Handbook of the Band Structure of Elemental Solids

Preface

This second edition of the handbook incorporates data for the whole periodic table up to the element copernicium. That is while the first edition contained 53 elements in the periodic table, this volume presents data up to $Z = 112$. There are major additions to the material contained in the first edition. First, results are given for the equation of state of the elements together with the parameters of a Birch fit so that the user can regenerate the results and also derive other information such as pressure-volume relations and the variation of the bulk modulus with pressure. For each element in addition to the equation of state, the energy bands, the densities of states and a set of tight-binding parameters is given. For most elements, the tight-binding parameters are presented for both a two-center and a three-center approximation. For the hcp structure new three-center tight-binding results are given. Other new materials included in this book are the energy bands and densities of states of all the rare earth metals, a discussion of the McMillan-Gaspari-Gyorffy theories and a tabulation of the electron-ion interaction matrix elements. The evaluation of the Stoner criterion for ferromagnetism is discussed and results are tabulated. This edition also contains two new appendices discussing the effects of spin-orbit interaction and a modified version of Harrison's tight-binding theory for metals which puts the theory on a quantitative basis.

The author believes that the accumulation of these results together in one source and their generation by the same methodology will be very useful for researchers and students to be able to quickly obtain the basic information of the electronic structure of all the elements in the periodic table.

The author is indebted to Michael J. Mehl for his assistance in resolving several problems that occurred in doing the LAPW calculations and continuous advice over the years. I also thank Noam Bernstein for useful discussions. The author's students Alex Koufos and Joseph McGrady contributed to the generation of many of the results. Special thanks go to Joseph McGrady for contributing results as well as technical support with figures, tables, formatting, and organizing the book. I want to thank my students Mak Keegan, Brahim Akdim, Lei Shi, Lane Nixon, Mohammed Assadullah, Joel Durgavich, and Mazhalai Chellathurai for providing tests and calculations they performed. I also wish to thank the late Mohammed Lach-hab for his contribution to the appendix on spin-orbit interaction. The encouragement and patience of Gail, Lia, Rea, Ben and Eleni is gratefully acknowledged.

I thank Robert Korec of Springer for his outstanding editorial assistance.

Fairfax, VA, USA
April 2014

Dimitris A. Papaconstantopoulos

Preface of the First Edition

This handbook consists of tabulations of the Slater-Koster parameters of 53 elements in the periodic table including metals, semiconductors, and insulators. These parameters are derived by fitting to first principles self-consistent and scalar-relativistic band structure calculations using in all cases (except for Si, Ge, and Sn) the augmented plane-wave method. For the cubic materials, four different sets of parameters are presented in this work corresponding to the following representations: (1) three-center nonorthogonal, (2) three-center orthogonal, (3) two-center nonorthogonal, and (4) two-center orthogonal.

For each element, figures showing the energy bands and densities of states resulting from nonorthogonal three-center and two-center calculations for the cubic and hexagonal materials, respectively, are also included.

I am grateful to Joel D. Shore for organizing and improving my computer codes and writing new plotting routines. I am indebted to W. E. Pickett and A. C. Switendick for numerous discussions and suggestions and to L. F. Mattheiss for his help in building an APW program for the hexagonal structure and for his valuable comments. I also wish to thank my colleagues, L. L. Boyer, J. L. Feldman, B. M. Klein, P. M. Laufer, and C. S. Wang at NRL and N. C. Bacalis at the Research Center of Crete for their continuous help. I also received valuable assistance and comments from J. R. Anderson, E. N. Economou, J. L. Fry, D. J. Gillespie, D. U. Gubser, W. A. Harrison, D. D. Koelling, N. C. Koon, D. J. Nagel, P. C. Pattnaik, D. G. Pettifor, and M. M. Saffren.

I am particularly thankful to Morrel H. Cohen for encouraging me to pursue this work and to V. L. Moruzzi, J. F. Janak, and A. R. Williams for the use of several of their potential functions. Finally, I wish to thank L. N. Blohm for technical assistance and my wife Gail for editorial assistance. The calculations on which this book was based were partially supported by the Office of Naval Research.

Fairfax, VA, USA
1985

Dimitris A. Papaconstantopoulos

Contents

Chapter 1
Introduction

1.1 Introduction to the New Edition

In the first edition of this Handbook we emphasized tight-binding (TB) parametrizations of augmented-plane wave (APW) results and presented energy bands and densities of states (DOS) for 53 elements in the periodic table. In this second edition we have added results for the equation of state for the above elements and added total energy and band structure data for the remaining elements in the periodic table, including the rare-earth elements and the over Z = 103 elements.

Over the years I realized that many users of this book were most interested in looking up the energy bands and DOS conveniently compiled in one source rather than getting into the details of TB parametrizations. For this reason and for the additional elements presented in this volume, I have included only one set of TB parameters. In addition, total energy results and calculations of the electron–phonon coupling and the Stoner criterion for ferromagnetism are also presented.

1.2 Introduction to the First Edition

The results of first-principles band structure calculations using elaborate augmented plane wave (APW), Green's function (KKR), linear combinations of Gaussian orbitals, or pseudopotential approaches can be recast on a linear combination of atomic orbitals (LCAO) basis using the Slater-Koster (SK) method as an interpolation scheme [1].

There are at least three important virtues in presenting the band structure of solids in this SK-LCAO form. First, using the SK parameters one can easily diagonalize a small matrix (9×9 for a typical transition metal) to obtain the energy bands and densities of states for a given material. This avoids the complexity of the APW or KKR techniques, which are used reliably and efficiently only by experts.

D.A. Papaconstantopoulos, *Handbook of the Band Structure of Elemental Solids*,
DOI 10.1007/978-1-4419-8264-3_1

Hence, the SK method is readily usable for the analysis of experiments by experimentalists themselves. Second, a great variety of theories utilize SK parameters to construct tight-binding (TB) Hamiltonians for the study of, for example, defects, disordered materials, surfaces, interfaces, and phonon spectra. The availability of such parameters is particularly useful to scientists working in these areas. At present, those who study such phenomena spend considerable time creating these TB Hamiltonians. In determining SK parameters they often invoke approximations that limit the accuracy of their results. Third, there is a clear educational value in this approach; it will enable students in solid-state physics courses to easily generate the band structure of a material and gain a better understanding of the concepts involved.

We present SK-LCAO parametrizations for 53 elements in the periodic table. We fit to self-consistent APW calculations that include the mass velocity and Darwin relativistic effects but not the spin-orbit coupling. We treat the following crystal structures: *fcc*, *bcc*, *hcp*, and *diamond.*

We tabulate *four* sets of SK-LCAO parameters: (1) in terms of three-center integrals using an orthogonal basis set, (2) in terms of three-center integrals using a nonorthogonal basis set, (3) in terms of two-center integrals using an orthogonal basis set, and (4) in terms of two-center integrals using a nonorthogonal basis set.

In addition to the SK-LCAO parameters we tabulate the following quantities: (1) the Fermi-level values of the density of states and its angular momentum decomposition, (2) the Fermi velocity and plasmon energy, and (3) the integrated densities of states, i.e., the number of electrons of s, p, t_{2g}, and e_g character.

Finally, we include two figures per element presenting the energy versus wave vector dispersion curve and the total density of states graph together with its angular momentum decomposition.

1.3 The Tight-Binding Formalism

In the LCAO method the one-electron wave function is expressed as a linear combination of Bloch sums. When such an expansion of the wave function is substituted into the Schrödinger equation one obtains a set of simultaneous linear algebraic equations which has a nonzero solution if the determinant of the coefficients vanishes, i.e.,

$$\ddot{H} - E\ddot{S} = 0 \tag{1.1}$$

The matrix elements in Eq. 1.1 have the form:

$$H_{nm} = \sum_{R_j} \exp i\mathbf{k} \cdot (\mathbf{R}_j - \mathbf{R}_i) \int \varphi_n^*(\mathbf{r} - \mathbf{R}_i) H \varphi_m(\mathbf{r} - \mathbf{R}_j) dv \tag{1.2}$$

$$S_{nm} = \sum_{R_j} \exp i\mathbf{k} \cdot (\mathbf{R}_j - \mathbf{R}_i) \int \varphi_n^*(\mathbf{r} - \mathbf{R}_i)\varphi_m(\mathbf{r} - \mathbf{R}_j)dv \tag{1.3}$$

where $\mathbf{R}_i$ and $\mathbf{R}_j$ denote the positions of atoms located on orbitals φ_n and φ_m respectively. The size of the matrices $\widetilde{H}$ and $\widetilde{S}$ is determined by the number of atoms per unit cell and the number of atomic orbitals taken on each atomic site. So for fcc and bcc structures with one atom per unit cell, $\widetilde{H}$ and $\widetilde{S}$ are 9×9 matrices representing one s function, three p functions, and five d functions. For crystal structures with two atoms per unit cell like the hcp the size of the matrices doubles to 18×18. In the diamond structure, which also has two atoms per unit cell, we have omitted the d orbitals and hence we work with 8×8 matrices. In Eqs. 1.2 and 1.3 the summations over $\mathbf{R}_j$ indicate summing over all neighbors of the original atom. In the present work for the bcc, hcp, and diamond structures we have carried out the summation to three neighboring atoms, and for the fcc structure to two neighbors.

The integrals in Eqs. 1.2 and 1.3 are three-center integrals since they are a product of an atomic function $\varphi^*_n(\mathbf{r}-\mathbf{R}_i)$ centered on the atom at position $\mathbf{R}_i$, an atomic function $\varphi_m(\mathbf{r}-\mathbf{R}_j)$ located on the atom at position $\mathbf{R}_j$, and a potential function (included in the Hamiltonian H) centered on a third atom.

The integrals in Eqs. 1.2 and 1.3 can be calculated from first principles in modern LCAO calculations. In this work we replace these integrals by adjustable parameters that are determined by fitting to APW band structure calculations. For Si, Ge, and Sn, instead of using APW calculations, we have fitted to empirical pseudopotential calculations [2] in order to obtain the correct energy gap.

In the classic paper of Slater and Koster [1] this method was first proposed with the simplification of using Löwdin functions [3] instead of atomic orbitals. These Löwdin functions are taken to be ortho-normal so that the integrals of Eq. 1.3 vanish unless $m = n$. In this case one has to solve the eigenvalue problem:

$$\widetilde{H} - E\widetilde{I} = 0 \tag{1.4}$$

Here we have performed the fitting to the APW results, both in a nonorthogonal representation (as was done by Mattheiss [4] in a number of materials) and on the more traditional orthogonal basis.

The integrals of Eqs. 1.2 and 1.3 are related by symmetry operations that reduce considerably the number of independent parameters to be determined in these calculations. These symmetry considerations have been worked out by Slater and Koster, who present tables of the matrix elements. Some typographical errors appear in these tables. For this reason and because phase factors may change the definition of these SK parameters by a sign, we have given here the full matrix for each structure as used in our calculations.

1.4 The Two-Center Approximation

The three-center integrals of Eq. 1.2 can be reduced to two-center integrals if we assume that the potential energy is a sum of spherical potentials located on the same two atoms where the atomic orbitals are centered. In this case the wavefunctions φ can be expressed as sums of functions that are space-quantized about the vector $\mathbf{R}_j - \mathbf{R}_i$, exactly as in a diatomic molecule. For example, if π is a p orbital, it becomes a linear combination of pσ, $p\pi_+$, and $p\pi\_$ functions and if φ is a d orbital it becomes a linear combination of dσ, $d\pi_+$, $d\pi\_$, $d\delta_+$, and $d\delta\_$ functions. The relationships between three- and two-center integrals have been worked out by Slater and Koster and are given in Table 1.1 of their paper. For the cubic structures we wrote our computer codes in the three-center approximation, so we have used this table to derive the corresponding expressions for first-, second-, and third-neighbor interactions and incorporated them into our codes. These relationships are given below for the bcc, fcc, and diamond structures. For the hcp structure we have done the fitting only in the two-center approximation using Miasek's tables [5], and therefore it was not necessary to derive such expressions for this structure.

1.4.1 bcc Structure

First-neighbor interactions:

$$E_{s,s}(111) = (ss\sigma)_1$$

$$E_{s,x}(111) = (sp\sigma)_1/\sqrt{3}$$

$$E_{s,xy}(111) = (sd\sigma)_1/\sqrt{3}$$

$$E_{x,x}(111) = \tfrac{1}{3}[(pp\sigma)_1 + 2(pp\pi)_1]$$

$$E_{x,y}(111) = \tfrac{1}{3}[(pp\sigma)_1 - (pp\pi)_1]$$

$$E_{x,xy}(111) = \frac{1}{3}\left[(pd\sigma)_1 + \frac{(pd\pi)_1}{\sqrt{3}}\right]$$

$$E_{x,yz}(111) = \frac{1}{3}\left[(pd\sigma)_1 - 2\frac{(pd\pi)_1}{\sqrt{3}}\right]$$

$$E_{x,x^2-y^2}(111) = \frac{(pd\pi)_1}{\sqrt{3}}$$

$$E_{xy,xy}(111) = \frac{(dd\sigma)_1}{3} + \frac{2}{9}(dd\pi)_1 + \frac{4}{9}(dd\delta)_1$$

Table 1.1 Block-diagonalization of the 9 × 9 TB Hamiltonian at high-symmetry points of the fcc lattice. The ordering of the states is s, p(x), p(y), p(z), d(yz), d(zx), d(xy), d(x^2-y^2), d($3z^2-r^2$)

State	k-point	Matrix	Degeneracy
Γ_1		$[H_{11}]$	1
Γ_{12}	(0, 0, 0)	$[H_{88}]$	2
Γ_{15}		$[H_{22}]$	3
Γ'_{25}		$[H_{55}]$	3
X_1		$\begin{bmatrix} H_{11} & H_{19} \\ H^*_{19} & H_{99} \end{bmatrix}$	1
X_2		$[H_{88}]$	1
X_3	$\left(0, 0, \frac{2\pi}{a}\right)$	$[H_{77}]$	1
X'_4		$[H_{44}]$	1
X_5		$[H_{55}]$	2
X'_5		$[H_{22}]$	2
L_1		$\begin{bmatrix} H_{11} & \frac{1}{\sqrt{3}}[H_{15}+H_{16}+H_{17}] \\ \frac{1}{\sqrt{3}}[H^*_{15}+H^*_{16}+H^*_{17}] & \frac{1}{3}\left\{ \begin{matrix} H_{55}+H_{66}+H_{77} \\ +2(H_{56}+H_{57}+H_{67}) \end{matrix} \right\} \end{bmatrix}$	1
L'_2	$\left(\frac{\pi}{a}, \frac{\pi}{a}, \frac{\pi}{a}\right)$	$[(1/3)(H_{22}+H_{33}+H_{44})+(2/3)(H_{23}+H_{24}+H_{34})]$	1
L_3		$\begin{bmatrix} (1/2)[H_{55}-2H_{56}+H_{66}] & (1/\sqrt{2})[H_{58}-H_{68}] \\ (1/\sqrt{2})[H^*_{58}-H^*_{68}] & H_{88} \end{bmatrix}$	2
L'_3		$[(1/2)(H_{22}-2H_{23}+H_{33})]$	2
W_1	$\left(0, \frac{2\pi}{a}, \frac{\pi}{a}\right)$	$\begin{bmatrix} H_{11} & H_{19} \\ H^*_{19} & H_{99} \end{bmatrix}$	1
W'_2		$\begin{bmatrix} H_{44} & H_{48} \\ H^*_{48} & H_{88} \end{bmatrix}$	1
W'_1		$[H_{66}]$	1
W_3	$\left(0, \frac{\pi}{a}, \frac{2\pi}{a}\right)$	$\begin{bmatrix} H_{22} & H_{27} \\ H^*_{27} & H_{77} \end{bmatrix}$	2

$$E_{xy,xz}(111) = \frac{(\mathrm{dd}\sigma)_1}{3} - \frac{(\mathrm{dd}\pi)_1}{9} - \frac{2}{9}(\mathrm{dd}\delta)_1$$

$$E_{xy,3z^2-r^2}(111) = \frac{2}{3\sqrt{3}}\left[(\mathrm{dd}\delta)_1 - (\mathrm{dd}\pi)_1\right]$$

$$E_{3z^2-r^2,3z^2-r^2}(111) = \frac{1}{3}\left[2(\mathrm{dd}\pi)_1 - (\mathrm{dd}\delta)_1\right]$$

Second-neighbor interactions:

$$E_{s,s}(200) = (ss\sigma)_2$$
$$E_{s,x}(200) = (sp\sigma)_2$$
$$E_{s,3z^2-r^2}(002) = (sd\sigma)_2$$
$$E_{x,x}(200) = (pp\sigma)_2$$
$$E_{y,y}(200) = (pp\pi)_2$$
$$E_{x,xy}(020) = (pd\pi)_2$$
$$E_{z,3z^2-r^2}(002) = (pd\sigma)_2$$
$$E_{xy,xy}(200) = (dd\pi)_2$$
$$E_{xy,xy}(002) = (dd\delta)_2$$
$$E_{3z^2-r^2,3z^2-r^2}(002) = (dd\sigma)_2$$
$$E_{x^2-y^2,x^2-y^2}(002) = (dd\delta)_2$$

The third-neighbor interaction relationship corresponding to coordinates (220) are the same as those of the first-neighbor interaction of the fcc structure that are given below.

1.4.2 fcc Structure

First-neighbor interactions:

$$E_{s,s}(110) = (ss\sigma)_1$$
$$E_{s,x}(110) = \frac{1}{\sqrt{2}}(sp\sigma)_1$$
$$E_{s,xy}(110) = \frac{\sqrt{3}}{2}(sd\sigma)_1$$
$$E_{s,3z^2-r^2}(110) = -\frac{1}{2}(sd\sigma)_1$$
$$E_{x,x}(110) = \frac{1}{2}[(pp\sigma)_1 + (pp\pi)_1]$$
$$E_{x,x}(011) = (pp\pi)_1$$
$$E_{x,y}(110) = \frac{1}{2}[(pp\sigma)_1 - (pp\pi)_1]$$
$$E_{x,xy}(110) = \frac{\sqrt{3}}{2\sqrt{2}}(pd\sigma)_1$$

$$E_{x,xy}(011) = \frac{1}{\sqrt{2}}(\mathrm{pd}\pi)_1$$

$$E_{z,3z^2-r^2}(011) = \frac{\sqrt{2}}{4}\left[\frac{(\mathrm{pd}\sigma)_1}{2} + \sqrt{3}(\mathrm{pd}\pi)_1\right]$$

$$E_{z,x^2-y^2}(011) = \frac{\sqrt{2}}{4}\left[\frac{(\mathrm{pd}\pi)_1}{2} - \sqrt{3}(\mathrm{pd}\sigma)_1\right]$$

$$E_{xy,xy}(110) = \tfrac{1}{4}[3(\mathrm{dd}\sigma)_1 + (\mathrm{dd}\delta)_1]$$

$$E_{xy,xy}(011) = \tfrac{1}{2}[(\mathrm{dd}\pi)_1 + (\mathrm{dd}\delta)_1]$$

$$E_{xy,xz}(011) = \tfrac{1}{2}[(\mathrm{dd}\pi)_1 - (\mathrm{dd}\delta)_1]$$

$$E_{xy,3z^2-r^2}(110) = \frac{\sqrt{3}}{4}[(\mathrm{dd}\delta)_1 + (\mathrm{dd}\sigma)_1]$$

$$E_{3z^2-r^2,3z^2-r^2}(110) = \tfrac{1}{4}[(\mathrm{dd}\sigma)_1 + 3(\mathrm{dd}\delta)_1]$$

$$E_{x^2-y^2,x^2-y^2}(110) = (\mathrm{dd}\pi)_1$$

The second-neighbor relations are the same as those of the bcc structure.

1.4.3 The Diamond Structure

First-neighbor interactions:

$$E_{s,s}\left(\frac{1}{2}\frac{1}{2}\frac{1}{2}\right) = (\mathrm{ss}\sigma)_1$$

$$E_{s,x}\left(\frac{1}{2}\frac{1}{2}\frac{1}{2}\right) = \frac{1}{\sqrt{3}}(\mathrm{sp}\sigma)_1$$

$$E_{x,x}\left(\frac{1}{2}\frac{1}{2}\frac{1}{2}\right) = \tfrac{1}{3}[(\mathrm{pp}\sigma)_1 + 2(\mathrm{pp}\pi)_1]$$

$$E_{x,y}\left(\frac{1}{2}\frac{1}{2}\frac{1}{2}\right) = \tfrac{1}{3}[(\mathrm{pp}\sigma)_1 - (\mathrm{pp}\pi)_1]$$

Second-neighbor interactions:

$$E_{s,s}(110) = (ss\sigma)_2$$
$$E_{s,x}(011) = 0$$
$$E_{s,x}(110) = \frac{1}{\sqrt{2}}(sp\sigma)_2$$
$$E_{x,x}(110) = \tfrac{1}{2}[(pp\sigma)_2 + (pp\pi)_2]$$
$$E_{x,x}(011) = (pp\pi)_2$$
$$E_{x,y}(110) = \tfrac{1}{2}[(pp\sigma)_2 - (pp\pi)_2]$$
$$E_{x,y}(011) = 0$$

Third-neighbor interactions:

$$E_{s,s}\left(\frac{3}{2}\frac{1}{2}\frac{1}{2}\right) = (ss\sigma)_3$$
$$E_{s,x}\left(\frac{3}{2}\frac{1}{2}\frac{1}{2}\right) = \frac{3}{\sqrt{11}}(sp\sigma)_3$$
$$E_{s,x}\left(\frac{1}{2}\frac{1}{2}\frac{3}{2}\right) = \frac{1}{\sqrt{11}}(sp\sigma)_3$$
$$E_{x,x}\left(\frac{3}{2}\frac{1}{2}\frac{1}{2}\right) = \tfrac{1}{11}[9(pp\sigma)_3 + 2(pp\pi)_3]$$
$$E_{x,x}\left(\frac{1}{2}\frac{1}{2}\frac{3}{2}\right) = \tfrac{1}{11}[(pp\sigma)_3 + 10(pp\pi)_3]$$
$$E_{x,y}\left(\frac{3}{2}\frac{1}{2}\frac{1}{2}\right) = \tfrac{3}{11}[(pp\sigma)_3 - (pp\pi)_3]$$
$$E_{x,y}\left(\frac{1}{2}\frac{1}{2}\frac{3}{2}\right) = \tfrac{1}{11}[(pp\sigma)_3 - (pp\pi)_3]$$

1.5 Computational Details

The SK matrices were formed by using the tables in the SK article [1] for the cubic materials, and the tables given by Miasek [5] for the hexagonal structure. To ensure correct assignments of states it was necessary to use group theory to reduce the

Table 1.2 Block-diagonalization of the 9×9 TB Hamiltonian at high-symmetry points of the bcc lattice. The ordering of the states is s, p(x), p(y), p(z), d(yz), d(zx), d(xy), d(x^2-y^2), d($3z^2-r^2$)

State	k-point	Matrix	Degeneracy
Γ_1		$[H_{11}]$	1
Γ_{12}	(0,0,0)	$[H_{88}]$	2
Γ_{15}		$[H_{22}]$	3
Γ'_{25}		$[H_{55}]$	3
H_1		$[H_{11}]$	1
H_{12}	$\left(0,0,\frac{2\pi}{a}\right)$	$[H_{88}]$	2
H_{15}		$[H_{22}]$	3
H'_{25}		$[H_{77}]$	3
N_1	$\left(\frac{\pi}{a},\frac{\pi}{a},0\right)$	$\begin{bmatrix} H_{11} & H_{17} & H_{19} \\ H^*_{17} & H_{77} & H_{79} \\ H^*_{19} & H^*_{79} & H_{99} \end{bmatrix}$	1
N_4		$[H_{88}]$	1
N'_1		$[(1/2)(H_{33}+2H_{34}+H_{44})]$	1
N_2		$[(1/2)(H_{66}-2H_{67}+H_{77})]$	1
N_3	$\left(0,\frac{\pi}{a},\frac{\pi}{a}\right)$	$[(1/2)(H_{66}+2H_{67}+H_{77})]$	1
N'_3		$[H_{22}]$	1
N'_4		$[(1/2)(H_{33}-2H_{34}+H_{44})]$	1
P_1		$[H_{11}]$	1
P_3	$\left(\frac{\pi}{a},\frac{\pi}{a},\frac{\pi}{a}\right)$	$[H_{88}]$	2
P_4		$\begin{bmatrix} H_{22} & H_{26} \\ H^*_{26} & H_{66} \end{bmatrix}$	3

9×9 secular equation of the cubic structure and the 18×18 Hamiltonian for the hexagonal materials to smaller matrices at high symmetry points or lines in the irreducible Brillouin zone.

The block diagonalization for bcc, fcc and hcp is shown in Tables 1.1, 1.2 and 1.3. It should also be noted that although the SK Hamiltonian is generally a complex matrix, for the fcc and bcc structures it was possible, by appropriate sign changes, to convert it to an equivalent real matrix. The SK fit was done for the **bcc** structure at a uniform mesh of 55 k-points in 1/48th of the Brillouin zone, and for the fcc and diamond structures at 33 k-points corresponding to a uniformly distributed grid of 20 k-points plus extra points at symmetry directions. For the bcc and fcc structures, aside from the high symmetry points Γ, H, N, P, X, L, and W (where we fit nine bands) we fit the first six bands for the rest of the k-points.

The determination of the SK parameters demands a least-squares procedure, and since it is a nonlinear problem it requires a reasonably good set of starting values. As mentioned above, by using symmetry considerations one can reduce the original secular equation to smaller matrices or to a number of linear equations. These linear equations can be solved simultaneously to determine an initial set of starting parameters. We now give the linear equations obtained for the particular crystal structure.

Table 1.3 Block-diagonalization of the 18x18 hcp Hamiltonian for high symmetry points

State	*k*-point	Matrix	Degeneracy
Γ_{1+}	(0,0,0)	$\begin{bmatrix} (H_{1,1}+H_{10,10}+2H_{1,10})/2 & 0.5*(H_{1,9}+H_{1,18}+H_{9,10}+H_{10,18}) \\ 0.5*(H^*_{1,9}+H^*_{1,18}+H^*_{9,10}+H^*_{10,18}) & (H_{9,9}+H_{18,18}+2H_{9,18})/2 \end{bmatrix}$	1
Γ_{4-}		$\begin{bmatrix} (H_{1,1}+H_{10,10}-2H_{1,10})/2 & 0.5*(H_{1,9}-H_{1,18}-H_{9,10}+H_{10,18}) \\ 0.5*(H^*_{1,9}-H^*_{1,18}-H^*_{9,10}+H^*_{10,18}) & (H_{9,9}+H_{18,18}-2H_{9,18})/2 \end{bmatrix}$	1
Γ_{6-}		$\begin{bmatrix} (H_{3,3}+H_{12,12}+2H_{3,12})/2 & 0.5*(H_{3,8}-H_{3,17}+H_{8,12}-H_{12,17}) \\ 0.5*(H^*_{3,8}-H^*_{3,17}+H^*_{8,12}-H^*_{12,17}) & (H_{8,8}+H_{17,17}-2H_{8,17})/2 \end{bmatrix}$	2
Γ_{5+}		$\begin{bmatrix} (H_{3,3}+H_{12,12}-2H_{3,12})/2 & 0.5*(-H_{3,8}-H_{3,17}+H_{8,12}-H_{12,17}) \\ 0.5*(-H^*_{3,8}-H^*_{3,17}+H^*_{8,12}+H^*_{12,17}) & (H_{8,8}+H_{17,17}+2H_{8,17})/2 \end{bmatrix}$	2
Γ_{2-}		$[(H_{4,4}+H_{13,13}+2H_{4,13})/2]$	1
Γ_{3+}		$[(H_{4,4}+H_{13,13}-2H_{4,13})/2]$	1
Γ_{6+}		$[(H_{6,6}+H_{15,15}+2H_{6,15})/2]$	2
Γ_{5-}		$[(H_{6,6}+H_{15,15}-2H_{6,15})/2]$	2
A_3	(0,0,0.5)	$\begin{bmatrix} H_{6,6} & H_{6,12} & H_{6,17} \\ H^*_{6,12} & H_{12,12} & H_{12,17} \\ H^*_{6,17} & H^*_{12,17} & H^*_{17,17} \end{bmatrix}$	4
A_1		$\begin{bmatrix} H_{1,1} & H_{1,9} & H_{1,13} \\ H^*_{1,9} & H_{9,9} & H_{9,13} \\ H^*_{1,13} & H^*_{9,13} & H_{13,13} \end{bmatrix}$	2
M_{4+}	(1.0,0,0)	$\begin{bmatrix} 0.5*(H_{2,2}-H_{2,11}-H_{11,2}+H_{11,11}) & 0.5*(H_{2,5}+H_{2,14}-H_{11,5}+H_{11,14}) \\ 0.5*(H^*_{2,5}+H^*_{2,14}-H^*_{11,5}-H^*_{11,14}) & 0.5*(H_{5,5}+H_{5,14}+H_{14,5}+H_{14,14}) \end{bmatrix}$	1
M_{1-}		$[0.5*(H_{7,7}-H_{7,16}-H_{16,7}+H_{16,16})]$	1

1.5.1 The bcc Structure

Neglecting third nearest-neighbor interactions and using the abbreviations $d1$ for $x^2 - y^2$ and $d2$ for $3z^2 - r^2$, we have the following:

$$E(\Gamma_1) = E_{s,s}(000) + 8E_{s,s}(111) + 6E_{s,s}(200) \tag{1.5}$$

$$E(\Gamma_{15}) = E_{x,x}(000) + 8E_{x,x}(111) + 2E_{x,x}(200) + 4E_{y,y}(200) \tag{1.6}$$

$$E(\Gamma_{25'}) = E_{xy,xy}(000) + 8E_{xy,xy}(111) + 4E_{xy,xy}(200) + 2E_{xy,xy}(002) \tag{1.7}$$

$$E(\Gamma_{12}) = E_{d2,d2}(000) + 8E_{d2,d2}(111) + 3E_{d2,d2}(002) + 3E_{d1,d1}(002) \tag{1.8}$$

$$E(\mathrm{H}_1) = E_{s,s}(000) - 8E_{s,s}(111) + 6E_{s,s}(200) \tag{1.9}$$

$$E(\mathrm{H}_{15}) = E_{x,x}(000) - 8E_{x,x}(111) + 2E_{x,x}(200) + 4E_{y,y}(200) \tag{1.10}$$

$$E(\mathrm{H}_{25'}) = E_{xy,xy}(000) - 8E_{xy,xy}(111) + 4E_{xy,xy}(200) + 2E_{xy,xy}(002) \tag{1.11}$$

$$E(\mathrm{H}_{12}) = E_{d2,d2}(000) - 8E_{d2,d2}(111) + 3E_{d2,d2}(002) + 3E_{d1,d1}(002) \tag{1.12}$$

$$E(\mathrm{N}_2) = E_{xy,xy}(000) + 8E_{xy,yz}(111) - 2E_{xy,yz}(002) \tag{1.13}$$

$$E(\mathrm{N}_3) = E_{xy,xy}(000) - 8E_{xy,yz}(111) - 2E_{xy,yz}(002) \tag{1.14}$$

$$E(\mathrm{N}_4) = E_{d2,d2}(000) - 3E_{d2,d2}(002) + E_{d1,d1}(002) \tag{1.15}$$

$$E(\mathrm{N}_{1'}) = E_{x,x}(000) - 2E_{x,x}(200) - 8E_{x,y}(111) \tag{1.16}$$

$$E(\mathrm{N}_{3'}) = E_{x,x}(000) + 2E_{x,x}(200) - 4E_{y,y}(200) \tag{1.17}$$

$$E(\mathrm{N}_{4'}) = E_{x,x}(000) - 2E_{x,x}(200) + 8E_{x,y}(111) \tag{1.18}$$

$$E(\mathrm{P}_1) = E_{s,s}(000) - 6E_{s,s}(200) \tag{1.19}$$

$$E(\mathrm{P}_3) = E_{d2,d2}(000) - 3E_{d2,d2}(002) - 3E_{d1,d1}(002) \tag{1.20}$$

$$E(\Delta_{2'}) = E_{xy,xy}(000) + 4E_{xy,xy}(200) - 2E_{xy,xy}(002)\,\text{at}\,k = \frac{\pi}{a}(0,0,1) \tag{1.21}$$

From Eqs. 1.5, 1.9, and 1.19 we determine the following s-s interaction parameters:

$$E_{s,s}(000) = \frac{1}{4}[E(\Gamma_1) + E(\mathrm{H}_1) + 2E(\mathrm{P}_1)]$$

$$E_{s,s}(111) = \frac{1}{16}[E(\Gamma_1) - E(\mathrm{H}_1)]$$

$$E_{s,s}(200) = \frac{1}{24}[E(\Gamma_1) + E(\mathrm{H}_1) + 2E(\mathrm{P}_1)]$$

From Eqs. 1.6, 1.10, and 1.16, 1.17, and 1.18 we determine the following p-p interaction parameters:

$$E_{x,x}(000) = \frac{1}{8}[E(\Gamma_{15}) + E(\mathrm{H}_{15}) + 2E(\mathrm{N}_{1'}) + 2E(\mathrm{N}_{3'}) + 2E(\mathrm{N}_{4'})]$$

$$E_{x,x}(111) = \frac{1}{16}[E(\Gamma_{15}) - E(\mathrm{H}_{15})]$$

$$E_{x,y}(111) = \frac{1}{16}[E(\mathrm{N}_{4'}) - E(\mathrm{N}_{1'})]$$

$$E_{y,y}(200) = \frac{1}{16}[E(\Gamma_{15}) + E(\mathrm{H}_{15}) - 2E(\mathrm{N}_{3'})]$$

$$E_{x,x}(200) = \frac{1}{16}[E(\Gamma_{15}) + E(\mathrm{H}_{15}) - 2E(\mathrm{N}_{1'}) + 2E(\mathrm{N}_{3'}) - 2E(\mathrm{N}_{4'})]$$

From Eqs. 1.7, 1.8, 1.11, 1.12, 1.13, 1.14, and 1.15, 1.20, and 1.21 we find the d-d interaction parameters:

$$E_{xy,xy}(000) = \frac{1}{4}[E(\Gamma_{25'}) + E(\mathrm{H}_{25'}) + 2E(\mathrm{N}_2) + 2E(\mathrm{N}_3) + 2E(\Delta_{2'})]$$

$$E_{xy,xy}(111) = \frac{1}{16}[E(\Gamma_{25'}) - E(\mathrm{H}_{25'})]$$

$$E_{d2,d2}(111) = \frac{1}{16}[E(\Gamma_{12}) - E(\mathrm{H}_{12})]$$

$$E_{xy,yz}(111) = \frac{1}{16}[E(\mathrm{N}_2) - E(\mathrm{N}_3)]$$

$$E_{d1,d1}(002) = \frac{1}{4}[E(\mathrm{N}_4) - E(\mathrm{P}_3)]$$

$$E_{xy,xy}(002) = \frac{1}{8}[E(\Gamma_{25'}) + E(\mathrm{H}_{25'}) - 2E(\Delta_{2'})]$$

$$E_{xy,xy}(200) = \frac{1}{8}[2E(\Delta_{2'}) - E(\mathrm{N}_2) - 2E(\mathrm{N}_3)]$$

$$E_{d2,d2}(000) = \frac{1}{4}[E(\Gamma_{12}) + E(\mathrm{H}_{12}) + 2E(\mathrm{P}_3)]$$

$$E_{d2,d2}(002) = \frac{1}{12}[E(\Gamma_{12}) + E(\mathrm{H}_{12}) + E(\mathrm{P}_3) - 3E(\mathrm{N}_3)]$$

Solving the above equations, we obtain reliable starting values for 17 parameters (representing three s-s interactions, five p-p interactions, and nine d-d interactions). The total number of three-center parameters in the bcc structure, up to second nearest neighbors, is 27. The additional 10 parameters appear in that part of the Hamiltonian that couples the different orbitals and therefore cannot be determined from linear equations. We proceed by assigning small values to these parameters, and we then run our least-squares program by fitting the APW energy bands at a uniform mesh of 55 k-points in the irreducible Brillouin zone. The resulting set of 27 parameters is then used as input to a subsequent run in which we have increased the number of parameters to 44 by including the third-neighbor interactions. This completes the determination of the SK parameters for an orthogonal basis in the three-center approximation.

Next we considered the nonorthogonal Hamiltonian also in three centers. In this case we dropped the third-neighbor integrals and introduced the corresponding 23 parameters of the overlap matrix for a total of 50 parameters. These parameters were determined by using the 27 parameters of the orthogonal fit as starting values, setting the rest of the parameters to small values, and letting our least-squares program iterate until a satisfactory fit was obtained.

In the two-center approximation we obtained starting values for the two-center parameters using the relations in Sect. 1.4 and then iterating with our least-squares program. Thirty-four parameters were used for the two-center orthogonal basis and 44 for the nonorthogonal.

For many elements, instead of following the procedure outlined above, we found it convenient to use the final parameters of a neighboring element in the periodic table as starting values.

1.5.2 The fcc Structure

To obtain starting values for the SK parameters in the fcc structure we use the following linear equations, which involve d-like functions:

$$E(\Gamma_{25'}) = E_{xy,xy}(000) + 4E_{xy,xy}(110) + 8E_{xy,xy}(011) + 4E_{xy,xy}(200) + 2E_{xy,xy}(002) \tag{1.22}$$

$$E(\Gamma_{12}) = E_{d2,d2}(000) + 6E_{d2,d2}(110) + 6E_{d1,d1}(110) + 3E_{d2,d2}(002) + 3E_{d1,d1}(002) \tag{1.23}$$

$$E(X_3) = E_{xy,xy}(000) + 4E_{xy,xy}(110) - 8E_{xy,xy}(011) + 4E_{xy,xy}(200) + 2E_{xy,xy}(002) \tag{1.24}$$

$$E(\mathrm{X}_2) = E_{d2,d2}(000) - 6E_{d2,d2}(110) + 2E_{d1,d1}(110) + 3E_{d2,d2}(002) + 3E_{d1,d1}(002) \quad (1.25)$$

$$E(\mathrm{X}_5) = E_{xy,xy}(000) - 4E_{xy,xy}(110) + 4E_{xy,xy}(200) + 2E_{xy,xy}(002) \quad (1.26)$$

$$E(\mathrm{W}_{1'}) = E_{xy,xy}(000) - 4E_{xy,xy}(110) + 4E_{xy,xy}(200) - 2E_{xy,xy}(002) \quad (1.27)$$

$$E(\Delta_{2'}, k = 0,0,4) = E_{xy,xy}(000) + 4E_{xy,xy}(110) + 4E_{xy,xy}(200) - 2E_{xy,xy}(002) \quad (1.28)$$

$$E(\Delta_2, k = 0,0,4) = E_{d2,d2}(000) + 4E_{d1,d1}(110) + 3E_{d2,d2}(002) - E_{d1,d1}(002) \quad (1.29)$$

From Eqs. 1.22, .1.24, and 1.26, 1.27, and 1.28 we find

$$E_{xy,xy}(011) = \frac{1}{16}[E(\Gamma_{25'}) - E(\mathrm{X}_3)]$$

$$E_{xy,xy}(110) = \frac{1}{8}[E(\Delta_{2'}, k = 0,0,4) - E(\mathrm{W}_{1'})]$$

$$E_{xy,xy}(002) = \frac{1}{4}[E(\mathrm{X}_5) - E(\mathrm{W}_{1'})]$$

If we now make the simplifying assumption that

$$E_{d2,d2}(002) = 0$$

then using Eqs. 1.23, 1.25, and 1.29 we can determine

$$E_{d1,d1}(002) = \frac{1}{8}[E(\Gamma_{12} + E(\mathrm{X}_2) - 2E(\Delta_2, k = 0,0,4)]$$

More d-d parameters can be found if we use Eq. 1.23 together with the equations resulting from the evaluation of $E(\Delta_2)$ at $k = (0,0,2)$ and $k = (0,0,6)$ [in our units $k = (0,0,4)$ is the middle of the Δ direction]. This system of linear equations gives

$$E_{d1,d1}(110) = \frac{E(\Gamma_{12}) - E(\mathrm{X}_2) - E(\Delta_2, k = 0,0,2) + E(\Delta_2, k = 0,0,6)}{2(2 - \sqrt{2})}$$

$$E_{d2,d2}(110) = \frac{1}{12}[E(\Gamma_{12}) - E(\mathrm{X}_2) - 4E_{d1,d1}(110)]$$

$$E_{d2,d2}(000) = E(\Gamma_{12}) - 6E_{d2,d2}(110) - 6E_{d1,d1}(110) - 3E_{d1,d1}(002)$$

We now bring in the s-s interaction parameters by utilizing the energy values at Γ_1, X_1, and W_1. Symmetry reduces the 9×9 secular equation to a linear equation at Γ_1 and to a quadratic at X_1 and W_1. However, linear equations can be constructed for X_1 and W_1 if we form the sum of the two roots of these quadratic equations. Hence, we have

$$E(\Gamma_1) = E_{s,s}(000) + 12E_{s,s}(110) + 6E_{s,s}(200) \tag{1.30}$$

$$E(X_{11}) + E(X_{12}) = E_{s,s}(000) - 4E_{s,s}(110) + 6E_{s,s}(200) + C_1 \tag{1.31}$$

$$E(W_{11}) + E(W_{12}) = E_{s,s}(000) - 4E_{s,s}(110) + 2E_{s,s}(200) + C_2 \tag{1.32}$$

where the indices 11 and 12 symbolize the two lowest eigenvalues of X_1 and W_1 symmetry and the quantities C_1 and C_2 are given below:

$$C_1 = E_{d2,d2}(000) + 2E_{d2,d2}(110) - 6E_{d1,d1}(110) + 3E_{d1,d1}(002)$$
$$C_2 = E_{d2,d2}(000) + 2E_{d2,d2}(110) + 3E_{d1,d1}(002)$$

Equations 1.30, 1.31, and 1.32 give

$$E_{s,s}(110) = \frac{1}{16}[E(\Gamma_1) - E(X_{11}) - E(X_{12}) + C_1]$$

$$E_{s,s}(200) = \frac{1}{4}[E(X_{11}) + E(X_{12}) - E(W_{11}) - E(W_{12}) - C_1 + C_2]$$

$$E_{s,s}(000) = E(\Gamma_1) - 12E_{s,s}(110) - 6E_{s,s}(200)$$

Next we use Eq. 1.22 together with the equations obtained from the sum of roots of the quadratics at L_1 and L_3, i.e.,

$$E(L_{11}) + E(L_{12}) = E_{s,s}(000) - 6E_{s,s}(200) + E_{xy,xy}(000) - 4E_{xy,xy}(200) - 2E_{xy,xy}(002) - 8E_{xy,xy}(011) \tag{1.33}$$

$$E(L_{31}) + E(L_{32}) = E_{xy,xy}(000) - 4E_{xy,xy}(200) - 2E_{xy,xy}(002) + 4E_{xy,xy}(011) + E_{d2,d2}(000) - 3E_{d1,d1}(002) \tag{1.34}$$

In Eqs. 1.22, 1.33, and 1.34 the only unknowns are $E_{xy,xy}(000)$, $E_{xy,xz}(011)$, and $E_{xy,xy}(200)$. Solving this system we have

$$E_{xy,xz}(011) = \frac{1}{12}[E(L_{31}) + E(L_{32}) - E(L_{11}) - E(L_{12}) + E_{s,s}(000) - E_{d2,d2}(000) - 6E_{s,s}(200) + 3E_{d1,d1}(002)]$$

$$E_{xy,xz}(000) = \frac{1}{6}[E(L_{11}) + E(L_{12}) + 2E(L_{31}) + 2E(L_{32}) + 3E(\Gamma_{25'}) - E_{s,s}(000) - 2E_{d2,d2}(000) - 12E_{xy,xy}(110) - 24E_{xy,xy}(011) + 6E_{s,s}(200) + 6E_{d1,d1}(002)]$$

$$E_{xy,xy}(200) = \frac{1}{24}[E(\Gamma_{25'}) - E(L_{11}) - E(L_{12}) - 2E(L_{31}) - 2E(L_{32}) + E_{s,s}(000)] + 2E_{d2,d2}(000) - 12E_{xy,xy}(110) - 24E_{xy,xy}(011) - 6E_{s,s}(200) - 12E_{xy,xy}(002) - 6E_{d1,d1}(002)]$$

We now proceed with the determination of parameters involving p-p interactions. We can obtain the following linear equations:

$$E(\Gamma_{15}) = E_{x,x}(000) + 8E_{x,x}(110) + 4E_{x,x}(011) + 2E_{x,x}(200) + 4E_{y,y}(200) \quad (1.35)$$

$$E(X_{4'}) = E_{x,x}(000) - 8E_{x,x}(110) + 4E_{x,x}(011) + 2E_{x,x}(200) + 4E_{y,y}(200) \quad (1.36)$$

$$E(X_{5'}) = E_{x,x}(000) - 4E_{x,x}(011) + 2E_{x,x}(200) + 4E_{y,y}(200) \quad (1.37)$$

$$E(L_{2'}) = E_{x,x}(000) - 8E_{x,y}(110) - 2E_{x,x}(200) - 4E_{y,y}(200) \quad (1.38)$$

$$E\left(W'_{21}\right) + E\left(W'_{22}\right) = E_{x,x}(000) - 4E_{x,x}(011) + 2E_{d2,d2}(000) - 4E_{d1,d1}(110) - 2E_{x,x}(200) + 4E_{y,y}(200) + 3E_{d2,d2}(002) - E_{d1,d1}(002) \quad (1.39)$$

$$E(L_{3'}) = E_{x,x}(000) + 4E_{x,y}(110) - 2E_{x,x}(200) - 4E_{y,y}(200) \quad (1.40)$$

From Eqs. 1.35, 1.36, and 1.37 we have

$$E_{x,x}(110) = \frac{1}{16}[E(\Gamma_{15}) - E(X_{4'})]$$
$$E_{x,x}(011) = \frac{1}{16}[E(\Gamma_{15}) + E(X_{4'}) - 2E(X_{5'})]$$

From Eqs. 1.38 and 1.40 we have

$$E_{x,y}(110) = \frac{1}{12}[E(L_{3'}) - E(L_{2'1})]$$

and

$$E_{x,x}(000) - 2E_{x,x}(200) - 4E_{y,y}(200) = \frac{1}{3}[E(L_{2'}) + 2E(L_{3'})] \quad (1.41)$$

If we rewrite Eq. 1.35 as

$$E_{x,x}(000) + 2E_{x,x}(200) + 4E_{y,y}(200) = E(\Gamma_{15}) - 8E_{x,x}(110) - 4E_{x,x}(011)] \quad (1.42)$$

Then from Eqs. 1.41 and 1.42 we get

$$E_{x,x}(000) = \frac{1}{6}\left[E(L_{2'}) + 2E(L_{3'})\right] + \frac{1}{2}[E(\Gamma_{15}) - 8E_{x,x}(110) - 4E_{x,x}(011)]$$

and

$$2E_{x,x}(200) + 4E_{y,y}(200) = \frac{1}{2}[E(\Gamma_{15}) - 8E_{x,x}(110) - 4E_{x,x}(011)] - \frac{1}{6}[E(L_{2'}) + 2E(L_{3'})] = D_1 \tag{1.43}$$

We then rewrite Eq. 1.39 as

$$2E_{x,x}(200) - 4E_{y,y}(200) = E_{x,x}(000) - 4E_{x,x}(011) + E_{d2,d2}(000) - 4E_{d1,d1}(110) + 3E_{d2,d2}(002) - E_{d1,d1}(002) - E\left(W'_{21}\right) - E\left(W'_{22}\right) = D_2 \tag{1.44}$$

Now Eqs. 1.43 and 1.44 give

$$E_{x,x}(200) = \frac{1}{4}(D_1 + D_2)$$

$$E_{y,y}(200) = \frac{1}{8}(D_1 - D_2)$$

We have thus determined initial values for 19 parameters with the approximation of $E_{d1,d1}(002) = 0$. These parameters, as in the bcc case, represent pure s-s, p-p, and d-d interactions. The hybridization of the different orbitals is taken into account by the remaining 12 parameters (within the second nearest-neighbor approximation), which enter into the problem via nonlinear equations. Using the parameters determined above as starting values we proceed with the least-squares fitting of the APW results, utilizing APW eigenvalues at 33 k-points in 1/48th of the irreducible zone. For the nonorthogonal three-center fit we used as starting values the final 32 parameters of the orthogonal three-center fit plus small initial values for the parameters of the overlap matrix. The total number of parameters in this fit is 60, which includes up to second-neighbor interaction.

In the two-center approximation the starting values were chosen by utilizing the relations of Sect. 1.3. The numbers of parameters used were 24 and 44 for the two-center orthogonal and nonorthogonal cases, respectively.

1.5.3 The Diamond Structure

If we consider first and second nearest-neighbor interactions, the block-diagonalization of the 8 × 8 Hamiltonian leads to the following linear equations:

$$E(\Gamma_1) = E_{s,s}(000) + 4E_{s,s}\left(\frac{1}{2}\frac{1}{2}\frac{1}{2}\right) + 12E_{s,s}(110) \tag{1.45}$$

$$E(\Gamma_{2'}) = E_{s,s}(000) - 4E_{s,s}\left(\frac{1}{2}\frac{1}{2}\frac{1}{2}\right) + 12E_{s,s}(110) \tag{1.46}$$

$$E(\Gamma_{15}) = E_{x,x}(000) + 4E_{x,x}\left(\frac{1}{2}\frac{1}{2}\frac{1}{2}\right) + 8E_{x,x}(110) + 4E_{x,x}(011) \tag{1.47}$$

$$E(\Gamma_{25'}) = E_{x,x}(000) - 4E_{x,x}\left(\frac{1}{2}\frac{1}{2}\frac{1}{2}\right) + 8E_{x,x}(110) + 4E_{x,x}(011) \tag{1.48}$$

$$E(X_3) = E_{x,x}(000) + 4E_{x,y}\left(\frac{1}{2}\frac{1}{2}\frac{1}{2}\right) - 4E_{x,x}(011) \tag{1.49}$$

$$E(X_4) = E_{x,x}(000) - 4E_{x,y}\left(\frac{1}{2}\frac{1}{2}\frac{1}{2}\right) - 4E_{x,x}(011) \tag{1.50}$$

To solve the above system of simultaneous equations we make the additional approximation of setting the parameters $E_{s,s}(\frac{1}{2}\frac{1}{2}\frac{1}{2})$ and $E_{x,x}$ (1 10) equal to zero. Then the solution is

$$E_{s,s}(000) = \frac{1}{2}\left[E(\Gamma_1) + E(\Gamma_{2'})\right] \tag{1.51}$$

$$E_{x,x}(000) = \frac{1}{4}\left[E(\Gamma_{15}) + E(\Gamma_{25'}) + E(X_3) + E(X_4)\right] \tag{1.52}$$

$$E_{s,s}\left(\frac{1}{2}\frac{1}{2}\frac{1}{2}\right) = \frac{1}{8}\left[E(\Gamma_1) - E(\Gamma_{25'})\right] \tag{1.53}$$

$$E_{x,x}\left(\frac{1}{2}\frac{1}{2}\frac{1}{2}\right) = \frac{1}{8}\left[E(\Gamma_{15}) - E(\Gamma_{25'})\right] \tag{1.54}$$

$$E_{x,y}\left(\frac{1}{2}\frac{1}{2}\frac{1}{2}\right) = \frac{1}{8}\left[E(X_3) - E(X_4)\right] \tag{1.55}$$

$$E_{x,x}(011) = \frac{1}{16}\left[E(\Gamma_{15}) + E(\Gamma_{25'}) - E(X_3) - E(X_4)\right] \tag{1.56}$$

We have thus determined starting values for seven out of the 13 three-center parameters corresponding to first- and second-neighbor interactions. We then run our least-squares program to determine all 13 parameters. The next step is to add the seven additional third-neighbor interaction parameters and go through the least-squares procedure again to find all 20 parameters. These parameters are then used as starting values for our separate fit on a nonorthogonal basis. Finally, utilizing the relations between three- and two-center integrals, we obtain estimates of starting parameters for our two-center fits. The fitting is done here, as in the fcc structure, for 33 k-points for eight bands. Owing to the absence of the d orbitals in our scheme only the first six bands fit well.

1.5.4 *The hcp Structure*

In the hcp structure we present results in the two-center approximation (both orthogonal and not), and for the three-center only in an orthogonal representation. In the hcp structure, owing to the larger size of the original secular equation, it is not possible to obtain linear equations, but several quadratic expressions can be found as shown in Table 1.3. For this structure we have included three-neighbor interactions resulting in 36 orthogonal and 67 nonorthogonal two-center parameters. In the three-center parametrizations we have 54 first-neighbor parameters following Miasek's notation. The fit was done for 12 bands and 23 k-points in the irreducible zone.

1.6 Nonorthogonal Hamiltonian

The use of nonorthogonal orbitals provides a more physically meaningful representation and, since the number of parameters involved is almost double that of the orthogonal SK method, the fit is superior. This approach has been utilized extensively by Mattheiss [4] in the parametrizations of the band structure of many compounds. Following Mattheiss we first set up the Hamiltonian and overlap matrices for a particular wave vector k, and then apply the Löwdin symmetric orthogonalization scheme to convert back to an orthogonal basis. In this scheme the matrices $\widetilde{H}$ and $\widetilde{S}$ of Eq. 1.2 are replaced by an effective Hamiltonian $\widetilde{H}'$ with a unit overlap matrix, where $\widetilde{H}' = \widetilde{S}^{-1/2}\widetilde{H}\widetilde{S}^{-1/2}$ and $\widetilde{S}' = \widetilde{S}^{-1/2}\widetilde{S}\widetilde{S}^{-1/2} = 1$. To determine $\widetilde{S}^{-1/2}$ the overlap matrix S is diagonalized, i.e., $\widetilde{U}^{+}\widetilde{S}\widetilde{U} = \widetilde{D}$. Then a new matrix $\widetilde{D}^{-1/2}$ is formed by replacing each diagonal element by its inverse square root, which is used to find $\widetilde{S}^{-1/2} = \widetilde{U}\widetilde{D}^{-1/2}\widetilde{U}^{+}$.

1.7 Total Energy: Birch Fit

Total Energy calculations provide an evaluation of the equilibrium lattice parameter and the bulk modulus. The total energy computed for a given crystal structure and several volumes is expanded by the Birch-fit formula [6] according to the expression:

$$E(V) = \sum_{i}^{N} a_i V_i^{-\frac{2i}{3}}$$

Where a_i are the expansion coefficients and N is the order of the fit. A third order fit usually provides an accurate description of the total energy. The second derivative of E with respect to volume gives the bulk modulus:

$$B = -V\frac{d^2E}{dV^2}$$

The Birch fit can also be used to calculate the bulk modulus and the pressure as a function of volume.

1.8 NRL-Tight-Binding Method

The NRL-Tight-Binding method (TB) is a Slater-Koster-like TB method [1, 7, 8], which has been successfully applied to many single element, binary and ternary systems [9, 10]. The method performs very well in metals, insulators, and semiconductors [11, 12], and can be extended to clusters and molecules. It is built on a fitting process to Density Functional Theory results. It works in both an orthogonal and a nonorthogonal basis and computes both the energy bands and total energies. It has transferability far beyond the fitted DFT database. The NRL-TB differs from the TB parameterizations presented in this book in that it extends to total energy evaluations as well as energy bands. The NRL-TB formalism with the basic equations is outlined below. First, the total energy is determined using the fact that in DFT the total energy E[n(r)] is given by the expression:

$$E[n(r)] = \sum_i \in_i + F[n(r)] \tag{1.57}$$

where n(r) is the electron density and F[n(r)] contains the remaining parts of the DFT total energy minus the sum of the one-electron eigenvalues.

In other TB approaches F[n(r)] is a sum of pair potentials. However, based on the fact that in DFT one can uniformly shift the energy bands by a constant, in this method a shift $V_0 = F[n(r)]/N$ is applied to the first-principles eigenvalues ϵ_i, where N is the number of valence electrons in the system. This shifts the one-electron eigenvalues to new values,

$$\in_i' = \in_i + V_0 \tag{1.58}$$

and hence the total energy becomes the sum of the shifted DFT eigenvalues over the Brillouin zone, i.e.

$$E[n(r)] = \sum_i \in_i' \tag{1.59}$$

This is a two-center TB scheme, where the on-site terms have a polynomial form as a function of the atomic density. For a single element, the density of atom i is defined as

$$\rho_i = \sum_i exp(-\lambda^2 R_{ij}) F_C(R_{ij}) \quad (1.60)$$

where the sum is over all the neighboring atoms j within a range of cutoff distance R_c of atom i, is a fitting parameter, and F_C (Rij) is a smooth cutoff function. The angular-momentum-dependent on-site terms are defined by

$$h_\ell(R_i) = \alpha_{i\ell} + \beta_{i\ell}\rho^{2/3} + \gamma_{i\ell}\rho^{4/3} + \delta_{i\ell}\rho^2 \quad (1.61)$$

where ℓ represents the s, p, and d orbitals, and $\alpha_{i\ell}$, $\beta_{i\ell}$, $\gamma_{i\ell}$ and $\delta_{i\ell}$ are fitting coefficients.

The two-center s-p-d Slater-Koster (SK) Hamiltonian and overlap integrals consist of ten independent SK parameters, which are assumed to all have polynomial times exponential forms in terms of the neighbor distance R, given by

$$P_\gamma(R) = \sum_j \left(e_\gamma + f_\gamma R + g_\gamma R^2\right) exp([-q_\gamma{}^2] F_C(R) \quad (1.62)$$

where γ indicates the type of interactions, including ssσ, ppσ, spσ, ddσ, sdσ, pdσ, ppπ, ddπ, pdπ, and ddδ. e_γ, f_γ, g_γ, and q_γ are our fitting coefficients. The SK overlap functions, in a non-orthogonal calculation, are defined to have the same form as the hopping parameters. Overall, there are in total 93 fitting coefficients for a single element in the on-site, hopping and overlap terms in the TB Hamiltonian with s, p, and d orbitals.

In order to determine the above coefficients a least-squares procedure is used to fit to DFT total energies and energy bands as a function of volume for high symmetry structures as bcc, fcc, and simple cubic crystal structures with varying atomic volumes. In some cases, to improve the fit one uses selected volumes for the hcp structure, and for semiconductors fit the diamond lattice as well. The total energy is usually weighed at around 200–300 times over a single band energy. In general, the fitting RMS error is less than 10 and 0.2 mRy for the energy bands and total energy, respectively.

Having determined the above coefficients, the method is used to predict total energies accessible to standard DFT which were not fitted, such as elastic constants, phonon spectra and surface energies. Furthermore, large scale simulations can be performed which are not practical via DFT, such as static calculations for the energetics of systems containing up to 20,000 atoms, or calculations for very large number of k-points needed for mapping Fermi surfaces and evaluating susceptibilities.

The method also has capabilities to perform molecular dynamics (MD) simulations accommodating as many as 10,000 atoms, and 10,000 MD steps, which is an impossible task for standard DFT codes. Such calculations should yield mean square displacements, thermal expansion, and vacancy formation energies [7–9].

1.9 The Gaspari-Gyorffy-McMillan Theory of Superconductivity

McMillan [13] first proposed to write the electron–phonon coupling constant λ as the following ratio

$$\lambda = \frac{\eta}{M\langle\omega^2\rangle} = \frac{N(\in_f)\langle I^2\rangle}{M\langle\omega^2\rangle} \tag{1.63}$$

where $N(\varepsilon_f)$ is the density of electronic states per spin and $<I^2>$ is and electron-ion matrix element, $<\omega^2>$ is an average phonon frequency, and M is the atomic mass.

One calculates the numerator η, known as the Hopfield parameters, by the theory known as rigid-muffin-tin approximation. The matrix element $<I^2>$, derived from multiple-scattering theory, is given by the formula due to Gaspari and Gyorffy [14]

$$\langle I^2\rangle = \frac{\in_f}{\pi^2 N^2(\in_f)} \sum_l \frac{2(l+1)sin^2(\delta_{l+1} - \delta_l)N_l N_{l+1}}{N_l^{(1)} N_{l+1}^{(1)}} \tag{1.64}$$

This formula for $<I^2>$ depends on the results of band structure calculations, specifically angular momentum components (N_l) of the density of states at the Fermi level, the scattering phase shifts (δ_l), calculated from the radial wave functions u_l, spherical Bessel functions j_l and Neumann functions n_l at the muffin-tin radius R_s from the well-known formula

$$tan\,\delta_l(R_s, E) = \frac{j_l'(kR_s) - j_l(kR_s)L_l(R_s, E)}{n_l'(kR_s) - n_l(kR_s)L_l(R_s, E)} \tag{1.65}$$

Where $L_l = \frac{u_l'}{u_l}$ is the logarithmic derivative of the radial wavefunction.

The equation for the free-scatterer density of states is

$$N_l^{(1)} = \frac{\sqrt{\in_f}}{\pi}(2l+1)\int_0^{R_s} u_l^2(r, \in_f)r^2 dr \tag{1.66}$$

Since Eq. 1.58 is based on scattering theory, it should be applied using touching muffin-tin spheres, as the excess interstitial volume in the case of nontouching spheres introduces errors. The denominator of Eq. 1.57 is approximated by $<\omega^2> = (1/2)\Theta^2_D$ where Θ_D is the measured value of the Debye temperature. Knowing λ and $<\omega>$ McMillan proposed the following equation to evaluate the superconducting transition temperature T_c [13]:

$$T_c = \frac{\Theta_D}{1.45}\exp\left[\frac{-1.04(1+\lambda)}{\lambda - \mu*(1+0.62\lambda)}\right] \tag{1.67}$$

The McMillan equation also involves the parameter μ^* called the Coulomb pseudopotential. The value of μ^* is usually taken in the range 0.1–0.15 and introduces small changes to T_c. The above formalism can also be used to study the pressure variation of T_c.

In Table 1.4 we list the values of the Hopfield parameter η and for some elements the electron–phonon coupling constant lambda. For the known transition-metal superconductors V, Nb, Ta and Tc the GG theory gives large values of η, but for the sp metals such as Hg and Pb this underestimates η, λ and therefore the superconducting critical temperature. For some elements such as H and Li we show the enhanced values of η under pressure. We note that in their

Table 1.4 Hopfield parameter η, electron phonon coupling λ, and Stoner criterion

Element	Lat.con. (Bohr)	Structure	η (eV/a^2)	λ	Stoner criterion
H	4.4	fcc	6.478	–	0.20
H	3.2	fcc	12.144	–	0.16
Li	6.6	bcc	0.265	0.363	0.47
Li	6.00(pres)	fcc	4.667	–	0.36
Be	6.06	fcc	0.56	0.034	–
Be	a: 4.19, c: 6.66	hcp	–	–	–
B	5.37	fcc	7.824	–	0.05
B	4.60(pres)	fcc	14.588	–	0.12
C	3.3	sc	14.339	0.27	–
Na	7.7	bcc	0.017	0.033	0.42
Na	7.00(pres)	fcc	2.47	–	–
Mg	8.58	fcc	0.123	0.036	–
Mg	a: 5.88, c: 9.63	hcp	–	–	–
Al	7.6	fcc	0.59	0.135	0.19
Si	4.68	sc	2.637	0.254	–
P	4.56	sc	4.266	1.723	0.13
S	4.78	sc	5.766	–	0.2
K	9.9	bcc	0.011	0.04	–
K	9.00(pres)	fcc	2.268	–	–
Ca	10.4	fcc	0.239	0.128	0.47
Ca	7.94	bcc	0.189	–	0.17
Sc	a: 6.06, c: 9.46	hcp	2.636	–	0.58
Sc	6.7	bcc	3.309	0.639	0.6
Sc	8.43	fcc	2.584	0.499	0.47
Ti	a: 5.37, c: 8.68	hcp	2.751	–	0.27
Ti	7.74	fcc	4.84	0.645	0.51
Ti	6.25	bcc	4.147	0.553	0.55
V	5.72	bcc	8.286	1.268	0.65
V	7.06	fcc	6.222	–	0.47
Cr	5.38	bcc	3.887	0.212	0.26
Cr	6.8	fcc	6.791	0.371	0.65

(continued)

Table 1.4 (continued)

Element	Lat.con. (Bohr)	Structure	η (eV/a²)	λ	Stoner criterion
Mn	6.8	fcc	5.624	0.685	0.68
Mn	5.4	bcc	8.882	1.082	1.24
Fe	5.3	bcc	5.608	0.512	1.51
Fe	6.5	fcc	5.714	0.522	0.72
Co	6.54	fcc	3.9	0.376	–
Co	5.5	bcc	2.909	0.281	1.95
Ni	6.55	fcc	1.762	0.167	2.14
Ni	5.16	bcc	2.763	–	0.87
Cu	6.83	fcc	0.456	0.069	–
Cu	5.34	bcc	0.687	–	0.09
Zn	7.25	fcc	0.544	0.088	–
Ga	7.83	fcc	0.595	0.094	–
Ge	5	sc	2.709	0.3	–
As	4.9	sc	3.69	0.697	0.12
Se	5.13	sc	4.607	–	0.17
Se	6.4	bcc	4.405	–	0.26
Rb	10.6	bcc	0.016	0.068	–
Rb	10.00(pres)	fcc	1.36	–	–
Sr	11	fcc	0	0	–
Sr	8.62	bcc	0.22	–	0.17
Y	a: 6.65, c: 10.60	hcp	3.183	–	0.53
Y	9.34	fcc	2.337	0.378	0.38
Y	7.39	bcc	3.344	–	0.5
Zr	a: 6.00, c: 9.72	hcp	2.323	–	0.16
Zr	8.37	fcc	5.759	–	0.38
Zr	6.6	bcc	4.658	0.679	0.32
Nb	6.2	bcc	7.627	1.222	0.39
Nb	7.85	fcc	7.014	–	0.39
Mo	5.89	bcc	5.797	0.336	0.13
Mo	7.48	fcc	7.21	–	0.3
Tc	a: 5.15, c: 8.33	hcp	6.922	0.55	0.24
Tc	7.25	fcc	8.094	–	0.27
Tc	5.76	bcc	12.077	–	0.47
Ru	a: 5.07, c: 8.17	hcp	7.151	–	0.25
Ru	7.2	fcc	6.51	0.201	–
Ru	5.7	bcc	8.245	–	0.57
Rh	7.18	fcc	4.722	0.224	–
Rh	5.72	bcc	5.036	–	0.57
Pd	7.36	fcc	3.662	0.281	0.85
Pd	5.81	bcc	2.716	–	0.4
Ag	7.68	fcc	0.288	0.059	0.06
Ag	6.07	bcc	0.357	–	0.06
Cd	8.2	fcc	0.359	0.082	–
In	8.95	fcc	0.355	0.298	–

(continued)

Table 1.4 (continued)

Element	Lat.con. (Bohr)	Structure	η (eV/a^2)	λ	Stoner criterion
Sb	5.82	sc	2.16	0.448	0.13
Te	5.83	sc	3.056	1.152	0.15
Cs	11.4	bcc	0.032	–	0.54
Cs	11.20(pres)	fcc	1.083	–	0.37
Ba	9.49	bcc	0.286	0.194	0.22
Ba	11.46	fcc	0.62	–	0.36
Ba	a: 8.19, c: 13.16	hcp	0.44	–	0.23
La	a: 7.10, c: 11.79	hcp	2.903	–	0.52
La	10	fcc	3.036	–	0.44
La	7.69	bcc	5.666	–	1.04
Ce	9.75	fcc	0.674	0.284	0.84
Pr	9.75	fcc	0.162	0.056	2.62
Nd	9.74	fcc	0.032	0.01	6.82
Pm	9.7	fcc	0.029	0.009	13.42
Sm	9.7	fcc	0.027	0.007	13.51
Eu	8.6	bcc	0.012	0.006	26.6
Gd	9.64	fcc	0.077	0.018	20.57
Tb	9.5	fcc	0.053	0.011	12.92
Dy	9.48	fcc	0.024	0.005	13.49
Ho	9.45	fcc	0.02	0.004	20.77
Er	9.39	fcc	0.022	0.004	12.14
Tm	9.34	fcc	0.068	0.001	9.83
Yb	10.36	fcc	0.404	0.182	0.27
Lu	9.26	fcc	2.079	0.303	0.38
Hf	a: 6.01, c: 9.73	hcp	2.803	–	0.18
Hf	8.3	fcc	6.212	0.617	0.39
Hf	6.59	bcc	5.412	–	0.35
Ta	6.24	bcc	9.423	1.018	0.34
Ta	7.9	fcc	6.486	–	0.27
Ta	a: 5.55, c: 9.23	hcp	7.848	–	0.36
W	6	bcc	4.521	0.173	0.1
W	7.6	fcc	7.25	–	0.32
W	a: 5.34, c: 8.93	hcp	14.748	–	0.5
Re	a: 5.23, c: 8.50	hcp	5.827	–	0.15
Re	7.36	fcc	8.1	0.265	0.21
Re	5.86	bcc	14.179	–	0.33
Os	a: 5.15, c: 8.29	hcp	6.14	–	0.17
Os	7.24	fcc	7.933	0.188	0.17
Os	5.81	bcc	8.553	–	0.37
Ir	7.26	fcc	6.899	0.229	0.29
Ir	5.84	bcc	8.159	–	1.01
Ir	a: 5.16, c: 8.43	hcp	8.516	–	0.38
Pt	7.41	fcc	4.871	0.488	0.59
Pt	5.92	bcc	4.566	–	0.26

(continued)

Table 1.4 (continued)

Element	Lat.con. (Bohr)	Structure	η (eV/a^2)	λ	Stoner criterion
Pt	a: 5.24, c: 8.73	hcp	6.523	–	0.65
Au	7.72	fcc	0.519	0.109	0.05
Au	6.15	bcc	0.705	–	0.05
Au	a: 5.44, c: 8.97	hcp	0.813	–	0.06
Hg	8.3	fcc	0.665	0.722	0.08
Hg	6.85	bcc	0.367	0.399	0.04
Tl	9	fcc	1.306	1.167	0.18
Tl	7.18	bcc	0.722	0.645	0.1
Pb	9.36	fcc	1.17	0.577	0.14
Pb	7.33	bcc	0.858	0.423	0.51
Bi	6.02	sc	1.755	0.668	0.11
Bi	9.35	fcc	2.441	0.929	0.19
Bi	7.33	bcc	2.169	0.825	0.16
Po	6.11	sc	2.573	0.962	0.13
Po	9.25	fcc	3.038	1.137	0.22
Po	7.36	bcc	2.775	1.038	0.2
Fr	11.58	bcc	0.032	0.255	0.48
Fr	14.62	fcc	0.026	0.212	0.48
Ra	9.82	bcc	0.238	0.438	0.19
Ra	12.4	fcc	0.061	0.112	0.08
Ac	10.35	fcc	3.38	–	0.41
Th	9.6	fcc	1.925	–	0.2
Pa	8.83	fcc	0.844	0.183	0.3
U	8.3	fcc	0.363	0.039	0.75
Np	8.06	fcc	0.389	0.052	1.43
Pu	8.16	fcc	0.268	0.055	0.74
Am	9.24	fcc	0.169	0.035	3.56
Cm	9.34	fcc	0.164	0.033	3.93
Bk	9.11	fcc	0.264	0.053	5.45
Cf	9.06	fcc	0.139	0.028	4.29
Es	9	fcc	0.092	0.018	5.76
Fm	8.95	fcc	0.107	0.021	4.31
Md	8.9	fcc	0.183	0.036	1.48
No	8.85	fcc	1.555	0.306	0.13
Lr	8.83	fcc	3.014	–	0.21
Rf	8.49	fcc	–	–	0.31
Db	6.45	bcc	–	–	0.29
Sg	6.28	bcc	5.029	–	0.1
Bh	6.18	bcc	10.479	–	0.23
Bh	7.75	fcc	8.674	–	0.19
Hs	7.62	fcc	6.98	–	0.14
Mt	7.69	fcc	6.877	–	0.19
Ds	7.85	fcc	5.297	–	0.43
Rg	6.47	bcc	2.229	–	0.16

non-equilibrium structure bcc-Re and hcp-W show large values of η. Also among the high Z elements only Bohrium has a large η, suggesting possibilities for superconductivity.

1.10 The Stoner Criterion

Many years ago Stoner [15] proposed that the following inequality:

$$N(E_F)I_s > 1$$

can be used as a criterion for the occurrence of ferromagnetism. As was shown by Vosko and Perdew [16], the parameter I_s can be accurately calculated using the results of band theory as follows.

$$I_s = \int dr\gamma^2(r)|K(r)|, \quad and \quad \gamma = \frac{1}{N(E_F)}\sum_l N_l(E_F)u_l^2(E_F)$$

where $u_l(E_F)$ is the Fermi level value of the radial wave function and K(r) is a kernel giving the exchange and correlation enhancement of the external field due to magnetization. We note again the key role of the angular momentum decomposed DOS, $N_l(E_F)$. The first implementation of this theory was made by Janak [17] who demonstrated that the Stoner criterion is satisfied for Fe and Ni with values clearly exceeding one, while for all other elements the values obtained by Janak were well below 1.0.

The ferromagnetic elements Fe, Co and Ni, as expected, satisfy the Stoner criterion as well as bcc-Mn. Because of the very large f-like DOS at the Fermi level many of the lanthanides and actinides also have very large Stoner criterion values confirming their magnetic nature.

For Co in the *fcc* structure, Janak reports a value of 0.97. Sigalas and Papaconstantopoulos [18], doing the calculations in the *bcc* lattice, found a value of 1.35. It is safe to assume that in the *hcp* structure, which is the ground state for Co, a value over 1.0 is a reasonable expectation. This method has been extended to binary systems by Papaconstantopoulos [19] who performed calculations for all the *3d, 4d,* and *5d* monohydrides in the NaCl structure. These calculations predicted that CoH exceeds the Stoner criterion, which makes it a ferromagnet in agreement with experiment.

1.11 Description of the First-Principles Calculations

The data presented in this book are based on the results of first-principles band structure calculations. With the exception of Si, Ge, and Sn (for which we used the empirical pseudopotential method [2]), we have performed these calculations by the APW or LAPW method. We have performed self-consistent scalar-relativistic (without spin-orbit) calculations using in most cases the local density approximation for exchange and correlation as applied by Hedin and Lundqvist [20]. For elements with $z < 20$ our results are close to the non-relativistic band structures given by Moruzzi et al. [21]. For the transition metals our band structure and total energy results are based on Ref. [22].

It should be noted here that the relativistic effects create bands that are wider and deeper in energy (Bacalis et al. [23]). In the 3d series these effects lower the s-like state Γ_1 by approximately 30 mRy and in the 4d series by as much as 80 mRy.

In the APW/LAPW calculations the energy levels are divided into inner and outer levels. The inner levels were treated as atomic-like in each iteration with a modified Liberman relativistic code [24]. The outer levels were treated by the APW/LAPW method on a uniform mesh of k-points depending on the crystal structure. We carried out our iterations to self-consistency using 55 k-points (in the irreducible Brillouin zone) for the bcc structure, 89 k-points for the fcc structure, and 23 k-points for the hcp structure. In the diamond structure we used the empirical pseudopotential approach for Si, Ge, and Sn, and the APW method with overlapping spheres [25] for C.

1.12 Accuracy of the Tight-Binding Calculations

In most of the bcc and fcc calculations we fit the lowest six bands except for the high symmetry points (Γ, H, N, and P for bcc and Γ, X, L, and W for fcc), for which we fit all nine bands. The three-center nonorthogonal calculations fit the first six bands with an accuracy that is better than 1 mRy. This means that these bands are virtually indistinguishable from the APW results for most elements. We believe that this is due to the much larger number of disposable parameters that we have available in the nonorthogonal representation and also because such a representation is more meaningful physically. In the other three representations our rms fitting errors are increasing with decreasing number of parameters. Typical rms errors for the first six bands in transition metals are 3 mRy for three-center orthogonal, 1 mRy for two-center nonorthogonal, and 5 mRy for two-center orthogonal. We should draw the attention of the reader to the fact that the higher bands, seventh through ninth, do not fit as well because these bands contain *f*-character, which is not included in our basis set. However, by fitting the high symmetry points for these bands, we place them correctly and do not allow them to come down and interfere incorrectly with the lower bands, as can happen if we restrict our fit to six bands.

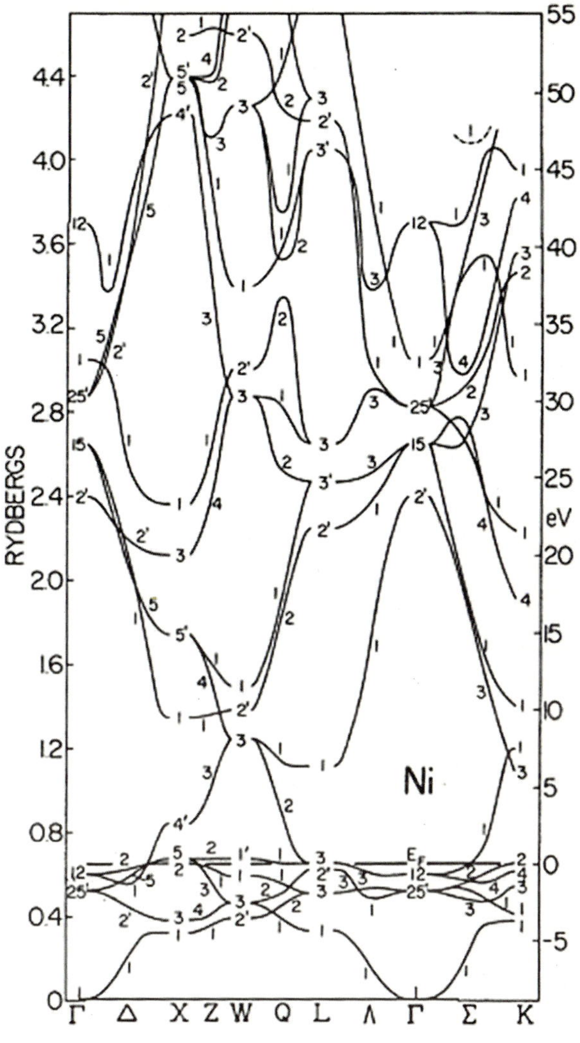

Fig. 1.1 First-principles APW energy bands for paramagnetic Ni covering an energy range of 5 Ry

In Fig. 1.1 we present the energy bands of an APW calculation for paramagnetic Ni, which shows an energy range of 5 Ry. This figure illustrates the only limitation of our SK fits. Specifically, the level $\Gamma_{2'}$ at about 2.4 Ry has purely *f*-character, which is not included in our nine-orbital SK basis set. Although we correctly position the p-like level Γ_{15} at about 2.7 Ry, our SK bands above 2 Ry are not reliable.

The situation is similar in the bcc structure where the level Γ_{25} of *f*-character is found to be significantly lower than the Γ_{15} state in the APW results.

For the semiconductors we have not included the d orbitals, so our results cannot be considered to be as accurate as those for the transition metals. Our rms errors are better than 0.2 eV for the valence bands and 0.4 eV for the fifth and sixth bands. The seventh and eighth bands have a much larger rms error (~0.9 eV) owing to the omission of the d orbitals. This difficulty in fitting the conduction bands in semiconductors with sufficient accuracy has led workers to generate widely differing sets of parameters depending on the number of neighbors used and on which conduction levels were fit.

This raises the question of the universality of the SK parameters. Our view is that our parameters are really unique for the metals for which we have achieved a high level of accuracy for the first six bands and the correct positioning of the high-energy p-like states. For semiconductors the same level of accuracy can only be achieved if the d orbitals are included. The work of Vogl et al. [26], who included a second s orbital instead of d orbitals, resulting in a 10×10 Hamiltonian, is certainly a step in the right direction. However, we have examined the wave functions of the pseudopotential calculations and we find that the d orbitals are located lower than the s orbitals of the conduction band, which means that inclusion of only s orbitals is not enough to give the desired accuracy.

Finally, we wish to comment on the combined interpolation scheme (CIS) of Hodges et al. [27] and of Mueller [28]. These works treat the d bands in exactly the same way as the SK method in an orthogonal representation does, but they treat the s- and p-like bands in the pseudopotential formalism. Our study shows that results obtained by the CIS are of comparable accuracy to our results when we use an orthogonal basis. However, our calculations utilizing the nonorthogonal SK Hamiltonian are clearly more accurate.

1.13 Wavefunctions

The Bloch eigenfunction $\psi_{kn}(r)$, which corresponds to an eigenvalue $E_n(k)$ of the SK Hamiltonian, is given by the following linear combination of LCAOs:

$$\psi_{kn}(r) = \sum_m v_{nm}(k) b_{km}(r)$$

where $v_{nm}(k)$ are the eigenvectors for band n and wave vector k, and the LCAOs are

$$b_{km}(r) = N^{-1/2} \sum_j \exp(ikR_j) \varphi_m(r - R_j)$$

where the sum is over the N atomic sites R_j and $\varphi_m(r - R_j)$ atomic wave-functions for $m = 1, \ldots, 9$. It is clear from the above equations that the wavefunctions of the solid can be calculated using the eigenvectors $v_{nm}(k)$ of the SK Hamiltonian together with atomic wavefunctions $\varphi_m(r - R_j)$. In this text we have used the SK eigenvectors as weights for obtaining the angular momentum components of the DOS. We have not calculated the wave-functions for the solid, but this is an easy task given our SK

Hamiltonian and parameters. Work along these lines has been done by [illegible] et al. [29], who used the CIS and obtained spherically averaged charge densities that were in good agreement with those of the first-principles calculations.

1.14 Scaling Laws

It is possible to include a lattice constant dependence in the SK parameters and thus obtain the band structure at different lattice spacings without performing new first-principles calculations. This can be done following Andersen et al. [30], who, using muffin-tin orbital theory, found that the distance-dependence of the tight-binding matrix elements is given by the expression

$$V_{ll'm} = C_{ll'm} d^{-(l+l'+1)}$$

where $C_{ll'm}$ is a constant that depends on the particular element and is volume independent, and d is the neighbor distance. This expression suggests that matrix elements for s-s interactions vary as d^{-1}, s-p interactions as d^{-2}, s-d and p-p interactions as d^{-3}, p-d interactions as d^{-4}, and d-d interactions as d^{-5}.

We have checked this scaling law [31], comparing its results to first-principles APW calculations at the same lattice spacing, and found very good agreement for variations of the lattice constant as large as 3 %.

Harrison [32], in his elegant theory of the physics of chemical bonds, has reformulated the above law to include only nearest neighbors. His bond-length dependences are d^{-2} for s-s, s-p, and p-p interactions, $d^{-7/2}$ for s-d and p-d interactions, and d^{-5} for d-d interactions. Harrison's theory provides a semiquantitative picture of the band structure of the elements that is remarkably successful considering the very small number of TB parameters used. On the other hand our work gives a very accurate band structure at the expense of introducing a large number of TB parameters. In Appendix 2 we present a modification to Harrison's theory for metals which gives a very substantial improvement to the band structure.

Beyond the scope of this Handbook is the NRL-TB methodology which we describe in Sect. 1.7. This reformulation of the SK approach reproduces the first-principles total energies and can give the lattice constant dependence near the equilibrium volume as well as having transferability to other structures.

1.15 Calculation of the Densities of States

To calculate the densities of states (DOS) we proceed as follows. Having determined the SK parameters for a particular element we then diagonalize the corresponding Hamiltonian for a large number of k-points. For the fcc and diamond structures our uniform k-mesh contains 505 k-points in the irreducible zone and for the bcc structure the same k-mesh results in 285 k-points. This diagonalization of the appropriate 9×9 (or 8×8 for the diamond structure) matrix gives, in addition to the eigenvalues, a quantity $Q_l = \sum_{j=1}^{9} v_{lj} * v_{lj}$ (with $l = s, p, t_{2g}, e_g$), which determines the angular momentum character of each state. The eigenvalues and Q_ls obtained by the above procedure are used to find the total DOS and its l-components by the tetrahedron method [33].

For the hcp materials we diagonalize an 18×18 matrix for 2,541 k-points in the irreducible hcp Brillouin zone. The reason we have used many more k-points in the hcp structure is that for hcp we have not employed the tetrahedron method but have calculated the DOS by a histogram technique directly from the eigenvalues and Q_ls of the 18×18 Hamiltonian. The computer codes that generate the eigenvalues and Q_ls are given in Appendix 3.

1.16 Systematics

In this section we consider the trends followed by the SK parameters as a function of atomic number Z, and also the systematic changes of the DOS at E_F across the periodic table, as reflected in the electronic specific heat coefficient.

We have studied as a function of Z the quantity $\Delta_l = E_l(000) - E_F$, where $E_l(00)$ denotes the on-site parameters for $l = s, p, t_{2_g}, e_g$ in the orthogonal three-center approximation and E_F is the Fermi energy. The quantities $\Delta_{t_{2g}}$ and Δ_{e_g} represent the approximate location of the d bands with respect to E_F. Both $\Delta_{t_{2g}}$ and Δ_{e_g} are positive at the beginning of the transition metal series and become negative around the middle of these series. The difference $\Delta_{t_{2g}} - \Delta_{e_g}$ which can be taken as the measure of the crystal-field splitting, has values of approximately 0.1 Ry at around Sc and Y and becomes almost 0 at the end of the transition metal series for Cu and Ag.

In Fig. 1.2 we have plotted the quantity $\Delta_d = \frac{3}{5}\Delta_{t_{2g}} + \frac{2}{5}\Delta_{e_g}$, which represents the center of gravity of the d bands with respect to E_F. We note the following: (1) The transition metals appear in three distinct groups corresponding to the 3d, 4d, and 5d series. The value of Δ_d is positive at the beginning of each series and continues to decrease as a function of Z until it takes negative values toward the middle of the series. Values of Δ_d that are greater than roughly 0.2 Ry represent elements with the entire d band unoccupied. On the other hand, when Δ_d is smaller than approximately -0.3 Ry, the d bands are completely filled. (2) The nontransition elements K, Ca, Rb, Sr, Cs, Ba, Ac, and Th have positive values of Δ_d that are very close to the Δ_ds of the early transition metals. (3) The free-electron-like elements Li, Na, Mg, Al, and Pb, as well as the noble gases, have large positive values of Δ_d, indicating that their d bands are far above E_F. (4) The elements Zn, Ga, Cd, In, and Tl have large negative values of Δ_d, which indicate that their d bands are far below E_F. The quantities Δ_s and Δ_p do not have a transparent physical

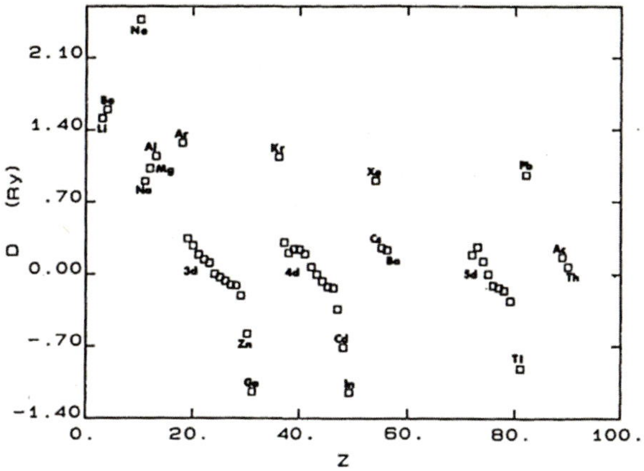

Fig. 1.2 The center of gravity of the d bands Δ_d plotted versus the atomic number Z

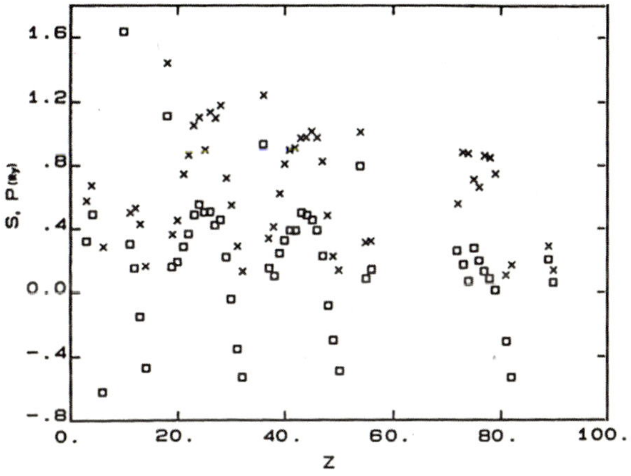

Fig. 1.3 The quantities Δ_s, (□) and Δ_p (X) plotted versus the atomic number Z

meaning, probably owing to hybridization effects with the d-like bands. It is, however, interesting to note from Fig. 1.3 that $\Delta_p > \Delta_s$, indicating that the p-like bands are located high above E_F.

We complete this section on systematics by considering the variation of the electronic specific heat coefficient γ across the periodic table. The coefficient γ is related to $N(E_F)$ by the formula

$$\gamma = 0.1734(1+\lambda)N(E_F)$$

where the numerical constant is selected in such a way that $N(E_F)$ is expressed in states/Ry per atom for both spins and γ is given in mJ mol^{-1} deg^{-2} The enhancement $1+\lambda$ has not been included in our calculation of γ but can be deduced from Table 1.5 if one compares the theoretical to the experimental value. An inspection of Table 1.5 shows that our calculated values of γ follow the trends seen in the measured values. As expected, the enhancement needed to match the experimental

Table 1.5 Electronic specific heat coefficients in mJ mol^{-1} deg^{-2}

	γ_{th}	γ_{exp} [a]
Li	1.17	1.63
Be	0.02	0.17
Na	1.17	1.38
Mg	1.03	1.30
AI	0.95	1.35
K	1.82	2.08
Ca	2.97	2.90
Sc	5.25	10.7
Ti	1.40	3.35
v	4.31	9.26
Cr	1.67	1.40
Mn	6.49	9.20
Fe	2.63	4.98
Co	2.08	4.73
Ni	4.13	7.02
Cu	0.70	0.695
Zn	0.42	0.64
Ga	0.93	0.596
Rb	2.17	2.41
Sr	0.0	3.6
Y	5.46	10.2
Zr	1.37	2.80
Nb	3.44	7.79
Mo	1.40	2.0
Ru	2.00	3.3
Rh	3.24	4.9
Pd	5.58	9.42
Ag	0.62	0.646
Cd	1.05	0.688
In	1.08	1.69
Cs	3.54	3.20
Ba	2.69	2.70
Hf	1.15	2.16
Ta	2.96	5.90
W	0.99	1.3
Re	1.91	2.3
Os	1.65	2.4
Ir	2.20	3.1
Pt	5.18	6.8
Au	0.69	0.729
Tl	0.648	1.47
Pb	1.19	2.98

[a]The experimental values are taken from Kittel, Ref. [34]

values is small for the free-electron-like metals and large for the transition metals. Serious disagreement between theory and experiment, in the sense that the unenhanced calculated value is larger than the measured, is found in Cr, Ga, and Cs. This discrepancy is understandable for Cr, which is antiferromagnetic; our calculations treated it as paramagnetic.

Our calculation of γ in Ga may be in error since we assumed the fcc structure instead of the tetragonal structure appropriate for this metal. In Cs our calculation is reliable, so we suggest the possibility of a 10 % uncertainty in the measured value. In Be we found a much smaller value of γ than the measured one. This is probably due to the poor fit we have obtained for Be. On the other hand there is a serious discrepancy in Sr, which we found to be a semimetal with $N(E_F)$ practically equal to zero.

1.17 Uses of This Handbook

The total energy results are presented in a graphical form and in terms of the coefficients of a Birch fit. These coefficients can be used to generate the bulk modulus as a function of volume or to obtain pressure versus volume plots. Labeling of the y-axis is not consistent throughout the book in that it ignores the absolute value of the total energy. This, of course, is not an issue since only the energy differences matter. In the energy band and density of states diagrams the Fermi level is set to a value with respect to the muffin-tin zero. The SK parameters given in this book can be used in a great variety of calculations in condensed matter physics. We will discuss a few of their applications here without describing the formalism of these theories. First, from the point of view of band theory of periodic materials, Fermi surface calculations can be performed by using the eigenvalues on a very dense grid of k-points as generated by diagonalizing our SK Hamiltonians, given in Appendix 3. We have found [35] that to obtain smooth cross-sectional areas of the Fermi surface of Nb it is necessary to use 1,785 k-points in the irreducible wedge of the Brillouin zone. This is slow if one uses first-principles calculations. This method also has the flexibility of adjusting a few of the SK parameters (usually the on-site parameters) to obtain a good fit to the Fermi surface for those cases in which the local density theory employed in the first-principles calculation is not in agreement with experiment. The SK Hamiltonian can also be used to obtain the DOS along symmetry directions in k-space, thus facilitating the comparison with photoemission data. Adjustment of SK parameters enables one to fit to various experiments and resolve discrepancies like that of the band structure of Ni, for which the angle-resolved photoemission data show a d-band width as well as an exchange splitting significantly smaller than those predicted by the first-principles calculations.

The SK parameters are particularly useful in the study of disordered materials. The TB form of the coherent-potential approximation (CPA) [36] leads to results that in many cases are nearly equivalent [37] to those obtained by more elaborate techniques. The TB-CPA has also been used to study the electronic structure of amorphous semiconductors [38].

SK parameters have been found to be the essential starting points for theories studying defects in solids and also for obtaining surface and interface electronic states. A study [39] of the band structure of metal–semiconductor interfaces using two-center parameters from this work appears to be very successful. Finally, we wish to refer to the work of Varma and Weber [40], who have used SK parametrizations to calculate both the electron–phonon (EP) interaction and phonon dispersion curves for Nb and Mo. Fry et al. [41] presented a systematic study of the EP interaction in most transition metals using two-center-orthogonal parameters from this work and Harrison's [32] scaling laws.

1.18 Description of the Slater-Koster Tables

The SK-parameter tables are arranged in the following way. For each element the first page displays the orthogonal and nonorthogonal parameters in the three-center approximation. The notation is self-explanatory with the exception of the abbreviations d1 for xy,yz,zx and d2 for x^2-y^2 and $3z^2-r^2$ for the two-center tables.

Different but obvious abbreviations are used for the hcp structure.

On the second page the rms errors and maximum deviations are given per band and a comparison is made of the eigenvalues resulting from the orthogonal and nonorthogonal fits to the APW eigenvalues. At the bottom of the second page a tabulation is made of the Fermi-level quantities, i.e., Fermi energy, total and angular momentum components of the DOS, Fermi velocity, plasmon energy, and integrated DOS up to E_F. These quantities correspond to the most accurate of the four calculations presented here, namely the one using a three-center nonorthogonal Hamiltonian. It should be noted that for the bcc materials the orthogonal Hamiltonians extend to three nearest neighbors while for the fcc structures only two nearest neighbors are used. For the hcp structure two-center parameters that include up to third-neighbor interactions are given, together with only orthogonal three-center parameters.

The diamond structure tables contain only s and p orbitals, unlike the cubic and hcp structures, which contain the d orbitals as well. In this structure the SK parameters are expressed in eV rather than Ry and the tables include, instead of Fermi-level quantities, the energy gap.

References

1. J.C. Slater, G.F. Koster, Phys. Rev. **94**, 1498 (1954)
2. J.R. Chelilkowsky, M.L. Cohen, Phys. Rev. **B14**, 556 (1976)
3. P.O. Löwdin, J. Chem. Phys. **18**, 1365 (1950); P.O. Löwdin, J. Chem. Phys. **19**, 1396 (1951)

4. L.F. Mattheiss, Phys. Rev. **B2**, 3918 (1970)
5. M. Miasek, Phys. Rev. **107**, 92 (1957); M. Miasek, Acta Physica Polonica **XVII**, 371 (1958)
6. F. Birch, J. Geophys. Res. **83**, 1257 (1978)
7. R.E. Cohen, M.J. Mehl, D.A. Papaconstantopoulos, Phys. Rev. **B 50**, 14694 (1994)
8. M.J. Mehl, D.A. Papaconstantopoulos, Phys. Rev. **B 54**, 4519 (1996)
9. D.A. Papaconstantopoulos, M.J. Mehl, J. Phys. Condens. Matter **15**, R413 (2003)
10. D.A. Papaconstantopoulos, M.J. Mehl, M.D. Johannes, Phys. Rev. **B 82**, 054503 (2010)
11. N. Bernstein, M.J. Mehl, D.A. Papaconstantopoulos, N.I. Papanicolaou, M.Z. Bazant, E. Kaxiras, Phys. Rev. **B 62**, 4477 (2000)
12. N. Bernstein, M.J. Mehl, D.A. Papaconstantopoulos, Phys. Rev. **B 66**, 075212 (2002)
13. W. McMillan, Phys Rev. **B 167**, 331 (1968)
14. G.D. Gaspari, B.L. Gyorffy, Phys. Rev. Lett. **28**, 801 (1972)
15. E.C. Stoner, Proc. R. Soc. A. **169**(938), 339 (1939)
16. S.H. Vosko, J.P. Perdev, Can. J. Phys. **53**, 1385 (1975)
17. J. F. Janak, Phys. Rev. **B 9**, (1974)
18. M.M. Sigalas, D.A. Papaconstantopoulos, Phys. Rev. **B 50**, 7255 (1994)
19. D.A. Papaconstantopoulos, Europhys. Lett. **15**, 621 (1991)
20. L. Hedin, B.I. Lundqvist, J. Phys. **C 4**, 2064 (1971)
21. V.L. Moruzzi, J.F. Janak, A.R. Williams, *Calculated Electronic Properties of Metals* (Pergamon Press, Elmsford, 1978)
22. M. Sigalas, D.A. Papaconstantopoulos, N.C. Bacalis, Phys. Rev. **B 45**, 5777 (1992)
23. N.C. Bacalis, K. Blathras, P. Thomaidis, D.A. Papaconstantopoulos, Phys. Rev. **B 32**, 4849 (1985)
24. D.A. Liberman, D.T. Cromer, J.T. Waber, Comp. Phys. Commun. **2**, 107 (1971)
25. D.A. Papaconstantopoulos, Phys. Rev. **B 15**, 2569 (1983)
26. P. Vogl, H.P. Hjalmarson, J.D. Dow, J. Phys. Chem. Solids **44**, 365 (1983)
27. L. Hodges, E. Ehrenreich, N.D. Lang, Phys. Rev. **152**, 505 (1965)
28. F.M. Mueller, Phys. Rev. **153**, 659 (1967)
29. N.A. Johnston, C.A. Sholl, P.V. Smith, J. Phys. **F 13**, 945 (1983)
30. O.K. Andersen, O. Jepsen, Physica **91B**, 317 (1977); O.K. Andersen, W. Close. H. Nohl, Phys. Rev. **B 17**, 1209 (1978)
31. J.D. Shore, D.A. Papaconstantopoulos, Phys. Rev. **B 35**, 1122 (1987)
32. W.A. Harrison, *Electronic Structure and the Properties of Solids* (W. H. Freeman, San Francisco, 1980)
33. G. Lehmann, M. Taut, Phys. Status Solidi **B 54**, 469 (1972)
34. C. Kittel, *Introduction to Solid State Physics* (Wiley, New York, 1971)
35. J.R. Anderson, D.A. Papaconstantopoulos, J.E. Schirber, Phys. Rev. **B 24**, 6790 (1981)
36. H. Ehrenreich, L.M. Schwartz, Solid State Phys. **31**, 150 (1976); J.S. Faulkner, Prog. Mat. Sci. **27**, 1 (1982)
37. D.A. Papaconstantopoulos, P.M. Laufer, A.C. Switendick, in *Hydrogen in Disordered and Amorphous Solids*, ed. by G. Bam bakidis, R.C. Bowman Jr. (Plenum Press, New York, 1986), pp. 139–152; P.M. Laufer, D.A. Papaconstantopoulos, Phys. Rev. **B 33**, 5134 (1986)
38. D.A. Papaconstantopoulos, E.N. Economou, Phys. Rev. **B 24**, 7233 (1981)
39. W.E. Pickett, D.A. Papaconstantopoulos, Phys. Rev. **B 34**, 8372 (1986)
40. C.M. Varma, W. Weber, Phys. Rev. Lett. **39**, 1094 (1977); C.M. Varma, E.I. Blount, P. Vashista, W. Weber, Phys. Rev. **B19**, 6130 (1979)
41. J.L. Fry, G. Fletcher, P.C. Pattnaik, D.A. Papaconstantopoulos, Physica **135B**, 473 (1985)

Chapter 2
Hydrogen and the Alkali Metals

2.1 Hydrogen

On the basis of a standard density functional theory calculation using the LAPW method, neglecting any zero point motion effects, we find that the diamond lattice gives the lowest total energy. However, the calculations shown in the figure below are not realized experimentally at present, since hydrogen is known to be in a molecular state under normal conditions. There have been numerous predictions of the formation of metallic hydrogen at very high pressures with a possible room temperature superconductivity [1, 2]. Our calculation for the H2 molecule gives a bond length of 0.75076 A. Two sets of energy bands and densities of states are presented at the fcc equilibrium a = 4.363 a.u. and at a lattice constant a = 3.2 a.u. corresponding to a pressure of 4.7 Mbar, where high temperature superconductivity is predicted.

D.A. Papaconstantopoulos, *Handbook of the Band Structure of Elemental Solids*,
DOI 10.1007/978-1-4419-8264-3_2

Hydrogen

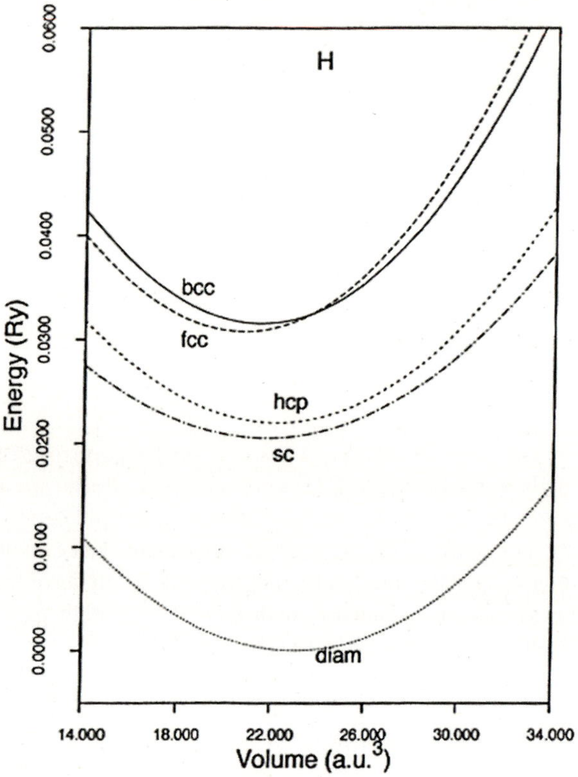

Fig. 2.1 Total energy of H

Table 2.1 Lattice constants and bulk modulus

	a (Bohr)	c (Bohr)	B (MBar)
fcc	4.363		1.403
bcc	3.426		1.585
diam	5.666		0.957
sc	2.715		1.106
hcp	3.749	3.374	1.078
$\Delta E = E_{diam} - E_{sc} = 20$ mRy			

Hydrogen fcc, a = 3.19857

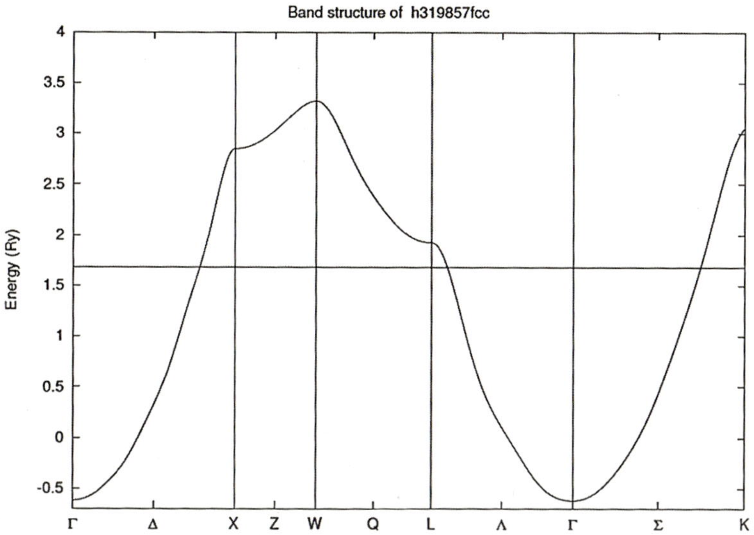

Fig. 2.2 Energy bands for H a = 3.19857

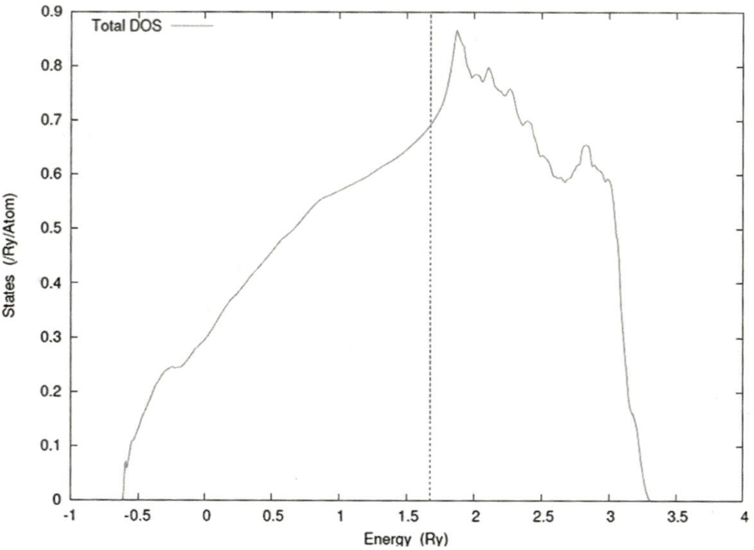

Fig. 2.3 Density of states for H a= 3.19857

Table 2.2 Fermi level quantities (non-orthogonal fit)

Energy		Densities of states					Velocity	Plasmon energy
	Total	s	p	t_{2g}	e_g	f		
Ry		States/Ry/atom					cm/s	eV
1.678	0.693	0.402	0.127	0.022	0.008	0.002	3.281×10E8	34.377

Hydrogen fcc, a = 4.4

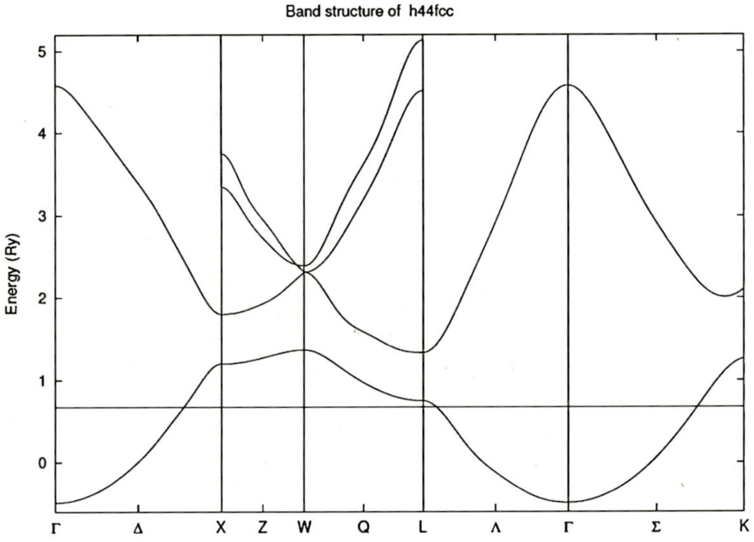

Fig. 2.4 Energy bands for H a = 4.4

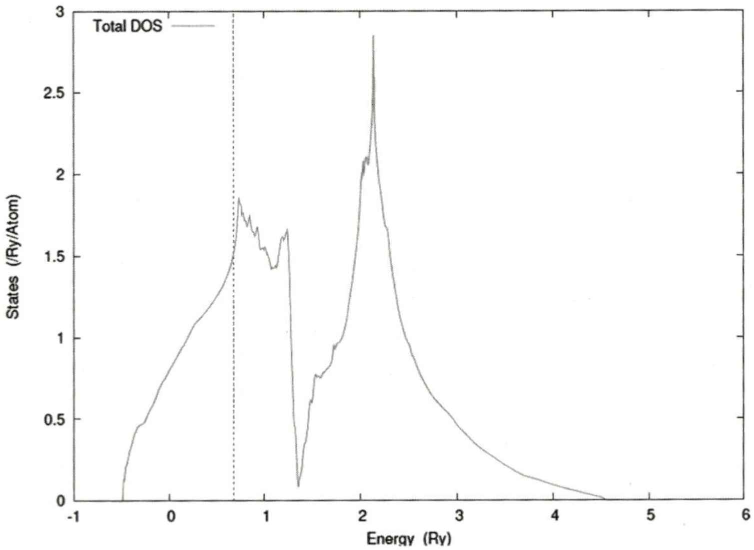

Fig. 2.5 Density of states for H a = 4.4

Table 2.3 Fermi level quantities (non-orthogonal fit)

Energy		Densities of states						
	Total	s	p	t_{2g}	e_g	f	Velocity	Plasmon energy
Ry		States/Ry/atom					cm/s	eV
0.674	1.506	1.011	0.199	0.044	0.013	0.005	2.048 × 10E8	19.611

Hydrogen fcc structure a=4.40 Bohr
Tight-binding parameters in Ry

Non-orthogonal
On-site
s
0.35589

sss
hopping dist= 3.11127
−0.09791
overlap dist = 3.11127
0.16403
hopping dist = 4.40
−0.04643
overlap dist= 4.40
0.02530
hopping dist= 5.38888
−0.01326
overlap dist= 5.38888
0.00133

Fitting rms error for 1 band 15 mRy

Orthogonal
On-site
s
0.60350

sss
hopping dist= 3.11127
−0.12723
hopping dist= 4.40
0.00994
hopping dist= 5.38888
0.02112

Fitting rms error for 1 band 70 mRy

2.2 The Alkali Metals

The alkali metals Li, Na, K, Rb, Cs and Fr all crystallize in the bcc structure. However, calculations of the total energy based on density functional theory with the exception of Li give incorrectly the fcc structure as the ground state. This is due to the very small energetic difference between the two structures. This problem is not resolved by the generalized gradient approximation used here, which improves the agreement with experiment only in the lattice constant value. The energy bands of these metals have a pronounced free-electron character and are similar to each other, particularly in the occupied region of states. Significant differences occur in the unoccupied states. We note changes in the ordering of states. For example, in Li the d-like states characterized by the symmetry points $\Gamma_{25'}$, and Γ_{12} are approximately 0.3 Ry above the p-like state Γ_{15}, while in Na the d states lie just below Γ_{15}. Going to K, Rb, and Cs, we note that both the d- and p-like states are lowered, with the d states moving at a much faster rate toward the Fermi level. The density of states at E_f is fairly small for Li and Na but becomes progressively larger for the heavier alkalis. The angular momentum decomposition of the DOS at E_f shows strong p-character for Li, about equal s and p contributions for Na, and a very strong s-like component for the heavier elements, including Fr which also has a small f-DOS contribution [3].

2.2.1 Lithium

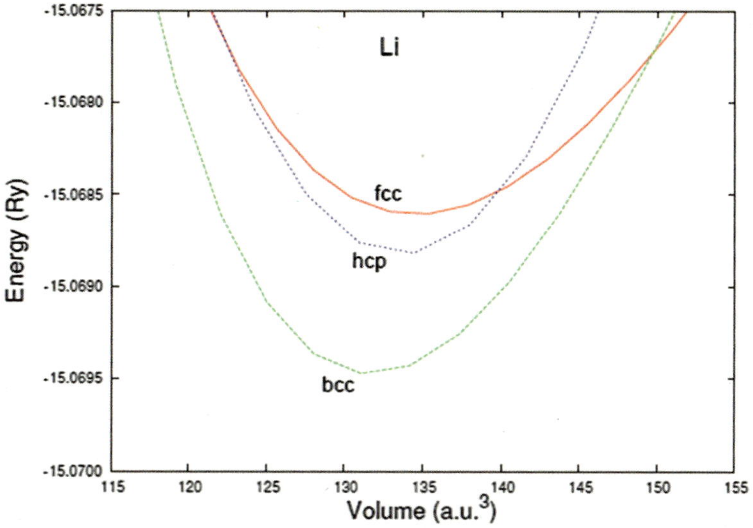

Fig. 2.6 Total energy of Li

Table 2.4 Lattice constants and bulk modulus

	a (Bohr)	c (Bohr)	B (MBar)
fcc	8.134		0.196
bcc	6.410		0.296
hcp	5.738	9.371	0.339
exp	6.597		0.116
$\Delta E_{bcc-hcp} = 0.654$ mRy			

Table 2.5 Birch fit coefficients

	A_1	A_2	A_3	A_4
fcc	–15.09897	7.67593	–340.29804	4192.88225
bcc	–15.24725	21.50653	–756.00191	8242.37038
hcp	–28.19296	–213.30253	7657.07755	–89401.46580

Table 2.6 Lithium bcc z = 3 lattice constant = 6.59700 a.u.

Slater-Koster 3-center parameters	Orthogonal	Non-orthogonal	
	Energy integrals (Ry)	Energy integrals (Ry)	Overlap integrals
On site			
s,s (000)	0.60662	0.59367	
x,x (000)	0.86181	0.71286	
xy,xy (000)	1.95620	1.60594	
d2,d2 (000)	1.56688	1.17802	
First neighbor			
s,s (111)	–0.07451	–0.06484	0.05237
s,x (111)	0.05369	0.02891	–0.05721
s,xy (111)	0.01458	0.03346	–0.03060
x,x (111)	0.03963	–0.00084	–0.03238
x,y (111)	0.05384	0.01155	–0.06674
x,xy (111)	0.02890	0.01496	–0.01242
x,yz (111)	0.04446	–0.00953	–0.07998
x,d1 (111)	0.01594	0.01607	–0.03733
xy,xy (111)	–0.05355	–0.06572	–0.01385
xy,xz (111)	–0.09364	0.06284	0.07927
xy,d2 (111)	–0.03736	–0.08553	–0.10995
d2,d2 (111)	0.04516	–0.08367	–0.13623
Second neighbor			
s,s (200)	–0.02172	–0.00252	0.05152
s,x (200)	0.03254	0.01742	–0.06560
s,d2 (002)	–0.05197	0.05380	0.13662
x,x (200)	0.05804	0.03961	–0.08825
y,y (200)	0.01578	–0.01600	–0.00320
x,xy (020)	0.02890	–0.02842	0.00328
z ,d2 (002)	–0.07330	0.04738	0.17079
xy,xy (200)	0.04772	–0.04483	–0.08137
xy,xy (002)	–0.07237	0.01076	–0.00313
d2,d2 (002)	–0.12519	0.22187	0.25525
d1,d1 (002)	–0.03933	–0.02737	–0.00994
Third neighbor			
s,s (220)	0.01239		
s,x (220)	–0.01325		
s,xy (220)	–0.01797		
s,d2 (220)	–0.01563		
x,x (220)	–0.00888		
x,x (022)	0.00468		
x,y (220)	–0.01543		
x,xy (220)	–0.01143		
x,xy (022)	–0.00107		
z,d2 (022)	0.00715		
z,d1 (022)	–0.01970		
xy,xy (220)	0.03458		
xy,xy (022)	–0.01429		
xy,xz (022)	–0.01701		
xy,d2 (220)	0.03421		
d2,d2 (220)	0.02619		
d1,d1 (220)	–0.00180		

Table 2.7 Lithium bcc

	Orthogonal			Non-orthogonal		
	RMS error	Maximum deviation		RMS error	Maximum	Deviation
Band	mRy	k	mRy	mRy	k	mRy
1	5.8	(044)	11.7	2.9	(044)	8.4
2	8.3	(350)	19.9	2.5	(008)	6.3
3	9.9	(442)	24.7	2.6	(044)	6.8
4	12.3	(044)	37.3	6.0	(004)	19.4
5	24.7	(000)	95.8	10.6	(010)	39.1
6	32.2	(004)	100.9	10.2	(353)	42.8
1–6	18.3			6.8		

Table 2.8 Energy values in Ry at selected k–points

	Orthogonal	APW	Non-orthogonal
GAMMA 1	0.02889	0.03681	0.03461
GAMMA 12	1.58090	1.73235	1.69030
GAMMA 15	1.30568	1.30778	1.30756
GAMMA 25′	1.59792	1.67675	1.65474
H1 (008)	1.22111	1.23414	1.23266
H12 (008)	0.85841	0.84628	0.86027
H15 (008)	0.67167	0.68054	0.68688
H25′(008)	2.45473	2.53647	2.53359
N1 (044)	0.50108	0.52042	0.51414
N1 (044)	1.46917	1.50042	1.50227
N1 (044)	2.12631	2.31314	2.30934
N2 (044)	1.21354	1.26749	1.27233
N3 (044)	2.71175	2.90831	2.90670
N4 (044)	1.74236	2.14378	2.16238
N1′ (044)	0.29627	0.30800	0.31641
N3′ (044)	1.00454	1.01691	1.02369
N4′ (044)	1.15774	1.12047	1.13000
P1 (444)	0.88566	0.87964	0.88120
P3 (444)	2.20676	2.31064	2.25140
P4 (444)	0.53957	0.53060	0.53128
P4 (444)	2.02479	1.95330	1.96269
(341)	0.28994	0.29742	0.29885
(341)	0.56062	0.56726	0.56892
(341)	0.92371	0.91992	0.91943
(341)	1.04696	1.05841	1.05814
(341)	1.43230	1.43512	1.45006
(341)	1.51663	1.54184	1.54127

Table 2.9 Fermi level quantities (non-orthogonal fit)

Energy		Densities of states				Velocity	Plasmon energy
	Total	s	p	t_{2g}	e_g		
Ry		States/Ry/atom				cm/s	eV
0.2925	6.73	1.47	5.01	0.11	0.14	0.84 × 10E8	6.53

Integrated densities of states Total Electrons	s	p	t_{2g}	e_g
1.00	0.52	0.46	0.02	0.01

Table 2.10 Lithium bcc

Slater-Koster 2-center parameters	Orthogonal	Non-orthogonal	
	Energy integrals (Ry)	Energy integrals (Ry)	Overlap integrals
On site			
s	0.58202	0.47162	
P	0.85622	0.66501	
dl	1.90161	1.28073	
d2	1.51783	1.26881	
First neighbor			
(sss)	–0.07394	–0.02923	0.10486
(pps)	0.14421	–0.03972	–0.23853
(ppp)	–0.01299	–0.02053	0.02532
(dds)	–0.21082	0.37081	0.28295
(ddp)	0.06821	0.01120	–0.10735
(ddd)	0.00278	–0.14875	–0.07580
(sps)	0.09031	–0.00724	0.15718
(sds)	–0.02153	0.09679	0.19923
(pds)	–0.08113	–0.16604	–0.26775
(pdp)	0.02489	–0.01934	0.06520
Second neighbor			
(sss)	–0.02221	–0.02631	0.04683
(pps)	0.05465	0.07571	–0.05651
(ppp)	0.02135	–0.02024	0.00327
(dds)	–0.11815	0.16333	0.21291
(ddp)	0.01402	–0.18027	–0.16930
(ddd)	–0.01384	–0.00398	–0.00015
(sps)	0.03215	–0.04033	0.05678
(sds)	–0.03691	0.01993	0.10741
(pds)	–0.05130	–0.00889	–0.12774
(pdp)	–0.03693	0.00251	0.05283
Third neighbor			
(sss)	0.01462		
(pps)	–0.03099		
(ppp)	0.00750		
(dds)	0.01856		
(ddp)	–0.00150		
(ddd)	0.01421		
(sps)	–0.02134		
(sds)	0.01719		
(pds)	0.03202		
(pdp)	–0.01111		

	Orthogonal			Non-orthogonal		
Band	RMS error mRy	Maximum k	Deviation mRy	RMS error mRy	Maximum k	Deviation mRy
1	6.3	(050)	16.0	2.4	(170)	6.2
2	8.9	(330)	20.6	4.7	(232)	13.2
3	6.1	(330)	13.1	4.4	(044)	9.9
4	10.4	(006)	22.0	5.1	(008)	17.3
5	27.2	(000)	100.9	12.8	(000)	44.1
6	29.2	(004)	107.3	14.4	(332)	47.7
1–6	17.6			8.6		

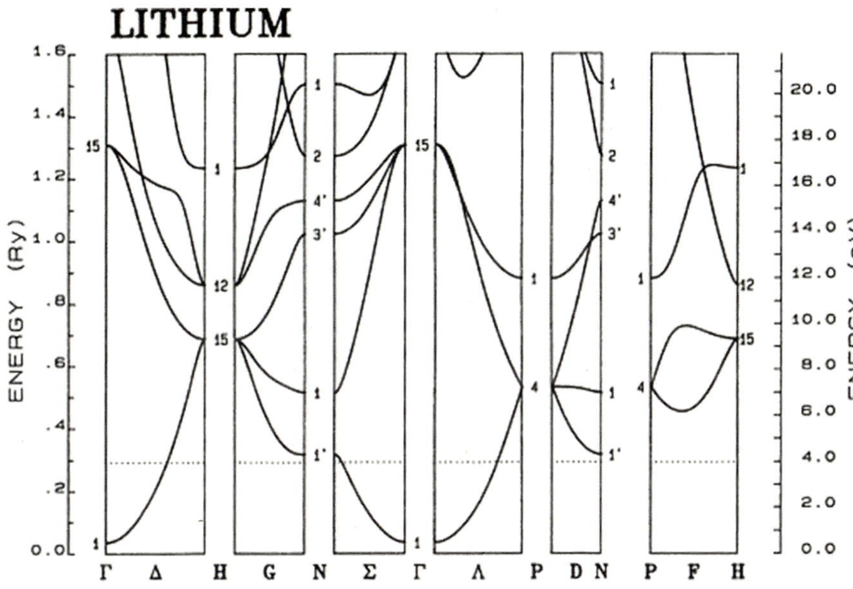

Fig. 2.7 Energy bands for Li

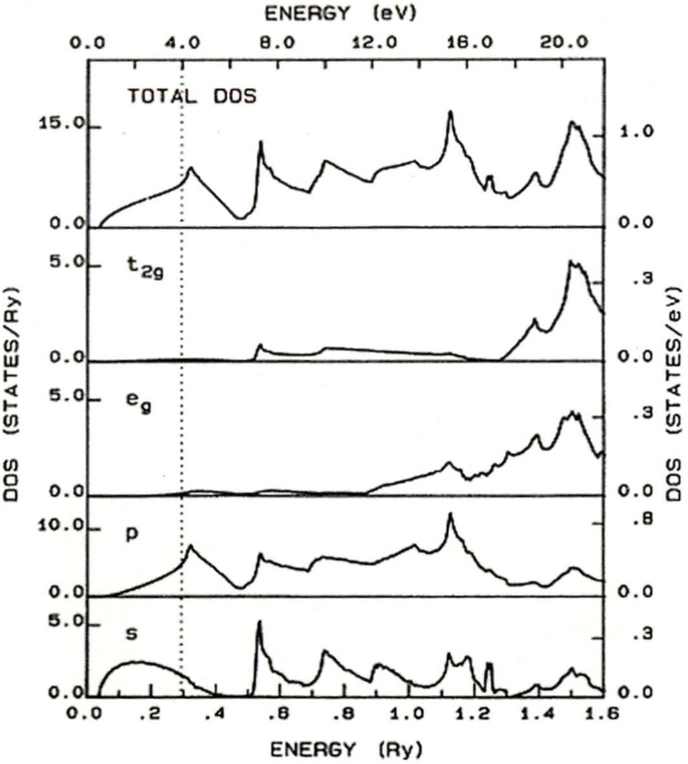

Fig. 2.8 Density of states for Li

2.2.2 Sodium

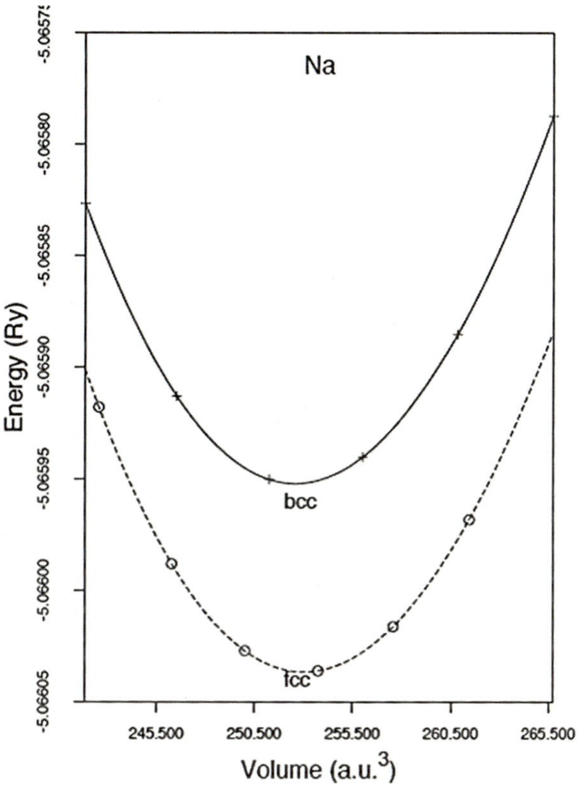

Fig. 2.9 Total energy of Na

Table 2.11 Lattice constants and bulk modulus

	a (Bohr)	B (MBar)
bcc	7.964	0.076
fcc	10.040	0.076
exp	7.984	0.068
$\Delta E = 0.08$ mRy		

Table 2.12 Birch fit coefficients

	A_1	A_2	A_3	A_4
bcc	–324.88318	–16.01031	404.03950	–2242.41563
fcc	–324.88638	–15.68708	392.58037	–2101.67260

Table 2.13 Sodium bcc z = 11 lattice constant = 8.10810 a.u.

Slater-Koster 3-center parameters	Orthogonal	Non-orthogonal	
	Energy integrals (Ry)	Energy integrals (Ry)	Overlap integrals
On site			
s,s (000)	0.50871	0.29767	
x,x (000)	0.70620	0.54175	
xy,xy (000)	1.15966	0.78557	
d2,d2 (000)	1.03615	0.87928	
First neighbor			
s,s (111)	–0.04811	–0.04383	0.06545
s,x (111)	0.04608	0.01656	–0.07334
s, x y (111)	0.05717	–0.01198	–0.09144
x,x (111)	0.03740	0.00662	–0.02553
x,y (111)	0.03966	–0.00929	–0.09702
x,x y (111)	0.03450	–0.03470	–0.06817
x,yz (111)	0.05465	–0.05852	–0.14463
x,d1 (111)	0.00072	–0.01796	–0.05916
x y, x y (111)	–0.03195	0.03922	0.04717
x y, x z (111)	–0.05185	0.09951	0.12583
xy,d2 (111)	0.01221	–0.01724	–0.06268
d2,d2 (111)	0.03333	–0.04639	–0.11442
Second neighbor			
s,s (200)	–0.01406	0.00232	0.07382
s,x (200)	0.01577	0.01169	–0.08027
s,d2 (002)	0.04125	–0.01967	0.04062
x,x (200)	0.07274	0.03301	–0.12325
y,y (200)	–0.01053	–0.00356	0.00543
x,xy (020)	–0.00652	–0.01948	0.01495
z,d2 (002)	–0.05256	–0.02861	0.09531
xy,x y (200)	0.01689	–0.00095	–0.08191
xy,xy (002)	–0.02287	–0.01697	–0.01001
d2,d2 (002)	–0.09741	0.01119	0.12276
d1,d1 (002)	–0.00020	–0.02393	–0.00638
Third neighbor			
s,s (220)	–0.00786		
s,x (220)	0.00300		
s,xy (220)	0.00760		
s,d2 (220)	0.00186		
x,x (220)	0.00460		
x,x (022)	–0.00165		
x,y (220)	–0.00402		
x,xy (220)	0.00310		
x,xy (022)	–0.00298		
z,d2 (022)	–0.00286		
z,d1 (022)	–0.00660		
xy,xy (220)	0.00303		
xy,xy (022)	0.01123		
x y,xz (022)	0.00956		
xy,d2 (220)	0.01148		
d2,d2 (220)	0.01104		
d1,d1 (220)	–0.00107		

Table 2.14 Sodium bcc

	Orthogonal			Non-orthogonal		
Band	RMS error mRy	Maximum k	Deviation mRy	RMS error mRy	Maximum k	Deviation mRy
1	11.6	(000)	32.5	2.5	(044)	9.2
2	10.6	(044)	35.4	3.8	(044)	11.0
3	6.1	(022)	24.0	4.8	(110)	16.0
4	15.1	(002)	59.8	5.3	(010)	21.5
5	18.1	(252)	37.9	7.2	(000)	21.2
6	17.8	(353)	61.4	7.5	(442)	21.8
1–6	13.9			5.5		

Table 2.15 Energy values in Ry at selected k-points

	Orthogonal	APW	Non-orthogonal
GAMMA 1	–0.05482	–0.02234	–0.01984
GAMMA 12	1.06981	1.10459	1.08337
GAMMA 15	1.13895	1.14548	1.13215
GAMMA 25′	1.02779	1.03976	1.03099
H1 (008)	0.71499	0.72029	0.72030
H12 (008)	0.53650	0.53248	0.53531
H15 (008)	0.54061	0.55119	0.55193
H25′(008)	1.53905	1.56962	1.57848
N1 (044)	0.25532	0.29074	0.27973
N1 (044)	0.89564	0.92789	0.92312
N1 (044)	1.35383	1.42573	1.39654
N2 (044)	0.77843	0.79440	0.79714
N3 (044)	1.60809	1.75602	1.75735
N4 (044)	1.25978	1.34353	1.31409
N1′ (044)	0.25002	0.26271	0.27196
N3′ (044)	0.85039	0.84264	0.84997
N4′ (044)	0.88462	0.85702	0.85342
P1 (444)	0.49875	0.49927	0.50935
P3 (444)	1.38878	1.42097	1.40964
P4 (444)	0.40424	0.40135	0.39996
P4 (444)	1.46858	1.48264	1.47754
(341)	0.20803	0.22037	0.21968
(341)	0.36143	0.36818	0.36503
(341)	0.64725	0.65242	0.65546
(341)	0.79801	0.80453	0.80323
(341)	0.94730	0.94031	0.94005
(341)	1.07938	1.08399	1.08036

Table 2.16 Fermi level quantities (non-orthogonal fit)

Energy		Densities of states				Velocity	Plasmon energy
	Total	s	p	t_{2g}	e_g		
Ry		States/Ry/atom				cm/s	eV
0.2090	6.72	3.60	2.95	0.11	0.06	$1.02 \times 10E8$	5.81

Integrated densities of states Total Electrons	s	p	t_{2g}	e_g
1.00	0.74	0.25	0.01	0.00

Table 2.17 Sodium bcc

Slater-Koster 2-center parameters	Orthogonal Energy integrals (Ry)	Non-orthogonal Energy integrals (Ry)	Overlap integrals
On site			
s	0.30461	0.29621	
P	0.70604	0.54786	
d1	1.23293	0.76245	
d2	1.03054	0.83689	
First neighbor			
(sss)	–0.04700	–0.03433	0.07336
(pps)	0.11504	–0.03206	–0.24002
(ppp)	–0.00340	0.01353	0.05972
(dds)	–0.13370	0.25074	0.30997
(ddp)	0.04807	–0.05542	–0.17671
(ddd)	0.00535	–0.02805	–0.00270
(sps)	0.05119	–0.01888	0.13579
(sds)	0.00536	0.02623	0.16935
(pds)	–0.09598	–0.15139	–0.30575
(pdp)	0.01270	0.02050	0.09858
Second neighbor			
(sss)	–0.01245	–0.00995	0.04968
(pps)	0.05772	0.06826	–0.06854
(ppp)	0.01386	–0.00786	0.00678
(dds)	–0.10155	0.05295	0.16414
(ddp)	–0.01391	0.00254	–0.08061
(ddd)	0.00162	–0.01432	–0.00087
(sps)	0.01736	–0.02971	0.04774
(sds)	–0.00519	–0.01810	0.05910
(pds)	–0.05729	0.02460	–0.10805
(pdp)	–0.02091	–0.01926	0.02030
Third neighbor			
(sss)	0.01002		
(pps)	–0.02022		
(ppp)	0.00507		
(dds)	0.02191		
(ddp)	0.00334		
(ddd)	0.00352		
(sps)	–0.01664		
(sds)	0.00549		
(pds)	0.01749		
(pdp)	–0.00645		

	Orthogonal			Non-orthogonal		
Band	RMS error mRy	Maximum k	Deviation mRy	RMS error mRy	Maximum k	Deviation mRy
1	4.9	(004)	14.3	3.4	(044)	12.5
2	7.5	(111)	24.3	5.2	(044)	13.2
3	9.9	(044)	42.7	5.3	(110)	17.6
4	16.6	(002)	64.1	6.2	(010)	21.5
5	21.8	(252)	56.0	8.8	(000)	26.3
6	26.9	(351)	74.5	8.0	(000)	26.3
1–6	16.6			6.4		

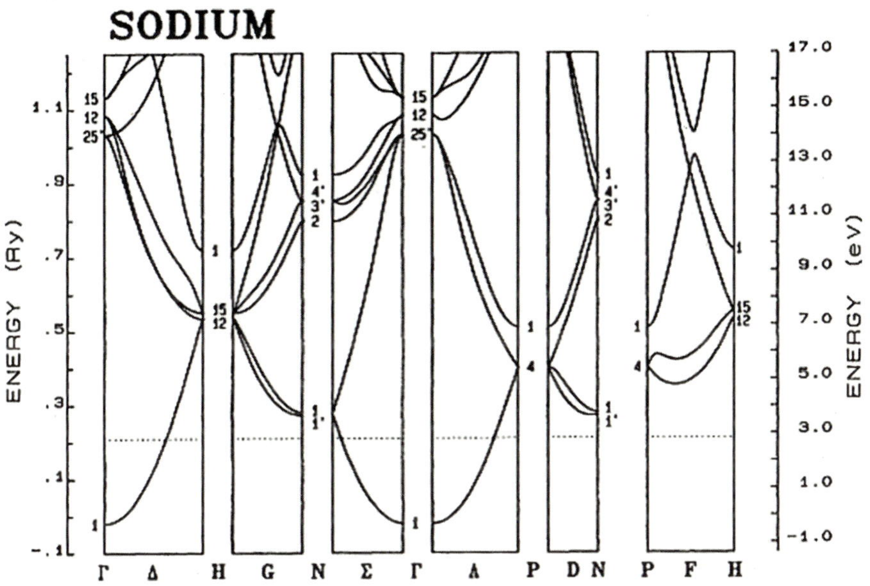

Fig. 2.10 Energy bands for Na

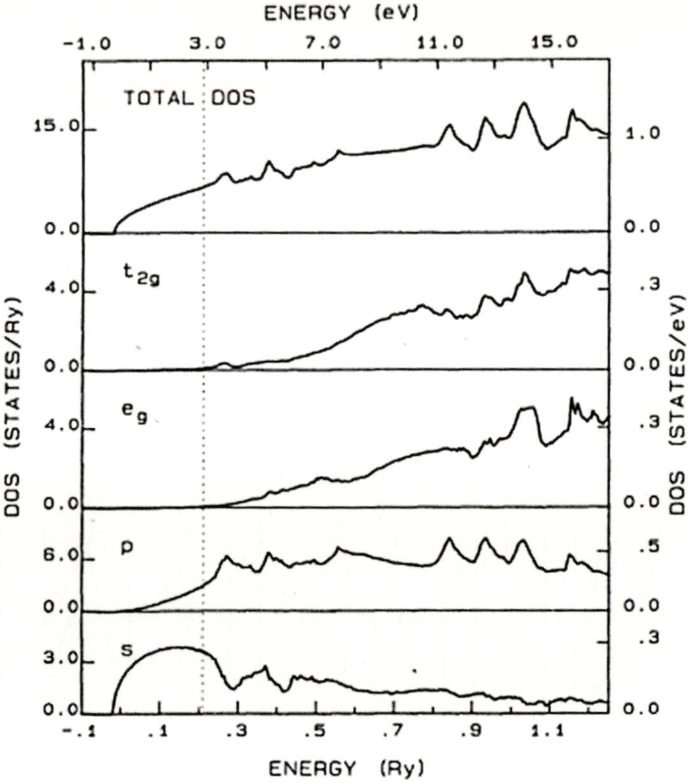

Fig. 2.11 Density of states for Na

2.2.3 Potassium

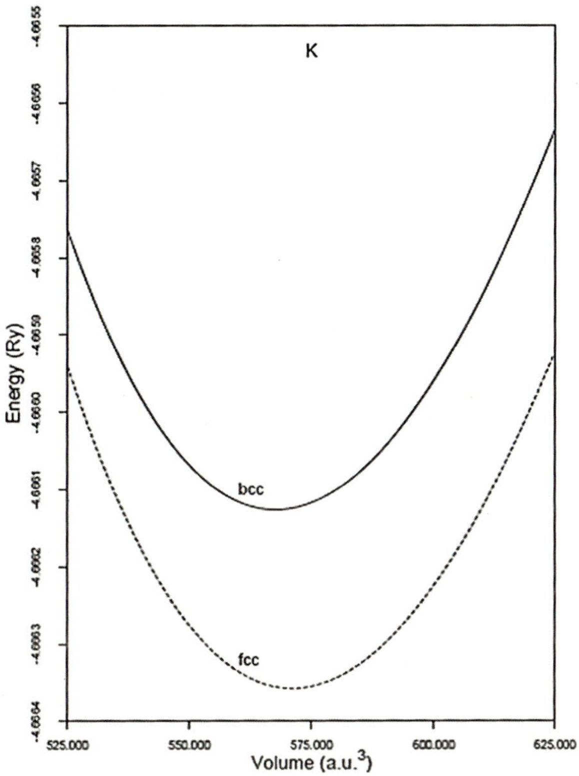

Fig. 2.12 Total energy of K

Table 2.18 Lattice constants and bulk modulus

	a (Bohr)	B (MBar)
bcc	10.432	0.02930
fcc	13.170	0.02901
exp	9.877	0.032
$\Delta E = 0.23$ mRy		

Table 2.19 Birch fit coefficients

	A_1	A_2	A_3	A_4
bcc	–1204.52676	–19.94321	769.44231	–3922.00214
fcc	–1204.51663	–22.19783	927.63600	1352.94743

Table 2.20 Potassium bcc z = 19 lattice constant = 9.90000 a.u.

Slater-Koster 3-center parameters	Orthogonal	Non-orthogonal	
	Energy integrals (Ry)	Energy integrals (Ry)	Overlap integrals
On site			
s,s (000)	0.31991	0.23658	
x,x (000)	0.52401	0.45486	
xy,xy (000)	0.51957	0.50876	
d2,d2 (000)	0.48266	0.43269	
First neighbor			
s,s (111)	–0.03559	–0.03855	0.05811
s,x (111)	0.03152	0.02845	–0.02626
s,xy (111)	0.02102	0.01246	–0.02029
x,x (111)	0.02970	0.01614	0.00863
x,y (111)	0.02419	0.03684	–0.00590
x,xy (111)	0.01636	0.01711	0.01100
x,yz (111)	0.02068	0.01950	–0.02522
x,d1 (111)	0.00538	–0.00084	–0.03379
xy,xy (111)	–0.00893	–0.02144	–0.02342
xy,xz (111)	–0.01549	–0.00954	0.01139
xy,d2 (111)	0.00525	–0.00464	–0.03608
d2,d2 (111)	0.01517	–0.00678	–0.06593
Second neighbor			
s,s (200)	–0.00813	0.01135	0.09306
s,x (200)	0.02244	–0.03633	–0.16525
s,d2 (002)	–0.01402	0.01154	0.09347
x,x (200)	0.06159	–0.09815	–0.30716
y,y (200)	0.00656	–0.00232	–0.00800
x,xy (020)	0.00570	–0.00650	–0.01175
z,d2 (002)	–0.04324	0.04704	0.19337
xy,xy (200)	–0.00191	0.00707	–0.00146
xy,xy (002)	–0.00552	0.00454	–0.00122
d2,d2 (002)	–0.03169	0.01222	0.10006
d1,d1 (002)	0.00337	–0.00923	–0.01313
Third neighbor			
s,s (220)	0.00074		
s,x (220)	–0.00180		
s,xy (220)	–0.00403		
s,d2 (220)	–0.00616		
x,x (220)	–0.00180		
x,x (022)	–0.00058		
x,y (220)	–0.00183		
x,xy (220)	–0.00412		
x,xy (022)	–0.00088		
z,d2 (022)	0.00097		
z,d1 (022)	–0.00383		
xy,xy (220)	0.00526		
xy,xy (022)	0.00189		
xy,xz (022)	0.00038		
xy,d2 (220)	0.00443		
d2,d2 (220)	0.00236		
dl,dl (220)	–0.00171		

Table 2.21 Potassium bcc

	Orthogonal			Non-orthogonal		
Band	RMS error mRy	Maximum k	deviation mRy	RMS error mRy	Maximum k	Deviation mRy
1	0.9	(008)	2.4	0.4	(260)	1.4
2	2.0	(030)	7.0	0.6	(242)	1.7
3	1.1	(442)	2.4	0.7	(151)	2.3
4	2.5	(002)	8.3	1.2	(044)	4.0
5	1.9	(000)	5.6	1.0	(004)	2.4
6	2.8	(008)	7.8	1.3	(008)	3.6
1–6	2.0			0.9		

Table 2.22 Energy values in Ry at selected k-points

	Orthogonal	APW	Non-orthogonal
GAMMA 1	–0.00465	–0.00309	–0.00184
GAMMA 12	0.52293	0.52854	0.52833
GAMMA 15	0.89437	0.89539	0.89528
GAMMA 25′	0.46559	0.46622	0.46577
H1 (008)	0.56480	0.55700	0.56063
H12(008)	0.28028	0.27793	0.27729
H15 (008)	0.41914	0.42093	0.42224
H25′(008)	0.60847	0.61014	0.60864
N1 (044)	0.18775	0.18888	0.18854
N1 (044)	0.50074	0.50583	0.50181
N1 (044)	0.62265	0.62346	0.62643
N2 (044)	0.38564	0.38694	0.38712
N3 (044)	0.63350	0.63093	0.63204
N4 (044)	0.56354	0.56553	0.56329
N1′ (044)	0.20965	0.21366	0.21455
N3′ (044)	0.63302	0.64050	0.64119
N4′ (044)	0.59662	0.60065	0.60355
P1 (444)	0.37759	0.38020	0.38151
P3 (444)	0.57152	0.57110	0.57322
P4 (444)	0.26847	0.26870	0.26796
P4 (444)	0.66385	0.66431	0.66139
(341)	0.16693	0.16722	0.16739
(341)	0.25054	0.25238	0.25260
(341)	0.36987	0.37056	0.37088
(341)	0.49385	0.49398	0.49370
(341)	0.55314	0.55519	0.55406
(341)	0.57950	0.57610	0.57606

Table 2.23 Fermi level quantities (non-orthogonal fit)

Energy		Densities of states				Velocity	Plasmon energy
	Total	s	p	t_{2g}	e_g		
Ry		States/Ry/atom				cm/s	eV
0.1605	10.49	6.72	3.04	0.68	0.04	0.80 × 10E8	4.21

Integrated densities of states Total Electrons	s	p	t_{2g}	e_g
1.00	0.80	0.17	0.03	0.00

Table 2.24 Potassium bcc

Slater-Koster 2-center parameters	Orthogonal	Non-orthogonal	
	Energy integrals (Ry)	Energy integrals (Ry)	Overlap integrals
On site			
s	0.27215	0.19115	
p	0.52399	0.46184	
d1	0.53054	0.45778	
d2	0.48660	0.44087	
First neighbor			
(sss)	–0.03593	–0.01746	0.10891
(pps)	0.07871	0.02533	–0.09507
(ppp)	0.00503	–0.04728	–0.05677
(dds)	–0.04036	0.02147	0.11630
(ddp)	0.02210	0.01139	–0.05714
(ddd)	0.00048	–0.01718	–0.01713
(sps)	0.04497	–0.00444	0.13311
(sds)	–0.02809	0.00606	0.11169
(pds)	–0.06075	0.00057	–0.10705
(pdp)	0.00848	–0.03728	–0.01593
Second neighbor			
(sss)	–0.00801	–0.00925	0.06002
(pps)	0.05477	–0.00667	–0.10384
(ppp)	0.00535	–0.01902	–0.00655
(dds)	–0.03133	0.00454	0.08125
(ddp)	–0.00319	0.00172	–0.02986
(ddd)	0.00262	0.00493	0.01126
(sps)	0.01797	–0.00424	0.06375
(sds)	–0.01193	0.00679	0.08793
(pds)	–0.03593	–0.01918	–0.12217
(pdp)	–0.00667	–0.00262	0.01945
Third neighbor			
(sss)	0.00484		
(pps)	–0.00410		
(ppp)	0.00214		
(dds)	0.00430		
(ddp)	–0.00265		
(ddd)	0.00196		
(sps)	–0.00659		
(sds)	0.00647		
(pds)	0.00350		
(pdp)	–0.00298		

	Orthogonal			Non-orthogonal		
	RMS error	Maximum	Deviation	RMS error	Maximum	Deviation
Band	mRy	k	mRy	mRy	k	mRy
1	1.6	(050)	3.8	0.7	(044)	1.8
2	2.1	(030)	8.2	1.0	(030)	3.4
3	2.2	(330)	4.7	1.4	(044)	3.0
4	4.4	(044)	16.4	2.4	(444)	5.3
5	2.9	(000)	9.5	1.9	(000)	6.2
6	4.3	(008)	12.6	2.7	(022)	6.4
1–6	3.1			1.8		

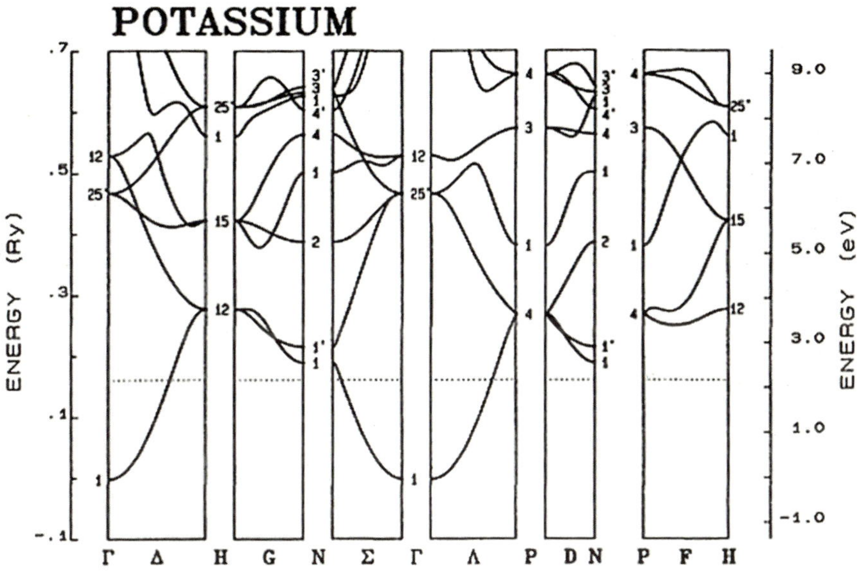

Fig. 2.13 Energy bands for K

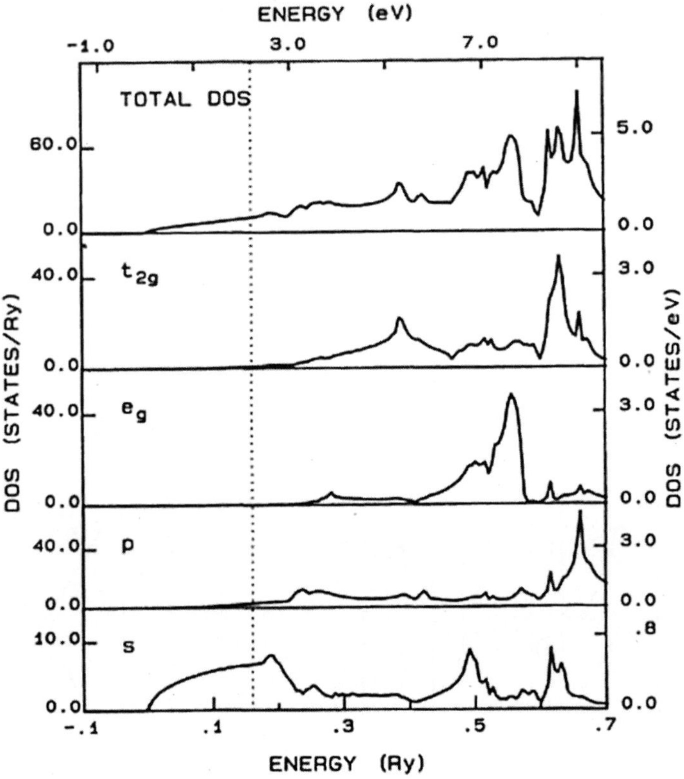

Fig. 2.14 Density of states for K

2.2.4 Rubidium

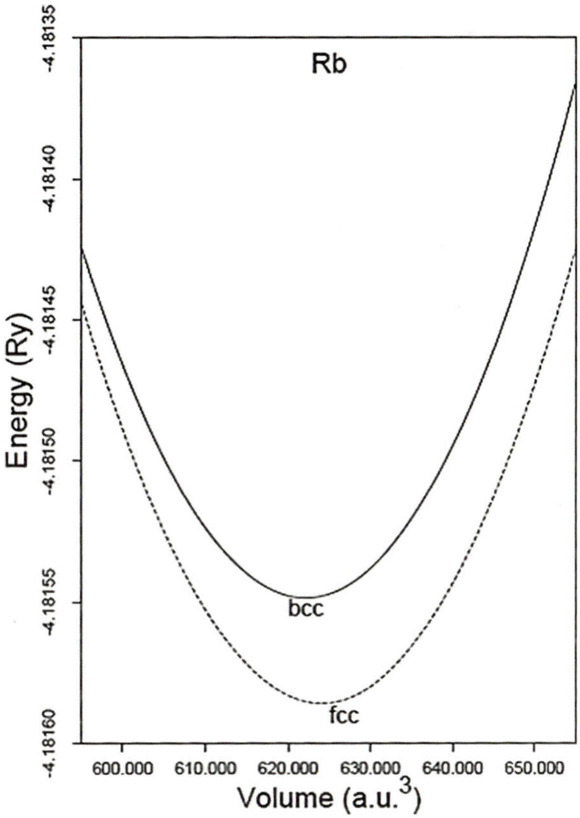

Fig. 2.15 Total energy of Rb

Table 2.25 Lattice constants and bulk modulus

	a (Bohr)	B (MBar)
bcc	10.753	0.03167
fcc	13.563	0.03114
exp	10.558	0.031
$\Delta E = 0.04$ mRy		

Table 2.26 Birch fit coefficients

	A_1	A_2	A_3	A_4
bcc	–5963.97658	–33.82265	1664.66250	–21015.22774
fcc	–5964.00170	–28.55059	1292.44782	–12181.83454

Table 2.27 Rubidium bcc z = 37 lattice constant = 10.60000 a.u.

Slater-Koster 3-center parameters	Orthogonal	Non-orthogonal	
	Energy integrals (Ry)	Energy integrals (Ry)	Overlap integrals
On site			
s,s (000)	0.28819	0.17799	
x,x (000)	0.47376	0.39837	
xy,xy (000)	0.45889	0.44251	
d2,d2 (000)	0.42899	0.37072	
First neighbor			
s,s (111)	–0.03103	–0.01249	0.09776
s,x (111)	0.02867	0.00479	–0.07177
s,xy (111)	0.02338	0.00712	–0.03627
x,x (111)	0.02734	0.00072	–0.02551
x,y (111)	0.02078	0.01789	–0.03200
x,xy (111)	0.01655	0.00931	–0.00891
x,yz (111)	0.02260	0.01110	–0.04217
x,d1 (111)	0.00448	0.00064	–0.03385
xy,xy (111)	–0.01010	–0.01284	–0.00752
xy,xz (111)	–0.01667	–0.00617	0.02312
xy,d2 (111)	0.00500	–0.00716	–0.04707
d2,d2 (111)	0.01419	–0.00768	–0.07749
Second neighbor			
s,s (200)	–0.00685	–0.01554	0.02821
s,x (200)	0.01642	0.01084	–0.07115
s,d2 (002)	–0.00493	–0.01167	0.05377
x,x (200)	0.05505	–0.03918	–0.21041
y,y (200)	0.00215	–0.00254	0.00258
x,xy (020)	0.00131	–0.00386	0.00064
z,d2 (002)	–0.03873	0.01359	0.14234
xy,xy (200)	0.00098	0.00031	–0.01720
xy,xy (002)	–0.00483	0.00351	–0.00198
d2,d2 (002)	–0.03135	0.01170	0.11390
d1,d1 (002)	0.00203	–0.01008	–0.01582
Third neighbor			
s,s (220)	–0.00086		
s,x (220)	–0.00007		
s,xy (220)	0.00001		
s,d2 (220)	–0.00515		
x,x (220)	0.00047		
x,x (022)	0.00079		
x,y (220)	–0.00041		
x,xy (220)	–0.00112		
x,xy (022)	–0.00063		
z,d2 (022)	0.00005		
z,d1 (022)	–0.00352		
xy,xy (220)	0.00275		
xy,xy (022)	0.00219		
xy,xz (022)	0.00058		
xy,d2 (220)	0.00352		
d2,d2 (220)	0.00298		
d1,d1 (220)	–0.00242		

Table 2.28 Rubidium bcc

	Orthogonal			Non-orthogonal		
Band	RMS error mRy	Maximum k	Deviation mRy	RMS error mRy	Maximum k	Deviation mRy
1	1.6	(000)	4.2	0.5	(260)	1.4
2	2.2	(030)	9.2	0.6	(030)	1.6
3	1.4	(006)	3.4	0.7	(006)	1.9
4	2.9	(002)	10.5	1.7	(350)	4.7
5	2.8	(000)	8.1	1.3	(004)	3.6
6	3.4	(008)	9.6	1.7	(044)	6.8
1–6	2.5			1.2		

Table 2.29 Energy values in Ry at selected k-points

	Orthogonal	APW	Non-orthogonal
GAMMA 1	–0.01149	–0.00729	–0.00776
GAMMA 12	0.45788	0.46593	0.46584
GAMMA 15	0.81806	0.81894	0.81882
GAMMA 25′	0.40089	0.40148	0.40143
H1 (008)	0.48502	0.47539	0.47700
H12 (008)	0.23086	0.22882	0.22832
H15 (008)	0.38063	0.38227	0.38319
H25′ (008)	0.56251	0.56335	0.56056
N1 (044)	0.15300	0.15445	0.15399
N1 (044)	0.43246	0.43985	0.43686
N1 (044)	0.57069	0.56921	0.57314
N2 (044)	0.32414	0.32516	0.32475
N3 (044)	0.59092	0.59157	0.59200
N4 (044)	0.50235	0.50926	0.50671
N1′ (044)	0.19426	0.19858	0.19895
N3′ (044)	0.57468	0.57949	0.58042
N4′ (044)	0.52675	0.52536	0.53216
P1 (444)	0.31902	0.32234	0.32647
P3 (444)	0.52029	0.51657	0.51839
P4 (444)	0.23523	0.23511	0.23508
P4 (444)	0.61992	0.62177	0.61874
(341)	0.14329	0.14386	0.14336
(341)	0.22023	0.22255	0.22280
(341)	0.31270	0.31318	0.31322
(341)	0.42780	0.42866	0.42883
(341)	0.49302	0.49520	0.49399
(341)	0.52197	0.51946	0.52012

Table 2.30 Fermi level quantities (non-orthogonal fit)

Energy		Densities of states				velocity	Plasmon energy
	Total	s	p	t_{2g}	e_g		
Ry		States/Ry/atom				cm/s	eV
0.1390	12.52	8.35	2.64	0.67	0.86	0.72 × 10E8	3.75

Integrated densities of states				
Total Electrons	s	p	t_{2g}	e_g
1.00	0.80	0.15	0.02	0.03

Table 2.31 Rubidium bcc

Slater-Koster 2-center parameters	Orthogonal	Non-orthogonal	
	Energy integrals (Ry)	Energy integrals (Ry)	Overlap integrals
On site			
s	0.22222	0.16075	
p	0.47173	0.38587	
d1	0.47861	0.39223	
d2	0.42944	0.37757	
First neighbor			
(sss)	–0.03119	–0.01326	0.11087
(pps)	0.06919	–0.00906	–0.17511
(ppp)	0.00626	–0.03651	–0.03955
(dds)	–0.04323	0.02531	0.14482
(ddp)	0.02041	0.00168	–0.08777
(ddd)	0.00125	–0.00958	–0.00524
(sps)	0.03783	0.00448	0.16476
(sds)	–0.02542	0.00709	0.12599
(pds)	–0.05942	–0.02304	–0.16963
(pdp)	0.00780	–0.02605	0.00726
Second neighbor			
(sss)	–0.00637	–0.01156	0.04986
(pps)	0.04970	–0.00237	–0.11575
(ppp)	0.00531	–0.00208	0.02434
(dds)	–0.03123	0.00531	0.09705
(ddp)	–0.00525	–0.00281	–0.04476
(ddd)	0.00217	–0.00020	0.00238
(sps)	0.01374	–0.01134	0.05867
(sds)	–0.00812	–0.00376	0.07112
(pds)	–0.03252	–0.01088	–0.12632
(pdp)	–0.00731	0.00393	0.03729
Third neighbor			
(sss)	0.00471		
(pps)	–0.00442		
(ppp)	0.00288		
(dds)	0.00458		
(ddp)	–0.00257		
(ddd)	0.00236		
(sps)	–0.00658		
(sds)	0.00536		
(pds)	0.00331		
(pdp)	–0.00316		

	Orthogonal			Non-orthogonal		
Band	RMS error mRy	Maximum k	Deviation mRy	RMS error mRy	Maximum k	Deviation mRy
1	1.9	(050)	4.4	0.6	(050)	1.5
2	2.3	(030)	9.3	1.0	(050)	2.4
3	2.2	(250)	4.7	1.1	(150)	2.9
4	4.9	(044)	17.6	2.3	(444)	6.0
5	3.7	(004)	10.0	1.8	(330)	6.9
6	5.5	(350)	20.0	2.6	(044)	7.8
1–6	3.7			1.7		

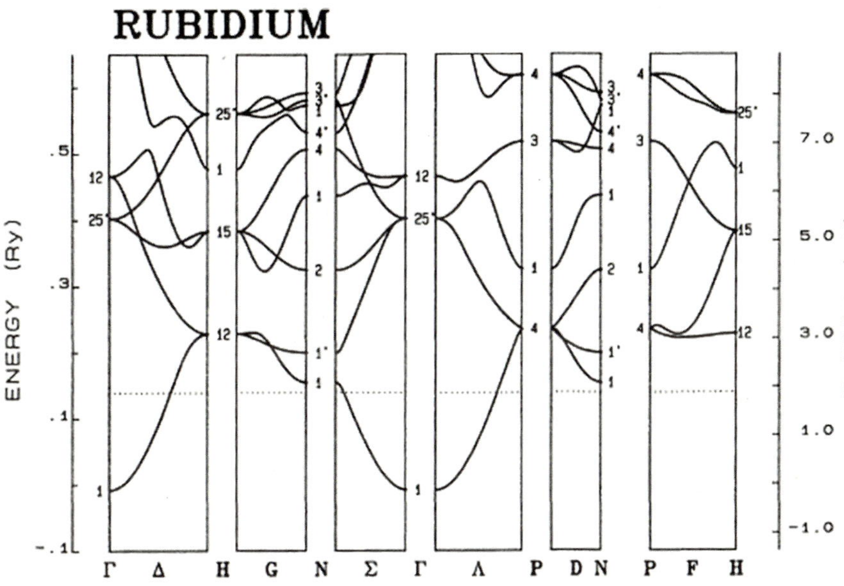

Fig. 2.16 Energy bands for Rb

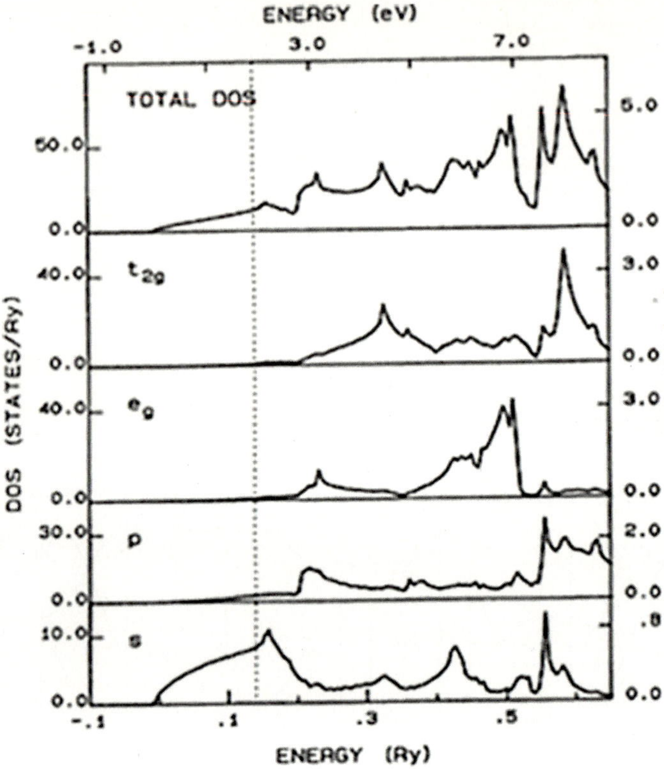

Fig. 2.17 Density of states for Rb

2.2.5 Cesium

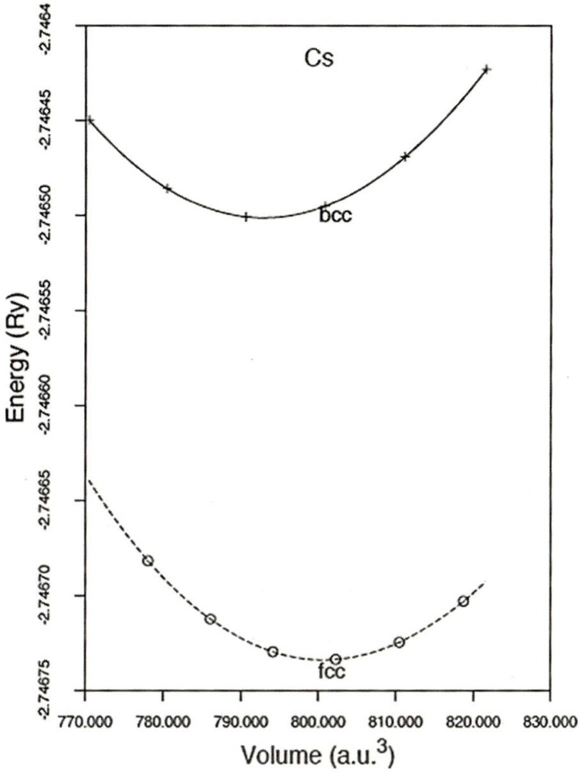

Fig. 2.18 Total energy of Cs

Table 2.32 Lattice constants and bulk modulus

	a (Bohr)	B (MBar)
bcc	11.661	0.02309
fcc	14.740	0.02311
exp	11.423	0.020
$\Delta E = 0.23$ mRy		

Table 2.33 Birch fit coefficients

	A_1	A_2	A_3	A_4
bcc	−15582.51423	−47.69720	3058.59737	−58004.91710
fcc	−15582.53067	−43.68709	2714.79413	−47787.68191

Table 2.34 Cesium bcc z = 55 lattice constant = 11.42340 a.u.

Slater-Koster 3-center parameters	Orthogonal	Non-orthogonal	
	Energy integrals (Ry)	Energy integrals (Ry)	Overlap integrals
On site			
s,s (000)	0.20421	0.14796	
x,x (000)	0.43484	0.32619	
xy,xy (000)	0.39653	0.31241	
d2,d2 (000)	0.33906	0.30870	
First neighbor			
s,s (111)	–0.02821	–0.01186	0.11157
s,x (111)	0.01959	–0.00416	–0.10578
s,xy (111)	0.01015	–0.00104	–0.06717
x,x (111)	0.02438	–0.01180	–0.05071
x,y (111)	0.01574	–0.01652	–0.11280
x,xy (111)	0.01640	–0.01022	–0.05277
x,yz (111)	0.01557	–0.01062	–0.09554
x,d1 (111)	0.00445	–0.00274	–0.04480
xy,xy (111)	–0.00913	0.00033	0.02198
xy,xz (111)	–0.01484	0.00767	0.06183
xy,d2 (111)	0.00075	–0.00065	–0.04026
d2, d2 (111)	0.01203	–0.00393	–0.06679
Second neighbor			
s,s (200)	–0.00570	–0.00960	0.05117
s,x (200)	0.01436	0.00778	–0.09159
s,d2 (002)	–0.01542	–0.00595	0.06169
x,x (200)	0.04815	0.00554	–0.15431
y,y (200)	0.01237	–0.01040	–0.00463
x,xy (020)	0.01246	–0.00943	0.00183
z,d2 (002)	–0.03567	–0.00655	0.10046
xy,xy (200)	–0.01079	0.00677	–0.02163
xy,xy (002)	0.00080	–0.00381	–0.01207
d2,d2 (002)	–0.02690	–0.00932	0.06077
d1,d1 (002)	0.00314	–0.00629	–0.01111
Third neighbor			
s,s (220)	0.00444		
s,x (220)	–0.00528		
s,xy (220)	–0.00699		
s,d2 (220)	–0.00465		
x,x (220)	–0.00438		
x,x (022)	0.00257		
x,y (220)	–0.00566		
x,xy (220)	–0.00569		
x,xy (022)	0.00105		
z,d2 (022)	0.00069		
z,d1 (022)	–0.00344		
xy,xy (220)	0.00626		
xy,xy (022)	0.00087		
xy,xz (022)	–0.00056		
xy,d2 (220)	0.00507		
d2,d2 (220)	0.00207		
d1,d1 (220)	–0.00103		

Table 2.35 Cesium bcc

	Orthogonal			Non-orthogonal		
Band	RMS error mRy	Maximum k	Deviation mRy	RMS error mRy	Maximum k	Deviation mRy
1	1.8	(260)	4.3	0.4	(000)	1.4
2	1.6	(030)	6.0	0.4	(443)	0.8
3	1.2	(442)	3.3	0.4	(006)	1.1
4	2.3	(002)	8.6	0.6	(444)	1.3
5	1.8	(000)	5.4	0.6	(131)	2.6
6	3.0	(350)	7.7	0.7	(030)	2.3
1–6	2.0			0.5		

Table 2.36 Energy values in Ry at selected k-points

	Orthogonal	APW	Non-orthogonal
GAMMA 1	–0.00239	–0.00347	–0.00206
GAMMA 12	0.37026	0.37561	0.37492
GAMMA 15	0.75095	0.75345	0.75340
GAMMA 25′	0.31393	0.31419	0.31406
H1 (008)	0.44897	0.44661	0.44685
H12 (008)	0.17783	0.17443	0.17425
H15 (008)	0.36081	0.36198	0.36166
H25′(008)	0.46000	0.46129	0.46145
N1 (044)	0.12466	0.12665	0.12664
N1 (044)	0.35986	0.36224	0.36102
N1 (044)	0.48577	0.48624	0.48712
N2 (044)	0.25116	0.25094	0.25112
N3 (044)	0.48859	0.48752	0.48845
N4 (044)	0.40844	0.41063	0.40962
N1′ (044)	0.20231	0.20212	0.20230
N3′ (044)	0.52698	0.53363	0.53369
N4′ (044)	0.45421	0.45006	0.45027
P1 (444)	0.29173	0.29539	0.29667
P3 (444)	0.41658	0.41645	0.41777
P4 (444)	0.20563	0.20646	0.20632
P4 (444)	0.52880	0.53002	0.52979
(341)	0.12828	0.12792	0.12785
(341)	0.20605	0.20632	0.20650
(341)	0.24604	0.24700	0.24697
(341)	0.35619	0.35572	0.35553
(341)	0.40364	0.40434	0.40383
(341)	0.44943	0.44580	0.44657

Table 2.37 Fermi level quantities (non-orthogonal fit)

Energy		Densities of states				Velocity	Plasmon energy
	Total	s	p	t_{2g}	e_g		
Ry		States/Ry/atom				cm/s	eV
0.1270	20.43	14.05	1.93	2.59	1.87	0.52 × 10E8	3.11

Integrated densities of states				
Total Electrons	s	p	t_{2g}	e_g
1.00	0.85	0.07	0.05	0.03

Table 2.38 Cesium bcc

Slater-Koster 2-center parameters	Orthogonal	Non-orthogonal	
	Energy integrals (Ry)	Energy integrals (Ry)	Overlap integrals
On site			
s	0.20957	0.14256	
p	0.43350	0.31104	
d1	0.38849	0.30768	
d2	0.34206	0.30603	
First neighbor			
(sss)	–0.02851	–0.01135	0.11717
(pps)	0.05671	–0.06847	–0.33248
(ppp)	0.00770	–0.00679	0.03905
(dds)	–0.03897	0.02077	0.15680
(ddp)	0.01698	0.00023	–0.09177
(ddd)	0.00174	–0.00593	–0.00076
(sps)	0.03375	0.01220	0.20530
(sds)	–0.02577	0.00683	0.13136
(pds)	–0.05391	–0.04446	–0.23055
(pdp)	0.00619	–0.00545	0.05598
Second neighbor			
(sss)	–0.00563	–0.00894	0.05658
(pps)	0.04577	0.00758	–0.14135
(ppp)	0.00497	–0.00103	0.02453
(dds)	–0.02669	–0.00475	0.07421
(ddp)	–0.00608	0.00302	–0.03309
(ddd)	0.00261	–0.00317	–0.00489
(sps)	0.01308	–0.00827	0.09006
(sds)	–0.00971	–0.00490	0.06718
(pds)	–0.03002	0.00071	–0.11388
(pdp)	–0.00628	–0.00305	0.02404
Third neighbor			
(sss)	0.00401		
(pps)	–0.00332		
(ppp)	0.00247		
(dds)	0.00429		
(ddp)	–0.00161		
(ddd)	0.00145		
(sps)	–0.00544		
(sds)	0.00514		
(pds)	0.00218		
(pdp)	–0.00255		

	Orthogonal			Non-orthogonal		
Band	RMS error mRy	Maximum k	Deviation mRy	RMS error mRy	Maximum k	Deviation mRy
1	1.8	(260)	4.1	0.6	(000)	2.6
2	1.7	(030)	7.6	0.4	(111)	1.1
3	2.0	(226)	5.0	0.7	(004)	1.2
4	4.6	(044)	15.9	1.0	(002)	3.0
5	2.9	(260)	7.5	0.8	(131)	2.7
6	5.1	(350)	14.8	1.1	(030)	3.2
1–6	3.3			0.8		

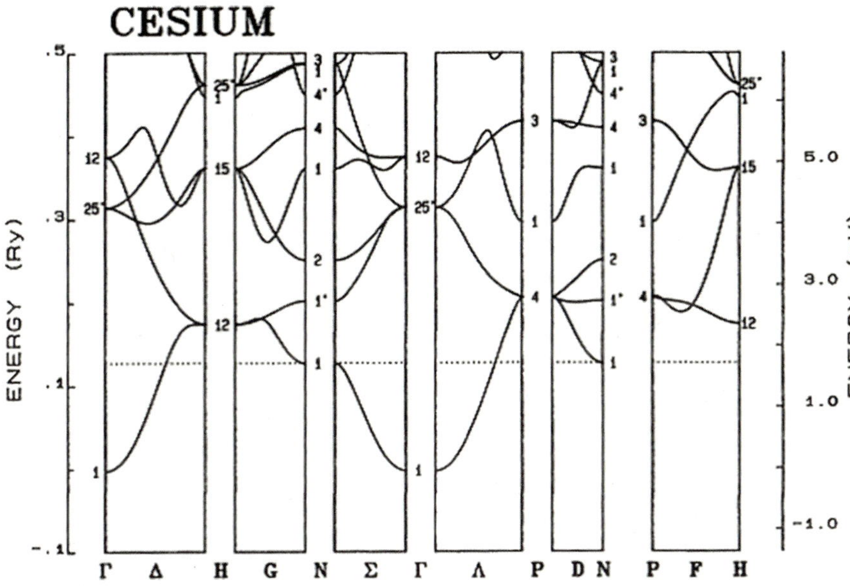

Fig. 2.19 Energy bands for Cs

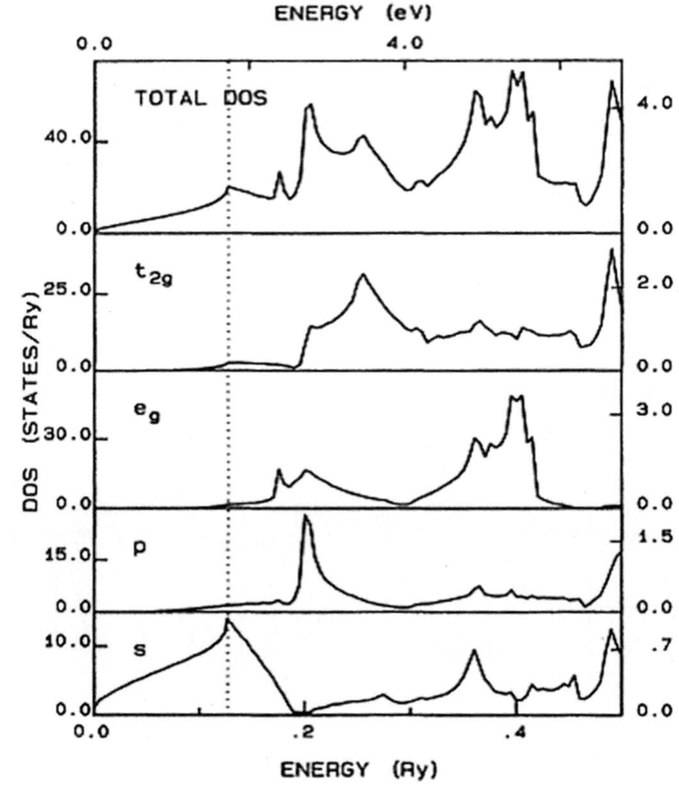

Fig. 2.20 Density of states for Cs

2.2.6 Francium

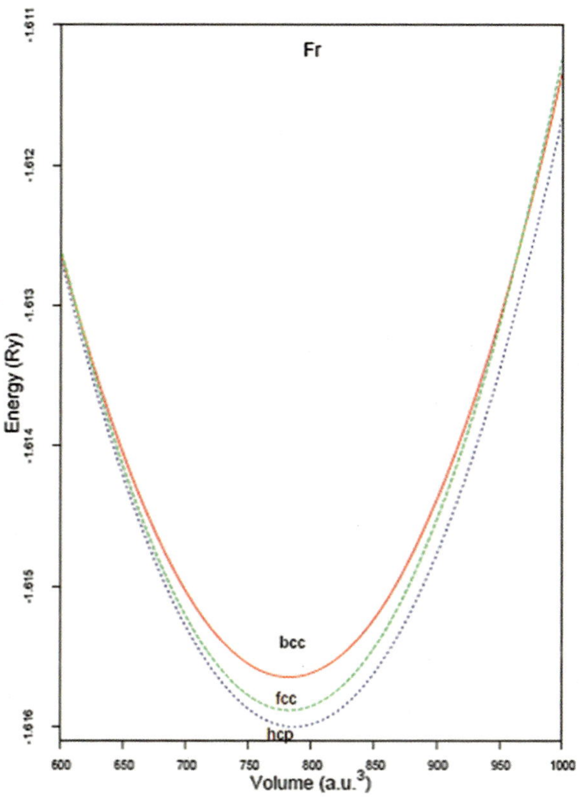

Fig. 2.21 Total energy of Fr

Table 2.39 Lattice constants and bulk modulus

	a (Bohr)	c (Bohr)	B (MBar)
bcc	11.578		0.01915
fcc	14.621		0.01919
hcp	10.307	17.006	0.01883
exp			(0.020)
$\Delta E_{fcc-hcp} = 0.10$ mRy			

Table 2.40 Birch fit coefficients

	A_1	A_2	A_3	A_4
bcc	–48661.45485	–31.16170	1823.70799	–28602.70449
fcc	–48661.48318	–24.04034	1213.99361	–10986.18652
hcp	–97323.00797	–60.19878	402.15876	3143.43348

Table 2.41 Density of states at E_F (States/Ry/atom)

	E_F	Total	s	p	d	f
bcc	0.14913	19.85549	5.60923	1.23482	2.90257	0.05151
fcc	0.13932	17.27354	5.68378	1.48802	2.58804	0.06354

Francium

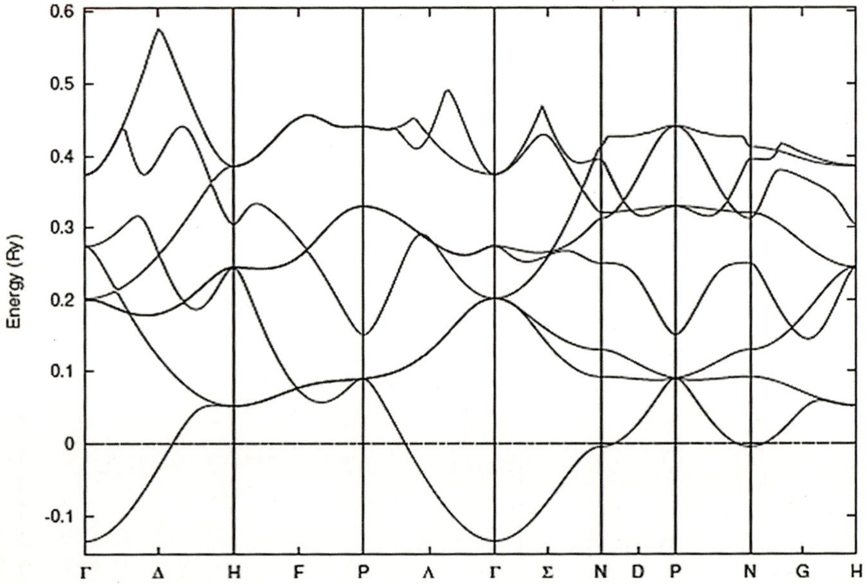

Fig. 2.22 Energy bands for Fr

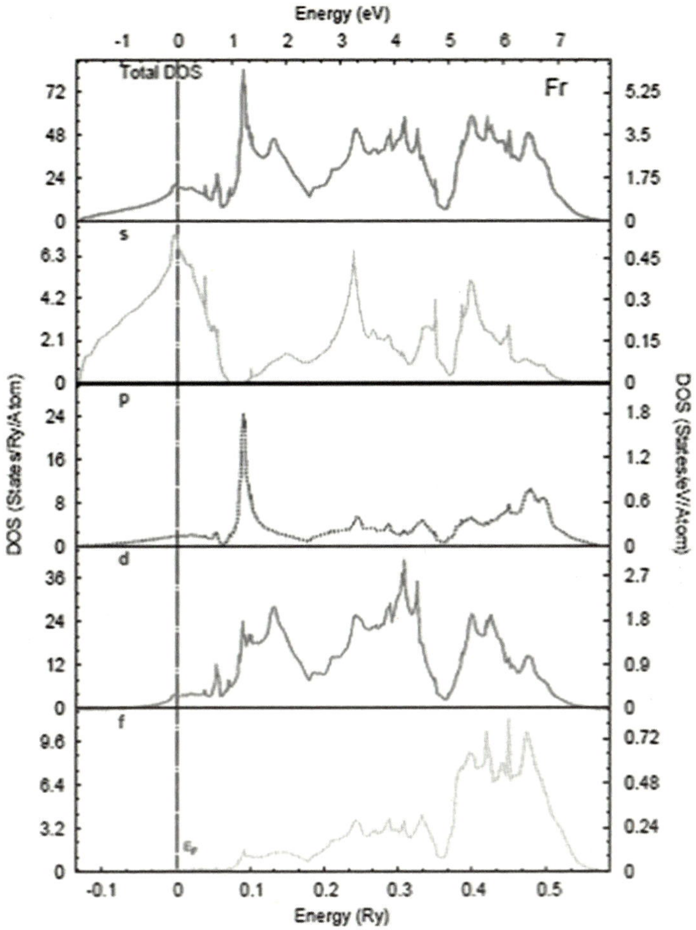

Fig. 2.23 Density of states for Fr

Table 2.42 DOS at Fermi level

E_{Fermi} (Ry)	Total (States/Ry)	s	p	d	f
0.13549	19.77186	7.19800	1.89658	3.90929	0.09406

Francium bcc structure a = 10.80 Bohr
Tight-binding parameters in Ry
Non-orthogonal

On-site									
s	p	t2g	eg						
0.35430	47.15784	0.41703	0.43817						
sss	sps	pps	ppp	sds	pds	pdp	dds	ddp	ddd
Hopping dist = 7.07107									
–0.01268	0.04110	0.24050	0.22087	–0.03252	–0.04721	–0.01670	0.14515	0.00509	–0.10044
Overlap dist = 7.07107									
0.03903	–0.10887	0.18627	0.21817	–0.15152	–0.06313	–0.07895	0.18368	–0.06484	–0.03820
Hopping dist = 10.00000									
–0.03031	0.00282	0.02273	0.01481	–0.00243	–0.00498	0.00000	0.01150	0.00196	–0.00678
Overlap dist = 10.00000									
0.00965	0.00259	0.00734	0.02127	–0.01660	–0.00742	–0.04224	0.01618	–0.01357	–0.00274
Hopping dist = 12.24745									
–0.00156	0.00003	0.00034	0.00017	–0.00003	–0.00008	0.00000	0.00015	0.00004	–0.00008
Overlap dist = 12.24745									
0.00011	0.00013	0.00006	0.00033	–0.00029	–0.00012	–0.00089	0.00022	–0.00032	–0.00003

Fitting rms error for 4 bands 13 mRy

Orthogonal

On-site									
s	p	d							
1.19419	1.13992	0.73800							
sss	sps	pps	ppp	sds	pds	pdp	dds	ddp	ddd
Hopping dist = 7.07107									
–0.05940	0.14955	0.23981	0.17580	–0.06973	–0.12695	–0.04101	0.05797	–0.03104	–0.04014
Hopping dist = 10.00000									
–0.04689	0.01435	0.07391	0.00970	0.04554	–0.05158	–0.00214	–0.00902	0.05152	–0.00056
Hopping dist = 12.24745									
–0.00223	0.00022	0.00205	0.00010	0.00272	–0.00168	0.00000	–0.00038	0.00228	0.00001

Fitting rms error for 4 bands 75 mRy

References

1. N.W. Ashcroft, Phys. Rev. Lett. **21**, 1748 (1968)
2. D.A. Papaconstantopoulos, B.M. Klein, Ferroelectrics **16**, 307 (1977)
3. A. Koufos, D.A. Papaconstantopoulos, Electronic structure of francium. Int. J. Quant. Chem. **133**(17), 2070–2077 (2013)

Chapter 3
The Alkaline Earth Metals

The elements of the second column of the periodic Table, known as the alkaline earths, are characterized by a full s shell. The two lighter of them, Be and Mg, crystallize in the hcp structure, Ca and Sr are fcc, and Ba and Ra are bcc. The total energy calculations predict the correct ground state for all of them. For Mg, Sr, and Ba the calculations are based on Hedin-Lundqvist LDA exchange and correlation, while for Be, Ca, and Ra the GGA form was used. The GGA gives lattice constants closer to the experimental values. Be has a very low value of $N(E_F)$. Ca and Sr have very similar energy bands i.e., free-electron-like occupied states and d states above E_F. Their main difference is in the value of $N(E_F)$ which is fairly high for Ca and practically zero for Sr, suggesting semi-metallic behavior. Finally, Ba and Ra resemble Cs in the shape of the energy bands and DOS.

D.A. Papaconstantopoulos, *Handbook of the Band Structure of Elemental Solids*,
DOI 10.1007/978-1-4419-8264-3_3

3.1 Beryllium

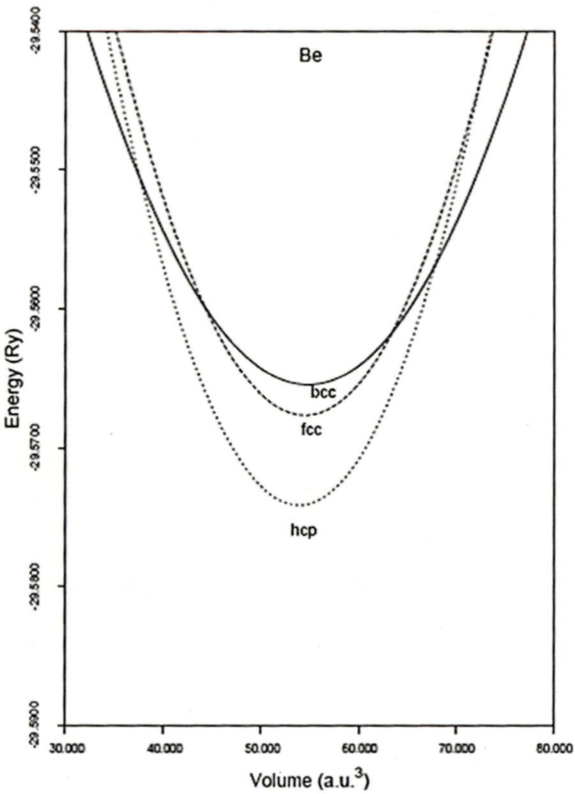

Fig 3.1 Total energy of Be

Table 3.1 Lattice constants and bulk modulus

	a (Bohr)	c (Bohr)	B (MBar)
bcc	4.723		1.241
fcc	5.966		1.224
hcp	4.267	6.742	1.244
exp	4.290	6.784	1.003
$E_{fcc}–E_{hcp} = \Delta E = 6.2$ mRy			

Table 3.2 Birch fit coefficients

	A_1	A_2	A_3	A_4
bcc	−28.93808	−19.50480	175.41096	−359.55137
fcc	−28.94239	−19.48243	176.08478	−362.57220

Table 3.3 Beryllium hcp Slater-Koster 3-center parameters for a = 4.32 a.u., c = 6.77 a.u.

	Orthogonal		
	Energy integrals (Ry)		Energy integrals (Ry)
On site			
s, s (0)	1.16373		
x, x (0)	1.51758		
z, z (0)	1.41444		
xy, xy (0)	2.72325		
yz, yz (0)	2.74154		
d2, d2 (0)	2.68968		
s, d2 (0)	– 0.06855		
x, xy (0)	– 0.19272		
First neighbor			
s, s (R)	–0.07750	s, s (T)	–0.11076
s, x (R)	0.13293	s, y (T)	–0.10893
s, y (R)	0.03181	s, z (T)	0.13147
s, xy (R)	–0.00448	s, d1 (T)	–0.02926
s, d1 (R)	0.10156	s, yz (T)	–0.07987
s, d2 (R)	–0.08464	s, d2 (T)	0.02007
x, x (R)	0.20766	x, x (T)	–0.01935
y, y (R)	–0.01374	y, y (T)	0.06735
x, y (R)	0.01242	y, z (T)	– 0.13183
x, xy (R)	–0.03434	x, xy (T)	0.02493
y, d1 (R)	0.03201	y, d1 (T)	–0.05315
x, d1 (R)	0.15753	x, xz (T)	– 0.08465
y, xy (R)	– 0.03741	y, yz (T)	0.08377
y, d2 (R)	0.00820	y, d2 (T)	– 0.08222
x, d2 (R)	–0.13050	z, z (T)	0.15055
z, z (R)	– 0.00936	z, d1 (T)	– 0.07279
z, yz (R)	– 0.01274	z, yz (T)	– 0.12591
z, xz (R)	– 0.01750	z, d2 (T)	0.03878
xy, xy (R)	0.12340	xy, xy (T)	0.02826
d1, d1 (R)	– 0.13099	d1, d1 (T)	0.14660
xy, d1 (R)	– 0.02860	xy, yz (T)	– 0.07170
d1, d2 (R)	0.07107	yz, d1 (T)	– 0.09022
xy, d2 (R)	0.03316	d1, d2 (T)	0.14252
yz, yz (R)	0.02727	yz, yz (T)	– 0.02825
xz, xz (R)	0.07429	xz, xz (T)	0.22046
yz, xz (R)	– 0.03852	yz, d2 (T)	0.17922
d2, d2 (R)	– 0.07342	d2, d2 (T)	0.06835

Table 3.4 Beryllium hcp

	Orthogonal		
Band	RMS error mRy	Maximum k	Deviation mRy
1	1.3	(8 0 12)	2.6
2	1.8	(24 0 9)	3.7
3	1.7	(12 0 0)	3.9
4	2.7	(0 12 0)	9.1
5	2.3	(0 12 12)	5.9
6	3.1	(0 18 0)	7.3
7	3.6	(0 6 0)	7.1
8	14.2	(0 12 0)	46.9
1–12	5.51		

Table 3.5 Energy values in Ry at selected k-points

	Orthogonal	APW
GAMMA 1+	−0.04271	−0.04304
GAMMA 3+	0.45493	0.45524
GAMMA 5+	1.89456	1.89516
GAMMA 4–	0.86490	0.86169
GAMMA 6–	2.08975	2.09180
M1+	0.40160	0.40331
Ml+	1.50942	1.50961
M3+	1.23735	1.23689
M4+	1.96553	1.96585
M2 –	0.55486	0.55405
M2 –	1.25388	1.25166
M3 –	1.42916	1.43105
M4 –	1.03879	1.03809
A1	0.11169	0.11339
A1	1.54969	1.54916
A3	2.01169	2.00870
L1	0.50742	0.50982
L1	0.88466	0.88549
L1	1.86978	1.86816
L2	1.62131	1.61821

Table 3.6 Beryllium hcp Slater-Koster 2-center parameters for a = 4.3210 a.u., c = 6.7715 a.u.

	Orthogonal	Non-orthogonal	
	Energy integrals (Ry)	Energy integrals (Ry)	Overlap integrals
On site			
s	1.25908	1.40584	
P	1.44105	1.22424	
d0	2.30436	1.50882	
d1	2.47063	2.37753	
d2	2.70933	1.85279	
pd	– 0.09439	0.09717	0.10566
First neighbor			
(sss)	– 0.11133	– 0.11560	0.21200
(sps)	0.16030	0.08987	– 0.15395
(sds)	0.11096	– 0.05956	– 0.23171
(pps)	0.22768	0.03027	– 0.26507
(ppp)	– 0.02182	0.07501	0.11711
(pds)	0.16909	– 0.09027	– 0.14695
(pdp)	– 0.04685	0.09730	0.19305
(dds)	– 0.15020	0.27013	0.15285
(ddp)	0.06830	– 0.23870	– 0.24651
(ddd)	– 0.03705	0.05909	0.08524
Second neighbor			
(sss)	0.00568	– 0.01895	– 0.06923
(sps)	– 0.00723	0.05966	– 0.04335
(sds)	– 0.04548	– 0.09895	– 0.09328
(pps)	– 0.00248	0.11105	– 0.04441
(ppp)	– 0.00362	0.04918	0.06324
(pds)	– 0.01610	0.09731	– 0.06081
(pdp)	0.00716	0.01690	0.06949

(continued)

Table 3.6 (continued)

	Orthogonal	Non-orthogonal	
	Energy integrals (Ry)	Energy integrals (Ry)	Overlap integrals
(dds)	– 0.18221	0.09509	0.12258
(ddp)	0.00306	0.02955	– 0.04012
(ddd)	– 0.00764	0.06735	0.06225
Third neighbor			
(sss)	– 0.00027	0.00374	– 0.26744
(sps)	– 0.00778	– 0.01299	– 0.14314
(sds)	– 0.04393	0.14749	0.13527
(pps)	– 0.01377	0.16270	0.20354
(ppp)	– 0.00457	– 0.04141	– 0.00730
(pds)	– 0.03567	0.26482	– 0.02641
(pdp)	– 0.00296	0.06846	– 0.00987
(dds)	0.10012	– 0.02543	– 0.03087
(ddp)	– 0.00226	– 0.10973	0.00278
(ddd)	0.01855	– 0.01462	– 0.02390

Table 3.7 Beryllium hcp

	Orthogonal			Non-orthogonal		
Band	RMS error mRy	Maximum k	Deviation mRy	RMS error mRy	Maximum k	Deviation mRy
1	21.2	(24 0 0)	55.8	29.0	(16 0 0)	77.8
2	18.1	(8 0 0)	37.4	24.8	(24 0 6)	51.5
3	17.8	(0 18 0)	36.5	22.3	(24 0 9)	47.3
4	26.5	(16 0 0)	72.8	42.3	(24 0 9)	121.5
1–4	20.8			29.5		

Table 3.8 Energy values in Ry at selected k-points

	Orthogonal	APW	Non-orthogonal
GAMMA 1+	−0.04630	−0.04304	−0.05012
GAMMA 3+	0.43495	0.45524	0.47256
GAMMA 5+	1.87723	1.89516	1.89607
GAMMA 4–	0.87339	0.86169	0.85079
GAMMA 6–	2.01501	2.09180	2.13300
M1+	0.45906	0.40331	0.38543
M1+	1.49824	1.50961	1.45233
M3+	1.21355	1.23689	1.21693
M4+	1.94467	1.96585	1.97010
M2 –	0.53202	0.55405	0.57121
M2 –	1.23687	1.25166	1.29946
M3 –	1.48860	1.43105	1.39770
M4 –	1.01612	1.03809	1.03277
A1	0.11679	0.11339	0.09385
A1	1.54314	1.54916	1.59400
A3	1.97262	2.00870	2.01131
L1	0.49872	0.50982	0.55515
L1	0.88259	0.88549	0.88614
L1	1.85627	1.86816	1.81684
L2	1.67207	1.61821	1.61090

Table 3.9 Fermi level quantities (non-orthogonal fit)

Energy		Densities of states			
	Total	s	p	t_{2g}	e_g
Ry		States/Ry/atom			
0.7731	0.10	0.00	0.09	0.01	0.00

Integrated densities of states				
Total	s	p	t_{2g}	e_g
Electrons				
2.00	0.45	1.48	0.01	0.06

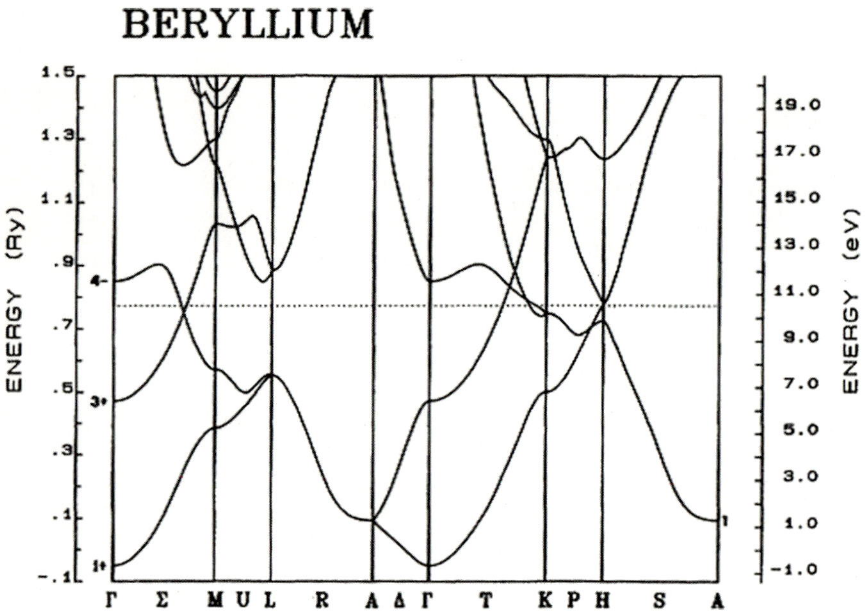

Fig. 3.2 Energy bands for Be

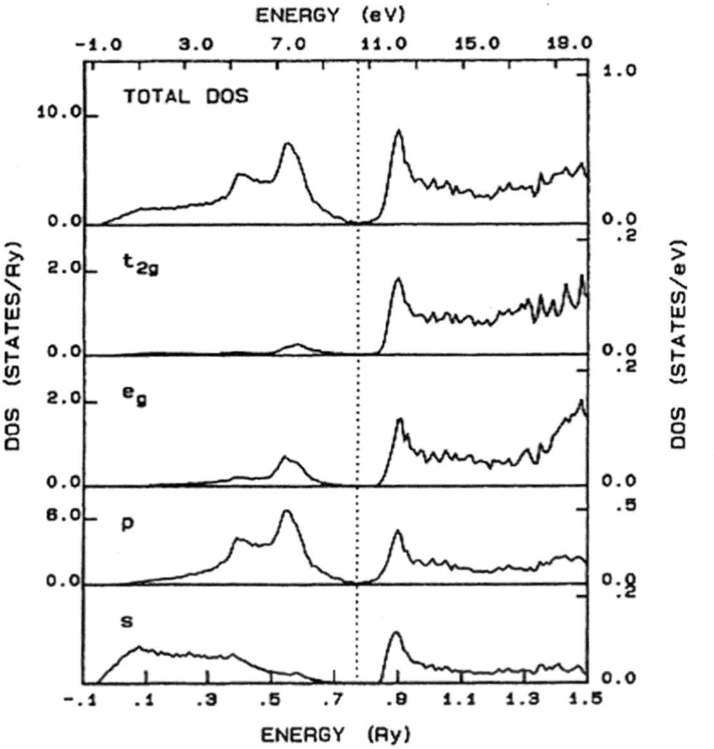

Fig. 3.3 Density of states for Be

3.2 Magnesium

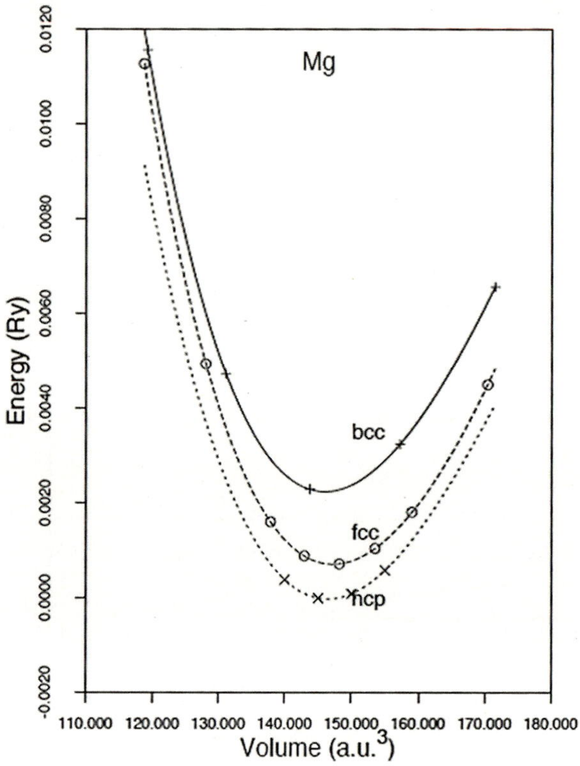

Fig. 3.4 Total energy of Mg

Table 3.10 Lattice constants and bulk modulus

	a (Bohr)	c (Bohr)	B (MBar)
bcc	6.636		0.383
fcc	8.383		0.392
hcp	5.957	9.531	0.392
exp	6.066	9.845	0.354
E_{hcp}–$E_{fcc} = \Delta E = 5.7$ mRy			

Table 3.11 Birch fit coefficients

	A_1	A_2	A_3	A_4
bcc	0.35457	−17.45867	155.26676	1607.18738
fcc	0.45507	−25.69558	373.07675	−274.76689
hcp	0.42779	−23.44687	312.16882	250.97414

Table 3.12 Magnesium hcp Slater-Koster 3-center parameters for a = 6.0647 a.u., c = 9.8460 a.u.

	Orthogonal		
	Energy integrals (Ry)		Energy integrals (Ry)
On site			
s, s (0)	0.44017		
x, x (0)	1.04205		
z, z (0)	0.82451		
xy, xy (0)	1.44296		
yz, yz (0)	1.42387		
d2, d2 (0)	1.45546		
s, d2 (0)	–0.04630		
x, xy (0)	–0.15888		
First neighbor			
s, s (R)	–0.03658	s, s (T)	–0.05124
s, x (R)	0.06708	s, y (T)	–0.04628
s, y (R)	0.02145	s, z (T)	0.06275
s, xy (R)	–0.03283	s, d1 (T)	–0.02541
s, d1 (R)	0.05536	s, yz (T)	–0.03364
s, d2 (R)	–0.04404	s, d2 (T)	0.01948
x, x (R)	0.12775	x, x (T)	–0.01650
y, y (R)	–0.02828	y, y (T)	0.05260
x, y (R)	–0.01300	y, z (T)	–0.08699
x, xy (R)	–0.02505	x, xy (T)	0.00049
y, d1 (R)	0.06269	y, d1 (T)	–0.01467
x, d1 (R)	0.10980	x, xz (T)	–0.02351
y, xy (R)	–0.02050	y, yz (T)	0.07344
y, d2 (R)	0.00088	y, d2 (T)	–0.08685
x, d2 (R)	–0.07122	z, z (T)	0.09237
z, z (R)	0.00108	z, d1 (T)	–0.03701
z, yz (R)	–0.01925	z, yz (T)	–0.07371
z, xz (R)	–0.01351	z, d2 (T)	0.03468
xy, xy (R)	0.07383	xy, xy (T)	0.03466
d1, d1 (R)	–0.11070	d1, d1 (T)	0.01351
xy, d1 (R)	0.04127	xy, yz (T)	–0.02316
d1, d2 (R)	0.03641	yz, d1 (T)	–0.00293
xy, d2 (R)	–0.02096	d1, d2 (T)	0.05686
yz, yz (R)	–0.03272	yz, yz (T)	–0.09451
xz, xz (R)	0.05263	xz, xz (T)	0.05293
yz, xz (R)	–0.04915	yz, d2 (T)	0.09355
d2, d2 (R)	–0.04601	d2, d2 (T)	0.00549

Table 3.13 Magnesium hcp

	Orthogonal		
Band	RMS error mRy	Maximum k	Deviation mRy
1	6.7	(12 0 0)	20.7
2	7.0	(12 0 0)	15.9
3	7.7	(12 0 0)	19.1
4	8.4	(0 0 9)	16.3
5	10.5	(0 12 0)	23.5
6	10.3	(4 0 0)	26.0
7	11.7	(0 6 0)	45.1
8	10.9	(0 0 0)	26.3
1–12	9.30		

Table 3.14 Energy values in Ry at selected k-points

	Orthogonal	APW
GAMMA 1+	−0.11504	−0.10480
GAMMA 1+	1.24057	1.26685
GAMMA 3+	0.27675	0.27880
GAMMA 5+	1.20352	1.51113
GAMMA 6+	1.35886	1.50168
GAMMA 4–	0.30979	0.30940
GAMMA 4–	1.36480	1.38191
GAMMA 5–	1.60829	1.63491
GAMMA 6–	1.18778	1.56715
M1+	0.23992	0.24166
M1+	0.66687	0.67309
M1+	0.98038	0.98963
M2+	1.91112	1.81844
M3+	0.62142	0.61575
M4+	1.24851	1.26674
M1–	1.23822	1.21815
M2–	0.25047	0.24935
M2–	0.64511	0.63459
M2–	1.30770	1.35948
M3–	0.89888	0.89639
M4–	0.58497	0.59939
A1	0.00716	−0.00317
A1	0.78010	0.76778
A3	1.24591	1.25461
A3	1.30260	1.31963
L1	0.33087	0.32469
L1	0.35986	0.36335
L1	1.00373	0.99496
L1	1.08868	1.09509
L2	0.98749	0.98179
L2	1.64242	1.67580

Table 3.15 Magnesium hcp Slater-Koster 2-center parameters for a = 6.0647 a.u., c = 9.8460 a.u.

	Orthogonal	Non-orthogonal	
	Energy integrals (Ry)	Energy integrals (Ry)	Overlap integrals
On site			
s	0.54102	0.54313	
p	0.91600	0.54265	
d0	1.48005	0.95976	
d1	1.29559	1.00093	
d2	1.46848	0.98900	
pd	−0.04237	0.02496	0.02343
First neighbor			
(sss)	−0.05217	−0.04980	−0.01209
(sps)	0.08743	0.06506	−0.08426
(sds)	0.07049	0.02129	−0.12290
(pps)	0.14634	−0.06420	−0.31478
(ppp)	−0.01519	0.00399	0.07295
(pds)	0.13185	−0.16103	−0.28241
(pdp)	−0.02540	0.03269	0.12338
(dds)	−0.14178	0.27314	0.28003
(ddp)	0.07357	−0.07809	−0.19603
(ddd)	−0.02016	0.01729	0.04596
Second neighbor			
(sss)	−0.00232	−0.00063	−0.01175
(sps)	0.00282	0.00597	0.00743
(sds)	−0.03247	0.01686	0.02688
(pps)	0.00897	0.04773	−0.01559
(ppp)	−0.00898	0.00519	0.00725
(pds)	−0.01643	0.04495	−0.02601
(pdp)	0.00239	0.00340	0.00889
(dds)	−0.04984	−0.07158	0.02911
(ddp)	−0.00963	0.03923	0.01289
(ddd)	0.02625	−0.01082	−0.00415
Third neighbor			
(sss)	0.00006	−0.00189	−0.00365
(sps)	0.00128	0.00370	0.00902
(sds)	−0.00881	0.00339	0.01172
(pps)	−0.00027	0.00278	−0.00506
(ppp)	−0.00855	0.00629	0.00519
(pds)	−0.00280	0.08088	0.06272
(pdp)	−0.01069	0.02444	0.01366
(dds)	0.01446	−0.05779	−0.02880
(ddp)	0.02991	0.00207	0.01073
(ddd)	−0.01594	−0.03402	−0.03693

Table 3.16 Magnesium hcp

	Orthogonal			Non-orthogonal		
Band	RMS error mRy	Maximum k	Deviation mRy	RMS error mRy	Maximum k	Deviation mRy
1	9.0	(16 0 6)	20.2	7.6	(0 0 0)	24.9
2	9.4	(0 24 12)	19.3	6.4	(24 0 0)	18.6
3	13.2	(8 0 12)	35.6	6.3	(0 0 6)	12.9
4	16.4	(16 0 0)	40.0	8.2	(0 12 0)	19.6
5	27.4	(16 0 9)	43.0	9.9	(0 0 12)	20.8
6	17.9	(16 0 12)	39.3	10.5	(0 6 0)	25.8
7	25.4	(0 12 0)	53.1	12.4	(0 6 0)	28.2
8	20.0	(8 0 0)	48.0	9.8	(0 0 12)	20.8
1–8	17.3			8.9		

Table 3.17 Energy values in Ry at selected k-points

	Orthogonal	APW	Non-orthogonal
GAMMA 1+	−0.09897	−0.10480	−0.07988
GAMMA 1+	1.27975	1.26685	1.26054
GAMMA 3+	0.28731	0.27880	0.27714
GAMMA 5+	1.20459	1.51113	1.22056
GAMMA 6+	1.32761	1.50168	1.34486
GAMMA 4–	0.31152	0.30940	0.30226
GAMMA 4–	1.35199	1.38191	1.39641
GAMMA 5–	1.70364	1.63491	1.67698
GAMMA 6–	1.19435	1.56715	1.20930
M1+	0.25600	0.24166	0.24551
M1+	0.66000	0.67309	0.66228
M1+	0.98898	0.98963	0.98974
M2+	1.76176	1.81844	1.59530
M3+	0.60221	0.61575	0.61180
M4+	1.23783	1.26674	1.26668
M1–	1.21718	1.21815	1.24445
M2–	0.24199	0.24935	0.26793
M2–	0.61875	0.63459	0.62211
M2–	1.32878	1.35948	1.33508
M3–	0.93915	0.89639	0.91223
M4–	0.57142	0.59939	0.60319
A1	−0.00594	−0.00317	0.00051
A1	0.77441	0.76778	0.77131
A3	1.25282	1.25461	1.23382
A3	1.30778	1.31963	1.30732
L1	0.30541	0.32469	0.32728
L1	0.37364	0.36335	0.36797
L1	0.98345	0.99496	1.00485
L1	1.09878	1.09509	1.08448
L2	0.99832	0.98179	0.98521
L2	1.48947	1.67580	1.47905

Table 3.18 Fermi level quantities (non-orthogonal fit)

Energy	Densities of states				
	Total	s	p	t_{2g}	e_g
Ry	States/Ry/atom				
0.3912	5.94	2.00	3.04	0.70	0.20

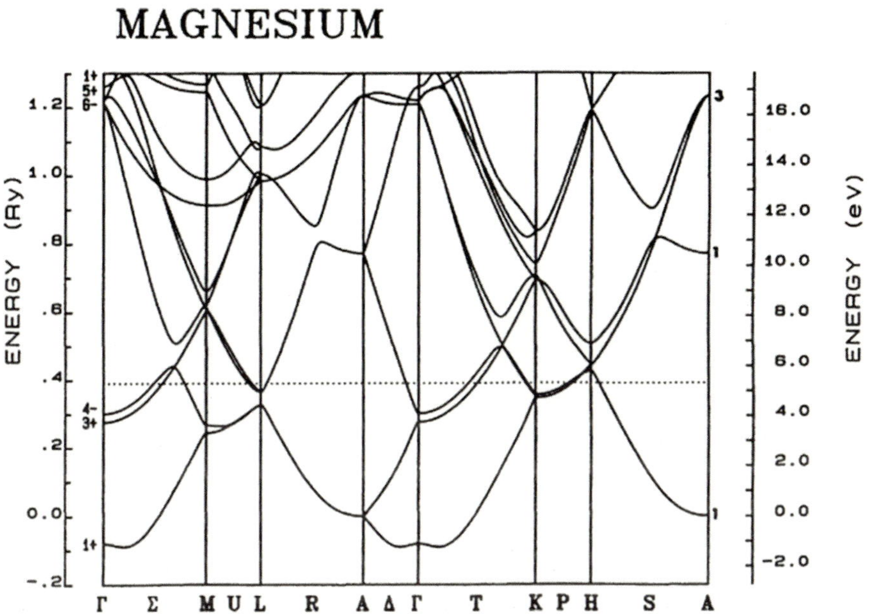

Fig 3.5 Energy bands for Mg

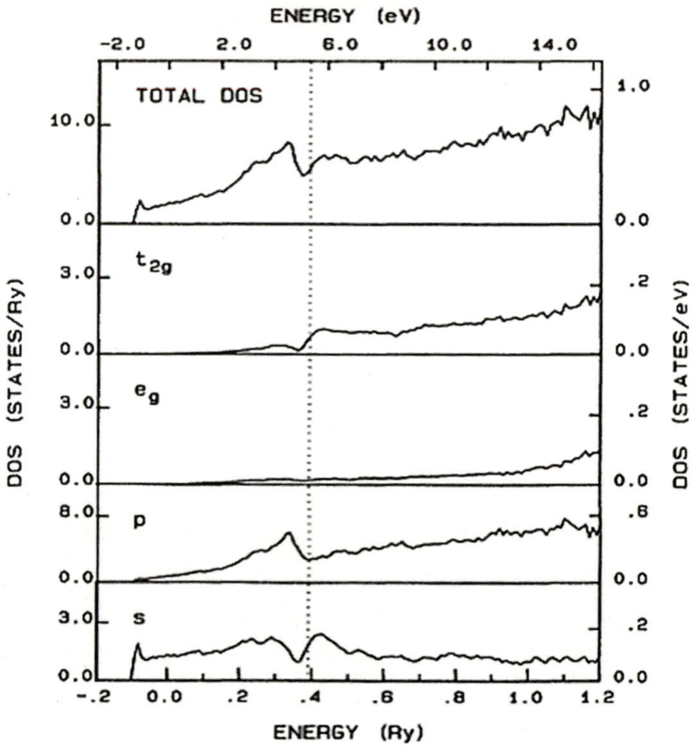

Fig 3.6 Density of states for Mg

3.3 Calcium

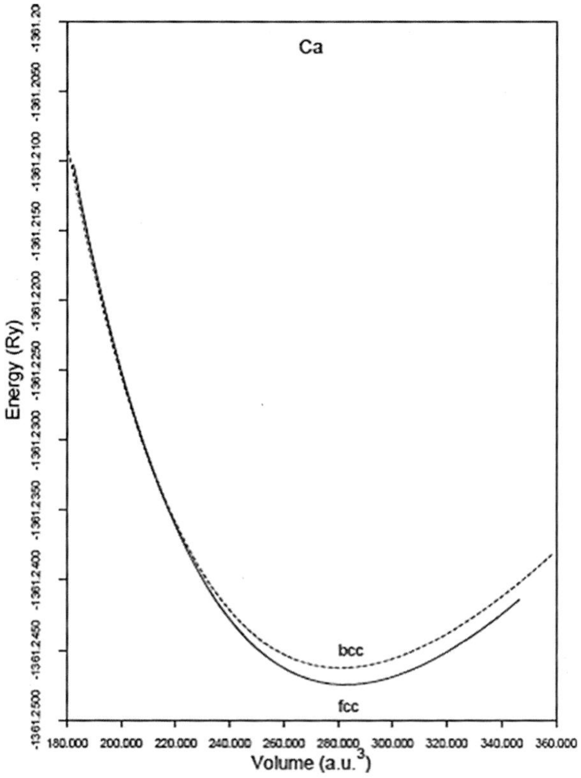

Fig. 3.7 Total energy of Ca

Table 3.19 Lattice constants and bulk modulus

	a (Bohr)	B (MBar)
fcc	10.420	0.168
bcc	8.251	0.161
exp	10.553	0.152
$\Delta E_{bcc\text{–}fcc} = 1.19$ mRy		

Table 3.20 Birch fit coefficients

	A_1	A_2	A_3	A_4
fcc	−1360.75885	−47.47408	1369.63161	−9964.885705
bcc	−1360.80613	−41.77205	1153.99445	−7381.704775

Table 3.21 Calcium fcc Slater-Koster 3-center parameters for a = 10.55300 a.u.

	Orthogonal	Non-orthogonal	
	Energy integrals (Ry)	Energy integrals (Ry)	Overlap integrals
On site			
s, s (000)	0.51021	0.42469	
x, x (000)	0.77165	0.69841	
xy, xy (000)	0.58160	0.59144	
d2, d2(000)	0.61800	0.55133	
First neighbor			
s, s (110)	−0.03573	−0.03673	0.01622
s, x (110)	0.04443	0.03470	−0.02950
s, xy (110)	−0.03633	−0.01437	0.05022
s, d2 (110)	0.03089	0.03811	0.01822
x, x (110)	0.04413	0.03477	−0.02230
x, x (011)	−0.00076	0.01958	0.03771
x, y (110)	0.06231	0.02822	−0.05998
x, xy (110)	−0.03676	−0.01952	0.03952
x, xy (011)	0.00845	−0.00790	−0.03032
z, d2 (011)	−0.01191	−0.02567	−0.04641
z, d1 (011)	0.03436	0.01943	−0.02115
xy, xy (110)	−0.03362	−0.02447	0.02166
xy, xy (011)	0.01011	−0.00566	−0.03553
xy, xz(011)	0.01580	−0.00433	−0.02608
xy, d2 (110)	0.01975	0.01455	−0.00185
d2, d2 (110)	−0.01152	0.00419	0.03351
d1, d1(110)	0.01823	−0.02056	−0.09661
Second neighbor			
s, s (200)	−0.01174	0.00501	−0.01964
s, x (200)	0.00825	0.00281	0.02349
s, d2 (002)	−0.00193	−0.01176	−0.01649
x, x (200)	0.01118	0.00815	0.00509
y, y (200)	−0.01192	0.00467	0.00300
x, xy (020)	0.00704	0.00207	0.00754
z, d2 (002)	−0.00488	−0.01619	−0.00291
xy, xy (200)	0.00532	0.00834	0.01670
xy, xy (002)	0.00111	−0.00221	−0.00273
d2, d2 (002)	−0.00316	−0.00845	0.01191
d1, d1 (002)	0.00067	0.00255	0.00361

Table 3.22 Calcium fcc

	Orthogonal			Non-orthogonal		
Band	RMS error mRy	Maximum k	Deviation mRy	RMS error mRy	Maximum k	Deviation mRy
1	2.6	(444)	7.8	0.1	(055)	0.3
2	4.3	(055)	12.8	0.3	(222)	0.6
3	6.9	(044)	22.2	0.3	(033)	0.7
4	4.3	(444)	8.4	0.6	(118)	2.5
5	2.6	(066)	5.8	0.4	(174)	0.9
6	2.9	(048)	8.5	0.5	(007)	1.3
1–6	4.2			0.4		

Table 3.23 Energy values in Ry at selected k-points

	Orthogonal	APW	Non-orthogonal
GAMMA 1	0.01102	0.01318	0.01300
GAMMA 12	0.65080	0.65104	0.65186
GAMMA 15	1.09637	1.09573	1.09574
GAMMA 25′	0.55154	0.55207	0.55250
X1 (008)	0.27780	0.28197	0.28220
X1 (008)	0.78296	0.79944	0.79927
X2 (008)	0.71610	0.71583	0.71656
X3 (008)	0.38971	0.39588	0.39642
X5 (008)	0.73958	0.73649	0.73687
X4′ (008)	0.39022	0.39463	0.39485
X5′ (008)	0.74936	0.75185	0.75178
L1 (444)	0.24370	0.25153	0.25156
L1 (444)	0.76864	0.77221	0.77262
L3 (444)	0.52663	0.53506	0.53522
L3 (444)	0.72016	0.71692	0.71683
L2′ (444)	0.29852	0.29994	0.30028
L3′ (444)	1.04620	1.05207	1.05202
W1 (048)	0.53194	0.52720	0.52721
W1 (048)	0.76696	0.75652	0.75692
W3 (048)	0.39582	0.39081	0.39063
W3 (048)	0.76502	0.76149	0.76211
W1′ (048)	0.73512	0.73695	0.73769
W2′ (048)	0.33209	0.33330	0.33352
W2′ (048)	0.90747	0.89749	0.89763
EVEN(224)	0.15946	0.15923	0.15908
EVEN(224)	0.40835	0.40563	0.40518
EVEN(224)	0.52206	0.51931	0.51981
EVEN(224)	0.65319	0.65386	0.65394
ODD (224)	0.55760	0.56177	0.56139
ODD (224)	0.70225	0.70116	0.70148

Table 3.24 Fermi level quantities (non-orthogonal fit)

Energy		Densities of states				Velocity	Plasmon energy
	Total	s	p	t_{2g}	e_g		
Ry		States/Ry/atom				cm/s	eV
0.3190	17.14	1.16	8.08	1.56	6.34	0.45 × 10E8	4.20

Integrated densities of states Total Electrons	s	p	t_{2g}	e_g
2.00	0.89	0.54	0.16	0.42

Table 3.25 Calcium fcc Slater-Koster 2-center parameters

	Orthogonal	Non-orthogonal	
	Energy integrals (Ry)	Energy integrals (Ry)	Overlap integrals
On site			
s	0.50375	0.26802	
p	0.75908	0.42308	
d1	0.59828	0.53056	
d2	0.57861	0.53196	
First neighbor			
(sss)	−0.03998	−0.01535	0.12109
(pps)	0.09955	−0.12553	−0.39701
(ppp)	−0.00803	0.01211	0.08928
(dds)	−0.04404	0.02585	0.11700
(ddp)	0.02589	−0.00216	−0.07680
(ddd)	−0.00346	−0.00387	0.00742
(sps)	−0.06314	−0.01451	−0.21891
(sds)	−0.04492	0.01168	0.11822
(pds)	0.06452	0.07018	0.21429
(pdp)	−0.01298	−0.00776	−0.08367
Second neighbor			
(sss)	−0.00212	−0.00830	0.00419
(pps)	0.00849	0.03386	−0.03611
(ppp)	−0.00544	0.00373	0.00943
(dds)	−0.00337	−0.02245	−0.00836
(ddp)	−0.00014	0.00607	0.00285
(ddd)	0.00107	−0.00065	−0.00076
(sps)	0.00014	0.01479	−0.01671
(sds)	0.00836	−0.00579	0.01091
(pds)	0.00610	−0.02672	0.00470
(pdp)	−0.00087	0.00245	−0.00222

	Orthogonal			Non-orthogonal		
Band	RMS error mRy	Maximum k	Deviation mRy	RMS error mRy	Maximum k	Deviation mRy
1	12.5	(084)	32.5	0.6	(264)	1.7
2	16.3	(008)	38.5	0.7	(002)	1.8
3	12.0	(044)	25.1	1.4	(042)	3.4
4	7.6	(280)	17.0	1.1	(222)	2.6
5	10.2	(066)	32.3	0.9	(000)	2.6
6	10.4	(044)	24.8	0.9	(000)	2.6
1–6	11.8			1.0		

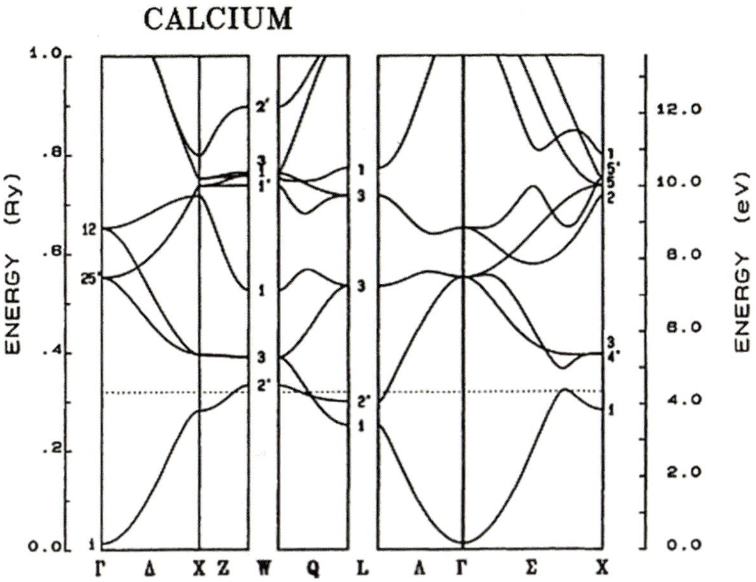

Fig. 3.8 Energy bands for Ca

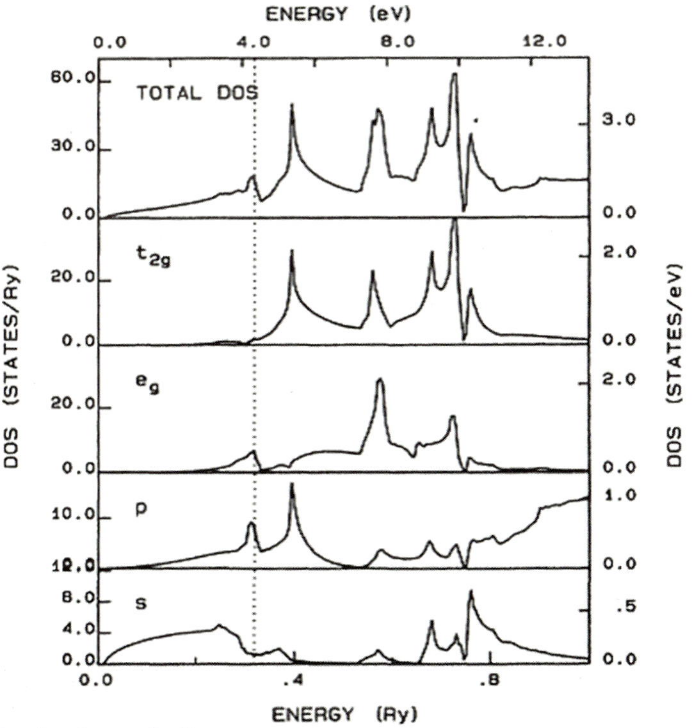

Fig 3.9 Density of states for Ca

3.4 Strontium

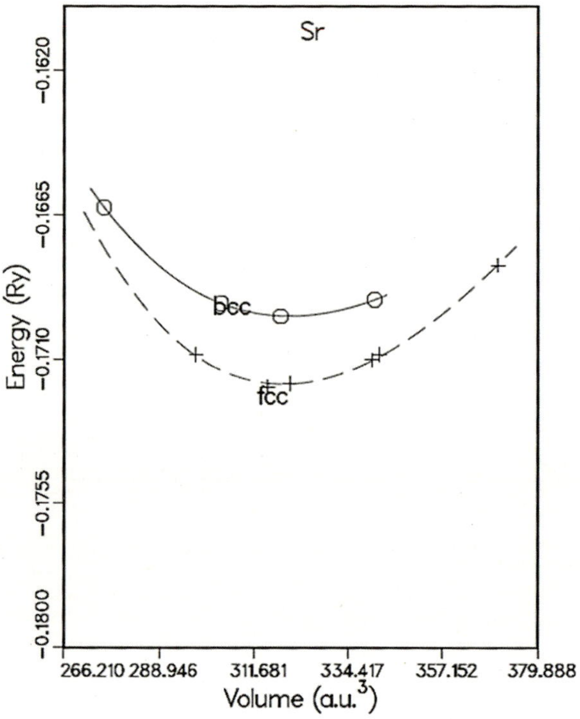

Fig. 3.10 Total energy of Sr

Table 3.26 Lattice constants and bulk modulus

	a (Bohr)	B (MBar)
bcc	8.619	0.126
fcc	10.84	0.169
exp	11.50	0.116
$\Delta E = 2.0$ mRy		

Table 3.27 Birch fit coefficients

	A_1	A_2	A_3
bcc	0.13870	−28.86488	675.47782
fcc	0.23981	−38.41015	896.09713

Table 3.28 Strontium fcc Slater-Koster 3-center parameters for a = 11.49840 a.u.

	Orthogonal	Non-orthogonal	
	Energy integrals (Ry)	Energy integrals (Ry)	Overlap integrals
On site			
s, s (000)	0.38720	0.40536	
x, x (000)	0.69674	0.66097	
xy, xy (000)	0.51737	0.48466	
d2, d2 (000)	0.45887	0.45590	
First neighbor			
s, s (110)	−0.03552	−0.03457	−0.01347
s, x (110)	0.03541	0.03763	−0.00086
s, xy (110)	−0.02578	−0.01767	0.03652
s, d2 (110)	0.01745	0.03582	0.02983
x, x (110)	0.03537	0.04362	0.00469
x, x (011)	−0.00178	0.01833	0.02828
x, y (110)	0.05213	0.02607	−0.04822
x, xy (110)	−0.03890	−0.01833	0.03834
x, xy (011)	0.00179	−0.00643	−0.02250
z, d2 (011)	0.00458	−0.02616	−0.05068
z, d1 (011)	0.02243	0.02120	−0.01169
xy, xy (110)	−0.02935	−0.01877	0.02774
xy, xy (011)	0.00789	−0.00306	−0.03220
xy, xz (01 l)	0.00062	−0.00253	−0.01987
xy, d2 (110)	0.01417	0.01789	0.00670
d2, d2 (110)	−0.01346	−0.00408	0.01823
d1, d1 (110)	0.02468	−0.01122	−0.07712
Second neighbor			
s, s (200)	0.01232	0.00567	−0.02452
s, x (200)	−0.01579	0.00088	0.02779
s, d2 (002)	0.00707	−0.00977	−0.02426
x, x (200)	−0.01552	0.00368	0.00598
y, y (200)	0.00266	0.00984	0.00685
x, xy (020)	−0.00515	−0.00299	0.00364
z, d2 (002)	0.00539	−0.01395	−0.01082
xy, xy (200)	−0.00751	0.00305	0.01342
xy, xy (002)	0.00163	−0.00100	−0.00128
d2, d2 (002)	−0.00242	−0.00834	0.00186
d1, d1 (002)	0.00179	0.00255	0.00362

Table 3.29 Strontium fcc

	Orthogonal			Non-orthogonal		
Band	RMS error mRy	Maximum k	Deviation mRy	RMS error mRy	Maximum k	Deviation mRy
1	1.0	(008)	2.4	0.1	(006)	0.3
2	2.2	(055)	5.8	0.2	(008)	0.6
3	2.4	(264)	5.7	0.2	(033)	0.5
4	1.6	(224)	4.1	0.5	(118)	2.2
5	1.8	(380)	4.9	0.4	(048)	0.8
6	2.1	(048)	5.6	0.5	(007)	1.4
1–6	1.9			0.3		

Table 3.30 Energy values in Ry at selected k-points

	Orthogonal	APW	Non-orthogonal
GAMMA 1	0.03487	0.03542	0.03539
GAMMA 12	0.52431	0.52215	0.52290
GAMMA 15	0.95212	0.94955	0.94958
GAMMA 25′	0.43631	0.43668	0.43701
X1 (008)	0.22977	0.23215	0.23231
X1 (008)	0.65539	0.67187	0.67103
X2 (008)	0.58710	0.58434	0.58526
X3 (008)	0.31008	0.31238	0.31294
X5 (008)	0.60798	0.60580	0.60619
X4′ (008)	0.38626	0.38754	0.38752
X5′ (008)	0.68345	0.68503	0.68488
L1 (444)	0.21487	0.21626	0.21640
L1 (444)	0.63758	0.64595	0.64610
L3 (444)	0.42207	0.42348	0.42364
L3 (444)	0.58534	0.58584	0.58618
L2′ (444)	0.30014	0.30114	0.30129
L3′ (444)	0.92567	0.93602	0.93588
W1 (048)	0.43770	0.43797	0.43835
W1 (048)	0.63671	0.63114	0.63116
W3 (048)	0.34295	0.34138	0.34143
W3 (048)	0.67035	0.66954	0.66948
W1′ (048)	0.60146	0.60636	0.60713
W2′ (048)	0.28166	0.28101	0.28110
W2′ (048)	0.81498	0.80087	0.80153
EVEN (224)	0.16087	0.16102	0.16097
EVEN (224)	0.34324	0.34281	0.34246
EVEN (224)	0.41580	0.41380	0.41408
EVEN (224)	0.53418	0.53398	0.53416
ODD (224)	0.44576	0.44982	0.44953
ODD (224)	0.57266	0.57161	0.57160

Table 3.31 Fermi level quantities (non-orthogonal fit)

Energy		Densities of states				Velocity	Plasmon energy
	Total	s	p	t_{2g}	e_g		
Ry		States/Ry/atom				cm/s	eV
0.2865	0.00	0.00	0.00	0.00	0.00	0.00 × 10E8	0.00

Integrated densities of states				
Total	s	p	t_{2g}	e_g
Electrons				
2.00	1.00	0.44	0.14	0.42

Table 3.32 Strontium fcc Slater-Koster 2-center parameters

	Orthogonal	Non-orthogonal	
	Energy integrals (Ry)	Energy integrals (Ry)	Overlap integrals
On site			
s	0.43114	0.26660	
p	0.66146	0.43018	
d1	0.48409	0.43183	
d2	0.47127	0.42999	
First neighbor			
(sss)	−0.02928	−0.01305	0.10296
(pps)	0.07667	−0.11453	−0.36874
(ppp)	−0.00637	0.00165	0.06787
(dds)	−0.03889	0.01664	0.11448
(ddp)	0.01983	0.00082	−0.06853
(ddd)	−0.00251	−0.00345	0.00496
(sps)	0.04860	−0.01323	−0.19487
(sds)	−0.03753	0.00679	0.10858
(pds)	−0.05387	0.05495	0.20334
(pdp)	0.01104	−0.00024	−0.06533
Second neighbor			
(sss)	−0.00746	−0.00505	0.00304
(pps)	0.02182	0.03021	−0.02275
(ppp)	−0.00651	0.00109	0.00753
(dds)	−0.00169	−0.02057	−0.01782
(ddp)	0.00025	0.00357	0.00185
(ddd)	0.00091	0.00043	0.00096
(sps)	0.00848	0.00962	−0.01140
(sds)	0.00564	−0.00384	0.00796
(pds)	−0.00716	−0.02502	−0.00589
(pdp)	0.00167	0.00199	−0.00193

	Orthogonal			Non-orthogonal		
Band	RMS error mRy	Maximum k	Deviation mRy	RMS error mRy	Maximum k	Deviation mRy
1	7.3	(084)	20.6	0.4	(264)	1.2
2	7.9	(008)	13.9	0.6	(002)	1.3
3	7.6	(044)	17.7	1.1	(044)	2.6
4	5.4	(066)	11.8	1.0	(224)	2.4
5	8.1	(066)	21.1	0.8	(000)	2.1
6	9.5	(280)	24.7	0.7	(000)	2.1
1–6	7.7			0.8		

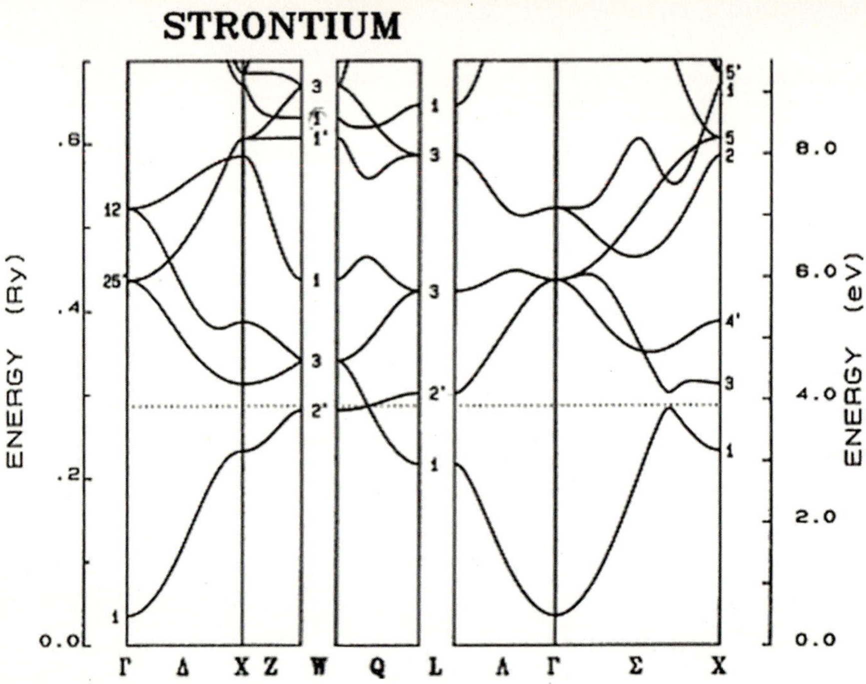

Fig. 3.11 Energy bands for Sr

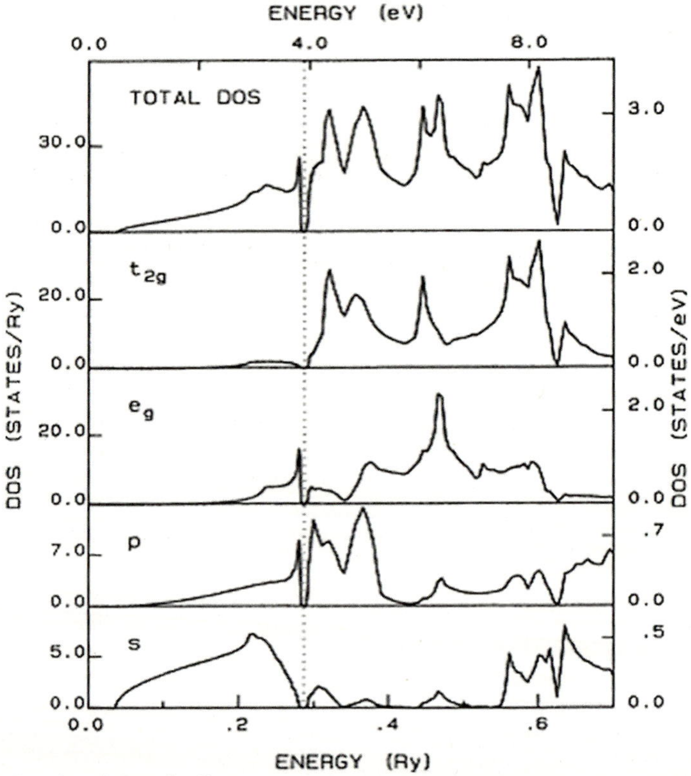

Fig. 3.12 Density of states for Sr

3.5 Barium

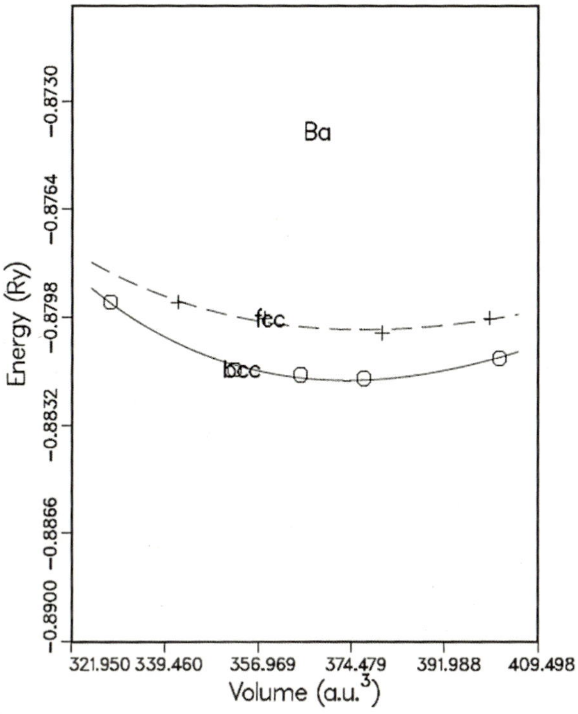

Fig. 3.13 Total energy of Ba

Table 3.33 Lattice constants and bulk modulus

	a (Bohr)	B (MBar)
fcc	11.46	0.070
bcc	9.073	0.111
exp	9.486	0.103
$\Delta E = 2.1$ mRy		

Table 3.34 Birch fit coefficients

	A_1	A_2	A_3
bcc	−0.56518	−32.84128	851.63253
fcc	−0.67640	−21.25447	554.19004

Table 3.35 Barium bcc Slater-Koster 3-center parameters for a = 9.48643 a.u.

	Orthogonal	Non-orthogonal	
	Energy integrals (Ry)	Energy integrals (Ry)	Overlap integrals
On site			
s, s (000)	0.38238	0.26979	
x, x (000)	0.56446	0.49654	
xy, xy (000)	0.49832	0.40009	
d2, d2 (000)	0.43440	0.37857	
First neighbor			
s, s (111)	−0.04502	−0.00855	0.11784
s, x (111)	0.02945	−0.01508	−0.10477
s, xy (111)	0.01953	0.00002	−0.06191
x, x (111)	0.01256	−0.01303	−0.03068
x, y (111)	0.01506	−0.01857	−0.08023
x, xy (111)	0.01799	−0.01025	−0.04265
x, yz (111)	0.02052	−0.01003	−0.07841
x, d1 (111)	0.00733	−0.00110	−0.03739
xy, xy (111)	−0.01272	−0.00200	0.02084
xy, xz (111)	−0.02073	0.00549	0.05782
xy, d2 (111)	−0.00006	0.00183	−0.03797
d2, d2 (111)	0.01638	−0.00302	−0.07025
Second neighbor			
s, s (200)	−0.00795	−0.01825	0.04035
s, x (200)	0.02128	0.01399	−0.07291
s, d2 (002)	−0.02576	−0.00237	0.07910
x, x (200)	0.04756	−0.00584	−0.12859
y, y (200)	0.00990	−0.03626	−0.04584
x, xy (020)	0.01007	−0.01811	−0.01473
z, d2 (002)	−0.04823	0.01788	0.14323
xy, xy (200)	−0.01272	0.01674	−0.00068
xy, xy (002)	−0.00081	−0.01025	−0.02823
d2, d2 (002)	−0.03600	−0.00035	0.09311
d1, d1 (002)	0.00453	−0.00767	−0.01052
Third neighbor			
s, s (220)	0.00561		
s, x (220)	−0.00795		
s, xy (220)	−0.00933		
s, d2 (220)	−0.00844		
x, x (220)	−0.00651		
x, x (022)	0.00106		
x, y (220)	−0.00770		
x, xy (220)	−0.00637		
x, xy (022)	−0.00043		
z, d2 (022)	0.00066		
z, d1 (022)	−0.00391		
xy, xy (220)	0.00755		
xy, xy (022)	0.00179		
xy, xz (022)	0.00054		
xy, d2 (220)	0.00745		
d2, d2 (220)	0.00303		
d1, d1 (220)	−0.00231		

Table 3.36 Barium bcc

	Orthogonal			Non-orthogonal		
Band	RMS error mRy	Maximum k	Deviation mRy	RMS error mRy	Maximum k	Deviation mRy
1	2.4	(222)	6.6	0.8	(353)	2.5
2	2.4	(030)	7.3	0.8	(260)	2.0
3	1.7	(161)	4.7	0.8	(151)	2.2
4	3.1	(330)	11.3	1.2	(044)	2.9
5	2.7	(131)	7.8	1.4	(131)	5.1
6	3.9	(010)	7.9	1.7	(004)	4.7
1–6	2.8			1.2		

Table 3.37 Energy values in Ry at selected k-points

	Orthogonal	APW	Non-orthogonal
GAMMA 1	0.04187	0.03999	0.04207
GAMMA 12	0.47533	0.48270	0.48173
GAMMA 15	0.75177	0.74979	0.75014
GAMMA 25′	0.38858	0.38903	0.38874
H1 (008)	0.76213	0.76484	0.76374
H12 (008)	0.21328	0.20868	0.20922
H15 (008)	0.55083	0.55230	0.55172
H25′ (008)	0.59204	0.59810	0.59751
N1 (044)	0.18175	0.18249	0.18277
N1 (044)	0.47853	0.48197	0.47904
N1 (044)	0.68378	0.71147	0.70823
N2 (044)	0.30389	0.30563	0.30577
N3 (044)	0.63559	0.63504	0.63430
N4 (044)	0.52416	0.52562	0.52377
N1′ (044)	0.34462	0.34494	0.34586
N3′ (044)	0.67631	0.68365	0.68010
N4′ (044)	0.58560	0.58601	0.58446
P1 (444)	0.49741	0.49757	0.50045
P3 (444)	0.53313	0.53280	0.53525
P4 (444)	0.29283	0.29286	0.29139
P4 (444)	0.68438	0.67977	0.68061
(341)	0.19641	0.19601	0.19581
(341)	0.30497	0.30591	0.30650
(341)	0.32552	0.32469	0.32514
(341)	0.47152	0.47125	0.47188
(341)	0.52367	0.52432	0.52300
(341)	0.61027	0.60608	0.60912

Table 3.38 Fermi level quantities (non-orthogonal fit)

Energy		Densities of states				Velocity	Plasmon energy
	Total	s	p	t_{2g}	e_g		
Ry		States/Ry/atom				cm/s	eV
0.2485	15.51	2.34	1.33	4.16	7.68	0.42 × 10E8	2.87

Integrated densities of states Total Electrons	s	p	t_{2g}	e_g
2.00	0.96	0.13	0.37	0.54

Table 3.39 Barium bcc Slater-Koster 2-center parameters

	Orthogonal	Non-orthogonal	
	Energy integrals (Ry)	Energy integrals (Ry)	Overlap integrals
On site			
s	0.38328	0.26261	
p	0.56001	0.48364	
d1	0.49391	0.39183	
d2	0.43399	0.38612	
First neighbor			
(sss)	−0.04524	−0.00877	0.12180
(pps)	0.04270	−0.09416	−0.27275
(ppp)	−0.00293	−0.00171	0.04107
(dds)	−0.05501	0.01658	0.14761
(ddp)	0.02304	0.00567	−0.08485
(ddd)	0.00318	−0.00748	0.00124
(sps)	0.04562	0.03478	0.20889
(sds)	−0.04276	0.00830	0.12538
(pds)	−0.06488	−0.05471	−0.20633
(pdp)	0.01111	−0.01024	0.04266
Second neighbor			
(sss)	−0.00843	−0.01539	0.04829
(pps)	0.04287	0.01195	−0.10084
(ppp)	0.00042	−0.02204	−0.01945
(dds)	−0.03639	−0.00560	0.07942
(ddp)	−0.00811	0.01058	−0.01684
(ddd)	0.00464	−0.00348	−0.00614
(sps)	0.01785	−0.01709	0.06699
(sds)	−0.01793	−0.00375	0.07272
(pds)	−0.04006	−0.01481	−0.13588
(pdp)	−0.00243	−0.01013	0.00765
Third neighbor			
(sss)	0.00569		
(pps)	−0.00310		
(ppp)	0.00169		
(dds)	0.00543		
(ddp)	−0.00185		
(ddd)	0.00132		
(sps)	−0.00838		
(sds)	0.00691		
(pds)	0.00101		
(pdp)	−0.00144		

Band	Orthogonal RMS error mRy	Maximum k	Deviation mRy	Non-orthogonal RMS error mRy	Maximum k	Deviation mRy
1	2.4	(250)	7.5	1.2	(000)	4.2
2	4.4	(044)	11.4	0.9	(002)	2.5
3	2.8	(151)	6.5	1.5	(030)	3.4
4	6.6	(044)	18.8	2.2	(002)	5.2
5	4.3	(260)	15.0	2.0	(131)	5.9
6	7.2	(008)	17.4	2.8	(341)	6.2
1–6	5.0			1.9		

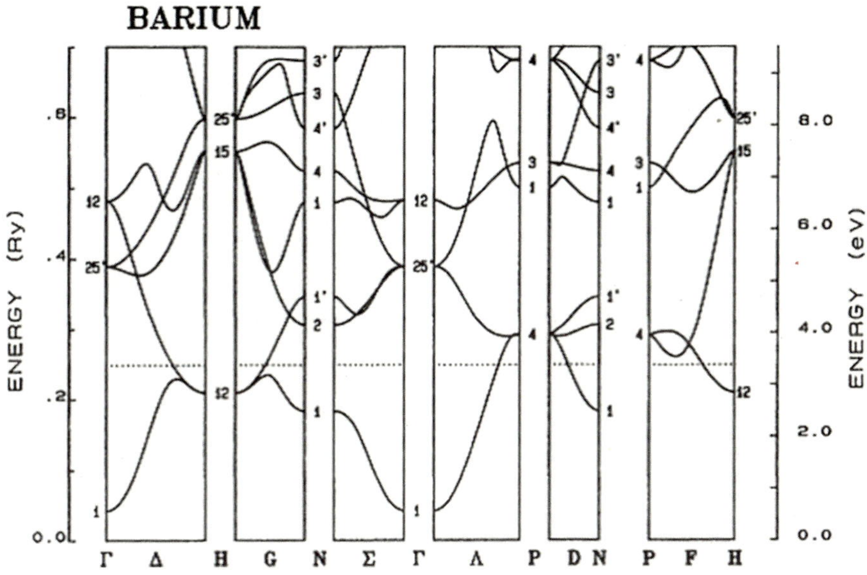

Fig. 3.14 Energy bands for Ba

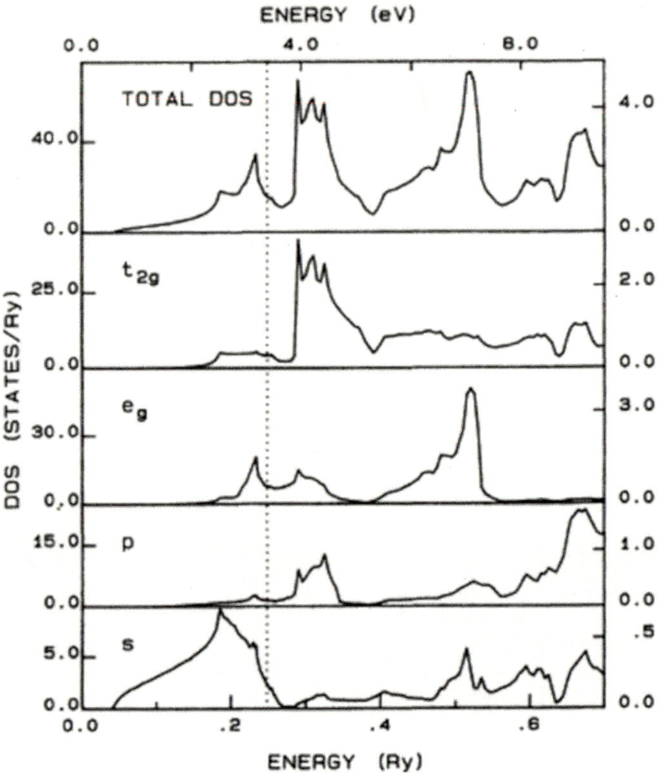

Fig. 3.15 Density of states for Ba

3.6 Radium

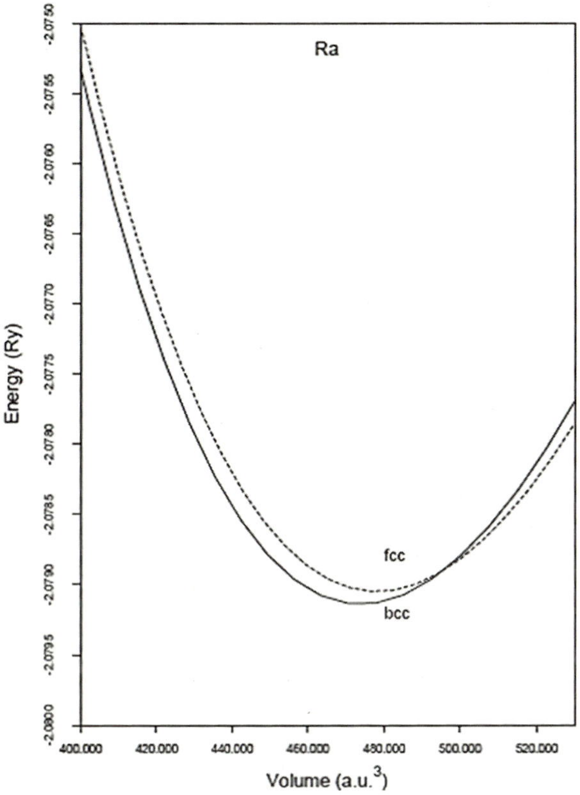

Fig. 3.16 Total energy of Ra

Table 3.40 Lattice constants and bulk modulus

	a (Bohr)	B (MBar)
bcc	9.823	0.076
fcc	12.413	0.073
exp	9.730	
$\Delta E_{bcc\ -fcc} = 0.09$ mRy		

Table 3.41 Birch fit coefficients

	A_1	A_2	A_3	A_4
bcc	−1.73421	−46.23133	1797.05356	−15884.30956
fcc	−1.71376	−50.79318	2114.21862	−22879.65398

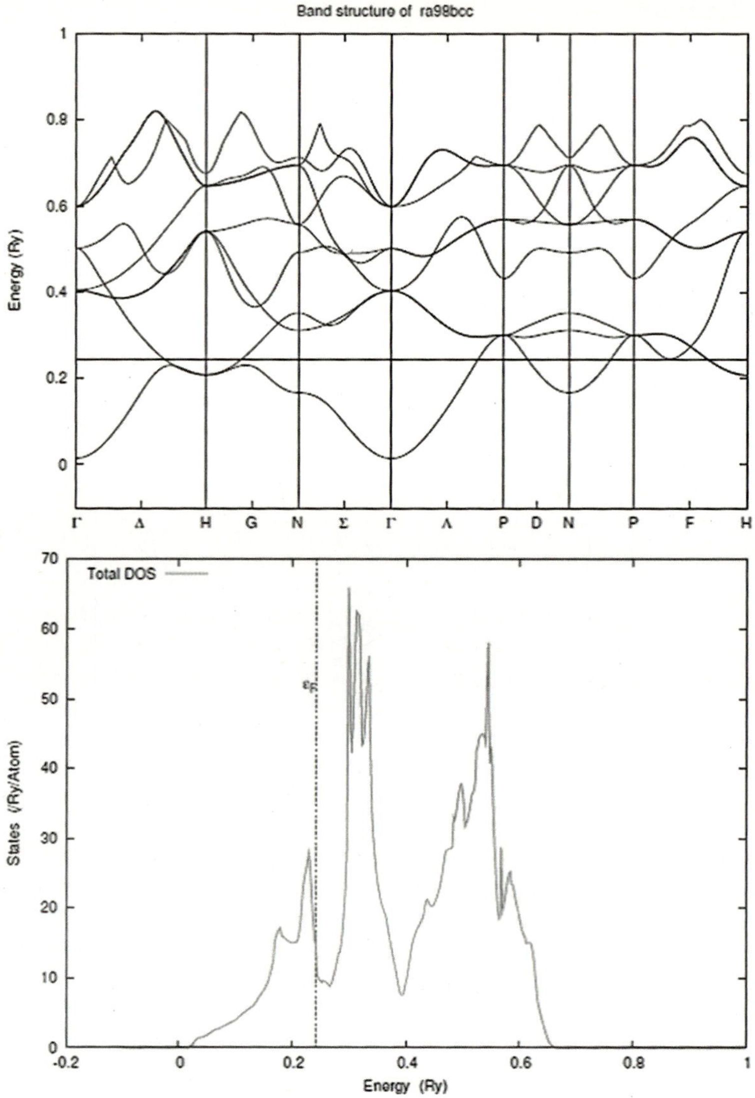

Fig. 3.17 Energy bands and density of states for Ra

Table 3.42 Density of states at the Fermi level (APW results)

Energy		s	p	e_g	t_{2g}	f	Velocity	Plasmon energy
Ry	Total	States/Ry/atom					cm/s	eV
0.243	13.287	1.289	0.978	4.268	1.909	0.238	0.487 × 10E8	2.892

Table 3.43 Radium bcc structure a = 9.80 Bohr
Tight-binding parameters in Ry
Non–orthogonal

On–site									
s	p	t_{2g}	e_g						
–0.04442	4.63298	0.16939	0.31848						
sss	sps	pps	ppp	sds	pds	pdp	dds	ddp	ddd
Hopping dist = 8.48705									
–0.01604	0.00656	0.30929	–0.03651	0.00239	0.11730	–0.03279	0.03019	0.00545	–0.02349
Overlap dist = 8.48705									
0.00002	0.00000	–0.27370	0.12964	–0.00023	–0.00075	0.02907	–0.00019	0.00113	0.00291
Hopping dist = 9.80000									
–0.00304	0.00126	0.02623	–0.02982	0.00046	0.02591	–0.00642	0.00506	0.00109	–0.00383
Overlap dist = 9.80000									
0.00000	0.00000	–0.04282	0.02099	0.00000	–0.00007	0.00618	–0.00002	0.00021	0.00050

Fitting rms error for 6 bands 38 mRy

Orthogonal

On-site									
s	p	t_{2g}	e_g						
–0.04389	4.63298	0.19546	0.32057						
sss	sps	pps	ppp	sds	pds	pdp	dds	ddp	ddd
Hopping dist = 8.48705									
–0.01649	0.00001	0.88758	0.36126	0.00000	0.22261	–0.01191	0.02369	0.00351	–0.02694
Hopping dist = 9.80000									
–0.00330	0.00000	0.09865	–0.03204	0.00000	0.05028	–0.00559	0.00384	0.00067	–0.00446

Fitting rms error for 6 bands 36 mRy

Chapter 4
The 3d Transition Metals

The first two elements of this series, i.e., Sc and Ti, and the last one, Zn, have the hcp structure as shown in the total energy plots. The total energy graphs for Cr and Mn are predicting the correct ground state i.e. antiferromagnetic for Cr and alpha-Mn for Mn. Sc and Ti have very similar energy bands, with Sc displaying a very high $N(E_F)$. Zn is very different, characterized by a set of narrow d bands located far below E_F. V and bcc Cr also have similar bands and DOS, differing in the position of E_F which results in a high $N(E_F)$ for V and low $N(E_F)$ for Cr. The tight-binding parameters of Cr and Mn were found for the bcc structure. The next three elements of the series, Fe, Co and Ni, are known to be ferromagnetic as it is confirmed in the total energy plots. Their energy bands and DOS are shown separately for the two spins and the tight-binding parameters are fitted to spin-polarized calculations in the bcc, hcp, and fcc structure and their integrated DOS give magneton numbers of 3.1, 1.64, and 0.67, respectively. Finally, Cu has an fcc ground state and d bands significantly below E_F and a low value of $N(E_F)$.

D.A. Papaconstantopoulos, *Handbook of the Band Structure of Elemental Solids*,
DOI 10.1007/978-1-4419-8264-3_4

4.1 Scandium

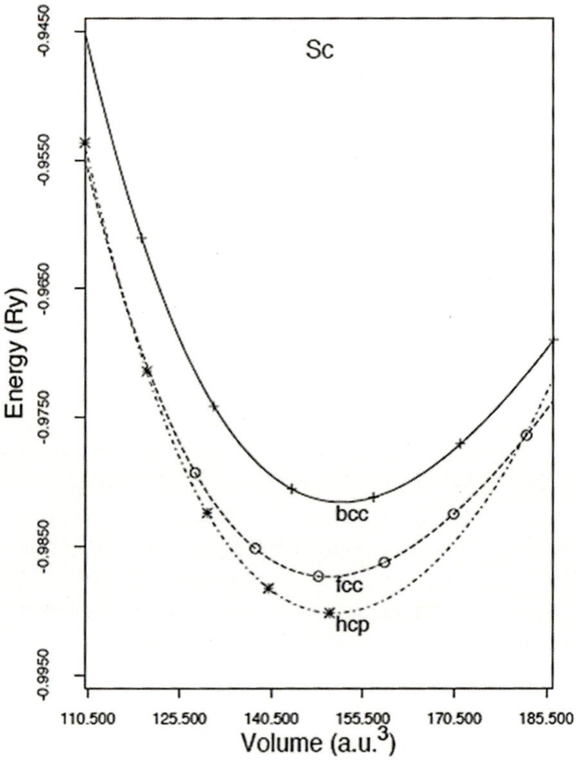

Fig. 4.1 Total energy of Sc

Table 4.1 Lattice constants and bulk modulus

	a (Bohr)	c (Bohr)	B (Mbar)
bcc	6.724		0.622
fcc	8.429		0.606
hcp	6.081	9.426	0.644
exp	6.255	9.959	0.435
$\Delta E = E_{hcp} - E_{fcc} = 2.8$ mRy			

Table 4.2 Birch fit coefficients

	A_1	A_2	A_3	A_4
bcc	0.10720	−72.42296	1475.77758	−8437.64310
fcc	−0.02793	−61.58144	1184.35943	−5942.67428
hcp	0.09994	−71.65320	1434.29109	−7911.49837

Table 4.3 Scandium hcp Z = 21 lattice constants = 6.2531 a.u. 9.9651 a.u.

Slater-Koster 3-center parameters	Orthogonal		
	Energy integrals (Ry)		Energy integrals (Ry)
On site			
s, s (0)	0.69748		
x, x (0)	0.97788		
z, z (0)	0.97039		
xy, xy (0)	0.59071		
yz, yz (0)	0.59887		
d2, d2 (0)	0.57530		
s, d2 (0)	−0.00733		
x, xy (0)	−0.01270		
First neighbor			
s, s (R)	−0.05036	s, s (T)	−0.05681
s, x (R)	0.07116	s, y (T)	−0.04588
s, y (R)	0.00986	s, z (T)	0.06727
s, xy (R)	0.00424	s, d1 (T)	−0.01845
s, d1 (R)	0.03865	s, yz (T)	−0.04750
s, d2 (R)	−0.02122	s, d2 (T)	0.02451
x, x (R)	0.09277	x, x (T)	−0.02941
y, y (R)	−0.02428	y, y (T)	0.03045
x, y (R)	0.02135	y, z (T)	−0.06463
x, xy (R)	0.01352	x, xy (T)	0.01440
y, d1 (R)	−0.00252	y, d1 (T)	0.00709
x, d1 (R)	0.05242	x, xz (T)	−0.01988
y, xy (R)	−0.01362	y, yz (T)	0.03337
y, d2 (R)	0.00106	y, d2 (T)	−0.02825
x, d2 (R)	−0.02354	z, z (T)	0.07548
z, z (R)	−0.00978	z, d1 (T)	−0.02760
z, yz (R)	−0.01038	z, yz (T)	−0.05836
z, xz (R)	−0.01105	z, d2 (T)	0.01858
xy, xy (R)	0.02071	xy, xy (T)	0.00720
d1, d1 (R)	−0.03066	d1, d1 (T)	−0.00017
xy, d1 (R)	−0.00382	xy, yz (T)	−0.01181
d1, d2 (R)	0.01476	yz, d1 (T)	−0.00509
xy, d2 (R)	0.00269	d1, d2 (T)	0.02041
yz, yz (R)	−0.00225	yz, yz (T)	−0.02975
xz, xz (R)	0.01601	xz, xz (T)	0.01703
yz, xz (R)	−0.00819	yz, d2 (T)	0.02611
d2, d2 (R)	−0.00913	d2, d2 (T)	0.00479

Table 4.4 Scandium hcp

	Orthogonal		
Band	RMS error mRy	Maximum k	Deviation mRy
1	1.5	(12 0 0)	4.7
2	2.7	(12 0 0)	6.9
3	1.4	(8 0 12)	4.4
4	3.0	(16 0 0)	7.2
5	1.5	(12 0 0)	4.0
6	3.4	(24 0 0)	9.9
7	3.1	(24 0 6)	8.5
8	3.9	(16 0 3)	12.4
9	2.9	(24 0 3)	8.0
10	4.8	(24 0 0)	17.2
11	4.4	(0 18 0)	14.6
12	3.5	(12 0 0)	8.4
1–12	3.21		

Table 4.5 Energy values in Ry at selected k-points

	Orthogonal	APW
GAMMA 1+	0.05412	0.05285
GAMMA 1+	0.54953	0.55001
GAMMA 3+	0.45882	0.45860
GAMMA 5+	0.57019	0.57010
GAMMA 6+	0.60198	0.59648
GAMMA 4−	0.30688	0.30764
GAMMA 4−	0.92109	0.92056
GAMMA 5−	0.67831	0.68486
GAMMA 6−	0.53677	0.53881
M1+	0.33726	0.33916
M1+	0.41594	0.41345
M1+	0.74752	0.74657
M2+	0.72791	0.71074
M3+	0.36972	0.36885
M4+	0.71110	0.71020
M1−	0.51533	0.52548
M2−	0.31815	0.31833
M2−	0.54972	0.55188
M2−	0.77652	0.77889
M3−	0.69498	0.69671
M4−	0.51558	0.51652
A1	0.16179	0.16162
A1	0.49240	0.49251
A3	0.53412	0.53529
A3	0.66346	0.66432
L1	0.31404	0.31411
L1	0.43402	0.43544
L1	0.51986	0.51864
L1	0.71481	0.71380
L2	0.59510	0.59756
L2	0.70353	0.70448

Table 4.6 Scandium hcp Z = 21 lattice constants = 6.2531 a.u. 9.9651 a.u.

Slater-Koster 2-center parameters	Orthogonal	Non-orthogonal	
	Energy integrals (Ry)	Energy integrals(Ry)	Overlap integrals
On site			
s	0.74560	0.62719	
p	0.91737	0.83913	
d0	0.57614	0.54730	
d1	0.57334	0.54839	
d2	0.58139	0.55662	
pd	−0.01716	0.01943	0.05031
First neighbor			
(sss)	−0.05610	−0.04058	0.04641
(sps)	0.07388	0.06326	−0.02128
(sds)	0.04311	0.03352	−0.01815
(pps)	0.09050	0.08107	−0.00935
(ppp)	−0.01454	0.05113	0.09125
(pds)	0.05269	0.03499	−0.00286
(pdp)	−0.01963	0.02048	0.07463
(dds)	−0.04407	−0.00160	0.06981
(ddp)	0.02646	0.00503	−0.04890
(ddd)	−0.00391	0.00177	0.01179
Second neighbor			
(sss)	−0.00319	−0.01018	0.01309
(sps)	0.00533	0.01336	−0.01273
(sds)	0.00197	0.01875	0.01500
(pps)	0.03071	0.03747	−0.03107
(ppp)	−0.00350	0.00749	0.02006
(pds)	0.01327	0.03663	0.02033
(pdp)	−0.00546	−0.00265	0.00233
(dds)	−0.00764	−0.02326	−0.01667
(ddp)	0.00172	0.00457	0.00434
(ddd)	0.00099	−0.00009	−0.00175
Third neighbor			
(sss)	−0.00140	0.00470	0.02031
(sps)	0.00129	−0.01396	−0.03852
(sds)	−0.00032	0.00388	0.02137
(pps)	0.00486	0.00551	−0.00965
(ppp)	−0.00011	−0.02518	−0.02823
(pds)	−0.00552	0.01251	0.02983
(pdp)	0.00028	−0.01713	−0.03021
(dds)	0.00136	−0.00247	−0.00403
(ddp)	0.00087	−0.00236	−0.00437
(ddd)	−0.00027	0.00558	0.00851

Table 4.7 Scandium hcp

	Orthogonal			Non-orthogonal		
Band	RMS error mRy	Maximum k	Deviation mRy	RMS error mRy	Maximum k	Deviation mRy
1	2.0	(0 24 12)	4.3	0.8	(16 0 12)	2.2
2	4.4	(0 18 0)	9.5	1.4	(16 0 9)	2.8
3	3.5	(24 0 9)	10.0	1.5	(0 18 0)	3.8
4	4.1	(0 24 12)	9.8	1.8	(0 0 0)	4.3
5	1.8	(0 24 12)	3.5	1.8	(0 0 0)	4.3
6	2.3	(8 0 0)	5.8	1.8	(8 0 0)	3.5
7	5.0	(8 0 12)	11.2	1.9	(8 0 12)	5.3
8	5.1	(8 0 12)	11.2	2.1	(8 0 12)	5.3
9	4.7	(0 0 12)	9.3	2.4	(16 0 12)	5.8
10	3.8	(0 0 12)	9.3	2.3	(16 0 12)	5.8
11	6.1	(24 0 0)	12.4	2.3	(16 0 12)	5.4
12	7.1	(8 0 0)	18.8	2.8	(0 18 0)	6.8
1–12	4.43			1.97		

Table 4.8 Energy values in Ry at selected k-points

	Orthogonal	APW	Non-orthogonal
GAMMA 1+	0.05048	0.05285	0.05280
GAMMA 1+	0.54702	0.55001	0.55093
GAMMA 3+	0.45952	0.45860	0.45821
GAMMA 5+	0.56914	0.57010	0.56764
GAMMA 6+	0.59943	0.59648	0.59908
GAMMA 4−	0.30621	0.30764	0.30849
GAMMA 4−	0.90951	0.92056	0.91696
GAMMA 5−	0.68598	0.68486	0.68426
GAMMA 6−	0.53544	0.53881	0.53454
M1 +	0.33139	0.33916	0.33845
M1 +	0.41083	0.41345	0.41363
M1 +	0.75896	0.74657	0.74640
M2 +	0.70119	0.71074	0.70924
M3+	0.36799	0.36885	0.36595
M4 +	0.70744	0.71020	0.70880
M1−	0.52536	0.52548	0.52753
M2−	0.31811	0.31833	0.31917
M2−	0.54640	0.55188	0.55130
M2−	0.78153	0.77889	0.78203
M3−	0.69713	0.69671	0.69961
M4−	0.51442	0.51652	0.51660
A1	0.16225	0.16162	0.16036
Al	0.49553	0.49251	0.49219
A3	0.53569	0.53529	0.53535
A3	0.65499	0.66432	0.66369
L1	0.31842	0.31411	0.31441
L1	0.42565	0.43544	0.43285
L1	0.52215	0.51864	0.51955
LI	0.71209	0.71380	0.71209
L2	0.60201	0.59756	0.59614
L2	0.70546	0.70448	0.70661

Table 4.9 Fermi level quantities (non-orthogonal fit)

Energy	Densities of states				
	Total	s	p	t_{2g}	e_g
Ry	States/Ry/atom				
0.4280	30.30	0.34	6.06	13.78	10.12

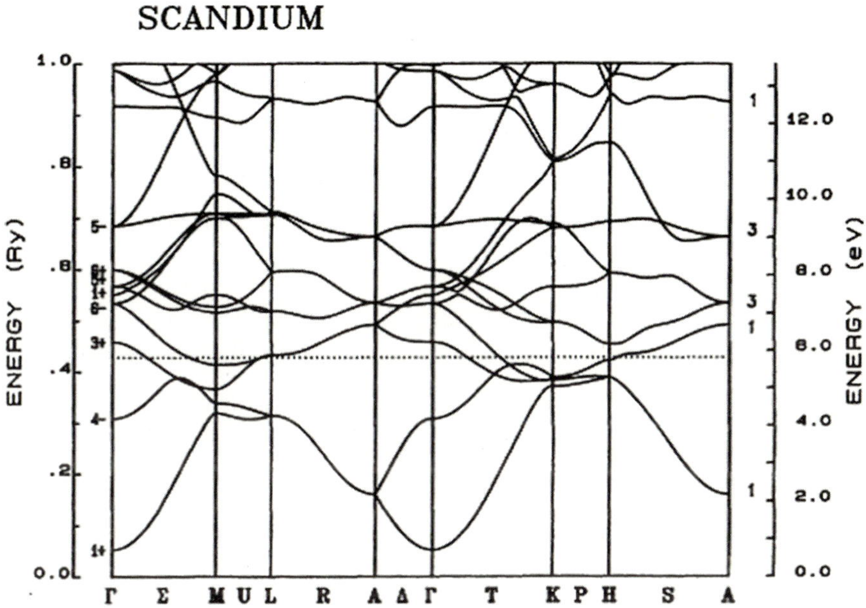

Fig. 4.2 Energy bands for Sc

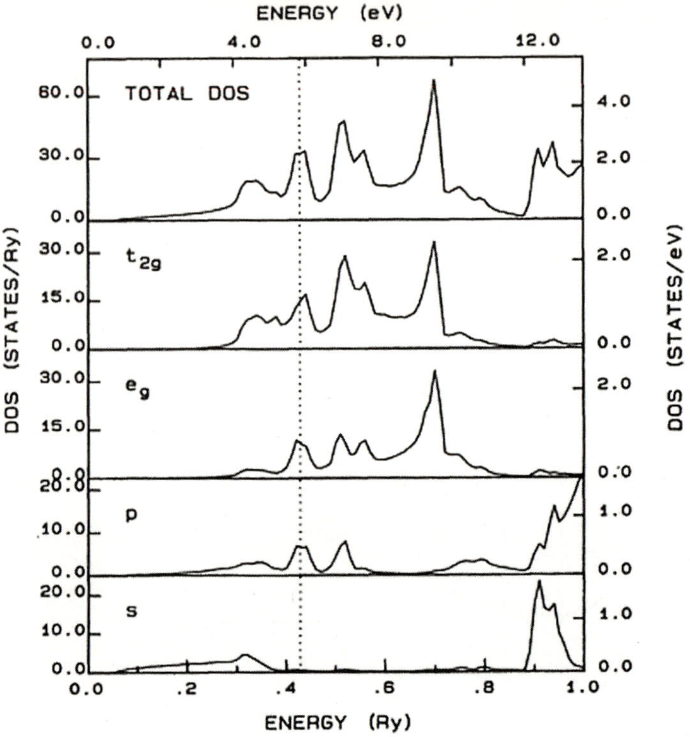

Fig. 4.3 Density of states for Sc

4.2 Titanium

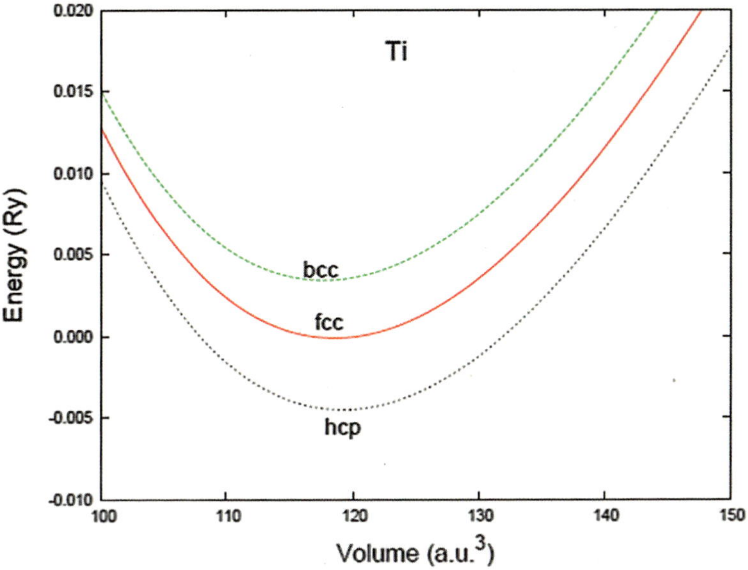

Fig. 4.4 Total energy of Ti

Table 4.10 Lattice constants and bulk modulus

	a (Bohr)	c (Bohr)	B (MBar)
bcc	6.175		1.052
fcc	7.798		1.076
hcp	5.583	8.832	1.095
exp hcp	5.576	8.847	
$\Delta E_{hcp\text{–}fcc} = 8.9$ mRy			

Table 4.11 Birch fit coefficients

	A_1	A_2	A_3	A_4
bcc	1.58855	−91.47905	1651.05074	−8846.26241
fcc	1.72537	−101.37442	1878.12553	−10536.04329
hcp	3.48029	−325.76082	9574.84683	−84897.02061

Table 4.12 Titanium hcp Z = 22 lattice constants =5.5769 a.u. 8.8523 a.u.

Slater-Koster 3-center parameters	Orthogonal		
	Energy integrals (Ry)		Energy integrals (Ry)
On site			
s, s (0)	0.90891		
x, x (0)	1.27537		
z, z (0)	1.17633		
xy, xy (0)	0.70397		
yz, yz (0)	0.71287		
d2, d2 (0)	0.68366		
s, d2 (0)	−0.02902		
x, xy (0)	−0.00709		
First neighbor			
s, s (R)	−0.05827	s, s (T)	−0.07130
s, x (R)	0.08514	s, y (T)	−0.05187
s, y (R)	0.01991	s, z (T)	0.07689
s, xy (R)	0.00466	s, d1 (T)	−0.02143
s, d1 (R)	0.04516	s, yz (T)	−0.05389
s, d2 (R)	−0.02255	s, d2 (T)	0.02911
x, x (R)	0.11447	x, x (T)	−0.02687
y, y (R)	−0.03667	y, y (T)	0.03840
x, y (R)	0.03594	y, z (T)	−0.07727
x, xy (R)	0.01828	x, xy (T)	0.02013
y, d1 (R)	−0.00936	y, d1 (T)	0.00601
x, d1 (R)	0.06400	x, xz (T)	−0.02133
y, xy (R)	−0.01714	y, yz (T)	0.03963
y, d2 (R)	0.00042	y, d2 (T)	−0.03585
x, d2 (R)	−0.01850	z, z (T)	0.07183
z, z (R)	−0.01806	z, d1 (T)	−0.02956
z, yz (R)	−0.01194	z, yz (T)	−0.06721
z, xz (R)	−0.01547	z, d2 (T)	0.01922
xy, xy (R)	0.02406	xy, xy (T)	0.00835
d1, d1 (R)	−0.03654	d1, d1 (T)	−0.00049
xy, d1 (R)	−0.00220	xy, yz (T)	−0.01453
d1, d2 (R)	0.01531	yz, d1 (T)	−0.00595
xy, d2 (R)	0.00264	d1, d2 (T)	0.02436
yz, yz (R)	−0.00228	yz, yz (T)	−0.03584
xz, xz (R)	0.01842	xz, xz (T)	0.01922
yz, xz (R)	−0.00887	yz, d2 (T)	0.03381
d2, d2 (R)	−0.01038	d2, d2 (T)	0.00557

Table 4.13 Titanium hcp

	Orthogonal		
Band	RMS error mRy	Maximum k	Deviation mRy
1	2.8	(16 0 0)	8.7
2	3.0	(12 0 0)	12.7
3	2.0	(12 0 0)	4.5
4	2.5	(4 0 0)	5.0
5	4.9	(24 0 0)	12.0
6	2.2	(24 0 9)	6.1
7	2.0	(0 12 12)	5.0
8	4.2	(16 0 3)	13.3
9	3.0	(24 0 3)	7.9
10	5.4	(24 0 0)	18.6
11	4.3	(0 18 0)	16.0
12	3.3	(12 0 0)	11.5
1–12	3.48		

Table 4.14 Energy values in Ry at selected k-points

	Orthogonal	APW
GAMMA 1+	0.13128	0.13044
GAMMA 1+	0.65499	0.65532
GAMMA 3+	0.63703	0.63704
GAMMA 5+	0.67804	0.67792
GAMMA 6+	0.71144	0.70749
GAMMA 4−	0.39419	0.39535
GAMMA 4−	1.18091	1.18170
GAMMA 5−	0.81118	0.81535
GAMMA 6−	0.63913	0.64063
M1+	0.41930	0.42263
M1+	0.54561	0.54273
M1+	0.90558	0.90446
M2+	0.86488	0.84626
M3+	0.45106	0.44725
M4+	0.84529	0.84412
M1−	0.61139	0.62338
M2−	0.42155	0.42289
M2−	0.67068	0.67287
M2−	0.97972	0.97859
M3−	0.82524	0.82759
M4−	0.63593	0.63594
A1	0.26578	0.26576
A1	0.59834	0.59844
A3	0.63430	0.63532
A3	0.78926	0.79034
L1	0.39696	0.39656
L1	0.53144	0.53238
L1	0.68502	0.68495
L1	0.86174	0.86295
L2	0.71230	0.71365
L2	0.83637	0.83722

Table 4.15 Titanium hcp Z = 22 lattice constants = 5.5769 a.u. 8.8523 a.u.

Slater-Koster 2-center parameters	Orthogonal	Non-orthogonal	
	Energy integrals (Ry)	Energy integrals (Ry)	Overlap integrals
On site			
s	0.90402	0.76848	
p	1.15021	1.00414	
d0	0.68539	0.65586	
d1	0.68485	0.65714	
d2	0.69467	0.66938	
pd	−0.01965	0.02711	0.06210
First neighbor			
(sss)	−0.06104	−0.03109	0.08966
(sps)	0.07517	0.04198	−0.06365
(sds)	0.04732	0.02655	−0.03504
(pps)	0.09812	0.08665	0.02131
(ppp)	−0.02239	0.06295	0.11650
(pds)	0.06046	0.04798	0.02360
(pdp)	−0.02427	0.02080	0.07173
(dds)	−0.05195	−0.00828	0.06154
(ddp)	0.03080	0.01087	−0.04001
(ddd)	−0.00442	−0.00013	0.00842
Second neighbor			
(sss)	−0.00552	−0.02083	0.01698
(sps)	0.01560	0.02053	−0.01927
(sds)	0.00626	0.02338	0.01172
(pps)	0.04627	0.02340	−0.04905
(ppp)	−0.00763	−0.00039	0.02019
(pds)	0.01487	0.02360	−0.00919
(pdp)	−0.00532	−0.00591	0.00224
(dds)	−0.00894	−0.02521	−0.01503
(ddp)	0.00190	0.00945	0.01013
(ddd)	0.00098	−0.00211	−0.00464
Third neighbor			
(sss)	−0.00435	0.01442	0.05065
(sps)	0.00718	−0.02513	−0.03452
(sds)	−0.00178	0.00541	0.02545
(pps)	0.01455	0.01473	0.00452
(ppp)	0.00344	−0.03190	−0.02485
(pds)	−0.00102	0.02261	0.03579
(pdp)	−0.00197	−0.02225	−0.03149
(dds)	0.00065	−0.01164	−0.01704
(ddp)	0.00026	−0.00264	−0.00412
(ddd)	0.00007	0.00599	0.00699

Table 4.16 Titanium hcp

	Orthogonal			Non-orthogonal		
Band	RMS error mRy	Maximum k	Deviation mRy	RMS error mRy	Maximum k	Deviation mRy
1	4.1	(24 0 0)	10.3	1.4	(12 0 0)	3.6
2	4.2	(0 18 0)	9.1	1.8	(0 18 0)	5.6
3	5.2	(24 0 9)	15.7	2.1	(24 0 0)	5.4
4	5.7	(0 24 12)	15.6	2.6	(24 0 9)	4.7
5	3.3	(8 0 0)	7.0	2.6	(8 0 0)	4.4
6	3.5	(0 12 0)	7.8	2.0	(0 12 0)	4.8
7	5.6	(0 12 12)	13.2	2.8	(8 0 12)	7.3
8	5.9	(0 12 12)	13.2	3.1	(8 0 12)	7.3
9	5.3	(24 0 0)	11.8	3.0	(0 12 12)	6.7
10	4.7	(0 18 0)	8.8	2.6	(0 12 12)	6.7
11	6.5	(24 0 0)	13.7	3.7	(16 0 6)	8.9
12	5.6	(12 0 0)	17.2	7.2	(16 0 0)	24.8
1–12	5.07			3.25		

Table 4.17 Energy values in Ry at selected k-points

	Orthogonal	APW	Non-orthogonal
GAMMA 1+	0.12970	0.13044	0.13131
GAMMA 1+	0.65081	0.65532	0.65620
GAMMA 3+	0.63526	0.63704	0.63504
GAMMA 5+	0.67640	0.67792	0.67356
GAMMA 6+	0.71239	0.70749	0.71201
GAMMA 4−	0.39156	0.39535	0.39672
GAMMA 4−	1.08241	1.18170	1.19487
GAMMA 5−	0.81663	0.81535	0.81494
GAMMA 6−	0.63755	0.64063	0.63630
M1+	0.41235	0.42263	0.42489
M1+	0.54206	0.54273	0.54257
M1+	0.91819	0.90446	0.90052
M2+	0.83236	0.84626	0.84385
M3+	0.44748	0.44725	0.44185
M4+	0.84109	0.84412	0.84336
M1−	0.62653	0.62338	0.62701
M2−	0.42364	0.42289	0.42462
M2−	0.66605	0.67287	0.67151
M2−	0.97882	0.97859	0.97676
M3−	0.82912	0.82759	0.83421
M4−	0.63065	0.63594	0.63371
A1	0.26568	0.26576	0.26564
Al	0.59792	0.59844	0.59863
A3	0.63679	0.63532	0.63566
A3	0.78197	0.79034	0.78898
L1	0.40288	0.39656	0.39600
L1	0.51680	0.53238	0.52817
L1	0.68784	0.68495	0.68494
LI	0.86123	0.86295	0.86181
L2	0.72018	0.71365	0.71225
L2	0.83701	0.83722	0.83913

Table 4.18 Fermi level quantities (non-orthogonal fit)

Energy	Densities of states				
	Total	s	p	t_{2g}	e_g
Ry	States/Ry/atom				
0.5953	8.06	0.07	0.80	4.61	2.59

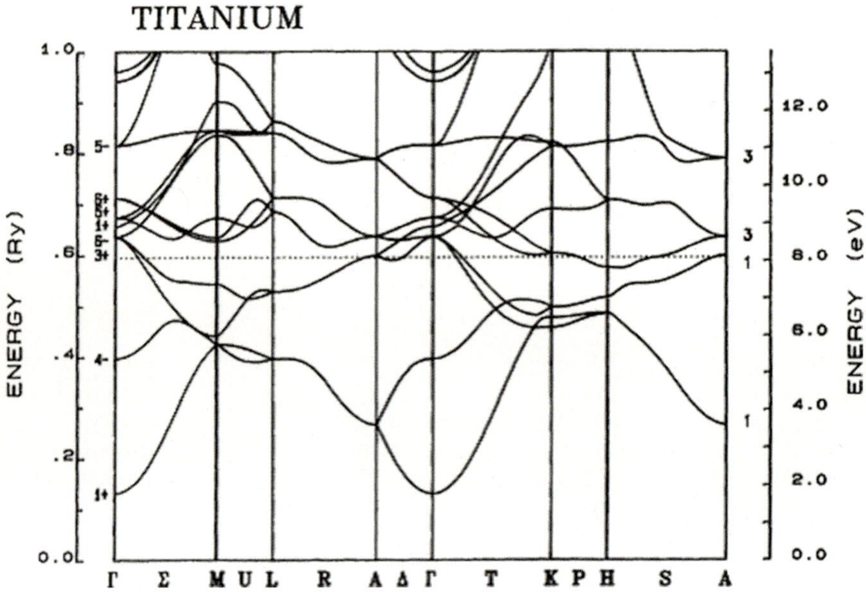

Fig. 4.5 Energy bands for Ti

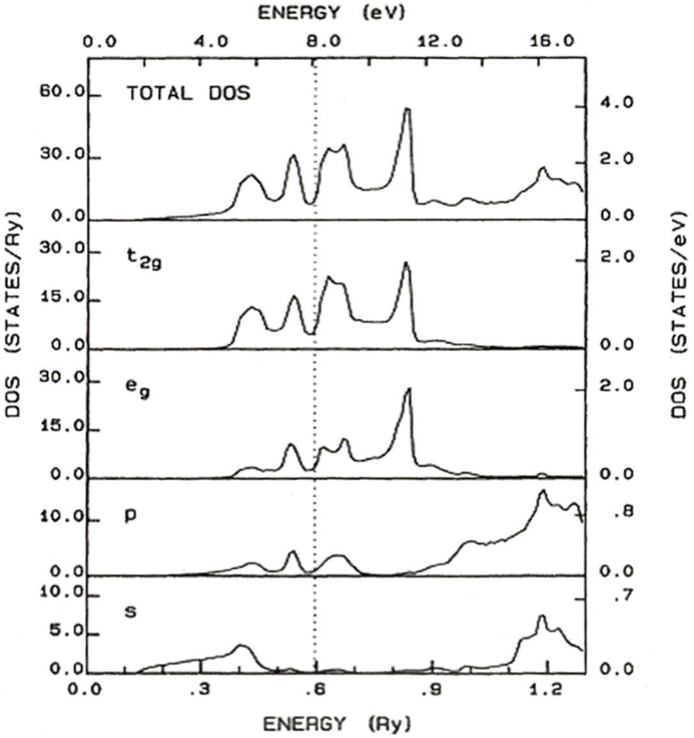

Fig. 4.6 Density of states for Ti

4.3 Vanadium

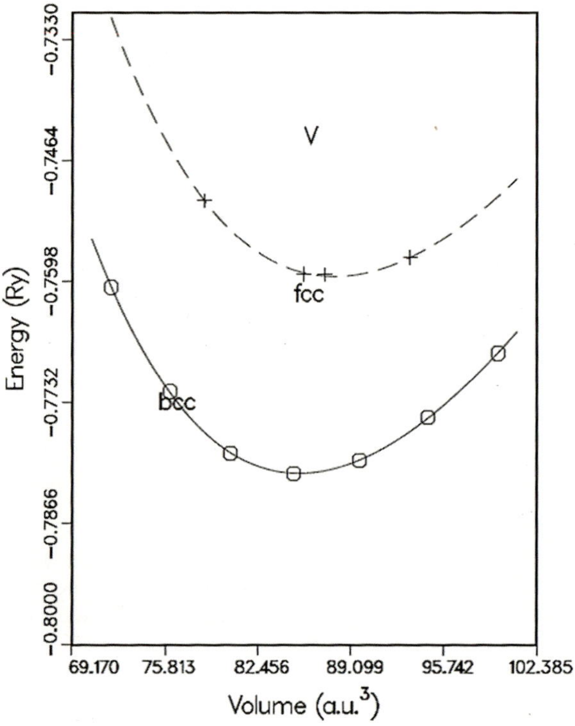

Fig. 4.7 Total energy of V

Table 4.19 Lattice constants and bulk modulus

	a (Bohr)	B (MBor)
bcc	5.548	2.155
fcc	7.059	2.081
exp	5.726	1.619
$\Delta E = 21.8$ mRy		

Table 4.20 Birch fit coefficients

	A_1	A_2	A_3
bcc	0.62618	−54.57822	529.20179
fcc	0.64010	−55.33772	547.08390

Table 4.21 Vanadium bcc Z = 23 lattice constant = 5.71453 a.u.

Slater-Koster 3-center parameters	Orthogonal	Non-orthogonal	
	Energy integrals (Ry)	Energy integrals (Ry)	Overlap integrals
On site			
s, s (000)	1.16730	0.91409	
x, x (000)	1.73196	1.34567	
xy, xy (000)	0.80516	0.76339	
d2, d2 (000)	0.76526	0.74155	
First neighbor			
s, s (111)	−0.12583	−0.04271	0.10056
s, x (111)	0.09358	0.02014	−0.07300
s, xy (111)	0.04475	0.02481	−0.01989
x, x (111)	0.09807	−0.02814	−0.03762
x, y (111)	0.06374	0.01658	−0.06020
x, xy (111)	0.02660	0.00436	−0.01188
x, yz (111)	0.04614	0.01786	−0.03262
x, d1 (111)	0.00824	0.01944	−0.01531
xy, xy (111)	−0.01442	−0.00851	0.00647
xy, xz (111)	−0.02327	−0.00937	0.01718
xy, d2 (111)	0.01188	0.01440	−0.00798
d2, d2 (111)	0.02404	0.01619	−0.01564
Second neighbor			
s, s (200)	−0.02754	−0.02566	0.05044
s, x (200)	0.06461	−0.02860	−0.11287
s, d2 (002)	−0.04963	0.00265	0.05549
x, x (200)	0.16614	−0.17782	−0.24412
y, y (200)	0.04314	0.00273	0.01216
x, xy (020)	0.00700	0.00131	0.01579
z, d2 (002)	−0.08335	0.02285	0.08821
xy, xy (200)	0.00042	0.00613	−0.01091
xy, xy (002)	−0.00191	0.00207	0.00284
d2, d2 (002)	−0.03981	−0.01720	0.03108
d1, d1 (002)	0.00241	−0.00132	−0.00073
Third neighbor			
s, s (220)	0.01670		
s, x (220)	−0.01720		
s, xy (220)	−0.00676		
s, d2 (220)	−0.00071		
x, x (220)	−0.01580		
x, x (022)	0.02307		
x, y (220)	−0.02222		
x, xy (220)	−0.00837		
x, xy (022)	0.00397		
z, d2 (022)	−0.00113		
z, d1 (022)	−0.00338		
xy, xy (220)	0.00495		
xy, xy (022)	0.00074		
xy, xz (022)	0.00044		
xy, d2 (220)	0.00326		
d2, d2 (220)	0.00136		
d1, d1 (220)	−0.00172		

Table 4.22 Vanadium bcc

	Orthogonal			Non-orthogonal		
Band	RMS error mRy	Maximum k	Deviation mRy	RMS error mRy	Maximum k	Deviation mRy
1	1.7	(330)	5.0	0.5	(333)	1.4
2	1.2	(350)	4.1	0.4	(131)	1.6
3	1.9	(242)	4.7	0.5	(161)	2.2
4	1.9	(044)	5.1	0.6	(140)	2.9
5	1.5	(131)	4.3	1.0	(131)	6.0
6	4.4	(353)	10.0	1.1	(140)	4.5
1–6	2.3			0.7		

Table 4.23 Energy values in Ry at selected k-points

	Orthogonal	APW	Non-orthogonal
GAMMA 1	0.19583	0.19759	0.19862
GAMMA 12	0.84325	0.84378	0.84428
GAMMA 15	2.98725	2.99017	2.99016
GAMMA 25′	0.71339	0.71374	0.71412
H1 (008)	2.20907	2.21196	2.21189
H12 (008)	0.45858	0.45676	0.45753
H15 (008)	1.41820	1.42595	1.42340
H25′ (008)	0.94415	0.94438	0.94489
N1 (044)	0.45178	0.45070	0.45018
N1 (044)	0.84362	0.84872	0.85001
N1 (044)	1.40495	1.40062	1.40096
N2 (044)	0.60304	0.60464	0.60460
N3 (044)	0.97534	0.97298	0.97360
N4 (044)	0.87549	0.87254	0.87395
N1′ (044)	0.79751	0.79605	0.79637
N3′ (044)	2.11030	2.11439	2.11413
N4′ (044)	1.81731	1.82199	1.82185
P1 (444)	1.53288	1.53205	1.53147
P3 (444)	0.87535	0.87693	0.87697
P4 (444)	0.60235	0.60293	0.60247
P4 (444)	1.42371	1.42134	1.42176
(341)	0.47710	0.47706	0.47692
(341)	0.60985	0.61059	0.61068
(341)	0.70790	0.70695	0.70736
(341)	0.84237	0.84292	0.84324
(341)	0.88898	0.88890	0.88941
(341)	1.02206	1.02315	1.02345

Table 4.24 Fermi level quantities (non-orthogonal fit)

Energy	Densities of states					Velocity	Plasmon energy
	Total	s	p	t_{2g}	e_g		
Ry	States/Ry/atom					cm/s	eV
0.6815	24.83	0.43	2.46	18.00	3.95	0.42 × 10E8	7.80

Integrated densities of states				
Total Electrons	s	p	t_{2g}	e_g
5.00	0.61	0.41	2.52	1.46

Table 4.25 Vanadium bcc

Slater-Koster 2-center parameters	Orthogonal	Non-orthogonal	
	Energy integrals (Ry)	Energy integrals (Ry)	Overlap integrals
On site			
s	1.18662	0.89757	
p	1.73152	1.31666	
d1	0.79925	0.75941	
d2	0.75975	0.74665	
First neighbor			
(sss)	−0.12571	−0.05055	0.10167
(pps)	0.22655	−0.00372	−0.16218
(ppp)	0.03351	−0.07638	0.00783
(dds)	−0.06196	−0.02230	0.04721
(ddp)	0.03633	0.02870	−0.02066
(ddd)	−0.00223	−0.01118	−0.00434
(sps)	0.14761	−0.04087	0.12778
(sds)	−0.07643	−0.03670	0.04012
(pds)	−0.10435	0.00564	−0.07585
(pdp)	0.02223	−0.03980	0.02022
Second neighbor			
(sss)	−0.02752	−0.00974	0.06260
(pps)	0.19000	−0.22995	−0.26740
(ppp)	0.02636	0.02248	0.02564
(dds)	−0.04002	−0.02812	0.01656
(ddp)	0.00361	0.00541	−0.01255
(ddd)	0.00184	−0.00130	−0.00067
(sps)	0.07469	0.06188	0.13135
(sds)	−0.04307	0.00446	0.05808
(pds)	−0.06916	−0.01217	−0.07782
(pdp)	−0.00643	0.00840	0.02464
Third neighbor			
(sss)	0.01492		
(pps)	−0.02241		
(ppp)	0.00922		
(dds)	0.00430		
(ddp)	−0.00004		
(ddd)	−0.00043		
(sps)	−0.01795		
(sds)	0.00180		
(pds)	0.00268		
(pdp)	−0.00253		

	Orthogonal			Non-orthogonal		
Band	RMS error mRy	Maximum k	Deviation mRy	RMS error Ry	Maximum k	Deviation mRy
1	3.3	(330)	10.4	0.5	(333)	1.7
2	4.9	(044)	15.1	0.4	(131)	1.4
3	4.2	(226)	12.4	0.5	(161)	2.3
4	4.2	(260)	14.9	0.7	(140)	3.0
5	3.0	(070)	8.1	1.1	(131)	6.2
6	6.1	(226)	15.6	1.1	(140)	4.5
1–6	4.4			0.8		

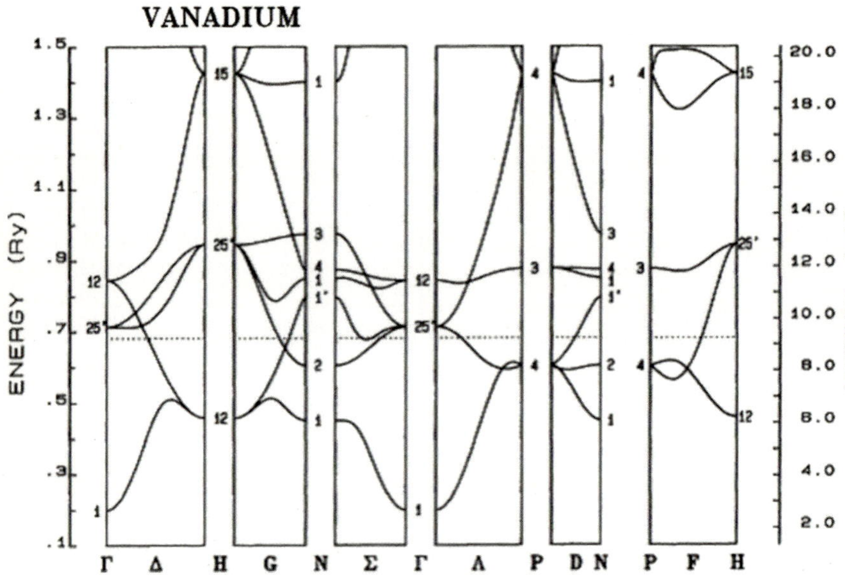

Fig. 4.8 Energy bands for V

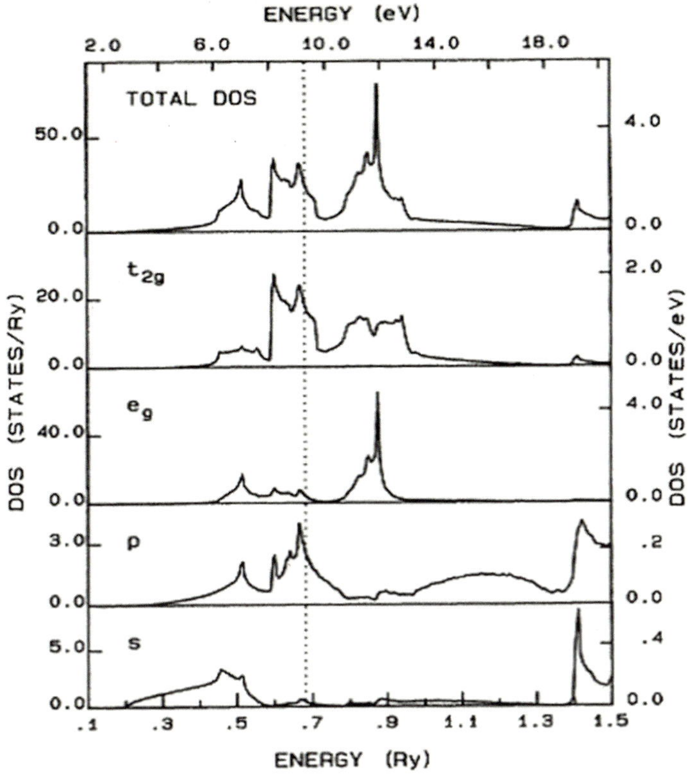

Fig. 4.9 Density of states for V

4.4 Chromium

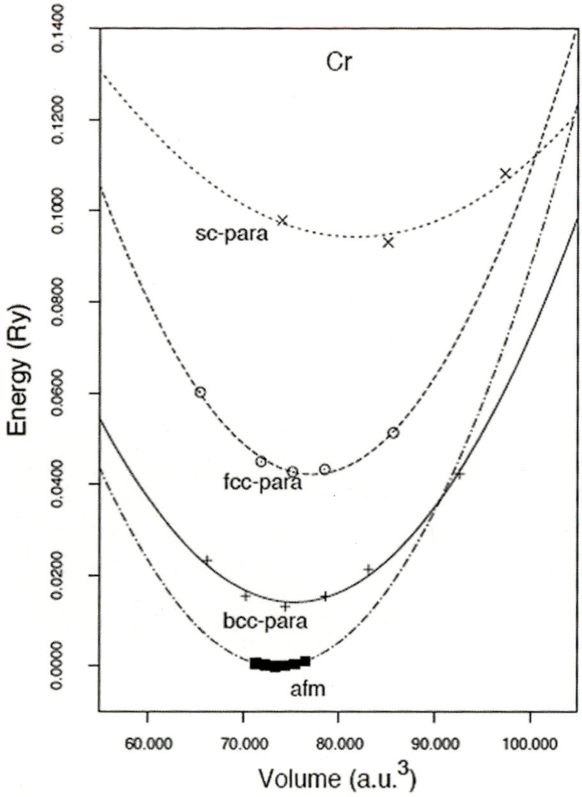

Fig. 4.10 Total energy of Cr

Table 4.26 Lattice constants and bulk modulus

	a (Bohr)	B (MBar)
sc	4.337	2.146
bcc	5.299	2.799
fcc	6.728	2.637
afm	5.279	2.820
exp	5.442	1.901
$\Delta E = E_{bcc} - E_{afm} = 13.2$ mRy		

Table 4.27 Birch fit coefficients

	A_1	A_2	A_3	A_4
sc	1.30129	−43.05919	336.10276	864.77672
bcc	1.48834	−50.11432	388.33227	646.64754
fcc	1.08811	−28.75841	21.00437	2842.34590
afm	0.24556	14.92299	−751.10131	7258.11132

Table 4.28 Chromium bcc Z = 24 lattice constant = 5.44979 a.u.

Slater-Koster 3-center parameters	Orthogonal	Non-orthogonal	
	Energy integrals (Ry)	Energy integrals (Ry)	Overlap integrals
On site			
s, s (000)	1.33055	0.94293	
x, x (000)	1.88031	1.40332	
xy, xy (000)	0.77732	0.77153	
d2, d2 (000)	0.77890	0.75199	
First neighbor			
s, s (111)	−0.13765	−0.03840	0.11005
s, x (111)	0.10971	0.00851	−0.08515
s, xy (111)	0.03907	0.02569	−0.01905
x, x (111)	0.10743	−0.03785	−0.04135
x, y (111)	0.07112	−0.01019	−0.08177
x, xy (111)	0.02832	0.00799	−0.00851
x, yz (111)	0.03218	0.01163	−0.04038
x, d1 (111)	0.00695	0.02314	−0.01224
xy, xy (111)	−0.01453	−0.01156	0.00334
xy, xz (111)	−0.02323	−0.01036	0.01579
xy, d2 (111)	0.01612	0.01516	−0.00591
d2, d2 (111)	0.02417	0.01734	−0.01300
Second neighbor			
s, s (200)	−0.03038	−0.02997	0.05418
s, x (200)	0.06450	−0.01976	−0.11535
s, d2 (002)	−0.04315	0.00606	0.06085
x, x (200)	0.17347	−0.16474	−0.24118
y, y (200)	0.03432	0.00216	0.01331
x, xy (020)	−0.01273	−0.00369	0.01358
z, d2 (002)	−0.07818	0.02284	0.09176
xy, xy (200)	0.01746	0.01073	−0.00493
xy, xy (002)	−0.00097	0.00398	0.00455
d2, d2 (002)	−0.03905	−0.02096	0.02465
d1, d1 (002)	0.00093	−0.00212	0.00166
Third neighbor			
s, s (220)	0.01291		
s, x (220)	−0.01191		
s, xy (220)	0.00353		
s, d2 (220)	0.00439		
x, x (220)	−0.01094		
x, x (022)	0.02775		
x, y (220)	−0.02190		
x, xy (220)	0.00256		
x, xy (022)	0.00357		
z, d2 (022)	−0.00466		
z, d1 (022)	0.00157		
xy, xy (220)	−0.00379		
xy, xy (022)	0.00045		
xy, xz (022)	0.00064		
xy, d2 (220)	−0.00039		
d2, d2 (220)	0.00176		
d1, d1 (220)	−0.00320		

Table 4.29 Chromium bcc

	Orthogonal			Non-orthogonal		
Band	RMS error mRy	Maximum k	Deviation mRy	RMS error mRy	Maximum k	Deviation mRy
1	2.0	(330)	5.6	0.5	(000)	1.3
2	1.3	(232)	3.0	0.5	(131)	1.8
3	1.9	(442)	5.0	0.6	(161)	2.1
4	1.5	(441)	4.0	0.7	(140)	3.3
5	1.4	(131)	4.6	1.0	(131)	5.6
6	5.4	(008)	11.7	1.3	(140)	4.5
1–6	2.7			0.8		

Table 4.30 Energy values in Ry at selected k-points

	Orthogonal	APW	Non-orthogonal
GAMMA 1	0.20202	0.20541	0.20672
GAMMA 12	0.84930	0.85064	0.85126
GAMMA 15	3.24749	3.24800	3.24799
GAMMA 25′	0.71738	0.71799	0.71840
H1 (008)	2.40435	2.40670	2.40664
H12 (008)	0.46253	0.46282	0.46380
H15 (008)	1.52853	1.54024	1.53632
H25′ (008)	0.94990	0.94981	0.95034
N1 (044)	0.46126	0.46057	0.46013
N1 (044)	0.85048	0.85523	0.85686
N1 (044)	1.47447	1.47104	1.47138
N2 (044)	0.60856	0.60927	0.60927
N3 (044)	0.98028	0.97817	0.97901
N4 (044)	0.88006	0.87766	0.87923
N1′ (044)	0.85345	0.84858	0.84917
N3′ (044)	2.28847	2.29395	2.29368
N4′ (044)	1.99131	1.99389	1.99387
P1 (444)	1.66772	1.66407	1.66361
P3 (444)	0.88462	0.88190	0.88208
P4 (444)	0.61543	0.61424	0.61389
P4 (444)	1.50198	1.49945	1.49994
(341)	0.48835	0.48860	0.48848
(341)	0.61573	0.61620	0.61627
(341)	0.73314	0.73414	0.73479
(341)	0.84879	0.84973	0.85008
(341)	0.89683	0.89749	0.89789
(341)	1.04692	1.04679	1.04731

Table 4.31 Fermi level quantities (non-orthogonal fit)

Energy	Densities of states					Velocity	Plasmon energy
	Total	s	p	t_{2g}	e_g		
Ry	States/Ry/atom					cm/s	eV
0.7795	9.63	0.08	0.72	6.52	2.31	0.52 × 10E8	6.41

Integrated densities of states				
Total Electrons	s	p	t_{2g}	e_g
6.00	0.60	0.44	3.32	1.64

Table 4.32 Chromium bcc

Slater-Koster 2-center parameters	Orthogonal	Non-orthogonal	
	Energy integrals (Ry)	Energy integrals (Ry)	Overlap integrals
On site			
s	1.27653	0.96894	
p	1.87795	1.43466	
d1	0.80247	0.76756	
d2	0.76581	0.75630	
First neighbor			
(sss)	−0.13748	−0.04768	0.10377
(pps)	0.24993	0.01550	−0.15539
(ppp)	0.03579	−0.07768	0.01081
(dds)	−0.06208	−0.02642	0.04213
(ddp)	0.03675	0.03093	−0.01625
(ddd)	−0.00268	−0.01148	−0.00503
(sps)	0.16377	−0.04396	0.12658
(sds)	−0.07784	−0.04029	0.03617
(pds)	−0.10654	0.01680	−0.06383
(pdp)	0.02331	−0.04186	0.01924
Second neighbor			
(sss)	−0.03008	−0.02343	0.05548
(pps)	0.20229	−0.24608	−0.26724
(ppp)	0.03149	0.02331	0.02382
(dds)	−0.03935	−0.03042	0.01225
(ddp)	0.00466	0.00658	−0.01035
(ddd)	0.00152	−0.00152	−0.00101
(sps)	0.07923	0.04950	0.12259
(sds)	−0.04396	−0.00210	0.04977
(pds)	−0.06943	−0.00800	−0.07150
(pdp)	−0.00627	0.00918	0.02575
Third neighbor			
(sss)	0.01715		
(pps)	−0.02985		
(ppp)	0.01202		
(dds)	0.00366		
(ddp)	0.00014		
(ddd)	−0.00049		
(sps)	−0.02166		
(sds)	0.00107		
(pds)	0.00215		
(pdp)	−0.00280		

	Orthogonal			Non-orthogonal		
Band	RMS error mRy	Maximum k	Deviation mRy	RMS error mRy	Maximum k	Deviation mRy
1	3.5	(330)	10.6	0.5	(008)	1.2
2	4.5	(044)	13.8	0.5	(131)	1.6
3	4.3	(226)	12.0	0.6	(161)	2.7
4	3.8	(260)	13.7	0.7	(140)	3.2
5	2.7	(070)	7.4	1.1	(131)	5.8
6	6.5	(226)	17.0	1.5	(008)	4.9
1–6	4.4			0.9		

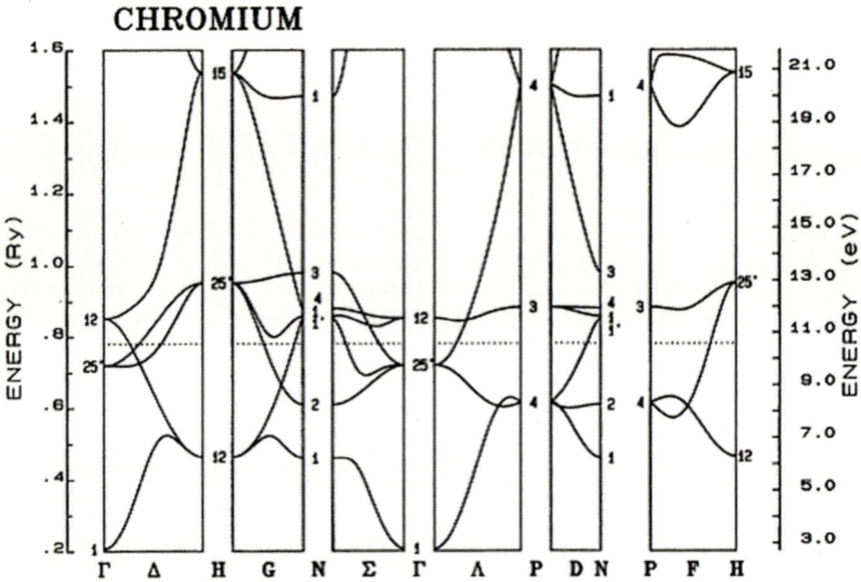

Fig. 4.11 Energy bands for Cr

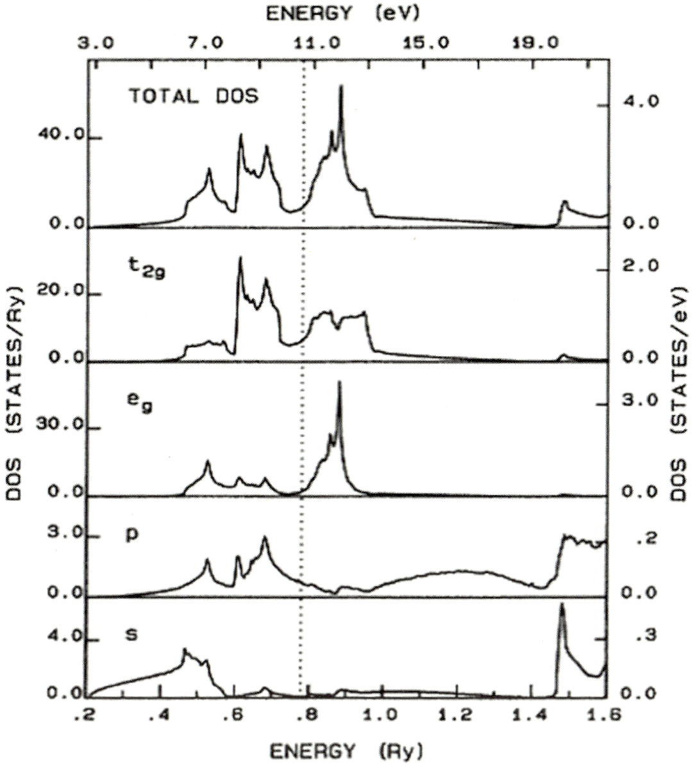

Fig. 4.12 Density of states for Cr

4.5 Manganese

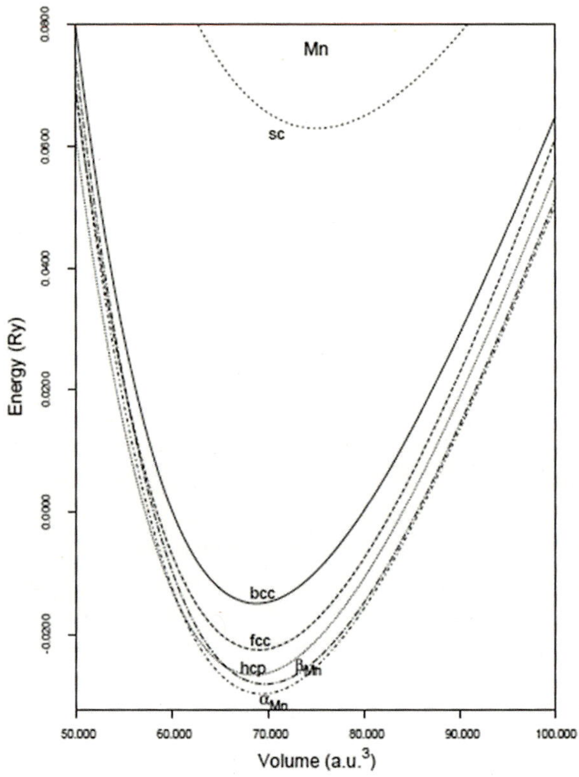

Fig. 4.13 Total energy of Mn

Table 4.33 Lattice constants and bulk modulus

	a (Bohr)	c (Bohr)	B (MBar)
sc	4.217		1.982
bcc	5.160		3.170
fcc	6.509		3.180
hcp	5.812	9.357	3.134
β_{Mn}(A13)	11.165		3.214
α_{Mn}(A12)	15.898		3.210
exp	16.849		1.200
$\Delta E_{\alpha Mn-\beta Mn} = 1.6$ mRy			

Table 4.34 Birch fit coefficients

	A_1	A_2	A_3	A_4
sc	1.73144	−68.80332	863.98904	−2989.37225
bcc	1.48417	−47.50070	328.12432	785.86489
fcc	1.93560	−70.57625	712.69000	−1338.80290
hcp	3.27660	−178.19805	2419.32910	−942.59354
β_{Mn}	35.37800	−9176.36481	612259.29932	−3349547.39778
α_{Mn}	48.73502	−15838.20988	1271475.43647	−1130712.60370

Table 4.35 Manganese bcc Z = 25 lattice constant = 5.39700 a.u.

Slater-Koster 3-center parameters	Orthogonal	Non-orthogonal	
	Energy integrals (Ry)	Energy integrals (Ry)	Overlap integrals
On site			
s, s (000)	1.15152	0.89521	
x, x (000)	1.84444	1.35172	
xy, xy (000)	0.74172	0.70062	
d2, d2 (000)	0.70887	0.68710	
First neighbor			
s, s (111)	−0.13558	−0.05023	0.10519
s, x (111)	0.10063	0.03018	−0.07571
s, xy (111)	0.04360	0.02556	−0.01468
x, x (111)	0.10621	−0.00050	−0.02033
x, y (111)	0.07479	0.00737	−0.08114
x, xy (111)	0.02696	0.00661	−0.00278
x, yz (111)	0.04560	0.01169	−0.04209
x, d1 (111)	0.00602	0.02368	−0.00967
xy, xy (111)	−0.01274	−0.00653	0.00825
xy, xz (111)	−0.02087	−0.00822	0.01679
xy, d2 (111)	0.00924	0.01593	−0.00320
d2, d2 (111)	0.02156	0.01924	−0.00800
Second neighbor			
s, s (200)	−0.02967	−0.03070	0.05352
s, x (200)	0.06511	−0.03416	−0.13221
s, d2 (002)	−0.05138	0.01312	0.07045
x, x (200)	0.17254	−0.24471	−0.29751
y, y (200)	0.05174	0.00968	0.00974
x, xy (020)	0.01002	−0.00124	0.01581
z, d2 (002)	−0.08636	0.05575	0.13556
xy, xy (200)	−0.00017	0.01117	−0.00365
xy, xy (002)	−0.00031	0.00135	0.00338
d2, d2 (002)	−0.03424	−0.00735	0.03870
d1, d1 (002)	0.00128	0.00023	0.00063
Third neighbor			
s, s (220)	0.02104		
s, x (220)	−0.02225		
s, xy (220)	−0.00488		
s, d2 (220)	−0.00116		

(continued)

Table 4.35 (continued)

Slater-Koster 3-center parameters	Orthogonal	Non-orthogonal	
	Energy integrals (Ry)	Energy integrals (Ry)	Overlap integrals
x, x (220)	−0.02012		
x, x (022)	0.02523		
x, y (220)	−0.03529		
x, xy (220)	−0.00743		
x, xy (022)	0.00509		
z, d2 (022)	−0.00137		
z, d1 (022)	−0.00507		
xy, xy (220)	0.00449		
xy, xy (022)	0.00014		
xy, xz (022)	0.00067		
xy, d2 (220)	0.00309		
d2, d2 (220)	0.00089		
d1, d1 (220)	−0.00169		

Table 4.36 Maganese bcc

	Orthogonal			Non-orthogonal		
Band	RMS error mRy	Maximum k	Deviation mRy	RMS error mRy	Maximum k	Deviation mRy
1	2.5	(330)	7.3	0.7	(008)	2.2
2	2.2	(353)	5.5	0.7	(008)	2.2
3	2.7	(230)	6.7	0.7	(044)	2.6
4	2.3	(044)	6.8	0.9	(330)	2.5
5	1.8	(044)	4.9	1.0	(131)	4.6
6	5.3	(070)	13.4	1.3	(140)	4.7
1–6	3.0			0.9		

Table 4.37 Energy values in Ry at selected k-points

	Orthogonal	APW	Non-orthogonal
GAMMA 1	0.14126	0.14181	0.14296
GAMMA 12	0.77767	0.77706	0.77761
GAMMA 15	3.18608	3.18919	3.18920
GAMMA 25′	0.65750	0.65706	0.65755
H1 (008)	2.31057	2.32056	2.32064
H12 (008)	0.43269	0.43087	0.43303
H15 (008)	1.48677	1.49349	1.49202
H25′ (008)	0.86142	0.86338	0.86400
N1 (044)	0.42105	0.42084	0.42043
N1 (044)	0.77737	0.77955	0.78211
N1 (044)	1.36973	1.37414	1.37485
N2 (044)	0.55743	0.56065	0.56062
N3 (044)	0.89130	0.88808	0.88912
N4 (044)	0.80413	0.79927	0.80197
N1′ (044)	0.80009	0.79327	0.79413
N3′ (044)	2.24443	2.25012	2.25006

(continued)

Table 4.37 (continued)

	Orthogonal	APW	Non-orthogonal
N4′ (044)	1.99681	2.00956	2.00884
P1 (444)	1.58200	1.59022	1.58988
P3 (444)	0.80298	0.80277	0.80325
P4 (444)	0.56321	0.56292	0.56247
P4 (444)	1.43122	1.42762	1.42806
(341)	0.44529	0.4458	0.44580
(341)	0.56580	0.56689	0.56695
(341)	0.67702	0.67517	0.67619
(341)	0.77526	0.77522	0.77580
(341)	0.81835	0.81785	0.81856
(341)	0.96169	0.96234	0.96317

Table 4.38 Fermi level quantities (non-orthogonal fit)

Energy	Densities of states					Velocity	Plasmon energy
	Total	s	p	t_{2g}	e_g		
Ry	States/Ry/atom					cm/s	eV
0.7600	37.42	0.19	0.31	17.88	19.05	0.27 × 10E8	7.38

Integrated densities of states				
Total	s	p	t_{2g}	e_g
Electrons				
7.00	0.60	0.41	2.01	3.98

Table 4.39 Manganese bcc

Slater-Koster 2-center parameters	Orthogonal	Non-orthogonal	
	Energy integrals (Ry)	Energy integrals (Ry)	Overlap integrals
On site			
s	1.16833	0.84661	
p	1.84211	1.35281	
d1	0.73531	0.70096	
d2	0.70341	0.69912	
First neighbor			
(sss)	−0.13490	−0.06272	0.11062
(pps)	0.25792	0.00720	−0.17223
(ppp)	0.03037	−0.09393	0.01411
(dds)	−0.05509	−0.02463	0.04069
(ddp)	0.03210	0.02937	−0.01027
(ddd)	−0.00173	−0.01337	−0.00949
(sps)	0.16844	−0.04958	0.13893
(sds)	−0.07524	−0.03315	0.03794
(pds)	−0.10344	0.01665	−0.05788
(pdp)	0.02040	−0.04574	0.01256

(continued)

Table 4.39 (continued)

Slater-Koster 2-center parameters	Orthogonal	Non-orthogonal	
	Energy integrals (Ry)	Energy integrals (Ry)	Overlap integrals
Second neighbor			
(sss)	−0.03047	−0.00138	0.07707
(pps)	0.18385	−0.28729	−0.29567
(ppp)	0.04128	0.01066	0.02043
(dds)	−0.03468	−0.03648	−0.00291
(ddp)	0.00347	0.01630	0.00319
(ddd)	0.00195	−0.00298	−0.00315
(sps)	0.06672	0.08811	0.15721
(sds)	−0.04100	0.01138	0.06883
(pds)	−0.06798	−0.00580	−0.07377
(pdp)	−0.01020	0.00132	0.02557
Third neighbor			
(sss)	0.01920		
(pps)	−0.04231		
(ppp)	0.01611		
(dds)	0.00346		
(ddp)	−0.00025		
(ddd)	−0.00071		
(sps)	−0.02683		
(sds)	−0.00013		
(pds)	0.00156		
(pdp)	−0.00461		

	Orthogonal			Non-orthogonal		
Band	RMS error mRy	Maximum k	Deviation mRy	RMS error mRy	Maximum k	Deviation mRy
1	4.3	(330)	10.6	0.8	(008)	2.1
2	4.4	(044)	11.3	0.8	(006)	2.2
3	3.9	(226)	10.4	0.7	(161)	2.6
4	3.7	(260)	10.8	0.6	(140)	2.9
5	3.4	(070)	8.7	1.0	(131)	4.8
6	6.2	(252)	16.1	1.9	(171)	6.6
1–6	4.4			1.1		

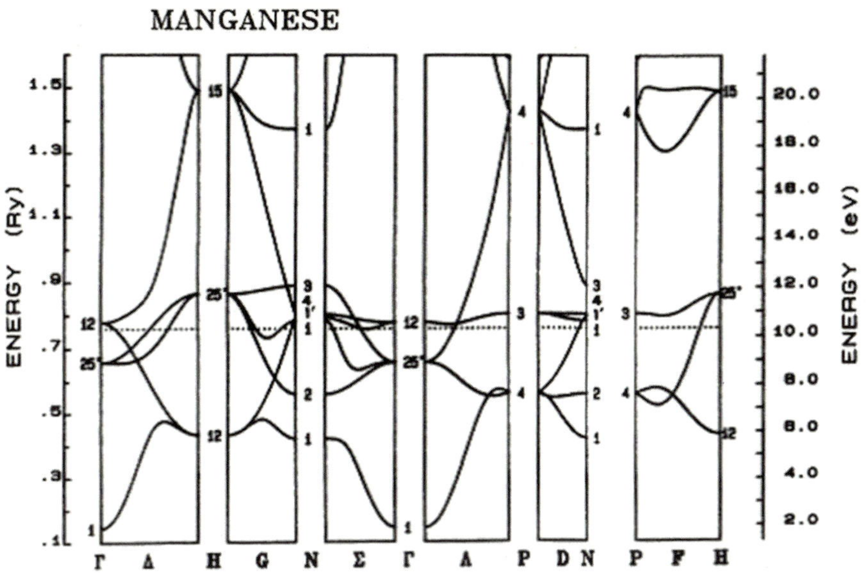

Fig. 4.14 Energy bands for Mn

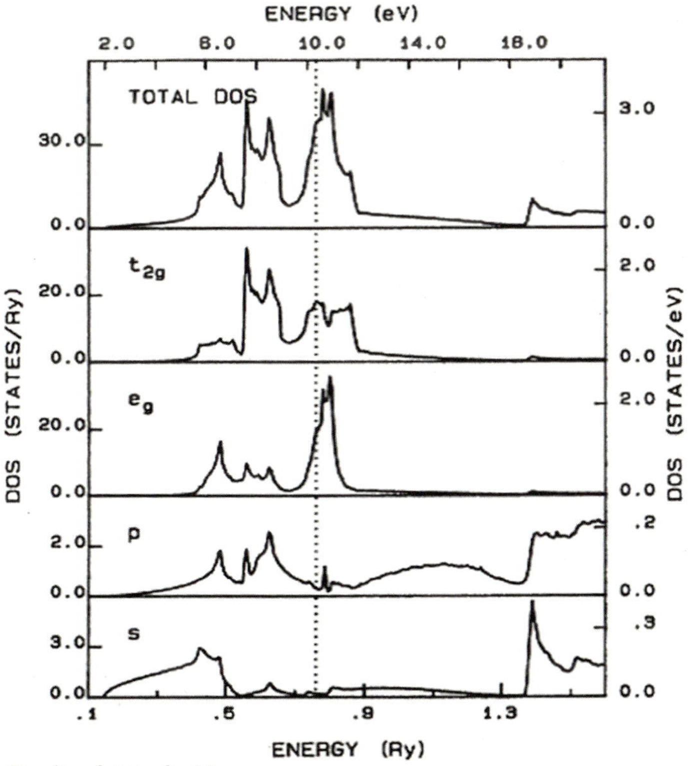

Fig. 4.15 Density of states for Mn

4.6 Iron

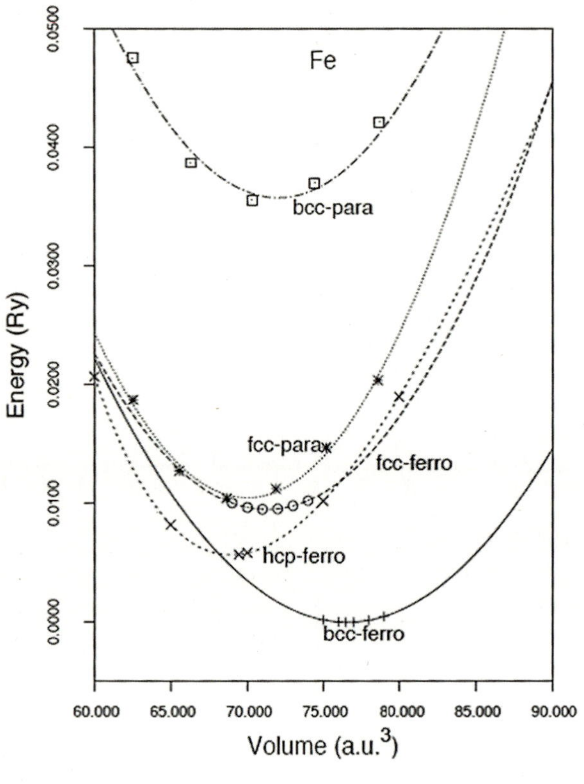

Fig. 4.16 Total energy of Fe

Table 4.40 Lattice constants and bulk modulus

	a (Bohr)	c (Bohr)	B (MBar)
bcc–para	5.215		2.769
fcc–para	6.521		3.005
bcc–ferro	5.348		1.898
fcc–ferro	6.581		2.187
hcp–ferro	4.663	7.375	2.927
exp	5.424		1.683

$E_{hcp-ferro} - E_{bcc-ferro} = \Delta E = 5.7$ mRy

Table 4.41 Birch fit coefficients

	A_1	A_2	A_3	A_4
bcc–para	0.94694	−21.11689	−79.12279	2970.51986
fcc–para	1.01738	−24.09172	−47.27591	2819.22474
bcc–ferro	−0.27085	34.66409	−985.65430	8090.16340
fcc–ferro	1.78727	−71.17948	871.25977	−2974.10286
hcp–ferro	1.01784	−25.10836	−14.03958	2523.14129

Table 4.42 IRON bcc spin up Z = 26 lattice constant = 5.40570 a.u.

Slater-Koster 3-center parameters	Orthogonal	Non-orthogonal	
	Energy integrals (Ry)	Energy integrals (Ry)	Overlap integrals
On site			
s, s (000)	1.15144	0.83255	
x, x (000)	1.81728	1.30038	
xy, xy (000)	0.63801	0.62784	
d2, d2 (000)	0.63647	0.62266	
First neighbor			
s, s (111)	−0.12944	−0.06346	0.10276
s, x (111)	0.10884	0.03434	−0.07844
s, xy (111)	0.03314	0.02169	−0.01429
x, x (111)	0.10187	−0.02649	−0.02992
x, y (111)	0.07787	0.02662	−0.07515
x, xy (111)	0.02339	0.00380	−0.00142
x, yz (111)	0.02742	0.01845	−0.03105
x, d1 (111)	0.00633	0.03017	0.00477
xy, xy (111)	−0.01064	−0.00288	0.01105
xy, xz (111)	−0.01705	−0.00745	0.01433
xy, d2 (111)	0.01104	0.01459	0.00063
d2, d2 (111)	0.01757	0.01659	−0.00407
Second neighbor			
s, s (200)	−0.02910	−0.00858	0.06849
s, x (200)	0.05525	−0.07102	−0.15114
s, d2 (002)	−0.03754	0.01211	0.07028
x, x (200)	0.14399	−0.34146	−0.34056
y, y (200)	0.04772	0.02868	0.02237
x, xy (020)	−0.00678	0.01191	0.03306
z, d2 (002)	−0.06401	0.06154	0.15119
xy, xy (200)	0.01140	0.00434	−0.00976
xy, xy (002)	−0.00006	0.00038	0.00324
d2, d2 (002)	−0.02659	−0.01034	0.02616
d1, d1 (002)	−0.00074	0.00011	0.00052
Third neighbor			
s, s (220)	0.01465		
s, x (220)	−0.01631		
s, xy (220)	0.00204		
s, d2 (220)	0.00215		
x, x (220)	−0.01931		
x, x (022)	0.03044		
x, y (220)	−0.03490		
x, xy (220)	0.00148		
x, xy (022)	0.00414		
z, d2 (022)	−0.00190		
z, d1 (022)	−0.00044		
xy, xy (220)	−0.00223		
xy, xy (022)	0.00000		
xy, xz (022)	0.00054		
xy, d2 (220)	0.00012		
d2, d2 (220)	0.00060		
d1, d1 (220)	−0.00143		

Table 4.43 IRON bcc spin up

	Orthogonal			Non-orthogonal		
Band	RMS error mRy	Maximum k	Deviation mRy	RMS error mRy	Maximum k	Deviation mRy
1	2.5	(222)	8.3	0.7	(008)	2.2
2	1.7	(353)	4.8	0.7	(008)	2.2
3	2.3	(442)	7.1	0.7	(044)	2.2
4	1.6	(330)	3.9	0.8	(140)	2.6
5	1.9	(044)	8.8	0.8	(131)	3.6
6	5.8	(170)	14.9	1.3	(170)	5.4
1–6	3.0			0.9		

Table 4.44 Energy values in Ry at selected k-points

	Orthogonal	APW	Non-orthogonal
GAMMA 1	0.11713	0.12119	0.12241
GAMMA 12	0.69008	0.69119	0.69184
GAMMA 15	3.07838	3.07849	3.07849
GAMMA 25′	0.58942	0.58951	0.58997
H1 (008)	2.18809	2.18876	2.18876
H12 (008)	0.40902	0.41054	0.41277
H15 (008)	1.44847	1.45932	1.45765
H25′ (008)	0.75968	0.76048	0.76106
N1 (044)	0.39854	0.39688	0.39677
N1 (044)	0.68885	0.69159	0.69383
N1 (044)	1.26893	1.26470	1.26491
N2 (044)	0.51063	0.51175	0.51209
N3 (044)	0.78346	0.78061	0.78130
N4 (044)	0.70902	0.70722	0.70910
N1′ (044)	0.78458	0.77469	0.77567
N3′ (044)	2.19061	2.19138	2.19134
N4′ (044)	2.03049	2.03382	2.03371
P1 (444)	1.50178	1.50070	1.50073
P3 (444)	0.71352	0.70977	0.71020
P4 (444)	0.52222	0.51962	0.51938
P4 (444)	1.36714	1.35991	1.36027
(341)	0.41864	0.41896	0.41868
(341)	0.51735	0.51783	0.51801
(341)	0.62590	0.62727	0.62839
(341)	0.68770	0.68872	0.68928
(341)	0.72555	0.72623	0.72636
(341)	0.87888	0.87887	0.87961

Table 4.45 Fermi level quantities (non-orthogonal fit)

Energy	Densities of states				
Ry	Total States/Ry/atom	s	p	t_{2g}	e_g
0.7505	11.89	0.22	0.14	10.45	1.07

Integrated densities of states Total Electrons	s	p	t_{2g}	e_g
5.05	0.30	0.16	2.68	1.91

Table 4.46 Iron bcc spin up

Slater-Koster 2-center parameters	Orthogonal	Non-orthogonal	
	Energy integrals (Ry)	Energy integrals (Ry)	Overlap integrals
On site			
s	1.13516	0.85674	
p	1.81739	1.36382	
d1	0.64840	0.63010	
d2	0.62960	0.62920	
First neighbor			
(sss)	−0.12950	−0.06479	0.09821
(pps)	0.25741	0.03700	−0.16189
(ppp)	0.02422	−0.06914	0.02686
(dds)	−0.04541	−0.02284	0.03372
(ddp)	0.02714	0.02648	−0.00430
(ddd)	−0.00260	−0.01242	−0.01105
(sps)	0.17363	−0.06822	0.12160
(sds)	−0.06115	−0.03943	0.02107
(pds)	−0.08485	0.03065	−0.03407
(pdp)	0.01778	−0.03899	0.01340
Second neighbor			
(sss)	−0.02915	−0.01214	0.06343
(pps)	0.16827	−0.26827	−0.28860
(ppp)	0.04112	0.00272	0.01255
(dds)	−0.02713	−0.02811	−0.00192
(ddp)	0.00589	0.01458	0.00565
(ddd)	0.00060	−0.00301	−0.00389
(sps)	0.06571	0.06506	0.14108
(sds)	−0.03560	0.01000	0.06231
(pds)	−0.05473	−0.01036	−0.07462
(pdp)	−0.00280	0.00142	0.02624
Third neighbor			
(sss)	0.01595		
(pps)	−0.04985		
(ppp)	0.01796		
(dds)	0.00112		
(ddp)	0.00034		
(ddd)	−0.00056		
(sps)	−0.02477		
(sds)	−0.00073		
(pds)	−0.00082		
(pdp)	−0.00241		

	Orthogonal			Non-orthogonal		
Band	RMS error mRy	Maximum k	Deviation mRy	RMS error mRy	Maximum k	Deviation mRy
1	3.4	(222)	8.6	0.9	(008)	2.7
2	2.8	(044)	7.9	0.9	(008)	2.7
3	3.3	(230)	8.9	0.8	(161)	2.7
4	2.3	(330)	8.3	0.7	(140)	2.3
5	2.3	(044)	10.3	0.9	(131)	3.3
6	6.6	(070)	17.1	2.2	(171)	9.2
1–6	3.8			1.2		

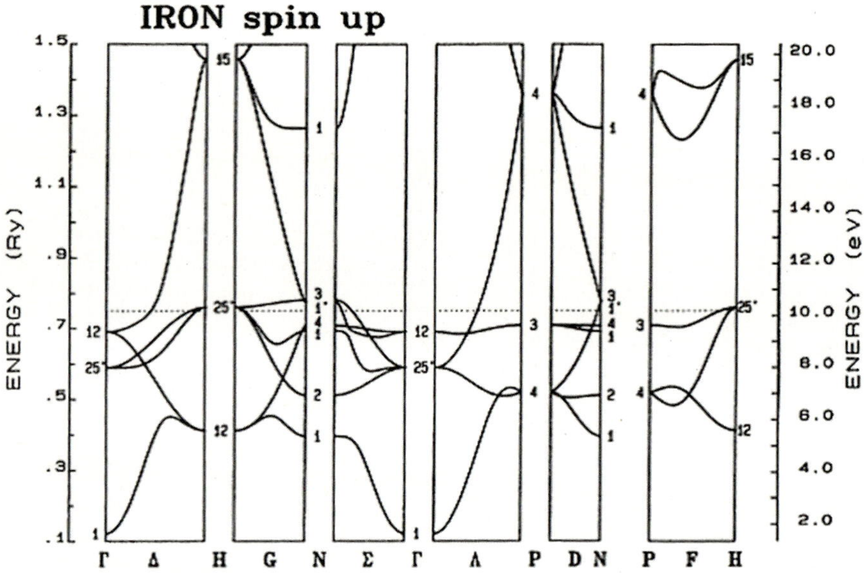

Fig. 4.17 Energy bands for Fe

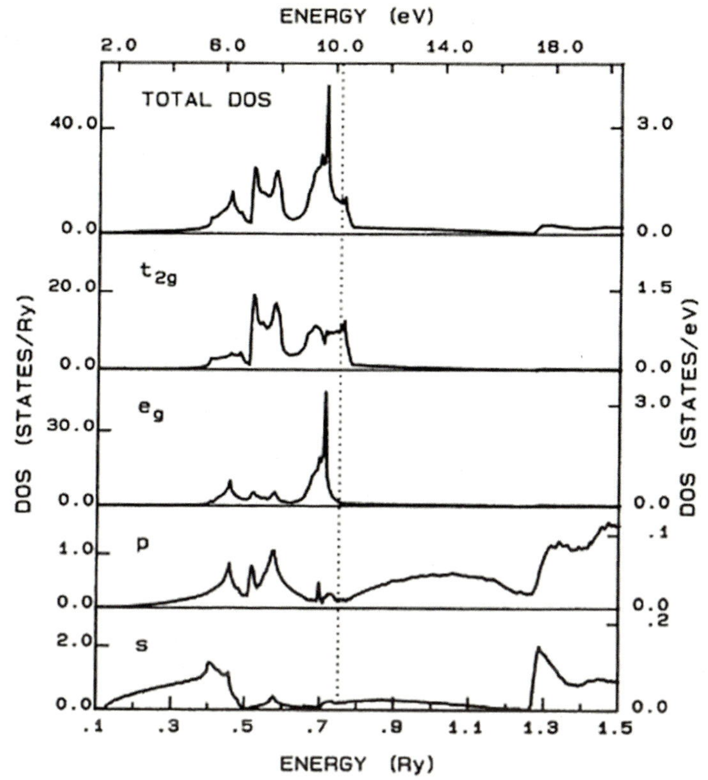

Fig. 4.18 Density of states for Fe

Table 4.47 Iron bcc spin down Z = 26 lattice constant =5.40570 a.u.

Slater-Koster 3-center parameters	Orthogonal	Non-orthogonal	
	Energy integrals (Ry)	Energy integrals (Ry)	Overlap integrals
On site			
s, s (000)	1.17753	0.85715	
x, x (000)	1.80796	1.29483	
xy, xy (000)	0.76816	0.75381	
d2, d2 (000)	0.76622	0.74556	
First neighbor			
s, s (111)	−0.13239	−0.07038	0.09727
s, x (111)	0.11124	0.04427	−0.07262
s, xy (111)	0.03831	0.02426	−0.01610
x, x (111)	0.10401	−0.02667	−0.03182
x, y (111)	0.07754	0.03597	−0.06989
x, xy (111)	0.02343	0.00001	−0.00834
x, yz (111)	0.03484	0.01723	−0.03568
x, d1 (111)	0.00669	0.03460	0.00262

(continued)

Table 4.47 (continued)

Slater-Koster 3-center parameters	Orthogonal	Non-orthogonal	
	Energy integrals (Ry)	Energy integrals (Ry)	Overlap integrals
xy, xy (111)	−0.01216	−0.00076	0.01325
xy, xz (111)	−0.01981	−0.00591	0.01732
xy, d2 (111)	0.01347	0.01859	0.00062
d2, d2 (111)	0.02142	0.01921	−0.00609
Second neighbor			
s, s (200)	−0.03001	−0.01218	0.06435
s, x (200)	0.05993	−0.06796	−0.14707
s, d2 (002)	−0.04215	0.01510	0.06955
x, x (200)	0.15789	−0.33238	−0.33665
y, y (200)	0.04305	0.04126	0.02879
x, xy (020)	−0.00744	0.01952	0.03686
z, d2 (002)	−0.07638	0.07263	0.15011
xy, xy (200)	0.01349	0.00170	−0.01509
xy, xy (002)	−0.00214	0.00177	0.00495
d2, d2 (002)	−0.03336	−0.01226	0.02848
d1, d1 (002)	0.00012	0.00109	0.00154
Third neighbor			
s, s (220)	0.01322		
s, x (220)	−0.01424		
s, xy (220)	0.00111		
s, d2 (220)	0.00269		
x, x (220)	−0.01670		
x, x (022)	0.02802		
x, y (220)	−0.02992		
x, xy (220)	−0.00100		
x, xy (022)	0.00377		
z, d2 (022)	−0.00314		
z, d1 (022)	−0.00119		
xy, xy (220)	−0.00188		
xy, xy (022)	0.00056		
xy, xz (022)	0.00096		
xy, d2 (220)	0.00038		
d2, d2 (220)	0.00113		
d1, d1 (220)	−0.00238		

Table 4.48 Iron bcc spin down

	Orthogonal			Non-orthogonal		
Band	RMS error mRy	Maximum k	Deviation mRy	RMS error mRy	Maximum k	Deviation mRy
1	2.4	(330)	6.6	0.7	(008)	1.8
2	1.6	(353)	4.8	0.6	(008)	1.8
3	2.4	(044)	7.9	0.5	(161)	2.0
4	1.7	(044)	4.3	0.8	(140)	2.2
5	1.4	(131)	3.7	1.0	(131)	4.7
6	5.2	(252)	13.4	1.3	(140)	4.6
1–6	2.8			0.9		

Table 4.49 Energy values in Ry at selected k-points

	Orthogonal	APW	Non-orthogonal
GAMMA 1	0.09701	0.10092	0.10212
GAMMA 12	0.83039	0.83089	0.83136
GAMMA 15	3.10645	3.10658	3.10658
GAMMA 25′	0.71751	0.71771	0.71817
H1 (008)	2.21528	2.21589	2.21589
H12 (008)	0.48774	0.48851	0.49030
H15 (008)	1.44235	1.44944	1.44804
H25′ (008)	0.91206	0.91275	0.91309
N1 (044)	0.45814	0.45701	0.45655
N1 (044)	0.82778	0.83205	0.83419
N1 (044)	1.34564	1.34159	1.34193
N2 (044)	0.62147	0.62276	0.62285
N3 (044)	0.93850	0.93582	0.93659
N4 (044)	0.85487	0.85276	0.85519
N1′ (044)	0.75974	0.75185	0.75159
N3′ (044)	2.19725	2.19828	2.19826
N4′ (044)	2.00044	2.00281	2.00266
P1 (444)	1.51627	1.51539	1.51530
P3 (444)	0.85848	0.85610	0.85621
P4 (444)	0.60368	0.60198	0.60140
P4 (444)	1.41024	1.40555	1.40610
(341)	0.47403	0.47490	0.47447
(341)	0.62588	0.62641	0.62652
(341)	0.68055	0.68058	0.68116
(341)	0.82732	0.82807	0.82848
(341)	0.86458	0.86510	0.86575
(341)	0.98259	0.98209	0.98307

Table 4.50 Fermi level quantities (non-orthogonal fit)

Energy Ry	Densities of states Total States/Ry/atom	s	p	t_{2g}	e_g
0.7505	3.27	0.02	0.04	2.80	0.41

Integrated densities of states Total Electrons	s	p	t_{2g}	e_g
2.95	0.34	0.27	1.62	0.72

Table 4.51 Iron bcc spin down

Slater-Koster 2-center Parameters	Orthogonal	Non-orthogonal	
	Energy integrals (Ry)	Energy integrals (Ry)	Overlap integrals
On site			
s	1.14481	0.83934	
p	1.80769	1.33360	
d1	0.78456	0.75962	
d2	0.75661	0.75455	
First neighbor			
(sss)	−0.13243	−0.06738	0.10106
(pps)	0.25911	0.01187	−0.17562
(ppp)	0.02653	−0.08700	0.01620
(dds)	−0.05266	−0.02266	0.03651
(ddp)	0.03276	0.03018	−0.00772
(ddd)	−0.00286	−0.01318	−0.00870
(sps)	0.17278	−0.06186	0.12949
(sds)	−0.07145	−0.04101	0.02996
(pds)	−0.09702	0.02645	−0.05081
(pdp)	0.02129	−0.04680	0.01062
Second neighbor			
(sss)	−0.03003	−0.01247	0.06612
(pps)	0.18256	−0.24059	−0.27472
(ppp)	0.03703	0.00721	0.01897
(dds)	−0.03396	−0.03625	−0.00328
(ddp)	0.00581	0.01076	−0.00279
(ddd)	0.00114	−0.00155	−0.00124
(sps)	0.07159	0.06053	0.13883
(sds)	−0.04075	0.00985	0.06217
(pds)	−0.06522	−0.00745	−0.07138
(pdp)	−0.00467	0.00696	0.02776
Third neighbor			
(sss)	0.01589		
(pps)	−0.04253		
(ppp)	0.01538		
(dds)	0.00233		
(ddp)	0.00013		
(ddd)	−0.00060		
(sps)	−0.02306		
(sds)	0.00016		
(pds)	0.00222		
(pdp)	−0.00351		

	Orthogonal			Non-orthogonal		
Band	RMS error mRy	Maximum k	Deviation mRy	RMS error mRy	Maximum k	Deviation mRy
1	3.4	(330)	9.8	0.8	(008)	1.9
2	3.8	(044)	11.3	0.7	(008)	1.9
3	4.0	(330)	11.7	0.7	(161)	2.5
4	3.1	(260)	10.9	0.6	(140)	2.3
5	2.6	(044)	6.3	1.0	(131)	4.4
6	6.2	(070)	16.4	1.9	(171)	6.8
1–6	4.0			1.1		

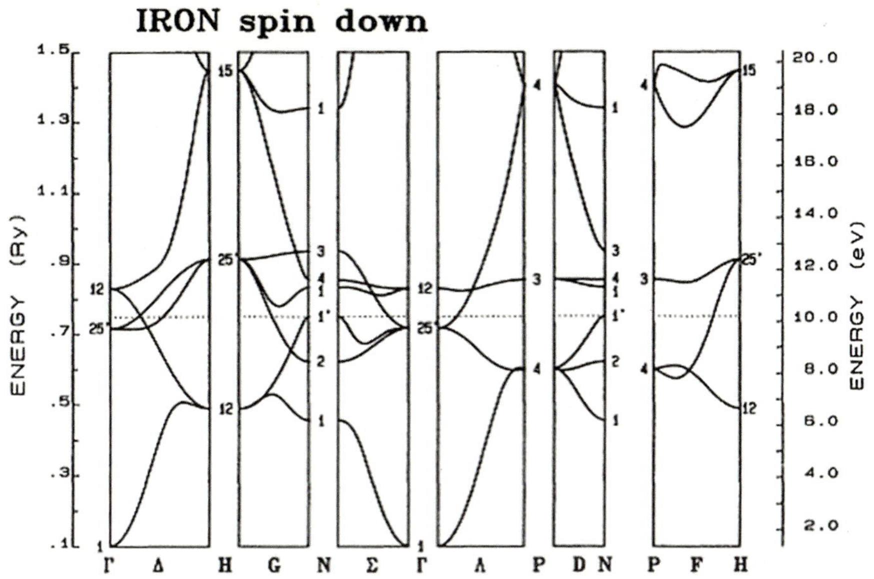

Fig. 4.19 Energy bands for Fe

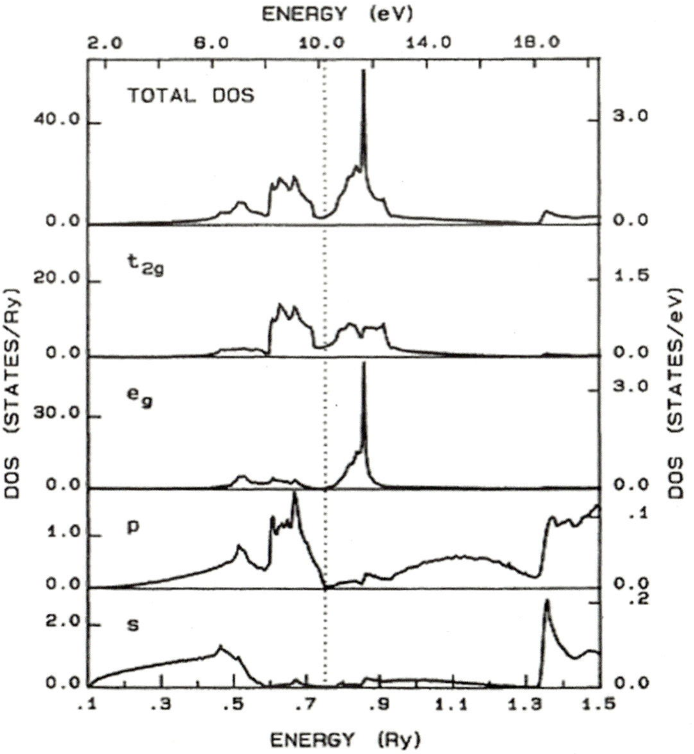

Fig. 4.20 Density of states for Fe

4.7 Cobalt

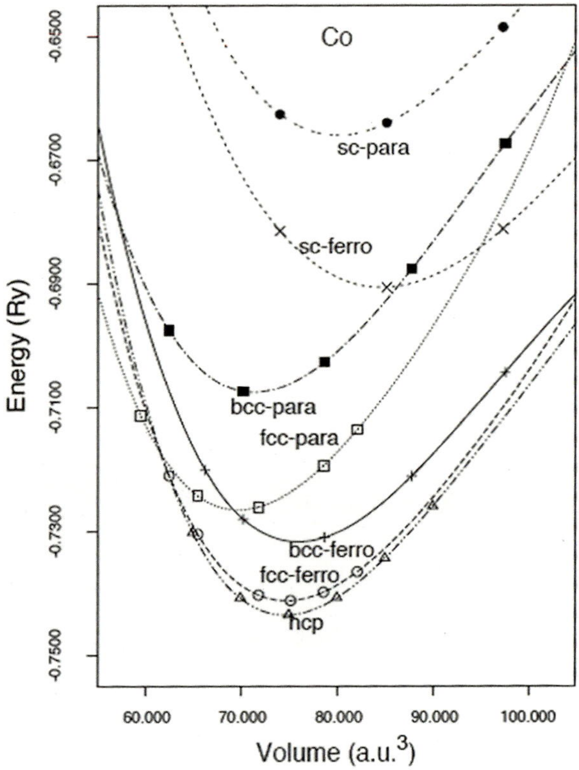

Fig. 4.21 Total energy of Co

Table 4.52 Lattice constants and bulk modulus

	a (Bohr)	c (Bohr)	B (MBar)
sc–para	4.306		1.874
bcc–para	5.219		2.229
fcc–para	6.536		2.287
sc–ferro	4.382		1.788
bcc–ferro	5.332		2.176
fcc–ferro	6.689		2.221
hcp	4.797	7.591	2.230
exp	4.743	7.691	1.914
$\Delta E = E_{fcc-ferro} - E_{hcp} = 2.27$ mRy			

Table 4.53 Birch fit coefficients

	A_1	A_2	A_3	A_4
sc–para	0.23031	−28.63632	137.83021	1576.82901
bcc–para	−0.03744	−13.70783	−121.40707	2733.34305
fcc–para	0.47843	−40.58622	337.25251	7615.62239
sc–ferro	0.05773	−21.03170	−20.15554	2842.36660
bcc–ferro	−0.27994	−1.68828	−374.08542	4646.43642
fcc–ferro	0.70382	−54.40561	565.33584	−974.75466
hcp	−0.17269	−6.25781	−315.90416	4386.82251

Table 4.54 Cobalt hcp spin up Z = 27 lattice constants = 4.7377 a.u. 7.6885 a.u.

Slater-Koster 3-center parameters	Orthogonal		
	Energy integrals (Ry)		Energy integrals (Ry)
On site			
s, s (0)	0.99140		
x, x (0)	1.49766		
z, z (0)	1.44433		
xy, xy (0)	0.53502		
yz, yz (0)	0.54702		
d2, d2 (0)	0.53022		
s, d2 (0)	−0.00311		
x, xy (0)	−0.04680		
First neighbor			
s, s (R)	−0.07978	s, s (T)	−0.08441
s, x (R)	0.11733	s, y (T)	−0.08609
s, y (R)	0.02579	s, z (T)	0.10146
s, xy (R)	−0.00431	s, d1 (T)	−0.02620
s, d1 (R)	0.03583	s, yz (T)	−0.06462
s, d2 (R)	−0.02652	s, d2 (T)	0.01806
x, x (R)	0.16695	x, x (T)	−0.00480
y, y (R)	−0.02013	y, y (T)	0.07559
x, y (R)	0.04058	y, z (T)	−0.10311
x, xy (R)	0.00463	x, xy (T)	0.02352
y, d1 (R)	−0.00511	y, d1 (T)	0.01115
x, d1 (R)	0.04790	x, xz (T)	−0.01189
y, xy (R)	−0.01773	y, yz (T)	0.04052
y, d2 (R)	0.00309	y, d2 (T)	−0.02764
x, d2 (R)	−0.02693	z, z (T)	0.13011
z, z (R)	−0.00506	z, d1 (T)	−0.02799
z, yz (R)	−0.01353	z, y2 (T)	−0.06936
z, xz (R)	−0.00997	z, d2 (T)	0.01554
xy, xy (R)	0.02217	xy, xy (T)	0.00341
d1, d1 (R)	−0.02846	d1, d1 (T)	−0.00122
xy, d1 (R)	−0.00418	xy, yz (T)	−0.01193
d1, d2 (R)	0.01650	yz, d1 (T)	−0.00410
xy, d2 (R)	0.00239	d1, d2 (T)	0.01598
yz, yz (R)	−0.00175	yz, yz (T)	−0.02589
xz, xz (R)	0.01479	xz, xz (T)	0.01410
yz, xz (R)	−0.00690	yz, d2 (T)	0.01902
d2, d2 (R)	−0.01376	d2, d2 (T)	0.00691

Table 4.55 Cobalt hcp spin up

Band	Orthogonal RMS error mRy	Maximum k	Deviation mRy
1	3.1	(16 0 0)	9.5
2	1.7	(12 0 0)	6.6
3	1.1	(0 12 12)	3.1
4	1.9	(24 0 0)	6.5
5	4.1	(24 0 3)	9.9
6	1.8	(0 12 0)	5.6
7	2.0	(8 0 0)	4.1
8	4.1	(16 0 3)	11.9
9	2.9	(24 0 3)	8.0
10	3.4	(24 0 0)	11.4
11	3.6	(0 18 0)	11.1
12	8.2	(12 0 0)	37.1
1–12	3.64		

Table 4.56 Energy values in Ry at selected k-points

	Orthogonal	APW
GAMMA 1+	0.00035	0.00108
GAMMA 1+	0.49505	0.49497
GAMMA 3+	0.63330	0.63356
GAMMA 5+	0.52014	0.52013
GAMMA 6+	0.55075	0.54701
GAMMA 4−	0.30381	0.30425
GAMMA 4−	1.12152	1.12003
GAMMA 5−	0.62149	0.62597
GAMMA 6−	0.49554	0.49628
M1+	0.32850	0.33119
M1+	0.48282	0.48180
M1+	0.76437	0.76327
M2+	0.65885	0.64270
M3+	0.35203	0.34945
M4+	0.64860	0.64747
M1−	0.47528	0.48545
M2−	0.33173	0.33332
M2−	0.52414	0.52449
M2−	0.91127	0.91004
M3−	0.63518	0.63669
M4−	0.49752	0.49955
A1	0.16612	0.16536
A1	0.46848	0.46866
A3	0.49164	0.49208
A3	0.60334	0.60388
L1	0.31085	0.30930
L1	0.41079	0.41095
L1	0.60380	0.60392
L1	0.77369	0.77406
L2	0.55180	0.55508
L2	0.64200	0.64312

Table 4.57 Cobalt hcp spin up Z = 27 lattice constants = 4.7377 a.u. 7.6885 a.u.

Slater-Koster 2-center parameters	Orthogonal	Non-orthogonal	
	Energy integrals (Ry)	Energy integrals (Ry)	Overlap integrals
On site			
s	0.97706	0.61407	
p	1.38919	1.25377	
d0	0.52737	0.51645	
d1	0.52598	0.51644	
d2	0.52898	0.51841	
pd	−0.01536	−0.00531	0.00540
First neighbor			
(sss)	−0.07777	−0.04320	0.10367
(sps)	0.11260	0.02958	−0.13124
(sds)	0.04575	0.01600	−0.06098
(pps)	0.16579	0.10255	−0.06868
(ppp)	0.00728	0.10943	0.10228
(pds)	0.05521	0.03678	0.00205
(pdp)	−0.01866	0.01807	0.08273
(dds)	−0.03880	−0.01844	0.03701
(ddp)	0.02387	0.01701	−0.01742
(ddd)	−0.00384	−0.00244	0.00325
Second neighbor			
(sss)	−0.00653	−0.01646	0.01355
(sps)	0.02260	0.01823	−0.05257
(sds)	0.00921	0.02302	0.01783
(pps)	0.06190	0.04436	−0.06180
(ppp)	0.00267	0.02474	0.01940
(pds)	0.01594	0.03483	0.02055
(pdp)	−0.00218	0.00339	0.01136
(dds)	−0.00669	−0.01394	−0.01099
(ddp)	0.00167	0.00391	0.00456
(ddd)	0.00023	−0.00048	−0.00098
Third neighbor			
(sss)	−0.00193	0.00427	0.01482
(sps)	0.00542	−0.03222	−0.05191
(sds)	−0.00112	0.00157	0.00506
(pps)	0.00525	−0.01029	−0.01758
(ppp)	0.00449	−0.00926	−0.00864
(pds)	−0.00317	0.00098	0.00760
(pdp)	0.00129	0.00247	0.00394
(dds)	−0.00068	−0.00063	0.00014
(ddp)	0.00036	−0.00195	−0.00393
(ddd)	0.00013	0.00160	0.00266

Table 4.58 Cobalt hcp spin up

	Orthogonal			Non-orthogonal		
Band	RMS error mRy	Maximum k	Deviation mRy	RMS error mRy	Maximum k	Deviation mRy
1	3.1	(24 0 0)	9.1	0.8	(0 18 0)	2.4
2	3.2	(24 0 3)	7.1	0.8	(0 18 0)	1.6
3	2.7	(0 24 12)	7.2	0.6	(0 24 12)	1.3
4	2.7	(0 24 12)	7.2	0.7	(24 0 9)	1.7
5	1.1	(8 0 12)	2.7	0.6	(0 12 0)	1.4
6	1.6	(24 0 9)	2.8	0.7	(8 0 0)	2.1
7	1.9	(8 0 12)	4.2	0.7	(8 0 12)	1.8
8	2.5	(16 0 9)	4.7	0.7	(8 0 12)	1.8
9	2.4	(0 12 0)	4.5	0.5	(8 0 0)	1.2
10	2.1	(0 0 12)	4.5	0.6	(8 0 0)	1.6
11	3.7	(16 0 12)	10.1	1.0	(16 0 12)	3.2
12	4.3	(12 0 0)	11.0	1.6	(8 0 0)	4.4
1–12	2.76			0.82		

Table 4.59 Energy values in Ry at selected k-points

	Orthogonal	APW	Non-orthogonal
GAMMA 1+	0.00084	0.00108	0,00233
GAMMA 1+	0.49468	0.49497	0.49560
GAMMA 3+	0.63115	0.63356	0.63316
GAMMA 5+	0.52180	0.52013	0.52001
GAMMA 6+	0.54800	0.54701	0.54751
GAMMA 4–	0.30186	0.30425	0.30479
GAMMA 4–	1.12011	1.12003	1.12249
GAMMA 5–	0.62563	0.62597	0.62562
GAMMA 6–	0.49592	0.49628	0.49523
M1 +	0.32213	0.33119	0.33083
M1 +	0.48426	0.48180	0.48191
M1 +	0.76846	0.76327	0.76316
M2 +	0.63838	0.64270	0.64270
M3+	0.34957	0.34945	0.34895
M4 +	0.64570	0.64747	0.64695
M1–	0.48580	0.48545	0.48615
M2–	0.33597	0.33332	0.33396
M2–	0.52330	0.52449	0.52382
M2–	0.91191	0.91004	0.90988
M3–	0.63615	0.63669	0.63847
M4–	0.49699	0.49955	0.49942
A1	0.16507	0.16536	0.16556
Al	0.46924	0.46866	0.46870
A3	0.49269	0.49208	0.49261
A3	0.59940	0.60368	0.60373
L1	0.31478	0.30930	0.30941
L1	0.40370	0.41095	0.40969
L1	0.60431	0.60392	0.60409
LI	0.77453	0.77406	0.77401
L2	0.55692	0.55508	0.55429
L2	0.64214	0.64312	0.64385

Table 4.60 Fermi level quantities (non-orthogonal fit)

Energy	Densities of states				
	Total	s	p	t_{2g}	e_g
Ry	States/Ry/atom				
0.7010	2.46	0.36	0.23	0.65	1.22

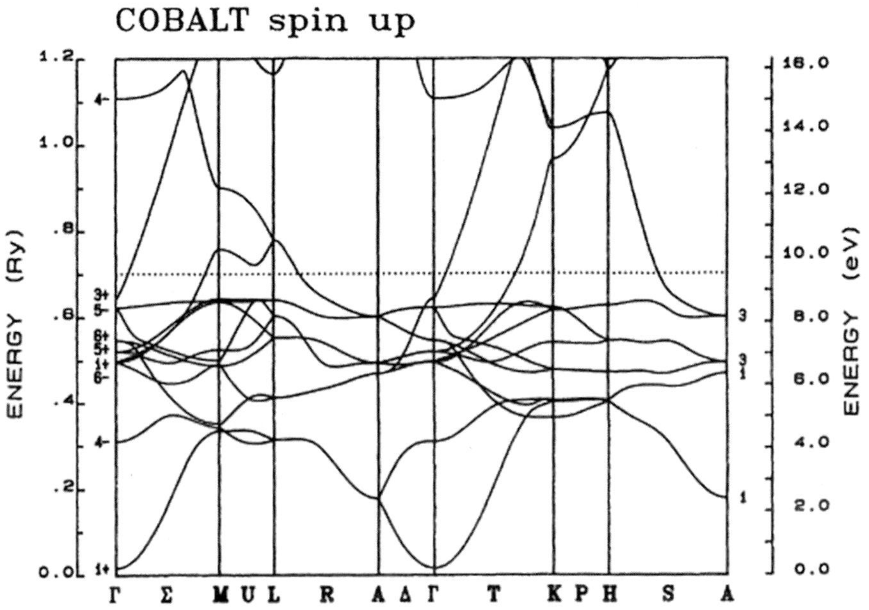

Fig. 4.22 Energy bands for Co

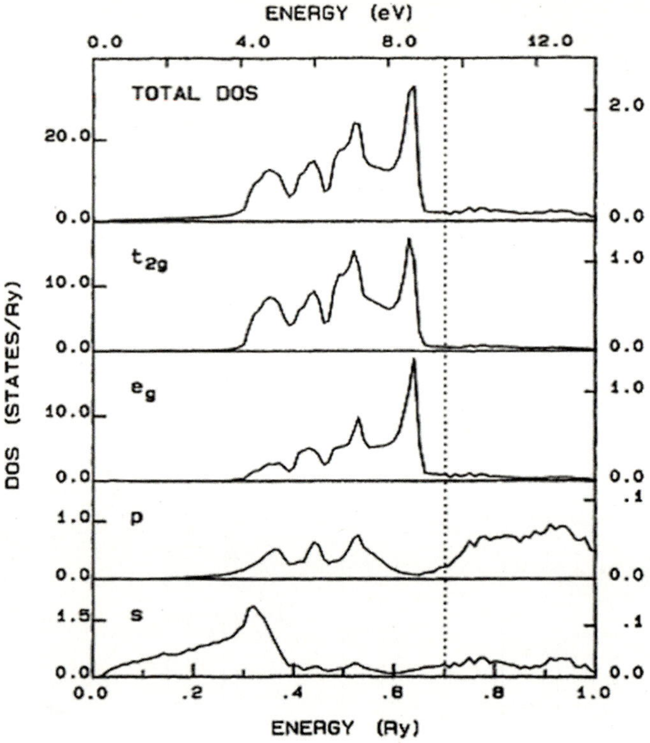

Fig. 4.23 Density of states for Co

Table 4.61 COBALT hcp spin down Z = 27 lattice constants = 4.7377 a.u. 7.6885 a.u.

Slater-Koster 3-center parameters	Orthogonal		
	Energy integrals (Ry)		Energy integrals (Ry)
On site			
s, s (0)	1.03033		
x, x (0)	1.53430		
z, z (0)	1.45442		
xy, xy (0)	0.67195		
yz, yz (0)	0.68501		
d2, d2 (0)	0.66068		
s, d2 (0)	−0.00876		
x, xy (0)	−0.03758		
First neighbor			
s, s (R)	−0.07849	s, s (T)	−0.08788
s, x (R)	0.11632	s, y (T)	−0.08416
s, y (R)	0.02761	s, z (T)	0.10301
s, xy (R)	−0.00335	s, d1 (T)	−0.02915
s, d1 (R)	0.04128	s, yz (T)	−0.06695

(continued)

Table 4.61 (continued)

Slater-Koster 3-center parameters			
	Orthogonal		
	Energy integrals (Ry)		Energy integrals (Ry)
s, d2 (R)	−0.02839	s, d2 (T)	0.02174
x, x (R)	0.16444	x, x (T)	−0.00753
y, y (R)	−0.02677	y, y (T)	0.07406
x, y (R)	0.04172	y, z (T)	−0.10382
x, xy (R)	0.00893	x, xy (T)	0.02473
y, d1 (R)	−0.00696	y, d1 (T)	0.01118
x, d1 (R)	0.05573	x, xz (T)	−0.01509
y, xy (R)	−0.01900	y, yz (T)	0.04345
y, d2 (R)	0.00201	y, d2 (T)	−0.03127
x, d2 (R)	−0.02823	z, z (T)	0.12338
z, z (R)	−0.00872	z, d1 (T)	−0.03322
z, yz (R)	−0.01479	z, yz (T)	−0.07439
z, xz (R)	−0.01206	z, d2 (T)	0.01648
xy, xy (R)	0.02432	xy, xy (T)	0.00466
d1, d1 (R)	−0.03189	d1, d1 (T)	−0.00110
xy, d1 (R)	−0.00402	xy, yz (T)	−0.01363
d1, d2 (R)	0.01799	yz, d1 (T)	−0.00416
xy, d2 (R)	0.00289	d1, d2 (T)	0.01891
yz, yz (R)	−0.00202	yz, yz (T)	−0.02945
xz, xz (R)	0.01662	xz, xz (T)	0.01648
yz, xz (R)	−0.00860	yz, d2 (T)	0.02317
d2, d2 (R)	−0.01440	d2, d2 (T)	0.00811

	Orthogonal		
Band	RMS error mRy	Maximum k	Deviation mRy
1	3.7	(16 0 0)	11.6
2	1.9	(12 0 0)	7.7
3	1.3	(0 12 12)	3.0
4	1.4	(8 0 0)	3.7
5	5.4	(24 0 0)	12.0
6	1.9	(0 12 0)	6.7
7	2.1	(0 12 12)	4.7
8	4.8	(16 0 3)	14.2
9	3.6	(24 0 3)	9.5
10	4.0	(24 0 0)	14.6
11	4.7	(0 18 0)	17.3
12	8.6	(12 0 0)	40.1
1–12	4.15		

Table 4.62 Energy values in Ry at selected k-points

	Orthogonal	APW
GAMMA 1+	0.02813	0.02850
GAMMA 1+	0.62690	0.62648
GAMMA 3+	0.66185	0.66165
GAMMA 5+	0.65466	0.65479
GAMMA 6+	0.68992	0.68550
GAMMA 4−	0.38836	0.38887
GAMMA 5−	0.76775	0.77326
GAMMA 6−	0.62579	0.62697
M1+	0.42732	0.42968
M1+	0.55304	0.55207
M1+	0.88081	0.88018
M2+	0.81252	0.79274
M3+	0.45365	0.45082
M4+	0.79967	0.79795
M1−	0.60284	0.61488
M2−	0.41144	0.41330
M2−	0.65048	0.65136
M2−	1.00426	1.00346
M3−	0.78397	0.78594
M4−	0.61836	0.61967
A1	0.19759	0.19693
A1	0.58441	0.58510
A3	0.62241	0.62276
A3	0.74882	0.74926
L1	0.39807	0.39721
L1	0.52509	0.52536
L1	0.70984	0.70987
L1	0.85131	0.85134
L2	0.68824	0.69210
L2	0.79192	0.79305

Table 4.63 Cobalt hcp spin down Z = 27 lattice constants = 4.7377 a.u. 7.6885 a.u.

Slater-Koster 2-center parameters	Orthogonal	Non-orthogonal	
	Energy integrals (Ry)	Energy integrals (Ry)	Overlap integrals
On site			
s	1.01504	0.72336	
p	1.41579	1.32097	
d0	0.66118	0.64560	
d1	0.65949	0.64550	
d2	0.66233	0.64716	
pd	−0.01691	0.00948	0.03011
First neighbor			
(sss)	−0.07700	−0.04168	0.12593
(sps)	0.11114	0.04156	−0.13035
(sds)	0.05120	0.02736	−0.04465
(pps)	0.16007	0.13835	−0.03869
(ppp)	−0.00357	0.13466	0.11965

(continued)

Table 4.63 (continued)

Slater-Koster 2-center parameters	Orthogonal	Non-orthogonal	
	Energy integrals (Ry)	Energy integrals (Ry)	Overlap integrals
(pds)	0.06229	0.05373	0.01943
(pdp)	−0.02129	0.03054	0.08684
(dds)	−0.04403	−0.01388	0.04331
(ddp)	0.02763	0.01381	−0.02595
(ddd)	−0.00442	0.00000	0.00736
Second neighbor			
(sss)	−0.01004	−0.02592	0.01100
(sps)	0.02275	0.02875	−0.03582
(sds)	0.00945	0.02250	0.00834
(pps)	0.05589	0.04042	−0.07810
(ppp)	−0.00168	0.02177	0.01751
(pds)	0.01825	0.02894	−0.00138
(pdp)	−0.00353	0.00065	0.00522
(dds)	−0.00793	−0.01939	−0.01359
(ddp)	0.00198	0.00245	0.00121
(ddd)	0.00026	0.00090	0.00131
Third neighbor			
(sss)	−0.00141	0.00569	0.01853
(sps)	0.00308	−0.02739	−0.03121
(sds)	−0.00005	−0.00429	−0.00635
(pps)	0.00248	−0.01183	−0.01685
(ppp)	0.00315	−0.01260	−0.00918
(pds)	−0.00410	−0.01260	−0.01439
(pdp)	0.00139	0.00032	−0.00065
(dds)	−0.00056	0.00483	0.00817
(ddp)	0.00034	−0.00440	−0.00671
(ddd)	0.00004	0.00178	0.00241

Table 4.64 Cobalt hcp spin down

	Orthogonal			Non-orthogonal		
Band	RMS error mRy	Maximum k	Deviation mRy	RMS error mRy	Maximum k	Deviation mRy
1	3.0	(16 0 0)	6.1	0.9	(0 18 0)	2.0
2	4.2	(24 0 0)	10.6	0.9	(0 18 0)	1.5
3	3.2	(24 0 9)	8.9	0.7	(4 0 0)	1.3
4	3.0	(0 24 12)	8.5	0.8	(12 0 0)	1.6
5	1.4	(8 0 12)	3.4	0.7	(8 0 12)	1.5
6	1.8	(8 0 12)	3.4	0.7	(8 0 12)	1.5
7	2.7	(8 0 12)	6.0	0.9	(8 0 12)	2.4
8	3.3	(16 0 9)	6.2	1.0	(8 0 12)	2.4
9	2.9	(0 12 0)	5.5	0.7	(8 0 0)	2.0
10	2.5	(0 0 12)	5.4	0.7	(16 0 12)	1.5
11	3.6	(24 0 0)	7.3	0.9	(16 0 12)	3.5
12	4.6	(12 0 0)	13.8	3.4	(12 0 0)	14.5
1–12	3.13			1.26		

Table 4.65 Energy values in Ry at selected k-points

	Orthogonal	APW	Non-orthogonal
GAMMA 1+	0.02795	0.02850	0.03021
GAMMA 1+	0.62596	0.62648	0.62763
GAMMA 3+	0.66117	0.66165	0.66117
GAMMA 5+	0.65669	0.65479	0.65423
GAMMA 6+	0.68633	0.68550	0.68581
GAMMA 4−	0.38685	0.38887	0.38980
GAMMA 4−	1.21020	1.21000	1.21738
GAMMA 5−	0.77331	0.77326	0.77283
GAMMA 6−	0.62628	0.62697	0.62575
M1 +	0.41910	0.42968	0.42854
M1 +	0.55297	0.55207	0.55134
M1 +	0.88750	0.88018	0.87971
M2 +	0.78730	0.79274	0.79361
M3+	0.45144	0.45082	0.45070
M4 +	0.79514	0.79795	0.79664
M1−	0.61486	0.61488	0.61534
M2−	0.41587	0.41330	0.41311
M2−	0.64871	0.65136	0.65059
M2−	1.00601	1.00346	1.00327
M3−	0.78517	0.78594	0.78792
M4−	0.61630	0.61967	0.61955
A1	0.19649	0.19693	0.19680
Al	0.58659	0.58510	0.58512
A3	0.62361	0.62276	0.62358
A3	0.74389	0.74926	0.74922
L1	0.40201	0.39721	0.39747
L1	0.51686	0.52536	0.52419
L1	0.71011	0.70987	0.70939
LI	0.85236	0.85134	0.85158
L2	0.69481	0.69210	0.69090
L2	0.79210	0.79305	0.79334

Table 4.66 Fermi level quantities (non-orthogonal fit)

Energy	Densities of states				
	Total	s	p	t_{2g}	e_g
Ry	States/Ry/atom				
0.7010	9.53	0.06	0.44	5.54	3.48

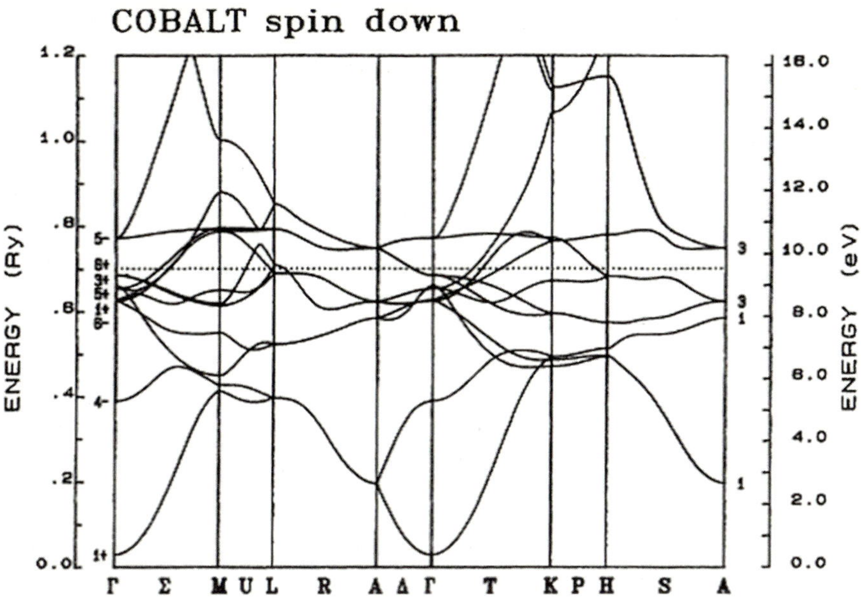

Fig. 4.24 Energy bands for Co

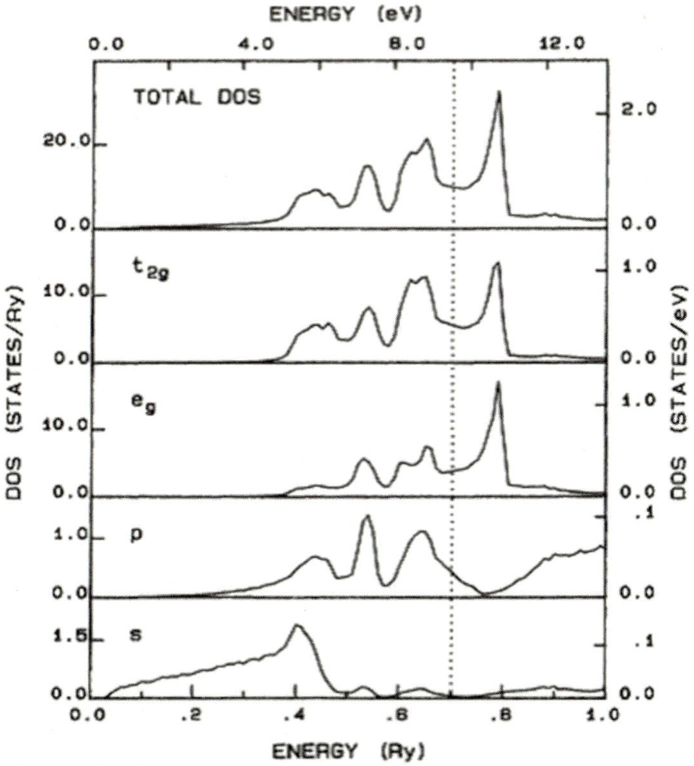

Fig. 4.25 Density of states for Co

4.8 Nickel

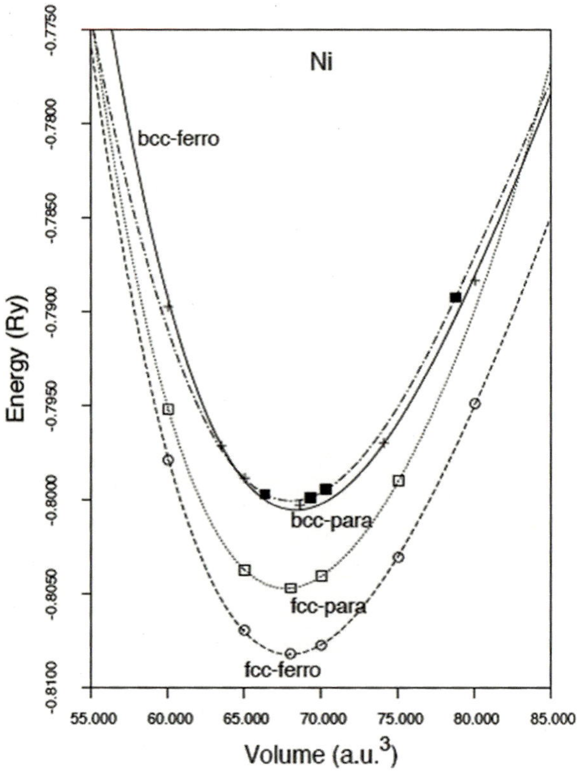

Fig. 4.26 Total energy of Ni

Table 4.67 Lattice constants and bulk modulus

	a (Bohr)	B (MBar)
bcc–ferro	5.151	2.418
fcc–ferro	6.479	2.525
bcc–para	5.143	2.446
fcc–para	6.469	2.574
exp	6.652	1.86
$E_{fcc-para} - E_{fcc-ferro} = \Delta E = 3.5$ mRy		

Table 4.68 Birch fit coefficients

	A_1	A_2	A_3	A_4
bcc–ferro	−0.06732	−15.64022	−91.43084	2474.76108
fcc–ferro	−0.07055	−14.99494	−114.42247	2657.42405
bcc–para	0.02666	−32.12109	181.97208	951.45119
fcc–para	0.01509	−25.48699	55.84384	1724.76015

Table 4.69 Nickel fcc spin up Z = 28 lattice constant = 6.65845 a.u.

Slater-Koster 3-center parameters	Orthogonal	Non-orthogonal	
	Energy integrals (Ry)	Energy integrals (Ry)	Overlap integrals
On site			
s, s (000)	0.98244	0.74407	
x, x (000)	1.42797	1.22609	
xy, xy (000)	0.48618	0.48712	
d2, d2 (000)	0.49080	0.48501	
First neighbor			
s, s (110)	−0.08599	−0.06503	0.08249
s, x (110)	0.08650	0.05131	−0.08550
s, xy (110)	−0.02721	−0.03504	0.00994
s, d2 (110)	−0.00783	0.03064	0.01013
x, x (110)	0.11245	0.02727	−0.07893
x, x (011)	0.00394	0.08801	0.09307
x, y (110)	0.07449	−0.04261	−0.15570
x, xy (110)	−0.02089	−0.01012	0.03467
x, xy (011)	0.01969	−0.00908	−0.05865
z, d2 (011)	0.00656	−0.02514	−0.06800
z, d1 (011)	−0.00337	0.01034	−0.02759
xy, xy (110)	−0.02676	−0.01124	0.03061
xy, xy (011)	0.00870	0.00607	−0.00772
xy, xz (011)	0.01616	0.00918	−0.00830
xy, d2 (110)	0.01393	0.01136	−0.00404
d2, d2 (110)	−0.01420	−0.00073	0.02099
d1, d1 (110)	0.02713	0.01066	−0.02726
Second neighbor			
s, s (200)	−0.00304	−0.01419	0.01052
s, x (200)	0.02362	0.02340	−0.01923
s, d2 (002)	−0.03739	−0.01474	−0.00461
x, x (200)	0.09389	0.06934	−0.02096
y, y (200)	0.01846	0.03977	0.02643
x, xy (020)	0.00741	−0.01122	−0.02179
z, d2 (002)	−0.05287	−0.04386	−0.04236
xy, xy (200)	0.00427	−0.00056	−0.00230
xy, xy (002)	0.00090	−0.00093	−0.00121
d2, d2 (002)	−0.01031	−0.00858	−0.00211
d1, d1 (002)	0.00088	0.00319	0.00721

Table 4.70 Nickel fcc spin up

	Orthogonal			Non-orthogonal		
Band	RMS error mRy	Maximum k	Deviation mRy	RMS error mRy	Maximum k	Deviation mRy
1	5.4	(444)	12.3	0.5	(044)	1.1
2	3.9	(033)	9.5	0.6	(008)	1.4
3	4.4	(008)	10.6	0.6	(007)	2.1
4	6.4	(044)	14.8	1.0	(118)	4.6
5	3.7	(224)	8.1	0.5	(000)	1.4
6	6.3	(174)	20.7	0.6	(000)	1.4
1–6	5.1			0.7		

Table 4.71 Energy values in Ry at selected k-points

	Orthogonal	APW	Non-orthogonal
GAMMA 1	−0.06767	−0.05993	−0.05913
GAMMA 12	0.54010	0.53906	0.54046
GAMMA 15	2.60495	2.60259	2.60259
GAMMA 25′	0.46757	0.46292	0.46388
X1 (008)	0.26754	0.27283	0.27375
X1 (008)	1.31196	1.27693	1.27726
X2 (008)	0.60197	0.59132	0.59238
X3 (008)	0.32841	0.33079	0.33220
X5 (008)	0.61210	0.60947	0.60975
X4′ (008)	0.80575	0.80251	0.80183
X5′ (008)	1.67381	1.69458	1.69453
L1 (444)	0.28813	0.27580	0.27611
L1 (444)	1.05053	1.03562	1.03576
L3 (444)	0.45699	0.45596	0.45660
L3 (444)	0.59405	0.59408	0.59465
L2′ (444)	0.57041	0.58230	0.58171
L3′ (444)	1.46433	2.41811	2.41809
W1 (048)	0.51968	0.53226	0.53383
W1 (048)	1.36120	1.41388	1.41386
W3 (048)	0.41795	0.41365	0.41449
W3 (048)	1.20119	1.19259	1.19287
W1′ (048)	0.60850	0.60962	0.61063
W2′ (048)	0.34968	0.34091	0.34119
W2′ (048)	1.29901	1.32808	1.32847
EVEN (224)	0.22702	0.23315	0.23394
EVEN (224)	0.42322	0.42002	0.42058
EVEN (224)	0.45874	0.45787	0.45816
EVEN (224)	0.59957	0.59358	0.59355
ODD (224)	0.48188	0.48703	0.48643
ODD (224)	0.57740	0.58547	0.58506

Table 4.72 Fermi level quantities (non-orthogonal fit)

Energy	Densities of states					Velocity	Plasmon energy
	Total	s	p	t_{2g}	e_g		
Ry	States/Ry/atom					cm/s	eV
0.6410	2.51	0.28	0.24	1.25	0.74	0.76 × 10E8	7.15

Integrated densities of states				
Total	s	p	t_{2g}	e_g
Electrons				
5.33	0.34	0.17	2.88	1.94

Table 4.73 Nickel fcc spin up

Slater-Koster 2-center parameters	Orthogonal	Non-orthogonal	
	Energy integrals (Ry)	Energy integrals (Ry)	Overlap integrals
On site			
s	0.99144	0.56544	
p	1.52711	1.00232	
d1	0.48464	0.48837	
d2	0.48665	0.48830	
First neighbor			
(sss)	−0.08657	−0.04783	0.13095
(pps)	0.21620	−0.19160	−0.34484
(ppp)	0.00965	−0.02827	0.05860
(dds)	−0.03441	−0.02037	0.02796
(ddp)	0.02534	0.01359	−0.01777
(ddd)	−0.00597	−0.00180	0.00286
(sps)	0.13148	−0.00561	−0.20766
(sds)	−0.03056	−0.02904	0.03895
(pds)	−0.03015	−0.01053	0.07146
(pdp)	0.02725	0.01666	−0.03554
Second neighbor			
(sss)	−0.00362	−0.02480	0.01250
(pps)	0.05322	0.06064	−0.04128
(ppp)	0.00842	−0.00408	0.00930
(dds)	−0.00693	−0.01467	−0.01672
(ddp)	0.00456	−0.00014	−0.00274
(ddd)	−0.00049	0.00016	0.00087
(sps)	0.01180	0.03069	−0.02607
(sds)	−0.00820	0.00002	0.02835
(pds)	−0.00776	−0.02877	0.00392
(pdp)	0.00782	−0.00191	−0.01270

	Orthogonal			Non-orthogonal		
Band	RMS error mRy	Maximum k	Deviation mRy	RMS error mRy	Maximum k	Deviation mRy
1	6.3	(444)	19.5	0.9	(005)	2.2
2	6.8	(007)	16.4	0.8	(008)	1.7
3	2.5	(000)	7.3	0.9	(007)	2.4
4	6.2	(444)	21.5	1.1	(118)	4.8
5	5.4	(444)	13.4	0.9	(000)	2.4
6	9.3	(174)	31.3	1.5	(174)	4.6
1–6	6.4			1.0		

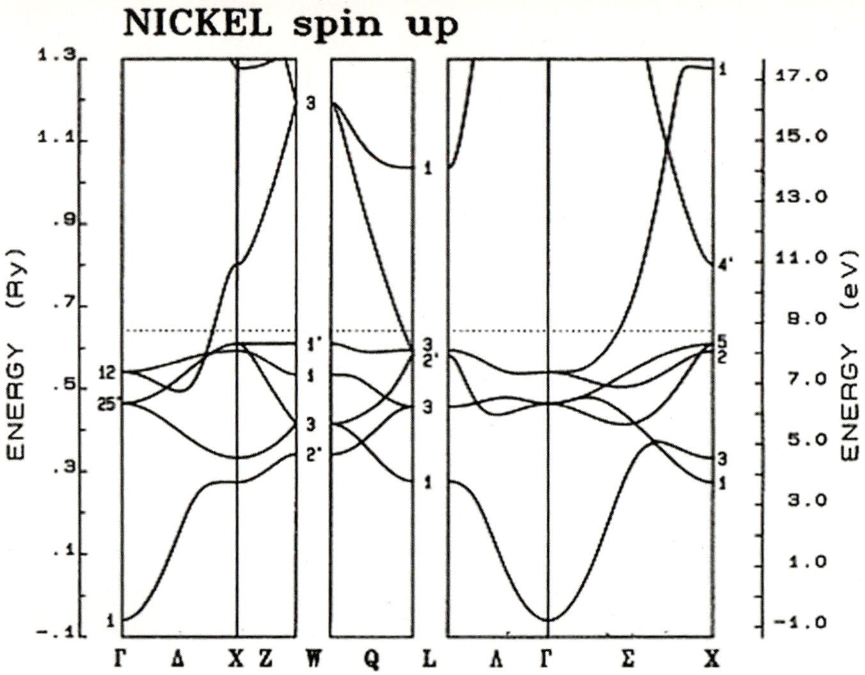

Fig. 4.27 Energy bands for Ni

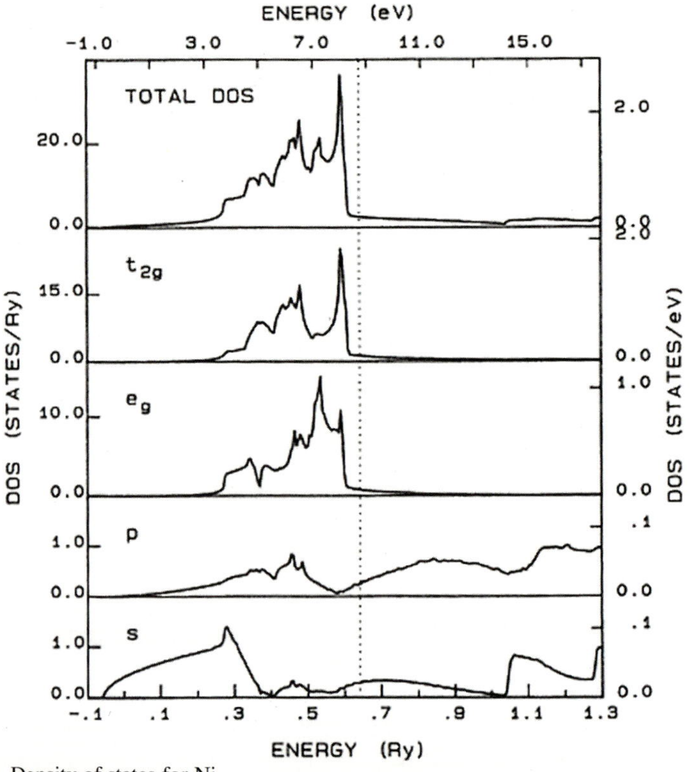

Fig. 4.28 Density of states for Ni

Table 4.74 Nickel fcc spin down Z = 28 lattice constant = 6.65845 a.u.

Slater-Koster 3-center parameters	Orthogonal	Non-orthogonal	
	Energy integrals (Ry)	Energy integrals (Ry)	Overlap integrals
On site			
s, s (000)	0.98172	0.64442	
x, x (000)	1.48057	1.15725	
xy, xy (000)	0.54277	0.54260	
d2, d2 (000)	0.54182	0.54087	
First neighbor			
s, s (110)	−0.08691	−0.03551	0.10730
s, x (110)	0.08742	0.02046	−0.11189
s, xy (110)	−0.03374	−0.03460	0.02465
s, d2 (110)	−0.00346	0.00558	−0.02635
x, x (110)	0.11012	0.00717	−0.08307
x, x (011)	0.01016	0.04464	0.06886
x, y (110)	0.08894	−0.08462	−0.18845
x, xy (110)	−0.02670	−0.00876	0.04431
x, xy (011)	0.01837	−0.00470	−0.04375
z, d2 (011)	0.00236	−0.00344	−0.04041
z, d1 (011)	−0.00065	0.00716	−0.03433
xy, xy (110)	−0.02789	−0.01945	0.01607
xy, xy (011)	0.00882	0.00563	−0.00814
xy, xz (011)	0.01594	0.00629	−0.01214
xy, d2 (110)	0.01466	0.01356	−0.00097
d2, d2 (110)	−0.01480	0.00656	0.03155
d1, d1 (110)	0.02878	0.02482	−0.00168
Second neighbor			
s, s (200)	0.00009	−0.05110	−0.02155
s, x (200)	0.00591	0.08731	0.02370
s, d2 (002)	−0.04055	−0.02964	−0.02976
x, x (200)	0.06463	0.09395	−0.01554
y, y (200)	0.01491	0.04263	0.02204
x, xy (020)	0.00485	−0.01301	−0.02418
z, d2 (002)	−0.05680	−0.03267	−0.01810
xy, xy (200)	0.00325	0.00324	0.00314
xy, xy (002)	0.00099	−0.00119	−0.00143
d2, d2 (002)	−0.01064	−0.00303	0.00691
d1, d1 (002)	0.00109	−0.00842	−0.01103

Table 4.75 Nickel fcc spin down

	Orthogonal			Non-orthogonal		
Band	RMS error mRy	Maximum k	Deviation mRy	RMS error mRy	Maximum k	Deviation mRy
1	7.7	(222)	21.3	0.7	(008)	1.7
2	3.1	(224)	7.3	0.6	(008)	1.7
3	4.5	(008)	10.9	0.8	(007)	2.3
4	7.9	(444)	22.3	1.1	(118)	5.5
5	3.9	(224)	10.4	0.5	(000)	1.2
6	4.7	(224)	12.7	0.9	(006)	1.8
1–6	5.6			0.8		

Table 4.76 Energy values in Ry at selected k-points

	Orthogonal	APW	Non-orthogonal
GAMMA 1	−0.06065	−0.04185	−0.04088
GAMMA 12	0.59706	0.59424	0.59542
GAMMA 15	2.59109	2.62229	2.62231
GAMMA 25′	0.51675	0.51453	0.51583
X1 (008)	0.31016	0.31358	0.31532
X1 (008)	1.33063	1.31414	1.31597
X2 (008)	0.65951	0.64865	0.65011
X3 (008)	0.37556	0.37611	0.37777
X5 (008)	0.66929	0.66760	0.66742
X4′ (008)	0.82918	0.81986	0.81963
X5′ (008)	1.62883	1.71133	1.71072
L1 (444)	0.31792	0.31209	0.31141
L1 (444)	1.06357	1.07326	1.07333
L3 (444)	0.50844	0.50721	0.50844
L3 (444)	0.65360	0.65155	0.65209
L2′ (444)	0.58011	0.60240	0.60225
L3′ (444)	1.64744	2.43392	2.43390
W1 (048)	0.57391	0.58583	0.58655
W1 (048)	1.37053	1.43440	1.43377
W3 (048)	0.46339	0.45910	0.45982
W3 (048)	1.21540	1.22043	1.21865
W1′ (048)	0.66535	0.66775	0.66842
W2′ (048)	0.39368	0.38430	0.38451
W2′ (048)	1.37035	1.36600	1.36597
EVEN (224)	0.25699	0.25741	0.25778
EVEN (224)	0.47165	0.46435	0.46431
EVEN (224)	0.50920	0.50841	0.50805
EVEN (224)	0.65817	0.64552	0.64615
ODD (224)	0.53605	0.53942	0.53906
ODD (224)	0.63207	0.64250	0.64184

Table 4.77 Fermi level quantities (non-orthogonal fit)

Energy		Densities of states				Velocity	Plasmon energy
	Total	s	p	t_{2g}	e_g		
Ry		States/Ry/atom				cm/s	eV
0.6410	21.28	0.16	0.09	13.07	7.94	0.25 × 10E8	6.76

Integrated densities of states				
Total	s	p	t_{2g}	e_g
Electrons				
4.66	0.35	0.17	2.42	1.73

Table 4.78 Nickel fcc spin down

Slater-Koster 2-center parameters	Orthogonal	Non-orthogonal	
	Energy integrals (Ry)	Energy integrals (Ry)	Overlap integrals
On site			
s	0.98738	0.57809	
p	1.50091	1.01983	
d1	0.54152	0.54087	
d2	0.53843	0.54070	
First neighbor			
(sss)	−0.08649	−0.04486	0.13224
(pps)	0.20673	−0.19683	−0.34555
(ppp)	0.01310	−0.02104	0.06151
(dds)	−0.03579	−0.02026	0.02813
(ddp)	0.02590	0.01214	−0.02054
(ddd)	−0.00633	−0.00073	0.00463
(sps)	0.12589	−0.00981	−0.20864
(sds)	−0.03632	−0.02878	0.04097
(pds)	−0.03814	−0.00807	0.07565
(pdp)	0.02894	0.01402	−0.03855
Second neighbor			
(sss)	−0.00171	−0.02492	0.01192
(pps)	0.05498	0.05861	−0.04197
(ppp)	0.01054	−0.00131	0.01032
(dds)	−0.00640	−0.01619	−0.01699
(ddp)	0.00339	−0.00076	−0.00355
(ddd)	−0.00002	0.00020	0.00086
(sps)	0.00683	0.03075	−0.02526
(sds)	−0.00600	0.00073	0.02676
(pds)	−0.00589	−0.03007	0.00165
(pdp)	0.00542	−0.00310	−0.01276

	Orthogonal			Non-orthogonal		
Band	RMS error mRy	Maximum k	Deviation mRy	RMS error mRy	Maximum k	Deviation mRy
1	7.7	(000)	18.9	0.9	(264)	2.1
2	6.1	(007)	14.4	0.9	(008)	2.0
3	2.6	(000)	5.1	0.9	(007)	2.4
4	7.7	(444)	28.1	1.2	(118)	5.2
5	5.4	(444)	11.6	0.9	(000)	2.4
6	6.1	(354)	17.3	1.4	(174)	4.5
1–6	6.2			1.0		

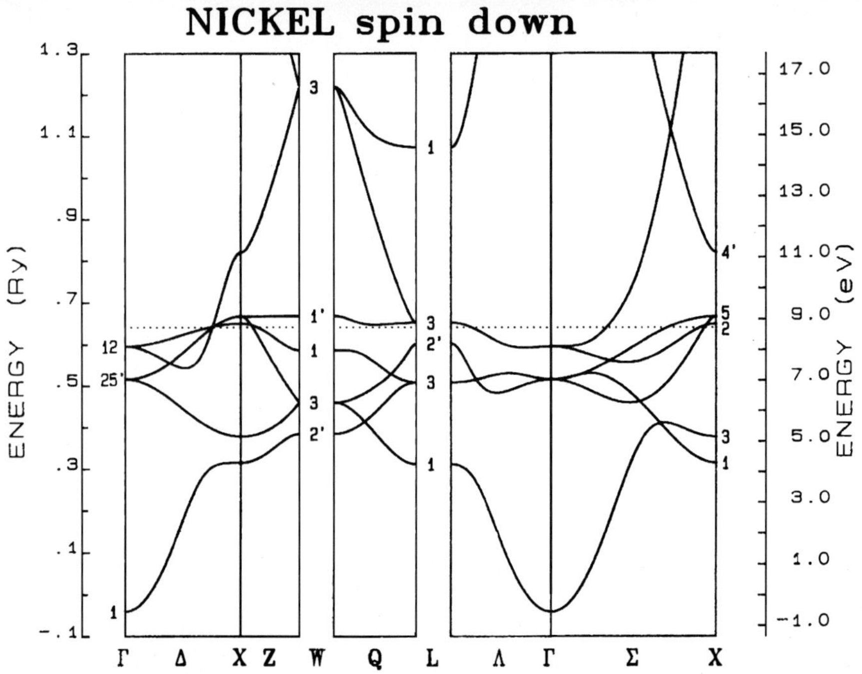

Fig. 4.29 Energy bands for Ni

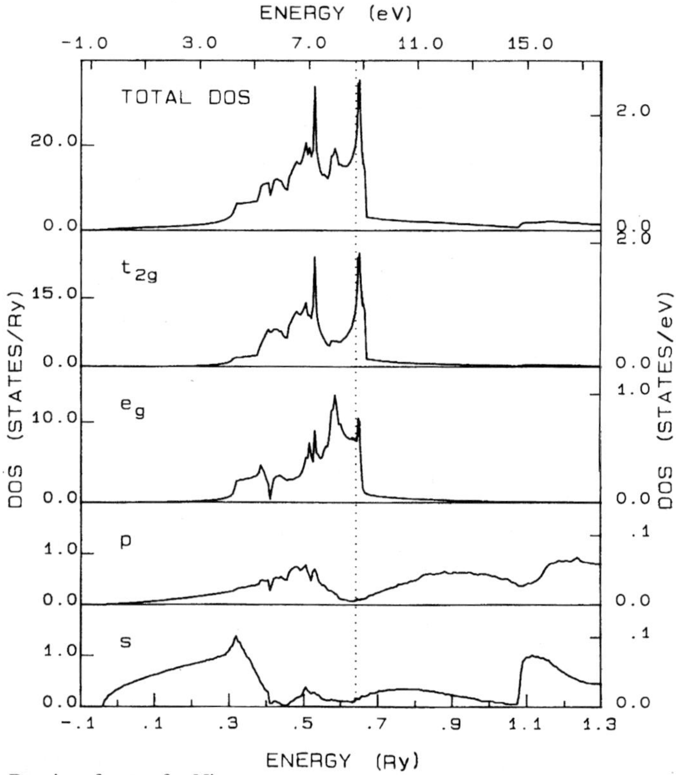

Fig. 4.30 Density of states for Ni

4.9 Copper

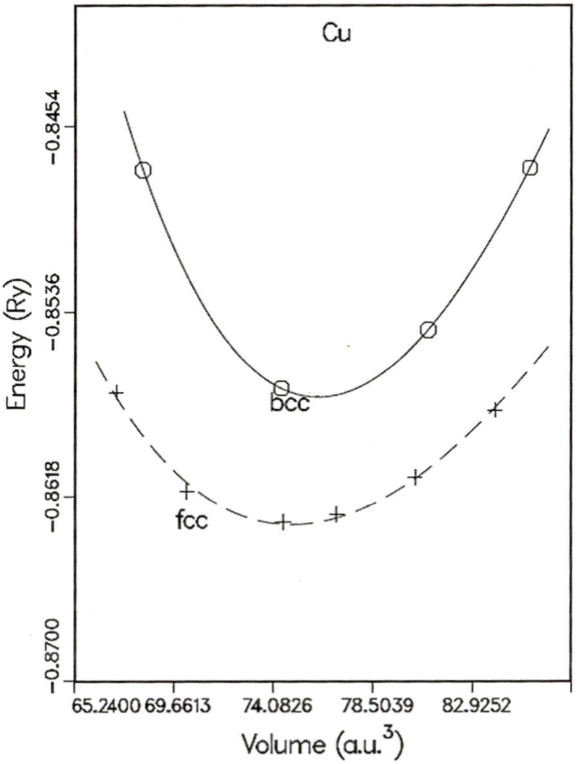

Fig. 4.31 Total energy of Cu

Table 4.79 Lattice constants and bulk modulus

	a (Bohr)	B (MBar)
fcc	6.692	1.698
bcc	5.338	3.122
exp	6.822	1.370
$\Delta E = 5.7$ mRy		

Table 4.80 Birch fit coefficients

	A_1	A_2	A_3
bcc	0.95847	−65.19925	585.24951
fcc	0.10975	−34.57379	307.19295

Table 4.81 Copper fcc Z = 29 lattice constant = 6.83000 a.u.

Slater-Koster 3-center parameters	Orthogonal	Non-orthogonal	
	Energy integrals (Ry)	Energy integrals (Ry)	Overlap integrals
On site			
s, s (000)	0.79651	0.60246	
x, x (000)	1.29808	1.22779	
xy, xy (000)	0.37266	0.37675	
d2, d2(000)	0.37320	0.37520	
First neighbor			
s, s (110)	−0.07583	−0.05801	0.08495
s, x (110)	0.07830	0.04580	−0.08397
s, xy (110)	0.02662	−0.03557	0.00492
s, d2 (110)	0.00731	0.02037	0.00966
x, x (110)	0.10756	0.07153	−0.05392
x, x (110)	0.01099	0.06477	0.06951
x, y (110)	0.07076	−0.01376	−0.12072
x, xy (110)	0.02077	−0.03287	−0.00862
x, xy (011)	−0.01276	−0.01085	−0.06060
z, d2 (011)	−0.01326	−0.01180	−0.06334
z, d1 (011)	0.00312	0.01291	−0.01291
xy, xy (110)	−0.02003	−0.01429	0.01400
xy, xy (011)	0.00643	0.00421	−0.00696
xy, xz (011)	0.01095	0.00315	−0.01384
xy, d2 (110)	0.01035	0.01009	0.00006
d2, d2 (110)	−0.01023	0.00069	0.02228
d1, d1 (110)	0.01971	0.01766	0.00209
Second neighbor			
s, s (200)	0.00003	−0.02041	−0.00003
s, x (200)	0.01779	0.05037	0.00468
s, d2 (002)	0.02286	−0.01651	−0.02370
x, x (200)	0.07993	0.13434	0.03163
y, y (200)	0.01761	0.03834	0.02838
x, xy (020)	−0.00439	−0.01329	−0.03333
z, d2 (002)	0.03134	−0.04240	−0.07768
xy, xy (200)	0.00273	−0.00150	−0.00499
xy, xy (002)	0.00029	−0.00126	−0.00235
d2, d2 (002)	−0.00726	−0.00484	−0.00218
d1, d1(002)	0.00040	0.00116	0.00519

Table 4.82 Copper fcc

	Orthogonal			Non-orthogonal		
Band	RMS error mRy	Maximum k	Deviation mRy	RMS error mRy	Maximum k	Deviation mRy
1	3.5	(444)	7.3	0.4	(174)	0.9
2	2.3	(033)	4.8	0.5	(008)	1.2
3	2.3	(008)	4.7	0.5	(007)	1.7
4	2.8	(118)	6.2	0.8	(118)	3.9
5	2.3	(226)	5.1	0.4	(000)	1.1
6	6.2	(174)	23.1	0.5	(000)	1.1
1–6	3.5			0.5		

Table 4.83 Energy values in Ry at selected k-points

	Orthogonal	Apw	Non-orthogonal
GAMMA 1	−0.11330	−0.10751	−0.10702
GAMMA 12	0.40947	0.40938	0.41048
GAMMA 15	2.43281	2.42427	2.42426
GAMMA 25′	0.35549	0.35257	0.35335
X1 (008)	0.21004	0.21247	0.21327
X1 (008)	1.10383	1.08946	1.08950
X2 (008)	0.45344	0.44873	0.44952
X3 (008)	0.25256	0.25396	0.25518
X5 (008)	0.46430	0.46247	0.46272
X4′ (008)	0.71185	0.70948	0.70867
X5′ (008)	1.48442	1.54749	1.54741
L1 (444)	0.21509	0.20777	0.20769
L1 (444)	0.85485	0.85002	0.85014
L3 (444)	0.34836	0.34800	0.34874
L3 (444)	0.45037	0.45091	0.45145
L2′ (444)	0.50168	0.51220	0.51194
L3′ (444)	1.35083	2.20587	2.20586
W1 (048)	0.40172	0.40613	0.40705
W1 (048)	1.12075	1.21673	1.21647
W3 (048)	0.32202	0.31933	0.32012
W3 (048)	1.05320	1.04828	1.04878
W1′ (048)	0.46313	0.46256	0.46345
W2′ (048)	0.27149	0.26720	0.26774
W2′ (048)	1.16539	1.18054	1.18075
EVEN (224)	0.16507	0.17002	0.17057
EVEN (224)	0.32934	0.32700	0.32737
EVEN (224)	0.34974	0.35021	0.35057
EVEN (224)	0.46102	0.45863	0.45850
ODD (224)	0.37052	0.37153	0.37090
ODD (224)	0.44083	0.44459	0.44453

Table 4.84 Fermi level quantities (non-orthogonal fit)

Energy	Densities of states					Velocity	Plasmon energy
	Total	s	p	t_{2g}	e_g		
Ry	States/Ry/atom					cm/s	eV
0.5805	4.03	0.88	1.16	1.38	0.61	1.13 × 10E8	9.11

Integrated densities of states Total Electrons	s	p	t_{2g}	e_g
11.00	0.74	0.35	6.00	3.91

Table 4.85 Copper fcc

Slater-Koster 2-center parameters	Orthogonal	Non-orthogonal	
	Energy integrals (Ry)	Energy integrals (Ry)	Overlap integrals
On site			
s	0.79466	0.57133	
p	1.35351	1.07554	
d1	0.37307	0.37380	
d2	0.37180	0.37395	
First neighbor			
(sss)	−0.07518	−0.05532	0.09758
(pps)	0.19669	−0.03520	−0.25299
(ppp)	0.01940	0.04690	0.07698
(dds)	−0.02566	−0.01749	0.02150
(ddp)	0.01800	0.01011	−0.01622
(ddd)	−0.00408	−0.00071	0.00429
(sps)	0.11571	0.04883	−0.14855
(sds)	−0.03107	−0.03636	0.00910
(pds)	−0.03289	−0.02911	0.02692
(pdp)	0.01753	−0.00332	−0.07058
Second neighbor			
(sss)	−0.00092	−0.02377	0.00366
(pps)	0.05389	0.08086	−0.02075
(ppp)	0.00846	0.01795	0.01857
(dds)	−0.00451	−0.00980	−0.01377
(ddp)	0.00241	0.00001	−0.00229
(ddd)	−0.00029	0.00031	0.00121
(sps)	0.01221	0.04105	−0.01159
(sds)	−0.00852	−0.01093	−0.00308
(pds)	−0.00536	−0.03070	−0.02682
(pdp)	0.00321	−0.00563	−0.01956

	Orthogonal			Non-orthogonal		
	RMS error	Maximum	Deviation	RMS error	Maximum	Deviation
Band	mRy	k	mRy	mRy	k	mRy
1	3.3	(444)	8.4	0.4	(066)	1.0
2	4.0	(007)	9.8	0.7	(008)	1.6
3	1.6	(000)	4.2	0.6	(007)	1.6
4	2.7	(444)	6.6	0.9	(118)	4.1
5	3.1	(444)	6.6	0.6	(000)	1.6
6	5.5	(174)	21.2	0.5	(000)	1.6
1–6	3.6			0.6		

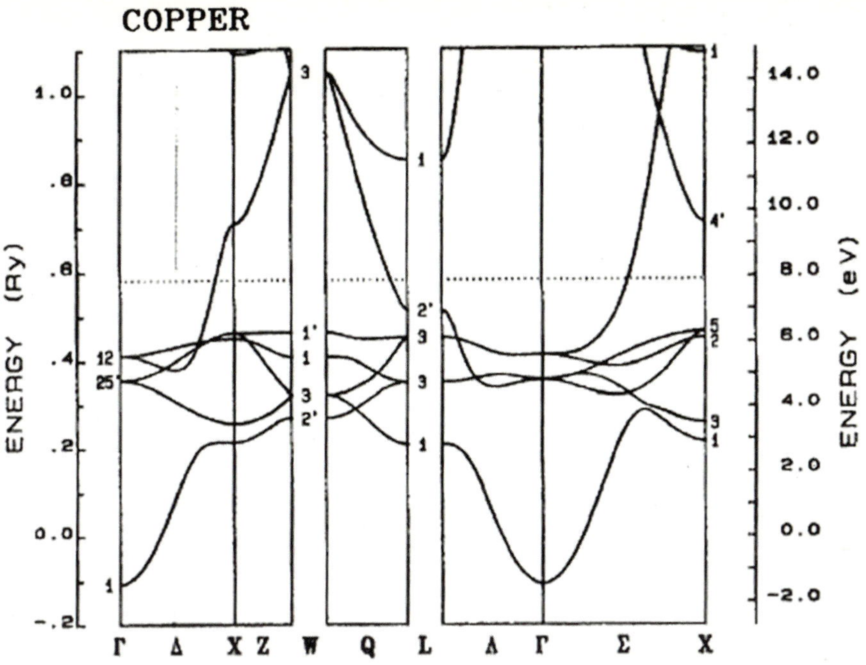

Fig. 4.32 Energy bands for Cu

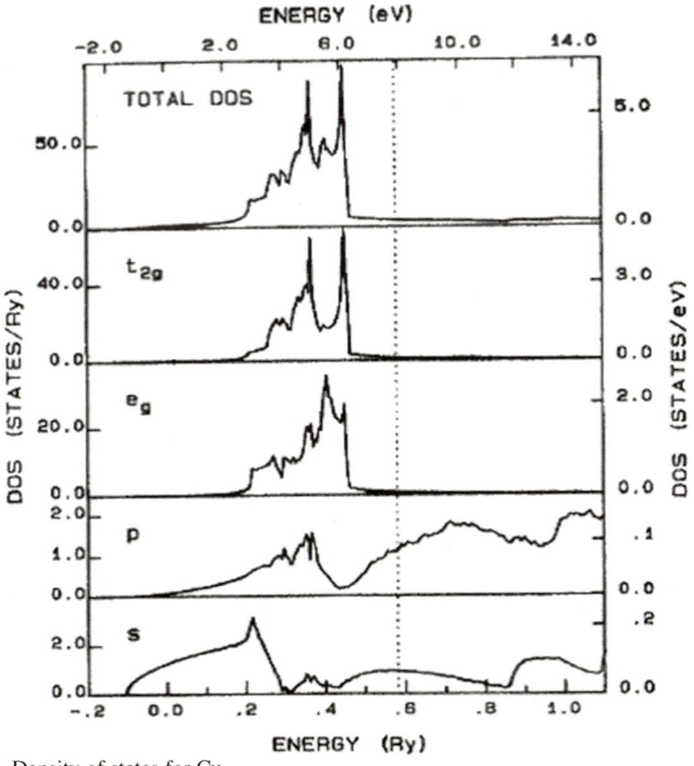

Fig. 4.33 Density of states for Cu

4.10 Zinc

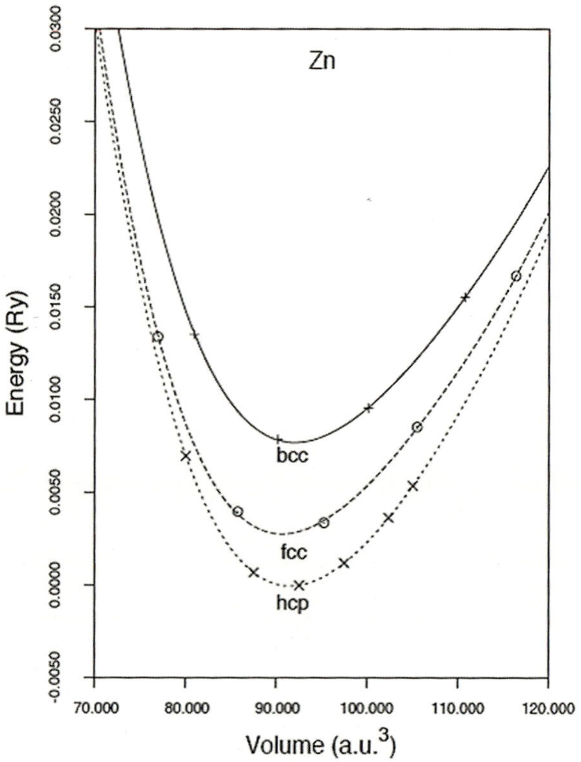

Fig. 4.34 Total energy of Zn

Table 4.86 Lattice constants and bulk modulus

	a (Bohr)	c (Bohr)	B (MBar)
bcc	5.694		0.928
fcc	7.139		0.986
hcp	4.861	9.046	1.069
exp	5.027	9.354	0.598
$\Delta E = E_{fcc} - E_{hcp} = 2.8$ mRy			

Table 4.87 Birch fit coefficients

	A_1	A_2	A_3	A_4
bcc	0.22314	0.17984	−276.90666	3745.24785
fcc	0.15996	4.34654	−368.69295	4378.95779
hcp	0.31888	−4.23175	−222.94247	3601.74125

Table 4.88 ZINC hcp Z = 30 lattice constants = 5.0357 a.u. 9.3479 a.u.

Slater-Koster 3-center parameters			
	Orthogonal		
	Energy integrals (Ry)		Energy integrals (Ry)
On site			
s, s (0)	0.26496		
x, x (0)	0.90594		
z, z (0)	0.69835		
xy, xy (0)	−0.29533		
yz, yz (0)	−0.30150		
d2, d2 (0)	−0.29565		
s, d2 (0)	0.04746		
x, xy (0)	0.07100		
First neighbor			
s, s (R)	−0.06698	s, s (T)	−0.04000
s, x (R)	0.10513	s, y (T)	−0.04905
s, y (R)	0.01114	s, z (T)	0.05285
s xy (R)	0.00285	s, d1 (T)	0.02050
s, d1 (R)	−0.00237	s, yz (T)	0.01067
s, d2 (R)	−0.01618	s, d2 (T)	0.00765
x, x (R)	0.17290	x, x (T)	0.01454
y, y (R)	0.00714	y, y (T)	0.06205
x, y (R)	0.02083	y, z (T)	−0.06524
x, xy (R)	0.01695	x xy (T)	0.01094
y, d1 (R)	0.01964	y, d1 (T)	−0.02219
x, d1 (R)	0.00758	x, xz (T)	−0.00289
y, xy (R)	0.02029	y, yz (T)	0.00481
y, d2 (R)	0.00258	y, d2 (T)	0.00494
x, d2 (R)	−0.01921	z, z (T)	0.07963
z, z (R)	−0.01605	z, d1 (T)	−0.01693
z, yz (R)	0.00617	z, yz (T)	−0.01338
z, xz (R)	0.00759	z, d2 (T)	0.00084
xy, xy (R)	0.00917	xy, xy (T)	0.00122
d1, d1 (R)	−0.00777	d1, d1 (T)	−0.00294
xy, d1 (R)	−0.00038	xy, yz (T)	−0.00020
d1, d2 (R)	−0.00488	yz, d1 (T)	0.00252
xy, d2 (R)	−0.00202	d1, d2(T)	−0.00041
yz, yz (R)	−0.00102	yz, yz (T)	−0.00424
xz, xz (R)	0.00717	xz, xz (T)	0.00296
yz, xz (R)	−0.00033	yz, d2 (T)	−0.00399
d2, d2 (R)	−0.00547	d2, d2 (T)	0.00041

Table 4.89 Zinc hcp

	Orthogonal		
Band	RMS error mRy	Maximum k	Deviation mRy
1	2.2	(0 12 12)	5.5
2	2.3	(0 12 12)	5.5
3	1.9	(0 12 12)	4.1
4	2.8	(16 0 6)	5.5
5	5.6	(0 0 9)	24.8
6	5.6	(0 0 9)	25.2
7	3.1	(8 0 0)	8.9
8	2.5	(12 0 0)	4.5
9	3.3	(24 0 0)	9.8
10	4.2	(0 0 9)	16.5
11	5.2	(0 0 9)	16.5
12	6.3	(12 0 0)	22.0
1–12	4.04		

Table 4.90 energy values in Ry at selected k-points

	Orthogonal	APW
GAMMA 1+	−0.37716	−0.37362
Gamma 1+	−0.32575	−0.32555
GAMMA 3+	0.12424	0.13536
GAMMA 5+	−0.31048	−0.31127
GAMMA 6+	−0.28688	−0.29078
GAMMA 4−	−0.35106	−0.35269
GAMMA 4−	0.12318	0.12411
GAMMA 5−	−0.27921	−0.27745
GAMMA 6−	−0.30916	−0.31382
M1+	−0.35596	−0.35471
M1+	−0.28040	−0.28051
M1+	0.18785	0.18328
M2+	−0.27560	−0.27921
M3+	−0.33782	−0.33957
M4+	−0.26042	−0.26154
M1−	−0.30696	−0.30450
M2−	−0.34492	−0.34487
M2−	−0.29823	−0.30002
M2−	0.21735	0.20955
M3−	−0.27492	−0.26512
M4−	−0.31778	−0.31280
A1	−0.34992	−0.34769
A1	−0.23336	−0.23494
A3	−0.31119	−0.31354
A3	−0.28163	−0.28301
L1	−0.34774	−0.35006
L1	−0.33486	−0.33644
L1	−0.27785	−0.27750
L1	0.25895	0.26362
L2	−0.29334	−0.29201
L2	−0.26166	−0.26362

Table 4.91 Zinc hcp Z = 30 lattice constants = 5.0357 a.u. 9.3479 a.u.

Slater-Koster 2-center parameters	Orthogonal	Non-orthogonal	
	Energy integrals (Ry)	Energy integrals (Ry)	Overlap integrals
On site			
s	0.24780	−0.08641	
p	0.61982	0.67115	
d0	−0.26757	−0.29942	
d1	−0.30101	−0.29840	
d2	−0.29806	−0.30347	
pd	0.16063	0.02761	−0.05658
First neighbor			
(sss)	−0.05768	−0.07426	0.14149
(sps)	0.06973	0.03698	−0.08575
(sds)	0.02402	0.05465	−0.10785
(pps)	0.08279	0.13625	0.02774
(ppp)	−0.02808	0.07834	0.14390
(pds)	0.02628	0.00682	0.05685
(pdp)	0.01164	−0.03501	0.11518
(dds)	−0.00895	0.00527	−0.04926
(ddp)	0.00391	0.02402	−0.06283
(ddd)	0.00000	−0.00276	0.00780
Second neighbor			
(sss)	0.01655	−0.00569	0.02238
(sps)	−0.00172	−0.00562	−0.01333
(sds)	0.02048	0.00418	0.01220
(pps)	0.04078	0.11073	0.11577
(ppp)	−0.00823	0.00383	−0.01229
(pds)	0.00820	−0.01656	0.10234
(pdp)	0.00127	0.00659	−0.04572
(dds)	−0.00140	0.00156	−0.01134
(ddp)	−0.00111	0.01569	−0.05034
(ddd)	0.00010	0.01756	−0.06000
Third neighbor			
(sss)	−0.00987	0.00635	0.00324
(sps)	−0.00943	−0.06477	−0.13175
(sds)	0.01227	0.02248	−0.08471
(pps)	−0.01251	−0.01889	−0.00685
(ppp)	0.01512	−0.01740	−0.03291
(pds)	0.03874	0.02415	−0.05173
(pdp)	−0.00644	−0.00656	0.00799
(dds)	−0.00218	−0.02407	0.08002
(ddp)	0.00004	0.00024	−0.00356
(ddd)	−0.00055	0.01034	−0.03564

Table 4.92 Zinc hcp

	Orthogonal			Non-orthogonal		
Band	RMS error mRy	Maximum k	Deviation mRy	RMS error mRy	Maximum k	Deviation mRy
1	10.6	(8 0 0)	25.9	5.5	(0 0 0)	16.1
2	8.6	(16 0 0)	19.7	5.2	(24 0 9)	8.7
3	9.9	(0 0 0)	18.1	4.5	(24 0 6)	9.9
4	6.4	(0 0 0)	11.8	3.4	(16 0 6)	6.1
5	9.1	(0 0 9)	33.2	5.6	(0 0 9)	23.4
6	9.0	(0 0 9)	33.6	5.7	(0 0 9)	23.9
7	7.8	(16 0 12)	14.8	3.9	(0 0 9)	7.6
8	6.9	(16 0 9)	16.7	3.8	(16 0 3)	6.9
9	7.1	(24 0 0)	14.3	5.3	(4 0 0)	10.7
10	5.5	(16 0 0)	11.9	5.0	(0 0 9)	13.7
11	21.5	(24 0 0)	47.6	6.9	(8 0 0)	18.4
12	32.9	(8 0 0)	93.2	9.1	(8 0 0)	23.0
1–12	13.63			5.52		

Table 4.93 Energy values in Ry at selected k-points

	Orthogonal	APW	Non-orthogonal
GAMMA 1+	−0.37534	−0.37362	−0.35747
GAMMA 1+	−0.30740	−0.32555	−0.32281
GAMMA 3+	0.10267	0.13536	0.13003
GAMMA 5+	−0.29907	−0.31127	−0.31255
GAMMA 6+	−0.30203	−0.29078	−0.29285
GAMMA 4−	−0.34405	−0.35269	−0.34562
GAMMA 4−	0.15904	0.12411	0.12818
GAMMA 5−	−0.27642	−0.27745	−0.28022
GAMMA 6−	−0.30143	−0.31382	−0.30985
M1+	−0.36162	−0.35471	−0.36032
M1+	−0.29275	−0.28051	−0.28440
M1+	0.20541	0.18328	0.17825
M2+	−0.28184	−0.27921	−0.26834
M3+	−0.33364	−0.33957	−0.34144
M4+	−0.26063	−0.26154	−0.26265
M1−	−0.31219	−0.30450	−0.30966
M2−	−0.35081	−0.34487	−0.33627
M2−	−0.29810	−0.30002	−0.30070
M2−	0.13570	0.20955	0.21703
M3−	−0.27940	−0.26512	−0.28067
M4−	−0.29941	−0.31280	−0.30146
A1	−0.34341	−0.34769	−0.34561
A1	−0.21898	−0.23494	−0.23149
A3	−0.32221	−0.31354	−0.31008
A3	−0.27417	−0.28301	−0.27725
L1	−0.34630	−0.35006	−0.34387
L1	−0.32880	−0.33644	−0.33290
L1	−0.27674	−0.27750	−0.27861
L1	0.25800	0.26362	0.25957
L2	−0.29845	−0.29201	−0.29112
L2	−0.26667	−0.26362	−0.27397

Table 4.94 Fermi level quantities (non-orthogonal fit)

Energy	Densities of states				
	Total	s	p	t_{2g}	e_g
Ry	States/Ry/atom				
0.3598	2.42	0.60	1.59	0.10	0.13

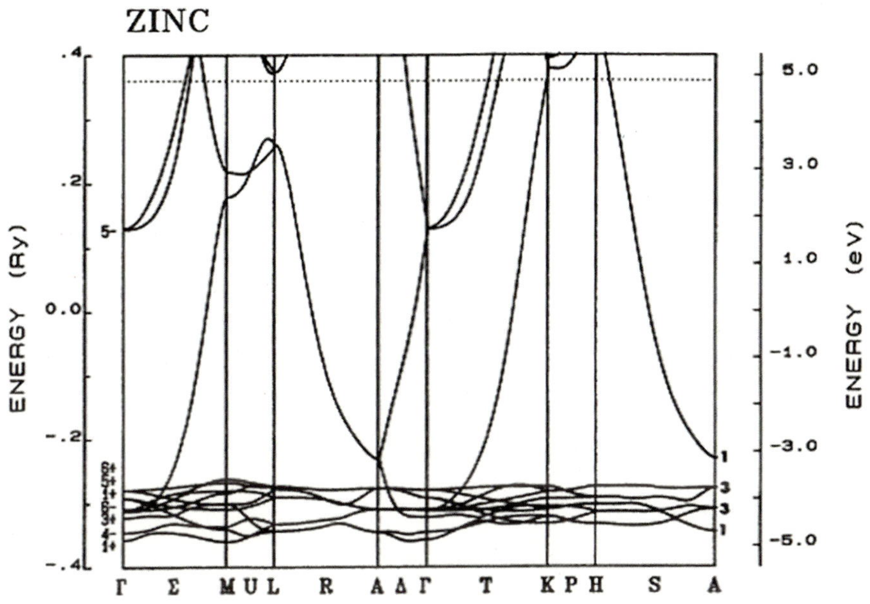

Fig. 4.35 Energy bands for Zn

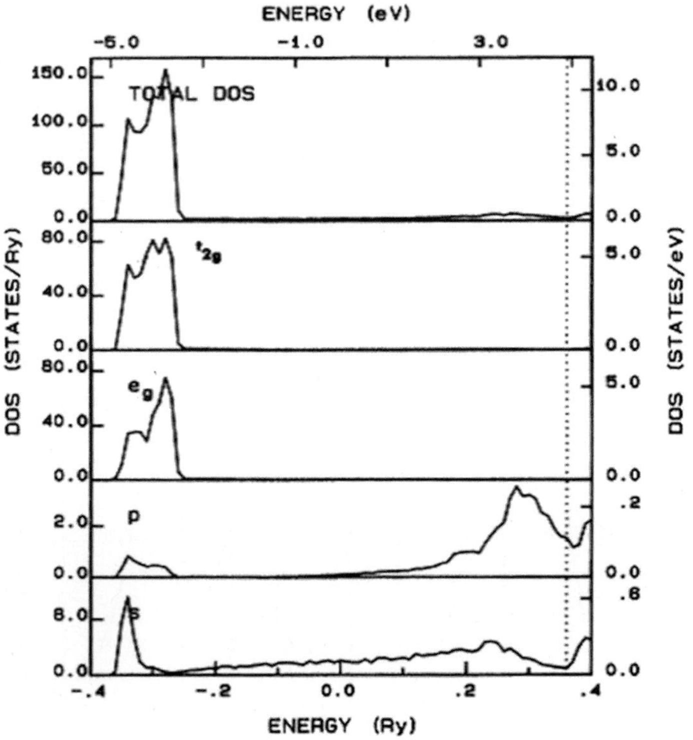

Fig. 4.36 Density of states for Zn

Chapter 5
The 4d Transition Metals

Similarly to the 3d elements, the first two of the 4d series, i.e. Y and Zr, as well as the last one, Cd, crystallize in the hcp structure as shown in the total energy plots. Most of the elements in the 4d transition metal series have band structures that resemble those of the corresponding 3d elements of the same column and structure. Specifically, the energy bands and DOS of Y and Zr look very much like those of Sc and Ti; Nb and Mo look like V and Cr; Ag looks like Cu; and Cd like Zn. In a more quantitative sense the most important difference between 3d and 4d elements is in the d-band widths, which are larger for the 4d elements. It is also interesting to note that the Ag d bands are located much deeper than in either Pd or Cu. Another important difference is that in Cd the d bands are well separated from the s-like band, unlike the situation in Zn, for which the s band close to the center of the Brillouin zone overlaps with the d bands.

D.A. Papaconstantopoulos, *Handbook of the Band Structure of Elemental Solids*,
DOI 10.1007/978-1-4419-8264-3_5

5.1 Yttrium

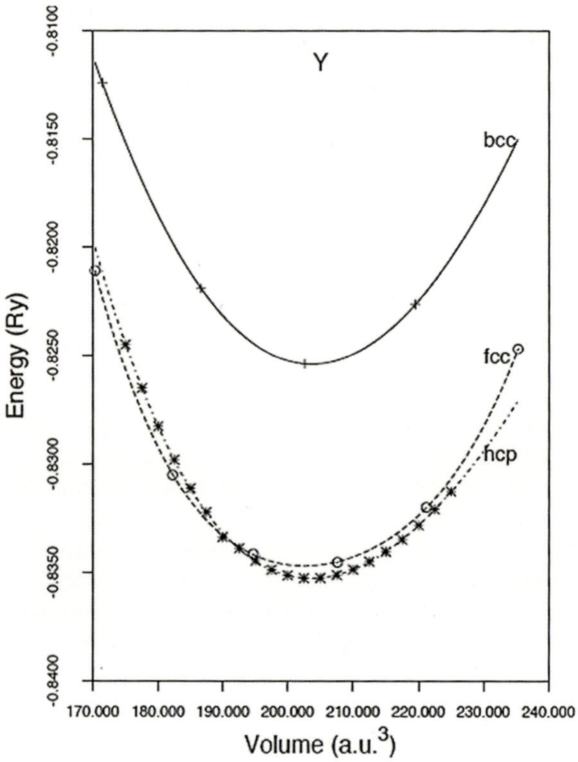

Fig. 5.1 Total energy of Y

Table 5.1 Lattice constants and bulk modulus

	a (Bohr)	c (Bohr)	B (Mbar)
bcc	7.412		0.689
fcc	9.317		0.673
hcp	6.690	10.504	0.616
exp	6.897	10.828	0.366
$\Delta E = E_{hcp} - E_{fcc} = 0.12$ mRy			

Table 5.2 Birch fit coefficients

	A_1	A_2	A_3	A_4
bcc	−6770.19779	−235.66296	6870.45964	−64431.23982
fcc	−6771.07939	−145.59629	3780.69822	−29235.41446
hcp	−13543.70958	−217.77828	6174.23196	−7054.34627

Table 5.3 Yttrium hcp Z = 39 lattice constants = 6.8926 a.u. 10.8293 a.u.

Slater-Koster 3-center parameters	Orthogonal		
	Energy integrals (Ry)		Energy integrals (Ry)
On site			
s, s (0)	0.61351		
x, x (0)	1.00595		
z, z (0)	0.90483		
xy, xy (0)	0.64414		
yz, yz (0)	0.63744		
d2, d2 (0)	0.55791		
s, d2 (0)	0.05244		
x, xy (0)	−0.00571		
First neighbor			
s, s (R)	−0.03485	s, s (T)	−0.05145
s, x (R)	0.05931	s, y (T)	−0.04184
s, y (R)	0.02748	s, z (T)	0.06836
s, xy (R)	0.02482	s, d1 (T)	−0.01643
s, d1 (R)	0.04642	s, yz (T)	−0.05198
s, d2 (R)	−0.02374	s, d2 (T)	0.03822
x, x (R)	0.10260	x, x (T)	−0.00025
y, y (R)	−0.00387	y, y (T)	0.06002
x, y (R)	0.02553	y, z (T)	−0.06985
x, xy (R)	0.02551	x, xy (T)	0.01105
y, d1 (R)	0.01773	y, d1 (T)	0.01461
x, d1 (R)	0.07662	x, xz (T)	−0.01366
y, xy (R)	0.00083	y, yz (T)	0.04799
y, d2 (R)	0.00220	y, d2 (T)	−0.02415
x, d2 (R)	−0.04275	z, z (T)	0.06486
z, z (R)	−0.01055	z, d1 (T)	−0.04372
z, yz (R)	−0.01252	z, yz (T)	−0.07799
z, xz (R)	−0.00418	z, d2 (T)	0.01873
xy, xy (R)	0.01356	xy, xy (T)	0.01929
d1, d1 (R)	−0.03643	d1, dl (T)	−0.00020
xy, d1 (R)	−0.00569	xy, yz (T)	−0.00915
d1, d2 (R)	0.02693	yz, d1 (T)	−0.01310
xy, d2 (R)	0.00130	d1, d2 (T)	0.02554
yz, yz (R)	0.00074	yz, yz (T)	−0.04409
xz, xz (R)	0.00774	xz, xz (T)	0.02229
yz, xz (R)	−0.01666	yz, d2 (T)	0.02676
d2, d2 (R)	−0.01052	d2, d2 (T)	0.00034

Table 5.4 Yttrium hcp

Band	Orthogonal RMS error mRy	Maximum k	Deviation mRy
1	1.8	(16 0 0)	4.1
2	3.1	(12 0 0)	7.3
3	3.1	(8 0 12)	8.5
4	3.8	(8 0 12)	8.5
5	3.4	(24 0 0)	8.6
6	2.1	(4 0 0)	6.2
7	3.2	(0 12 12)	8.3
8	3.9	(0 12 12)	8.3
9	4.0	(0 12 12)	11.5
10	5.5	(16 0 0)	14.3
11	4.4	(16 0 12)	10.2
12	4.6	(16 0 12)	10.2
1–12	3.72		

Table 5.5 Energy values in Ry at selected k-points

	Orthogonal	APW
GAMMA 1+	0.05204	0.05304
GAMMA 1+	0.54042	0.54043
GAMMA 3+	0.45238	0.45651
GAMMA 5+	0.56090	0,56057
GAMMA 6+	0.59750	0.59388
GAMMA 4–	0.26513	0.26085
GAMMA 4–	0.94071	0.93773
GAMMA 5–	0.72829	0.72669
GAMMA 6–	0.51595	0.51481
M1+	0.29256	0.29667
M1+	0.39501	0.39212
M1+	0.81704	0.81903
M2+	0.79751	0.78479
M3+	0.32470	0.32513
M4+	0.77849	0.77667
M1–	0.48840	0.49695
M2–	0.28066	0.28000
M2–	0.53799	0.53862
M2–	0.80868	0.80927
M3–	0.74709	0.75136
M4–	0.49685	0.49848
A1	0.15698	0.15651
A1	0.46339	0.46493
A3	0.50777	0.51050
A3	0.69766	0.70262
L1	0.26932	0.26792
L1	0.40182	0.40331
L1	0.49681	0.49544
L1	0.76871	0.76664
L2	0.59341	0.59158
L2	0.76507	0.76539

Table 5.6 Yttrium hcp Z = 39 lattice constants = 6.8926 a.u. 10.8293 a.u.

Slater-Koster 2-center parameters	Orthogonal	Non-orthogonal	
	Energy integrals (Ry)	Energy integrals (Ry)	Overlap integrals
On site			
s	0.70944	0.51819	
p	0.84896	0.80726	
d0	0.58906	0.52943	
d1	0.58323	0.53475	
d2	0.59697	0.54335	
pd	−0.02918	0.00643	0.01068
First neighbor			
(sss)	−0.05505	−0.03056	0.05661
(sps)	0.06975	0.04103	−0.07512
(sds)	0.04806	0.02355	−0.07109
(pps)	0.07071	0.10545	0.01500
(ppp)	−0.02610	0.11109	0.13998
(pds)	0.06285	0.01710	−0.05995
(pdp)	−0.02445	0.04437	0.10440
(dds)	−0.06331	0.00003	0.09952
(ddp)	0.03080	−0.00331	−0.08148
(ddd)	−0.00338	0.00932	0.02385
Second neighbor			
(sss)	0.00066	−0.01173	−0.01118
(sps)	−0.00892	0.03301	0.01823
(sds)	−0.00804	0.03456	0.03821
(pps)	0.01820	0.09469	0.05633
(ppp)	−0.00377	0.02128	0.02710
(pds)	0.00996	0.06818	0.06141
(pdp)	−0.00835	−0.00183	−0.00269
(dds)	−0.00693	−0.03537	−0.02475
(ddp)	0.00127	0.00209	−0.00014
(ddd)	0.00161	−0.00441	−0.00971
Third neighbor			
(sss)	−0.00099	0.00178	0.01607
(sps)	0.00242	0.00036	−0.01623
(sds)	−0.00545	0.00708	0.02503
(pps)	0.00744	0.01471	0.00325
(ppp)	−0.00446	−0.00667	0.00231
(pds)	−0.00645	0.01049	0.02797
(pdp)	0.00065	−0.01384	−0.01727
(dds)	0.00694	−0.01376	−0.02750
(ddp)	0.00036	−0.00623	−0.00828
(ddd)	−0.00057	0.01062	0.01578

Table 5.7 Yttrium hcp

	Orthogonal			Non-orthogonal		
Band	RMS error mRy	Maximum k	Deviation mRy	RMS error mRy	Maximum k	Deviation mRy
1	6.6	(16 0 0)	13.9	2.2	(16 0 12)	4.6
2	7.8	(24 0 0)	13.5	2.8	(0 18 0)	5.3
3	7.7	(24 0 9)	19.4	3.3	(12 0 0)	8.5
4	8.0	(0 24 12)	19.1	4.7	(8 0 0)	9.8
5	4.2	(0 18 0)	11.0	3.9	(16 0 12)	9.0
6	5.1	(8 0 0)	15.9	4.2	(16 0 12)	9.0
7	12.3	(0 12 12)	27.1	3.8	(8 0 12)	9.2
8	12.2	(0 12 12)	27.1	4.4	(8 0 12)	9.2
9	9.8	(24 0 0)	21.7	5.2	(0 12 12)	11.4
10	11.1	(24 0 0)	19.1	5.5	(0 12 12)	11.4
11	13.8	(0 18 0)	27.6	4.6	(8 0 12)	11.6
12	15.6	(8 0 0)	33.7	5.3	(8 0 12)	11.6
1–12	10.11			4.28		

Table 5.8 Energy values in Ry at selected k-points

	Orthogonal	APW	Non-orthogonal
GAMMA 1+	0.05064	0.05304	0.05134
GAMMA 1+	0.53579	0.54043	0.53992
GAMMA 3+	0.45530	0.45651	0.45492
GAMMA 5+	0.55698	0.56057	0.55729
GAMMA 6+	0.59796	0.59388	0.59788
GAMMA 4−	0.26264	0.26085	0.26023
GAMMA 4−	0.90668	0.93773	0.94672
GAMMA 5−	0.73446	0.72669	0.73101
GAMMA 6−	0.50822	0.51481	0.50886
M1 +	0.27908	0.29667	0.29387
M1 +	0.38641	0.39212	0.39407
M1 +	0.83451	0.81903	0.83006
M2+	0.75501	0.78479	0.78179
M3+	0.32852	0.32513	0.31874
M4+	0.76567	0.77667	0.77602
M1−	0.49474	0.49695	0.50432
M2−	0.28317	0.28000	0.27938
M2−	0.52569	0.53862	0.53413
M2−	0.80862	0.80927	0.81380
M3−	0.75239	0.75136	0.75233
M4−	0.49793	0.49848	0.49504
A1	0.16031	0.15651	0.15727
A1	0.46650	0.46493	0.46447
A3	0.51078	0.51050	0.51111
A3	0.68805	0.70262	0.70289
L1	0.27640	0.26792	0.26906
L1	0.38421	0.40331	0.39715
L1	0.49870	0.49544	0.49862
L1	0.75725	0.76664	0.76077
L2	0.60641	0.59158	0.59108
L2	0.76337	0.76539	0.77224

Table 5.9 Fermi level quantities (non-orthogonal fit)

Energy		Densities of states			
	Total	s	p	t_{2g}	e_g
Ry		States/Ry/atom			
0.3961	31.50	0.87	7.67	13.29	9.67

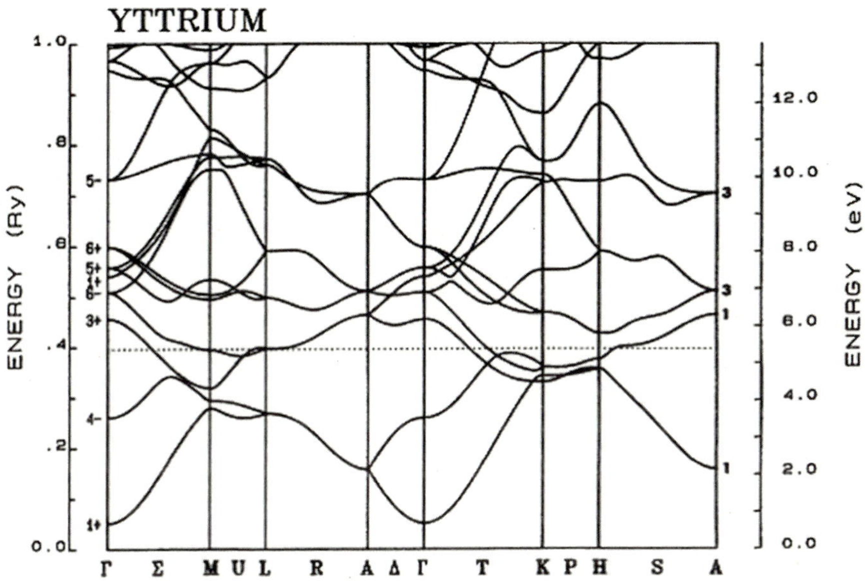

Fig. 5.2 Energy bands for Y

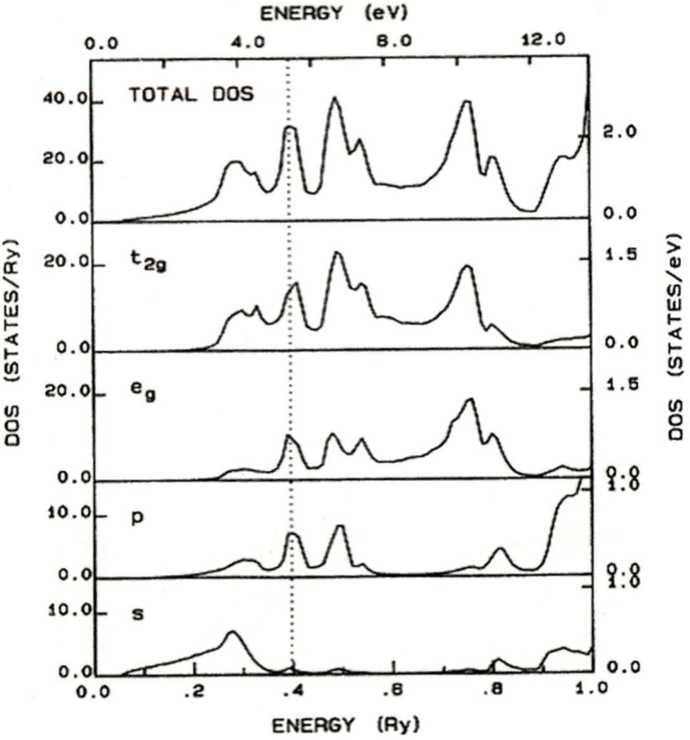

Fig. 5.3 Density of states for Y

5.2 Zirconium

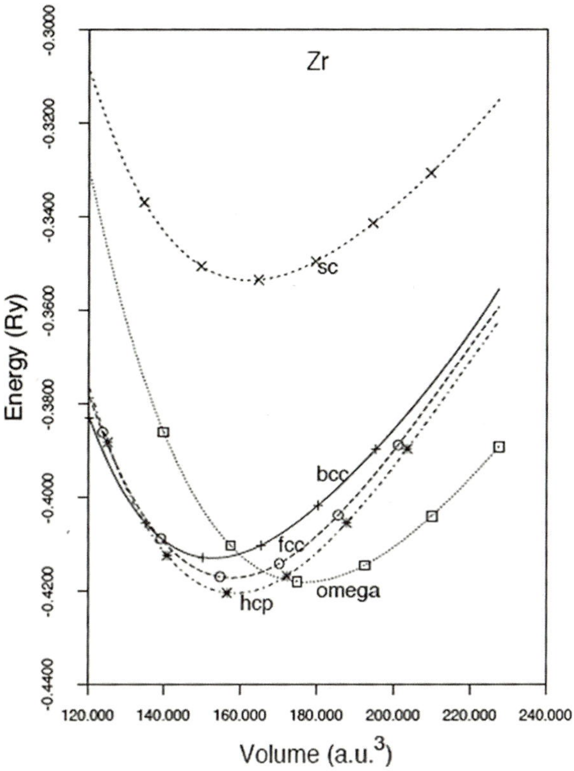

Fig. 5.4 Total energy of Zr

Table 5.10 Lattice constants and bulk modulus

	a (Bohr)	c (Bohr)	B (Mbar)
fcc	8.577		0.949
bcc	6.739		0.861
hcp	6.054	9.990	0.998
exp	6.104	9.732	0.833
$\Delta E = E_{hcp} - E_{fcc} = 3.4$ mRy			

Table 5.11 Birch fit coefficients

	A_1	A_2	A_3	A_4
sc	0.54953	−52.55506	728.27010	1070.64343
bcc	0.70474	−67.10190	1095.54193	−2589.61084
fcc	1.51894	−136.12597	2998.06429	−19675.62466
hcp	1.43279	−127.40370	2693.79826	−16164.75428
omega	1.10147	−106.93681	2211.23996	−11035.90122

Table 5.12 Zirconium hcp Z = 40 lattice constants = 6.1076 a.u. 9.7264 a.u.

Slater-Koster 3-center parameters	Orthogonal		
	Energy integrals (Ry)		Energy integrals (Ry)
On site			
s, s (0)	0.86190		
x, x (0)	1.07676		
z, z (0)	1.29992		
xy, xy (0)	0.72282		
yz, yz (0)	0.73752		
d2, d2 (0)	0.64164		
s, d2 (0)	0.01838		
x, xy (0)	−0.00942		
First neighbor			
s, s (R)	−0.04978	s, s (T)	−0.07157
s, x (R)	0.05893	s, y (T)	−0.04724
s, y (R)	0.01518	s, z (T)	0.09032
s, xy (R)	0.01580	s, d1 (T)	−0.01739
s, d1 (R)	0.06129	s, yz (T)	−0.06256
s, d2 (R)	−0.02694	s, d2 (T)	0.03480
x, x (R)	0.04050	x, x (T)	−0.03260
y, y (R)	−0.02171	y, y (T)	0.01501
x, y (R)	0.01683	y, z (T)	−0.09541
x, xy (R)	0.02255	x, xy (T)	0.01733
y, d1 (R)	−0.01013	y, d1 (T)	−0.00210
x, d1 (R)	0.08101	x, xz (T)	−0.02886
y, xy (R)	−0.01470	y, yz (T)	0.04068
y, d2 (R)	−0.00304	y, d2 (T)	−0.03473
x, d2 (R)	−0.03318	z, z (T)	0.10742
z, z (R)	−0.00578	z, d1 (T)	−0.03617
z, yz (R)	−0.01809	z, yz (T)	−0.10073
z, xz (R)	−0.00582	z, d2 (T)	0.01921
xy, xy (R)	0.02376	xy, xy (T)	0.01398
d1, d1 (R)	−0.05474	d1, d1 (T)	−0.00360
xy, d1 (R)	0.00975	xy, yz (T)	−0.01685
d1, d2 (R)	0.02569	yz, d1 (T)	−0.01092
xy, d2 (R)	0.00150	d1, d2 (T)	0.02740
yz, yz (R)	−0.00044	yz, yz (T)	−0.05215
xz, xz (R)	0.01437	xz, xz (T)	0.02430
yz, xz (R)	−0.01589	yz, d2 (T)	0.04370
d2, d2 (R)	0.01397	d2, d2 (T)	0.00832

Table 5.13 Zirconium hcp

Band	Orthogonal RMS error mRy	Maximum k	Deviation mRy
1	9.0	(16 0 0)	25.4
2	12.1	(16 0 0)	36.0
3	3.5	(12 0 0)	8.4
4	2.8	(0 18 0)	6.3
5	3.4	(0 12 0)	7.1
6	2.4	(8 0 12)	5.2
7	3.2	(0 12 12)	10.3
8	4.0	(0 12 12)	10.3
9	4.2	(0 0 12)	10.4
10	5.5	(16 0 0)	15.6
11	12.8	(16 0 12)	46.7
12	17.2	(12 0 0)	47.4
1–12	8.14		

Table 5.14 Energy values in Ry at selected k-points

	Orthogonal	APW
GAMMA 1+	0.12493	0.12509
GAMMA 1+	0.61658	0.61613
GAMMA 3+	0.62070	0.62324
GAMMA 5+	0.65092	0.65052
GAMMA 6+	0.69575	0.69529
GAMMA 4–	0.32287	0.32166
GAMMA 4–	1.17769	1.20477
GAMMA 5–	0.86284	0.85830
GAMMA 6–	0.59306	0.59745
M1 +	0.33964	0.34474
M1 +	0.52940	0.52530
Ml +	0.99690	0.99075
M2 +	0.93396	0.92393
M3+	0.37307	0.37302
M4 +	0.92222	0.92228
M1–	0.57247	0.57605
M2–	0.36153	0.36279
M2–	0.64162	0.64693
M2–	1.01435	1.01607
M3–	0.88424	0.89261
M4–	0.59930	0.60075
A1	0.25095	0.25006
A1	0.54735	0.54737
A3	0.58975	0.59190
A3	0.81512	0.82556
L1	0.32447	0.32334
L1	0.46487	0.46551
L1	0.66305	0.66311
L1	0.91874	0.92097
L2	0.70159	0.69902
L2	0.91049	0.90924

Table 5.15 Zirconium hcp Z = 40 lattice constants = 6.1076 a.u. 9.7264 a.u.

Slater-Koster 2-center parameters	Orthogonal	Non-orthogonal	
	Energy integrals (Ry)	Energy integrals (Ry)	Overlap integrals
On site			
s	0.89346	0.68413	
p	1.08881	1.01725	
d0	0.68679	0.61871	
d1	0.67297	0.61857	
d2	0.70076	0.63420	
pd	−0.03446	−0.01323	0.01341
First neighbor			
(sss)	−0.06206	−0.03370	0.08582
(sps)	0.07342	0.04219	−0.06861
(sds)	0.05436	0.01640	−0.06586
(pps)	0.07289	0.08818	0.00505
(ppp)	−0.04471	0.06903	0.10237
(pds)	0.07613	0.00475	−0.04728
(pdp)	−0.02926	0.02040	0.07506
(dds)	−0.07679	−0.00017	0.10670
(ddp)	0.03875	0.01552	−0.05754
(ddd)	−0.00526	−0.00020	0.01113
Second neighbor			
(sss)	−0.00231	−0.00261	0.02110
(sps)	0.00004	−0.00871	−0.03704
(sds)	−0.00131	0.00851	−0.00846
(pps)	0.02611	0.02094	−0.03534
(ppp)	−0.00791	0.02684	0.03667
(pds)	0.01146	0.04114	0.01513
(pdp)	−0.00803	0.01117	0.02210
(dds)	−0.00741	−0.02974	−0.01548
(ddp)	0.00271	0.00555	0.00571
(ddd)	0.00077	0.00159	−0.00051
Third neighbor			
(sss)	−0.00467	0.00778	0.03006
(sps)	0.00240	−0.04965	−0.08475
(sds)	−0.00053	0.00359	0.02119
(pps)	0.00982	0.00689	−0.00773
(ppp)	0.01178	−0.01719	−0.01585
(pds)	0.00632	0.01093	0.01675
(pdp)	0.00086	−0.00970	−0.01451
(dds)	−0.00062	−0.01312	−0.02051
(ddp)	0.00271	−0.00167	−0.00299
(ddd)	−0.00164	0.00594	0.00782

Table 5.16 Zirconium hcp

	Orthogonal			Non-orthogonal		
Band	RHS error mRy	Maximum k	Deviation mRy	RMS error mRy	Maximum k	Deviation mRy
1	8.8	(24 0 0)	21.4	2.2	(0 18 0)	5.2
2	5.0	(0 18 0)	14.7	2.4	(0 18 0)	5.8
3	7.7	(24 0 9)	18.6	3.1	(24 0 6)	6.6
4	7.0	(0 24 12)	17.8	3.4	(24 0 9)	8.6
5	3.8	(16 0 0)	7.5	2.7	(16 0 0)	8.3
6	4.4	(0 12 0)	7.8	3.1	(16 0 0)	8.3
7	8.6	(8 0 12)	19.1	4.2	(8 0 12)	10.3
8	10.8	(0 12 0)	19.2	4.3	(8 0 12)	10.3
9	14.1	(24 0 0)	30.2	4.9	(16 0 12)	12.1
10	12.6	(0 0 12)	27.3	6.1	(16 0 0)	15.9
11	19.2	(16 0 12)	35.2	6.5	(16 0 12)	21.7
12	21.8	(0 18 0)	47.4	8.8	(0 18 0)	26.3
1–12	11.67			4.70		

Table 5.17 Energy values in Ry at selected k-points

	Orthogonal	APW	Non-orthogonal
GAMMA 1+	0.12557	0.12509	0.12617
GAMMA 1+	0.61022	0.61613	0.61638
GAMMA 3+	0.61748	0.62324	0.62039
GAMMA 5+	0.64844	0.65052	0.64442
GAMMA 6+	0.69778	0.69529	0.69944
GAMMA 4–	0.32439	0.32166	0.32181
GAMMA 4–	1.08463	1.20477	1.20176
GAMMA 5–	0.85994	0.85830	0.86147
GAMMA 6–	0.58996	0.59745	0.59355
M1 +	0.32333	0.34474	0.34861
M1 +	0.52775	0.52530	0.52402
M1 +	1.00314	0.99075	0.99124
M2 +	0.89204	0.92393	0.91743
M3 +	0.36930	0.37302	0.36606
M4 +	0.90809	0.92228	0.91829
M1–	0.57384	0.57605	0.57863
M2–	0.36831	0.36279	0.36800
M2–	0.63658	0.64693	0.64289
M2–	1.01617	1.01607	1.00760
M3–	0.88290	0.89261	0.89815
M4–	0.59486	0.60075	0.59893
A1	0.25113	0.25006	0.25063
A1	0.54940	0.54737	0.54675
A3	0.59459	0.59190	0.59290
A3	0.79824	0.82556	0.82168
L1	0.32865	0.32334	0.32034
L1	0.44774	0.46551	0.46092
L1	0.66689	0.66311	0.66251
L1	0.91155	0.92097	0.91298
L2	0.70478	0.69902	0.69529
L2	0.90350	0.90924	0.91492

Table 5.18 Fermi level quantities (non-orthogonal fit)

Energy		Densities of states			
	Total	s	p	t_{2g}	e_g
Ry		States/Ry/atom			
0.5461	7.91	0.13	0.58	4.90	2.30

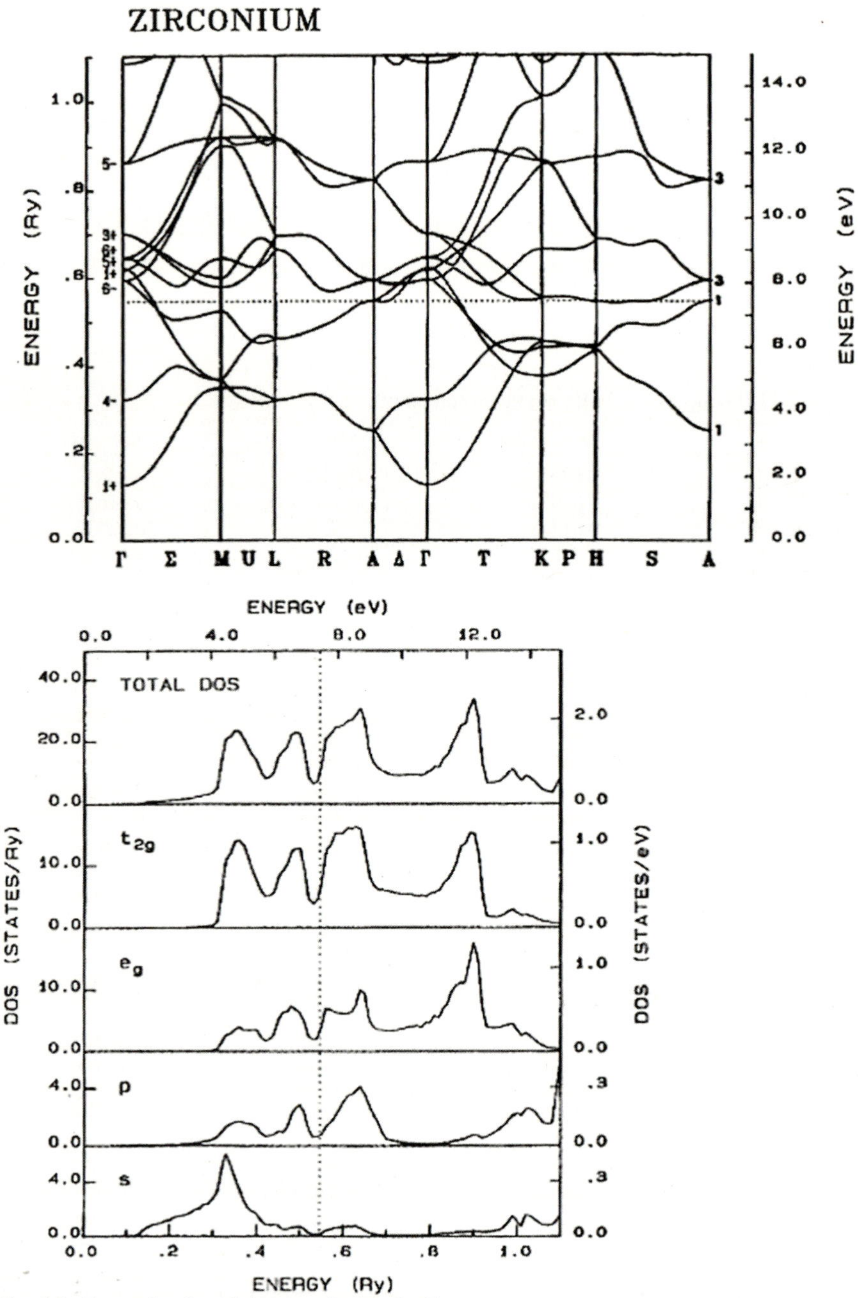

Fig. 5.5 Energy bands and density of states for Zr

5.3 Niobium

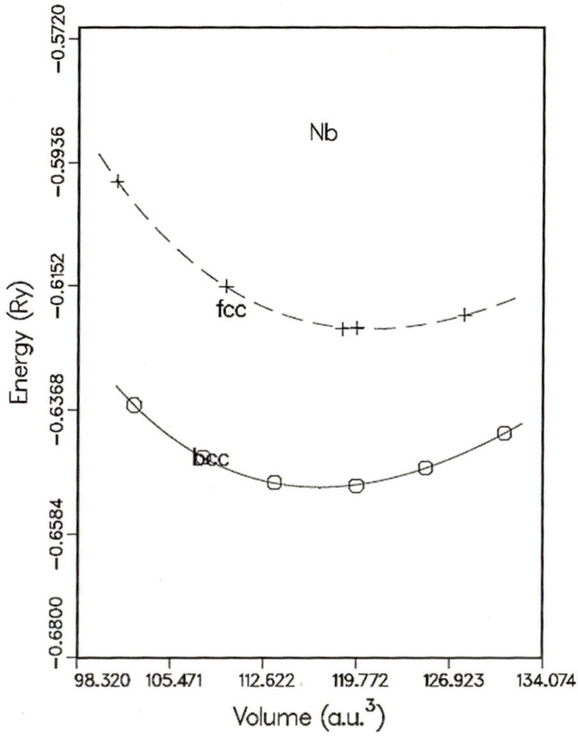

Fig. 5.6 Total energy of Nb

Table 5.19 Lattice constants and bulk modulus

	a (Bohr)	B (mbar)
bcc	6.161	1.947
fcc	7.847	1.798
exp	6.236	1.702
$\Delta E = 27.6$ mRy		

Table 5.20 Birch fit coefficients

	A_1	A_2	A_3
bcc	1.09131	−83.29348	995.90669
fcc	1.03897	−81.21771	992.42916

Table 5.21 Niobium bcc Z = 41 lattice constant = 6.23610 a.u.

Slater-Koster 3-center parameters	Orthogonal	Non-orthogonal	
	Energy integrals (Ry)	Energy integrals (Ry)	Overlap integrals
On site			
s, s (000)	1.05228	0.77626	
x, x (000)	1.56150	1.12855	
xy, xy (000)	0.90335	0.76571	
d2, d2 (000)	0.79760	0.70725	
First neighbor			
s, s (111)	−0.11028	−0.01014	0.12102
s, x (111)	0.07185	−0.03372	−0.11088
s, xy (111)	0.05645	0.01124	−0.04676
x, x (111)	0.08402	−0.05267	−0.04780
x, y (111)	0.04273	−0.08930	−0.13043
x, xy (111)	0.04372	−0.01126	−0.03515
x, yz (111)	0.05825	−0.02152	−0.08275
x, d1 (111)	0.00881	0.00336	−0.04196
xy, xy (111)	−0.02354	−0.01429	0.01000
xy, xz (111)	−0.03653	−0.00767	0.03533
xy, d2 (111)	0.00461	0.01148	−0.02680
d2, d2 (111)	0.03139	0.01157	−0.04528
Second neighbor			
s, s(200)	−0.02061	−0.02775	0.05223
s, x (200)	0.06077	−0.01404	−0.11830
s, d2 (002)	−0.06497	0.01397	0.08789
x, x (200)	0.17455	−0.13248	−0.24765
y, y (200)	0.03361	−0.01696	0.00420
x, xy (020)	0.01895	−0.02190	0.00006
z, d2 (002)	−0.11530	0.07370	0.16619
xy, xy (200)	−0.02231	0.02533	0.00079
xy, xy (002)	0.00260	−0.00580	−0.01110
d2, d2 (002)	−0.06120	0.00335	0.08892
d1, d1 (002)	0.00742	−0.00585	−0.00409
Third neighbor			
s, s (220)	0.01561		
s, x (220)	−0.01476		
s, xy (220)	−0.01416		
s, d2 (220)	−0.00788		
x, x (220)	−0.00940		
x, x (022)	0.01198		
x, y (220)	−0.00413		
x, xy (220)	−0.01109		
x, xy (022)	0.00282		
z, d2 (022)	0.00077		
z, d1 (022)	−0.00591		
xy, xy (220)	0.01457		
xy, xy (022)	0.00103		
xy, xz (022)	−0.00051		
xy, d2 (220)	0.00994		
d2, d2 (220)	0.00441		
d1, d1 (220)	−0.00518		

Table 5.22 Niobium bcc

	Orthogonal			Non-orthogonal		
Band	RMS error mRy	Maximum k	Deviation mRy	RMS error mRy	Maximum k	Deviation mRy
1	2.9	(260)	8.3	0.7	(000)	2.7
2	2.4	(030)	7.3	0.6	(232)	1.1
3	2.6	(044)	8.0	0.8	(161)	3.3
4	2.4	(330)	5.9	0.8	(140)	3.1
5	3.6	(131)	11.9	1.7	(131)	10.8
6	5.1	(171)	13.7	1.7	(140)	6.9
1–6	3.3			1.2		

Table 5.23 Energy values in Ry at selected k-points

	Orthogonal	APW	Non-orthogonal
GAMMA 1	0.23373	0.22898	0.23172
GAMMA 12	0.88272	0.88856	0.88798
GAMMA 15	2.68992	2.69151	2.69151
GAMMA 25′	0.69748	0.69906	0.69847
H1 (008)	1.99822	2.00121	2.00107
H12 (008)	0.38055	0.37484	0.37556
H15 (008)	1.34555	1.34585	1.34648
H25′ (008)	1.07415	1.07456	1.07636
N1 (044)	0.38832	0.38852	0.38836
N1 (044)	0.89825	0.90205	0.90343
N1 (044)	1.54635	1.54901	1.54875
N2 (044)	0.54765	0.54906	0.54871
N3 (044)	1.13211	1.13252	1.13401
N4 (044)	0.95181	0.94826	0.94817
NI′ (044)	0.82264	0.83062	0.83031
N3′ (044)	1.89931	1.90928	1.90918
N4′ (044)	1.50631	1.50273	1.50280
P1 (444)	1.36330	1.37222	1.37302
P3 (444)	0.95431	0.95793	0.95876
P4 (444)	0.58625	0.58728	0.58634
P4 (444)	1.51827	1.53049	1.52978
(341)	0.42572	0.42475	0.42429
(341)	0.55827	0.56027	0.56047
(341)	0.72706	0.72743	0.72761
(341)	0.88980	0.88997	0.88976
(341)	0.96940	0.97007	0.97039
(341)	1.16665	1.16591	1.16585

Table 5.24 Fermi level quantities (non-orthogonal fit)

Energy		Densities of states				Velocity	Plasmon energy
	Total	s	p	t_{2g}	e_g		
Ry		States/Ry/atom				cm/s	eV
0.6655	19.86	0.80	2.08	13.05	3.94	0.63 × 10E8	9.12

Integrated densities of states Total Electrons	s	p	t_{2g}	e_g
5.00	0.64	0.25	2.52	1.58

Table 5.25 Niobium bcc

Slater-Koster 2-center parameters	Orthogonal	Non-orthogonal	
	Energy integrals (Ry)	Energy integrals (Ry)	Overlap integrals
On site			
s	1.12855	0.79458	
p	1.54674	1.10503	
d1	0.87358	0.74857	
d2	0.78444	0.71756	
First neighbor			
(sss)	−0.10924	−0.02873	0.11271
(pps)	0.16935	−0.22210	−0.30660
(ppp)	0.03944	0.03037	0.07845
(dds)	−0.09786	−0.01147	0.09797
(ddp)	0.04451	0.02433	−0.06277
(ddd)	0.00214	−0.00953	0.00455
(sps)	0.10087	0.03433	0.18670
(sds)	−0.10570	−0.01108	0.09122
(pds)	−0.14396	−0.08920	−0.18796
(pdp)	0.02362	−0.00366	0.07679
Second neighbor			
(sss)	−0.02174	−0.00497	0.06646
(pps)	0.19817	−0.18680	−0.28131
(ppp)	0.00536	0.03176	0.03023
(dds)	−0.06165	−0.00745	0.07438
(ddp)	−0.00955	0.00755	−0.02252
(ddd)	0.00600	−0.00591	−0.00545
(sps)	0.07223	0.05402	0.14491
(sds)	−0.05088	0.00519	0.07601
(pds)	−0.09164	−0.04745	−0.13911
(pdp)	−0.00854	0.00361	0.02161
Third neighbor			
(sss)	0.00913		
(pps)	0.02321		
(ppp)	−0.00568		
(dds)	0.01259		
(ddp)	−0.00161		
(ddd)	0.00028		
(sps)	0.00170		
(sds)	0.00587		
(pds)	0.00201		
(pdp)	0.00040		

	Orthogonal			Non-orthogonal		
Band	RMS error mRy	Maximum k	Deviation mRy	RMS error mRy	Maximum k	Deviation mRy
1	4.9	(330)	12.8	1.0	(000)	3.6
2	10.2	(044)	29.8	1.0	(121)	2.4
3	7.1	(008)	20.2	1.5	(161)	4.6
4	10.1	(260)	26.3	2.2	(231)	4.9
5	8.0	(008)	20.2	2.4	(131)	11.1
6	10.2	(442)	26.0	2.3	(140)	6.0
1–6	8.7			1.8		

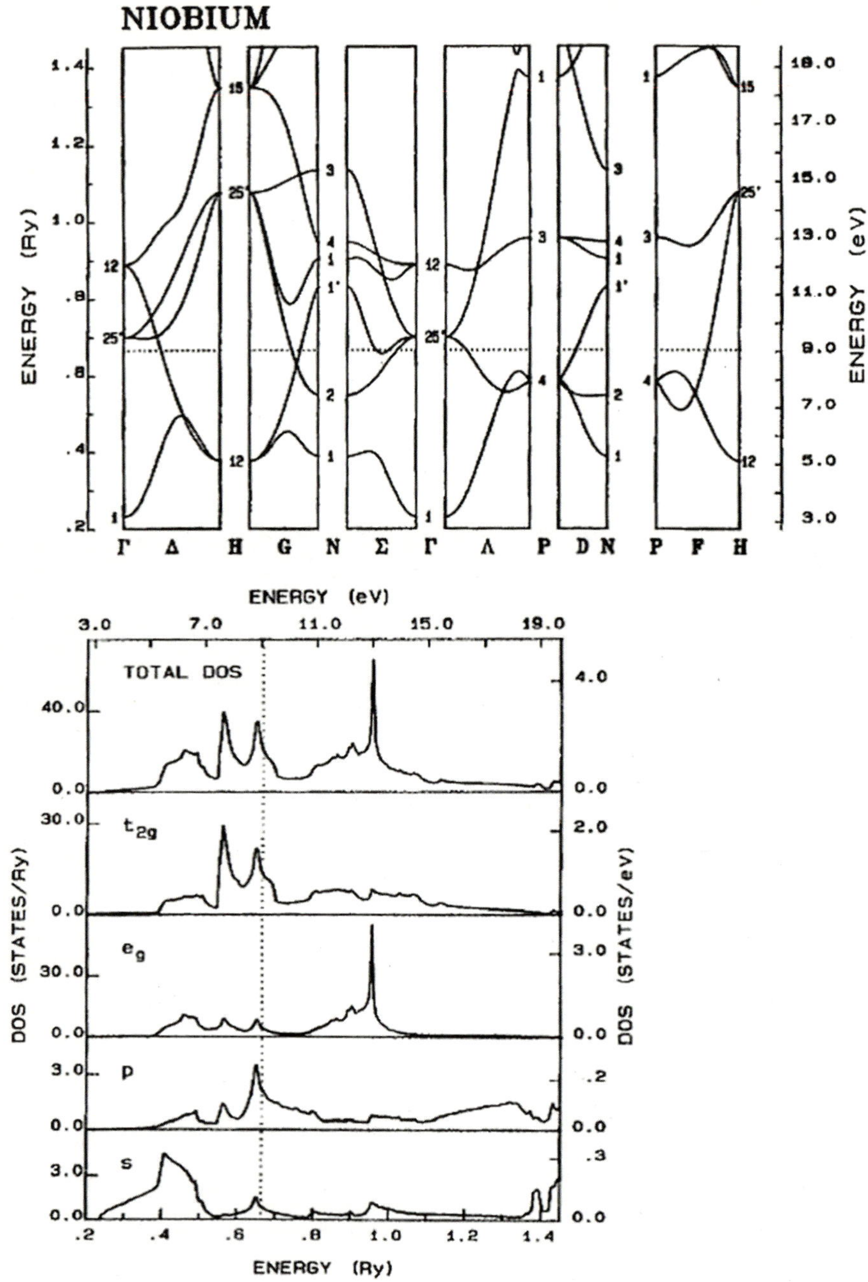

Fig. 5.7 Energy bands and density of states for Nb

5.4 Molybdenum

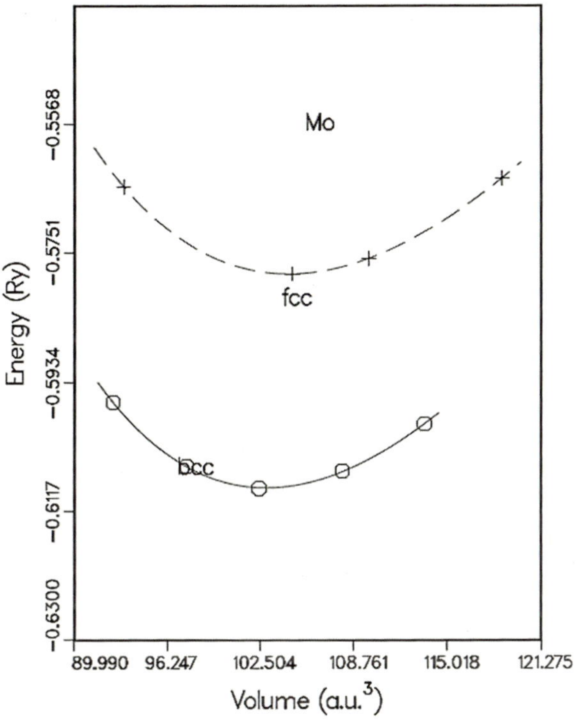

Fig. 5.8 Total energy of Mo

Table 5.26 Lattice constants and bulk modulus

	a (Bohr)	B (mbar)
fcc	7.475	2.564
bcc	5.904	2.884
exp	5.953	2.725
$\Delta E = 30.3$ mRy		

Table 5.27 Birch fit coefficients

	A_1	A_2	A_3
bcc	1.66166	−99.70914	1094.90895
fcc	1.46921	−90.79810	1006.73603

Table 5.28 Molybdenum bcc Z = 42 lattice constant = 5.95348 a.u.

Slater-Koster 3-center parameters	Orthogonal	Non-orthogonal	
	Energy integrals (Ry)	Energy integrals (Ry)	Overlap integrals
On site			
s, s (000)	1.16667	0.84380	
x, x (000)	1.69088	1.23852	
xy, xy (000)	0.88248	0.75992	
d2, d2 (000)	0.79511	0.70916	
First neighbor			
s, s (111)	−0.12088	−0.01057	0.12053
s, x (111)	0.08025	−0.03156	−0.10638
s, xy (111)	0.06125	0.01361	−0.04458
x, x (111)	0.09144	−0.04663	−0.04329
x, y (111)	0.04727	−0.08711	−0.12370
x, xy (111)	0.04312	−0.00969	−0.03333
x, yz (111)	0.06229	−0.01866	−0.07922
x, d1 (111)	0.00839	0.00606	−0.03863
xy, xy (111)	−0.02403	−0.01528	0.00996
xy, xz (111)	−0.03704	−0.01049	0.03290
xy, d2 (111)	0.00682	0.01361	−0.02352
d2, d2 (111)	0.03205	0.01515	−0.03982
Second neighbor			
s, s (200)	−0.02231	−0.03473	0.04900
s, x (200)	0.06534	−0.00775	−0.11333
s, d2 (002)	−0.06844	0.01132	0.08353
x, x (200)	0.19098	−0.13161	−0.24107
y, y (200)	0.03155	−0.02012	0.00179
x, xy (020)	0.01521	−0.02719	−0.00426
z, d2 (002)	−0.12115	0.06526	0.15471
xy, xy (200)	−0.01715	0.02812	0.00443
xy, xy (002)	0.00013	−0.00527	−0.00999
d2, d2 (002)	−0.06060	−0.00407	0.07860
d1, d1 (002)	0.00688	−0.00522	−0.00355
Third neighbor			
s, s (220)	0.01495		
s, x (220)	−0.01304		
s, xy (220)	−0.01157		
s, d2 (220)	−0.00673		
x, x (220)	−0.00785		
x, x (022)	0.01223		
x, y (220)	−0.00346		
x, xy (220)	−0.01035		
x, xy (022)	0.00234		
z, d2 (022)	−0.00026		
z, d1 (022)	−0.00620		
xy, xy (220)	0.01280		
xy, xy (022)	0.00153		
xy, xz (022)	0.00021		
xy, d2 (220)	0.00866		
d2, d2 (220)	0.00445		
d1, d1 (220)	−0.00606		

Table 5.29 Molybdenum bcc

	Orthogonal			Non-orthogonal		
Band	RMS error mRy	Maximum k	Deviation mRy	RMS error mRy	Maximum k	Deviation mRy
1	2.7	(260)	7.1	0.7	(000)	2.3
2	2.3	(131)	6.6	0.6	(131)	1.4
3	2.7	(044)	8.2	0.9	(161)	3.8
4	2.4	(332)	6.0	0.8	(140)	3.6
5	3.2	(131)	10.8	1.7	(131)	10.6
6	5.5	(171)	12.8	1.6	(140)	7.0
1–6	3.3			1.1		

Table 5.30 Energy values in Ry at selected k-points

	Orthogonal	APW	Non-orthogonal
GAMMA 1	0.24511	0.24163	0.24395
GAMMA 12	0.88073	0.88564	0.88520
GAMMA 15	2.91668	2.91980	2.91974
GAMMA 25′	0.68530	0.68668	0.68650
H1 (008)	2.17927	2.18385	2.18365
H12 (008)	0.36790	0.36238	0.36282
H15 (008)	1.45368	1.45427	1.45507
H25′ (008)	1.06984	1.06975	1.07182
N1 (044)	0.38340	0.38344	0.38339
N1 (044)	0.89283	0.89834	0.90001
N1 (044)	1.62464	1.62568	1.62558
N2 (044)	0.53472	0.53556	0.53499
N3 (044)	1.12737	1.12722	1.12908
N4 (044)	0.94496	0.94131	0.94149
N1′ (044)	0.88182	0.89002	0.88951
N3′ (044)	2.05834	2.06747	2.06736
N4′ (044)	1.63819	1.63391	1.63401
P1 (444)	1.47990	1.49079	1.49027
P3 (444)	0.94666	0.95033	0.95115
P4 (444)	0.58730	0.58799	0.58726
P4 (444)	1.59579	1.61218	1.61136
(341)	0.42330	0.42201	0.42166
(341)	0.54688	0.54894	0.54910
(341)	0.75329	0.75411	0.75424
(341)	0.88729	0.88760	0.88751
(341)	0.97047	0.97148	0.97208
(341)	1.18344	1.18315	1.18292

Table 5.31 Fermi level quantities (non-orthogonal fit)

Energy		Densities of states				Velocity	Plasmon energy
	Total	s	p	t_{2g}	e_g		
Ry		States/Ry/atom				cm/s	eV
0.7810	8.05	0.18	0.76	4.90	2.22	0.81 × 10E8	8.06

Integrated densities of states Total Electrons	s	p	t_{2g}	e_g
6.00	0.61	0.32	3.26	1.81

Table 5.32 Molybdenum bcc

Slater-Koster 2-center parameters	Orthogonal	Non-orthogonal	
	Energy integrals (Ry)	Energy integrals (Ry)	Overlap integrals
On site			
s	1.23823	0.86107	
p	1.67475	1.21343	
d1	0.85996	0.74577	
d2	0.77276	0.71692	
First neighbor			
(sss)	−0.11957	−0.03015	0.11301
(pps)	0.18564	−0.21887	−0.29295
(ppp)	0.04323	0.02897	0.07325
(dds)	−0.09948	−0.02091	0.09010
(ddp)	0.04647	0.02986	−0.05511
(ddd)	0.00093	−0.01047	0.00306
(sps)	0.10619	0.02966	0.17691
(sds)	−0.10802	−0.01589	0.08656
(pds)	−0.14707	−0.08248	−0.18119
(pdp)	0.02715	−0.00973	0.06931
Second neighbor			
(sss)	−0.02260	−0.01002	0.06355
(pps)	0.21961	−0.18001	−0.26745
(ppp)	0.00548	0.03248	0.02881
(dds)	−0.06055	−0.01165	0.06824
(ddp)	−0.00752	0.00996	−0.01885
(ddd)	0.00463	−0.00553	−0.00522
(sps)	0.08595	0.04719	0.13357
(sds)	−0.05454	0.00332	0.07287
(pds)	−0.09013	−0.04315	−0.13244
(pdp)	−0.00762	−0.00049	0.01689
Third neighbor			
(sss)	0.00859		
(pps)	0.02682		
(ppp)	−0.00745		
(dds)	0.01193		
(ddp)	−0.00031		
(ddd)	−0.00008		
(sps)	0.00482		
(sds)	0.00502		
(pds)	0.00248		
(pdp)	0.00124		

	Orthogonal			Non-orthogonal		
Band	RMS error mRy	Maximum k	Deviation mRy	RMS error mRy	Maximum k	Deviation mRy
1	5.4	(330)	13.8	1.0	(000)	3.3
2	9.9	(044)	28.8	1.0	(350)	2.0
3	7.3	(008)	17.1	1.5	(161)	4.9
4	10.0	(260)	30.7	1.9	(231)	4.3
5	6.9	(008)	17.1	2.2	(131)	11.1
6	9.7	(140)	20.0	2.1	(140)	6.0
1–6	8.4			1.7		

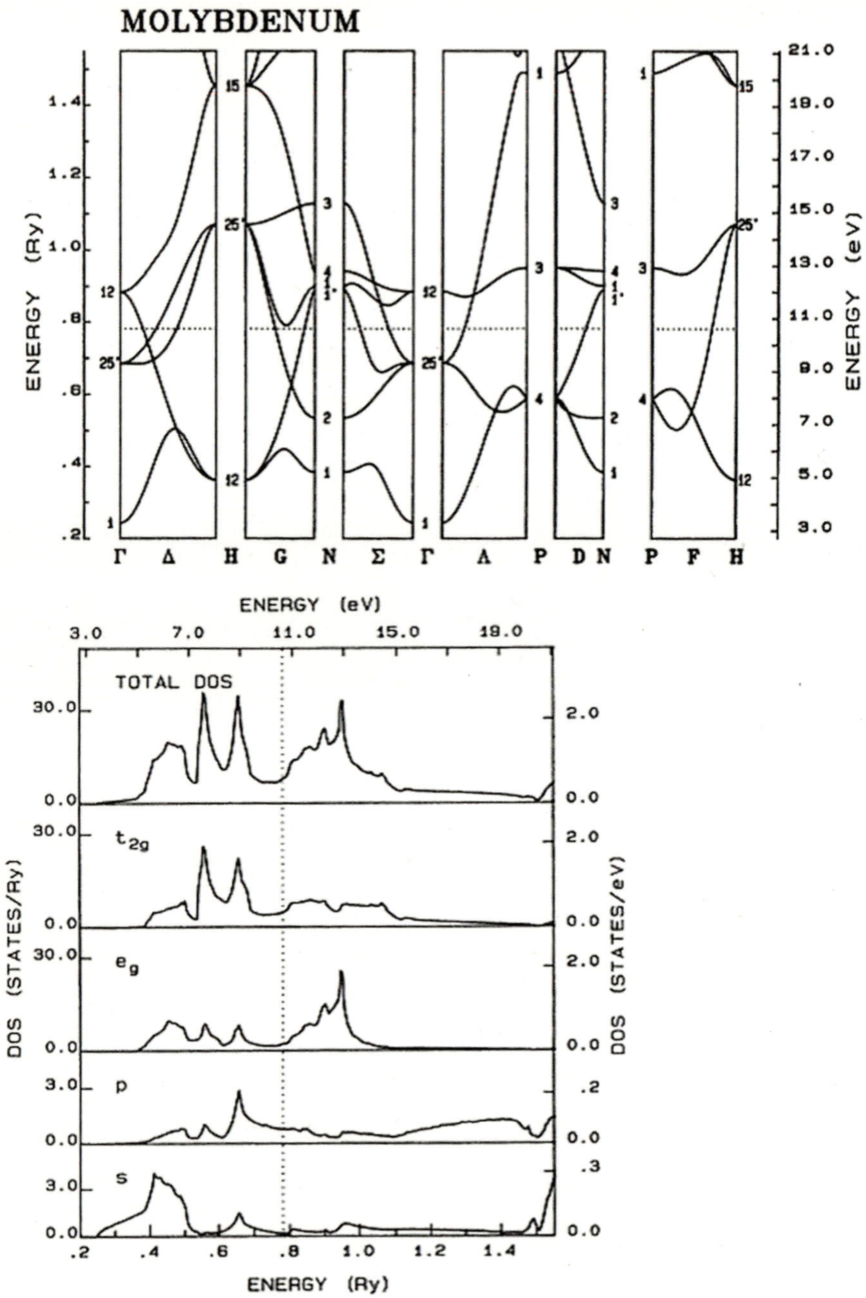

Fig. 5.9 Energy bands and density of states for Mo

5.5 Technetium

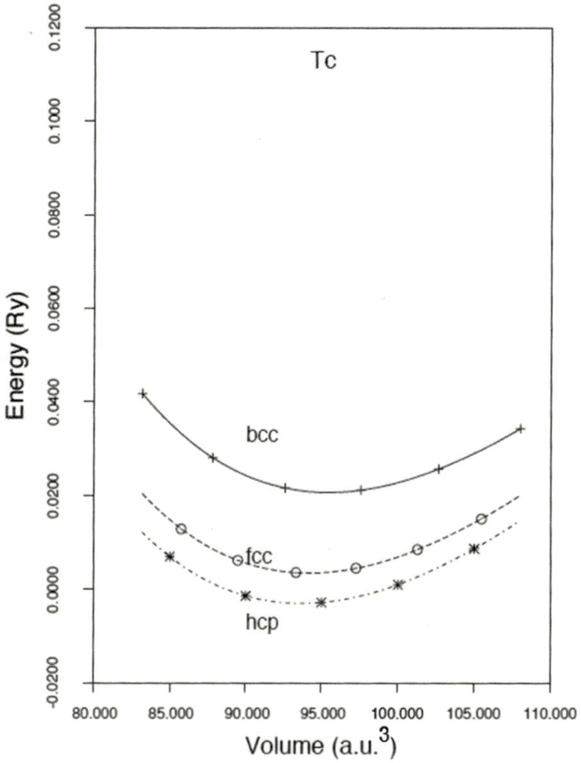

Fig. 5.10 Total energy of Tc

Table 5.33 Lattice constants and bulk modulus

	a (Bohr)	c (Bohr)	B (Mbar)
bcc	5.758		3.047
fcc	7.221		3.098
hcp	5.132	8.211	3.076
exp	5.178	8.315	2.97
$\Delta E = E_{hcp} - E_{fcc} = 6.5$ mRy			

Table 5.34 Birch fit coefficients

	A_1	A_2	A_3	A_4
bcc	1.57875	−51.16375	98.63349	6064.07774
fcc	1.52107	−48.05661	39.19879	6319.95720
hcp	1.57214	−52.01177	135.85977	5503.74794

Table 5.35 Technetium hcp Z = 43 lattice constants = 5.1779 a.u. 8.3148 a.u.

Slater-Koster 3-center parameters			
	Orthogonal		
	Energy integrals (Ry)		Energy integrals (Ry)
On site			
s, s (0)	1.28303		
x, x (0)	1.86617		
z, z (0)	1.48051		
xy, xy (0)	0.76103		
yz, yz (0)	0.74938		
d2, d2 (0)	0.70557		
s, d2 (0)	−0.09063		
x, xy (0)	−0.08135		
First neighbor			
s, s (R)	−0.03233	s, s (1)	−0.13907
s, x (R)	0.05162	s, y (T)	−0.10431
s, y (R)	0.06974	s, z (T)	0.11372
s, xy (R)	0.00424	s, d1 (T)	−0.03146
s, d1 (R)	0.07124	s, yz (T)	−0.07956
s, d2 (R)	−0.03437	s, d2 (T)	0.02599
x, x (R)	0.01067	x, x (T)	−0.04070
y, y (R)	−0.23489	y, y (T)	0.02905
x, y (R)	0.20308	y, z (T)	−0.16176
x, xy (R)	0.03221	x, xy (T)	0.00667
y, d1 (R)	−0.01026	y, d1 (T)	0.00566
x, d1 (R)	0.09523	x, xz (T)	−0.03964
y, xy (R)	−0.01908	y, yz (T)	0.04546
y, d2 (R)	0.01101	y, d2 (T)	−0.06292
x, d2 (R)	−0.02798	z, z (T)	0.06487
z, z (R)	−0.04278	z, d1 (T)	−0.03881
z, yz (R)	−0.00911	z, yz (T)	−0.09621
z, xz (R)	−0.02388	z, d2 (T)	0.02140
xy, xy (R)	0.02348	xy, xy (T)	0.01210
d1, d1 (R)	−0.05929	d1, d1 (T)	−0.00332
xy, d1 (R)	−0.02455	xy, yz (T)	−0.00991
d1, d2 (R)	0.02646	yz, d1 (T)	−0.02098
xy, d2 (R)	−0.00126	d1, d2 (T)	0.03989
yz, yz (R)	−0.00185	yz, yz (T)	−0.06004
xz, xz (R)	0.02926	xz, xz (T)	0.02913
yz, xz (R)	−0.01117	yz, d2 (T)	0.05521
d2, d2 (R)	−0.02926	d2, d2 (T)	0.00930

Table 5.36 Technetium hcp

Band	Orthogonal RMS error mRy	Maximum k	Deviation mRy
1	6.1	(16 0 0)	11.2
2	4.9	(0 12 0)	11.1
3	4.9	(0 12 12)	12.4
4	7.0	(24 0 0)	22.4
5	7.9	(0 12 12)	25.4
6	6.6	(0 12 12)	25.4
7	4.0	(24 0 9)	10.2
8	6.7	(24 0 0)	15.1
9	6.4	(24 0 3)	15.8
10	5.7	(16 0 0)	16.7
11	6.0	(16 0 0)	11.5
12	13.8	(0 18 0)	61.8
1-12	7.06		

Table 5.37 Energy values in Ry at selected k-points

	Orthogonal	APW
GAMMA 1+	0.20277	0.19481
GAMMA 1+	0.63762	0.63158
GAMMA 3+	0.83460	0.84289
GAMMA 5+	0.67908	0.67814
GAMMA 6+	0.73887	0.73148
GAMMA 4−	0.34435	0.34390
GAMMA 5−	0.92432	0.92195
GAMMA 6−	0.62211	0.62234
M1+	0.33886	0.33862
M1+	0.67105	0.66079
M1+	1.13621	1.12918
M2+	0.99345	0.98258
M3+	0.37767	0.37695
M4+	0.98397	0.98697
M1−	0.57495	0.59731
M2−	0.40304	0.40597
M2−	0.69946	0.70283
M2−	1.26171	1.26342
M3−	0.94027	0.95541
M4−	0.65236	0.65539
A1	0.33780	0.33762
A1	0.59939	0.59989
A3	0.61401	0.61399
A3	0.86956	0.87602
L1	0.33393	0.33804
L1	0.46972	0.46958
L1	0.84127	0.84156
L1	1.05741	1.05853
L2	0.75459	0.74757
L2	0.97458	0.97368

Table 5.38 technetium hcp z = 43 lattice constants = 5.1779 a.u. 8.3148 a.u.

Slater-Koster 2-center parameters	Orthogonal	Non-orthogonal	
	Energy integrals (Ry)	Energy integrals (Ry)	Overlap integrals
On site			
s	1.20940	0.82416	
p	1.33436	1.31908	
d0	0.72227	0.66333	
d1	0.71415	0.66474	
d2	0.72638	0.67268	
pd	−0.02880	−0.03383	−0.01136
First neighbor			
(sss)	−0.07919	−0.01576	0.13642
(sps)	0.08644	−0.03171	−0.16272
(sds)	0.06255	0.01979	−0.07706
(pps)	0.05472	−0.12672	−0.18122
(ppp)	−0.06303	0.03236	0.10306
(pds)	0.08304	−0.00462	−0.06780
(pdp)	−0.03658	−0.00253	0.05467
(dds)	−0.08814	−0.02571	0.08020
(ddp)	0.04391	0.02232	−0.04857
(ddd)	−0.00572	−0.00004	0.01063
Second neighbor			
(sss)	−0.00475	−0.01632	0.01968
(sps)	0.00238	0.00812	−0.02904
(sds)	−0.00069	0.00763	−0.02136
(pps)	0.05303	0.11498	0.00761
(ppp)	−0.01190	0.07656	0.08040
(pds)	0.01626	0.05258	0.00607
(pdp)	−0.01335	0.02004	0.03536
(dds)	−0.00814	−0.02555	−0.00978
(ddp)	0.00330	0.00331	0.00058
(ddd)	0.00048	−0.00013	−0.00193
Third neighbor			
(sss)	−0.01211	0.02051	0.06264
(sps)	0.01378	−0.07426	−0.10778
(sds)	−0.00306	0.01156	0.02989
(pps)	0.01835	0.09267	0.05672
(ppp)	0.02360	0.06200	0.01774
(pds)	0.00175	0.05988	0.06451
(pdp)	−0.00091	0.04190	0.04406
(dds)	−0.00027	−0.01541	−0.01975
(ddp)	0.00060	−0.00938	−0.01134
(ddd)	−0.00122	0.00583	0.00650

Table 5.39 Technetium hcp

	Orthogonal			Non-orthogonal		
Band	RMS error mRy	Maximum k	Deviation mRy	RMS error mRy	Maximum k	Deviation mRy
1	13.6	(24 0 0)	26.8	3.1	(0 18 0)	7.1
2	7.3	(0 24 12)	15.2	2.9	(24 0 9)	6.9
3	8.6	(24 0 9)	20.7	3.5	(24 0 6)	8.4
4	8.8	(0 24 12)	19.3	3.2	(24 0 9)	8.2
5	5.5	(8 0 0)	12.5	3.2	(16 0 0)	9.7
6	5.4	(0 6 0)	13.8	4.2	(16 0 0)	9.7
7	8.9	(0 12 12)	21.7	3.7	(8 0 12)	10.4
8	11.5	(16 0 9)	22.2	4.3	(8 0 12)	10.4
9	9.5	(24 0 0)	23.9	3.5	(0 12 12)	7.6
10	7.9	(0 18 0)	14.5	4.0	(12 0 0)	8.5
11	16.6	(16 0 12)	38.8	8.6	(16 0 9)	28.7
12	24.8	(0 18 0)	75.7	8.9	(16 0 9)	28.7
1–12	11.90			4.84		

Table 5.40 Energy values in Ry at selected k-points

	Orthogonal	APW	Non–orthogonal
GAMMA 1+	0.20626	0.19481	0.20061
GAMMA 1+	0.62273	0.63158	0.63017
GAMMA 3+	0.84158	0.84289	0.83844
GAMMA 5+	0.67840	0.67814	0.67387
GAMMA 6+	0.73598	0.73148	0.73690
GAMMA 4–	0.34557	0.34390	0.34623
GAMMA 5–	0.92386	0.92195	0.92293
GAMMA 6–	0.61532	0.62234	0.61739
M1+	0.31186	0.33862	0.34287
M1+	0.65807	0.66079	0.65815
M1+	1.15165	1.12918	1.12907
M2+	0.95867	0.98258	0.98194
M3+	0.37323	0.37695	0.37220
M4+	0.97903	0.98697	0.98419
M1–	0.59417	0.59731	0.59640
M2–	0.41119	0.40597	0.41350
M2–	0.69719	0.70283	0.70602
M2–	1.28172	1.26342	1.26405
M3–	0.94900	0.95541	0.95726
M4–	0.65417	0.65539	0.65624
A1	0.34058	0.33762	0.34005
A1	0.59654	0.59989	0.59657
A3	0.61668	0.61399	0.61545
A3	0.86246	0.87602	0.87375
L1	0.35323	0.33804	0.33534
L1	0.45026	0.46958	0.46691
L1	0.84594	0.84156	0.83978
L1	1.05102	1.05853	1.05672
L2	0.75800	0.74757	0.74477
L2	0.97239	0.97368	0.97629

Table 5.41 Fermi level quantities (non-orthogonal fit)

Energy		Densities of states			
	Total	s	p	t_{2g}	e_g
Ry		States/Ry/atom			
0.7572	13.63	0.27	0.92	7.26	5.17

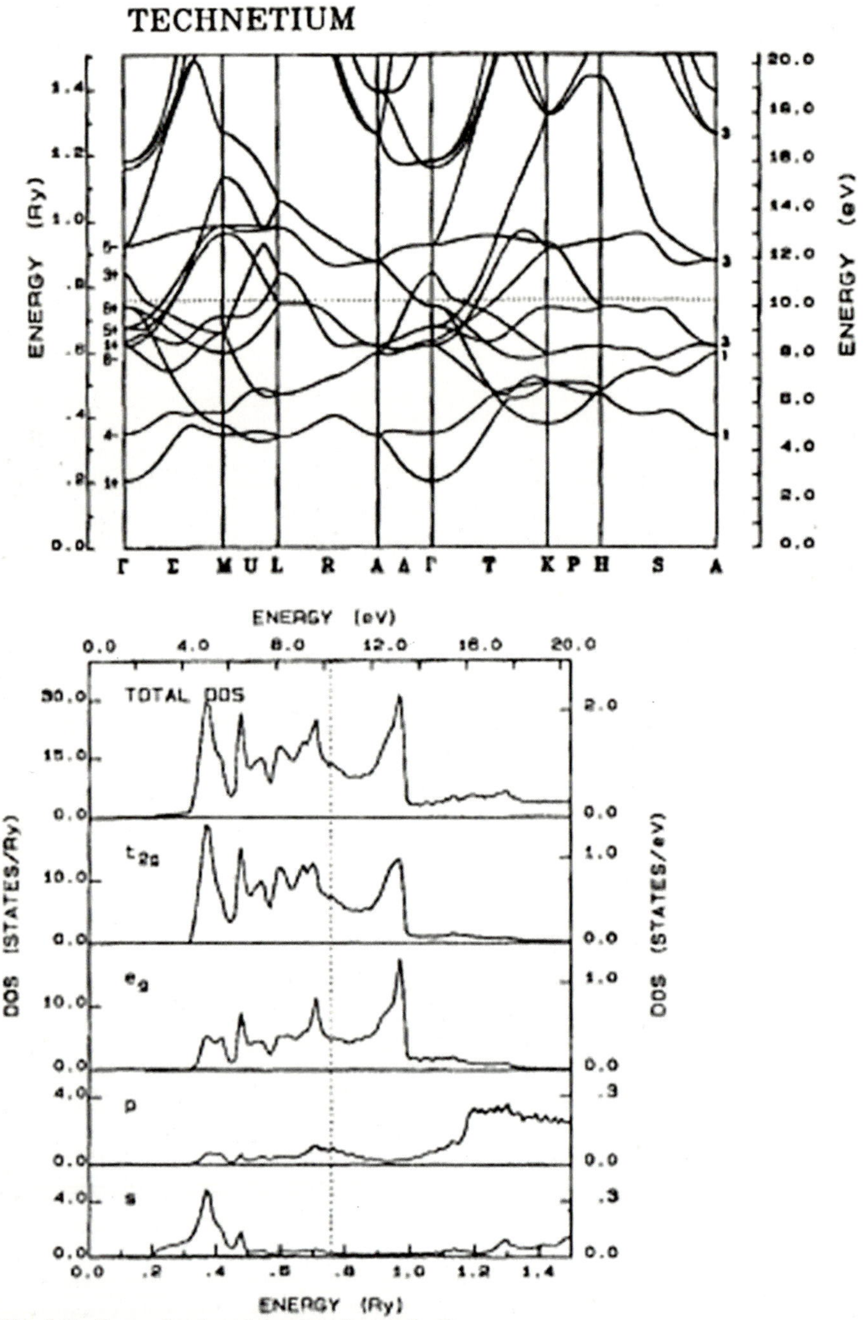

Fig. 5.11 Energy bands and density of states for Tc

5.6 Ruthenium

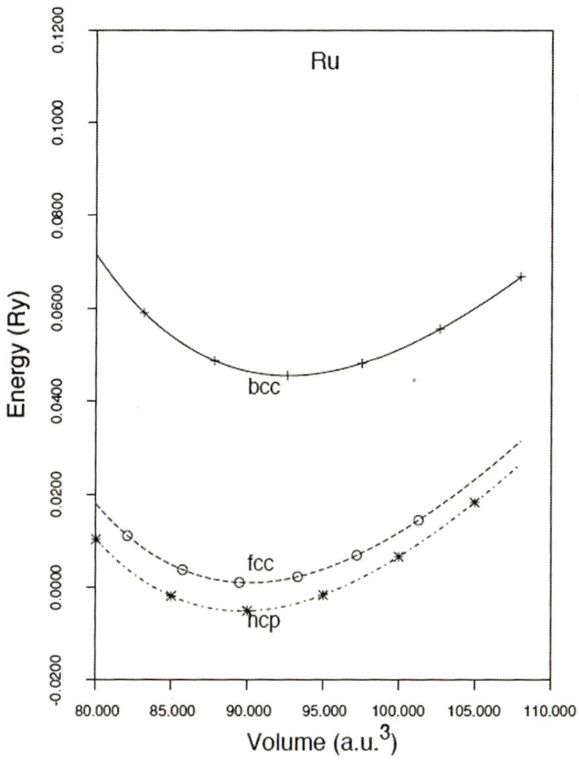

Fig. 5.12 Total energy for Ru

Table 5.42 Lattice constants and bulk modulus

	a (Bohr)	c (Bohr)	B (Mbar)
bcc	5.701		3.328
fcc	7.117		3.522
hcp	5.090	7.992	3.512
exp	5.121	8.088	3.208
$\Delta E = E_{hcp} - E_{fcc} = 11.1$ mRy			

Table 5.43 Birch fit coefficients

	A_1	A_2	A_3	A_4
bcc	1.28348	−27.76930	−419.76902	9609.93822
fcc	1.71536	−54.58832	116.72923	5786.33698
hcp	3.35614	−167.96106	470.10384	46649.06543

Table 5.44 Ruthenium hcp Z = 44 lattice constants = 5.1096 a.u. 8.0912 a.u.

Slater-Koster 3-center parameters			
	Orthogonal		
	Energy integrals (Ry)		Energy integrals (Ry)
On site			
s, s (0)	1.21181		
x, x (0)	2.02293		
z, z (0)	1.44639		
xy, xy (0)	0.68343		
yz, yz (0)	0.69691		
d2, d2 (0)	0.65073		
s, d2 (0)	−0.09944		
x, xy (0)	0.00059		
First neighbor			
s, s (R)	−0.03779	s, s (T)	−0.12820
s, x (R)	0.04906	s, y (T)	−0.11240
s, y (R)	0.06197	s, z (T)	0.11196
s, xy (R)	0.01553	s, d1 (T)	−0.03263
s, d1 (R)	0.06507	s, yz (T)	−0.08812
s, d2 (R)	−0.03429	s, d2 (T)	0.02951
x, x (R)	0.00532	x, x (T)	−0.05495
y, y (R)	−0.29033	y, y (T)	0.02067
x, y (R)	0.23742	y, z (T)	−0.17461
x, xy (R)	0.04002	x, xy (T)	0.01692
y, d1 (R)	−0.01621	y, d1 (T)	0.01398
x, d1 (R)	0.09781	x, xz (T)	−0.03414
y, xy (R)	−0.03426	y, yz (T)	0.05062
y, d2 (R)	0.00469	y, d2 (T)	−0.06950
x, d2 (R)	−0.03128	z, z (T)	0.06989
z, z (R)	−0.03355	z, d1 (T)	−0.03409
z, yz (R)	−0.01163	z, yz (T)	−0.10410
z, xz (R)	−0.01619	z, d2 (T)	0.02747
xy, xy (R)	0.02861	xy, xy (T)	0.01661
d1, d1 (R)	−0.05407	d1, d1 (7)	−0.00102
xy, d1 (R)	−0.01304	xy, yz (T)	−0.01333
d1, d2 (R)	0.02638	yz, d1 (T)	−0.01613
xy, d2 (R)	−0.00337	d1, d2 (T)	0.03494
yz, yz (R)	0.00089	yz, yz (T)	−0.05699
xz, xz (R)	0.01796	xz, xz (T)	0.02731
yz, xz (R)	−0.01437	yz, d2 (T)	0.05312
d2, d2 (R)	−0.02455	d2, d2 (T)	0.00674

Table 5.45 Ruthenium hcp

Band	Orthogonal RMS error mRy	Maximum k	Deviation mRy
1	6.0	(0 0 0)	10.5
2	4.4	(24 0 9)	8.2
3	4.1	(0 6 0)	8.5
4	6.8	(24 0 0)	20.2
5	7.4	(0 12 12)	22.8
6	6.4	(0 12 12)	22.8
7	4.1	(24 0 9)	10.1
8	7.0	(24 0 3)	15.4
9	5.7	(8 0 0)	10.3
10	6.2	(16 0 0)	15.1
11	4.8	(12 0 0)	11.8
12	16.1	(0 18 0)	74.9
1–12	7.26		

Table 5.46 Energy values in Ry at selected k-points

	Orthogonal	APW
GAMMA 1+	0.17172	0.16121
GAMMA 1+	0.58802	0.58148
GAMMA 3+	0.82577	0.83221
GAMMA 5+	0.61259	0.61571
GAMMA 6+	0.66441	0.65736
GAMMA 4−	0.30277	0.30218
GAMMA 5−	0.84246	0.84266
GAMMA 6−	0.55944	0.56254
M1+	0.29516	0.29729
M1+	0.62387	0.61410
M1+	1.05368	1.05409
M2+	0.91049	0.89934
M3+	0.34162	0.33857
M4+	0.89196	0.89373
M1−	0.51393	0.53415
M2−	0.35732	0.36494
M2−	0.64524	0.64639
M2−	1.20526	1.20758
M3−	0.84785	0.86294
M4−	0.60490	0.60549
A1	0.30212	0.30358
A1	0.55625	0.55772
A3	0.55338	0.55193
A3	0.79451	0.79833
L1	0.29846	0.30060
L1	0.42717	0.42504
L1	0.78786	0.78898
L1	1.00209	1.00239
L2	0.68672	0.67912
L2	0.88042	0.88118

Table 5.47 Ruthenium hcp Z = 44 lattice constants = 5.1096 a.u. 8.0912 a.u.

Slater-Koster 2-center parameters	Orthogonal	Non-orthogonal	
	Energy integrals (Ry)	Energy integrals (Ry)	Overlap integrals
On site			
s	1.21824	0.78265	
p	1.34284	1.45058	
d0	0.65178	0.60589	
d1	0.64968	0.60847	
d2	0.66649	0.62177	
pd	−0.02537	−0.00049	0.06862
First neighbor			
(sss)	−0.07589	−0.00629	0.12745
(sps)	0.09273	−0.04219	−0.15834
(sds)	0.06440	0.02653	−0.05689
(pps)	0.07019	−0.09797	−0.15998
(ppp)	−0.07048	−0.14582	−0.03236
(pds)	0.07927	0.00473	−0.05156
(pdp)	−0.03384	−0.04133	0.00820
(dds)	−0.08143	−0.03208	0.07081
(ddp)	0.04049	0.02333	−0.04362
(ddd)	−0.00497	−0.00183	0.00788
Second neighbor			
(sss)	−0.01803	−0.02918	0.01565
(sps)	0.00423	0.01228	−0.02754
(sds)	0.00000	0.00170	−0.03310
(pps)	0.03513	−0.02400	−0.13041
(ppp)	−0.02361	0.05081	0.07688
(pds)	0.01856	0.02885	−0.01852
(pdp)	−0.01297	0.01388	0.03782
(dds)	−0.00934	−0.02257	−0.00767
(ddp)	0.00306	0.00826	0.00625
(ddd)	0.00121	−0.00445	−0.00881
Third neighbor			
(sss)	−0.01112	−0.05792	−0.03653
(sps)	0.00631	0.05029	0.02166
(sds)	−0.00644	0.01029	0.02789
(pps)	0.01514	0.14495	0.10826
(ppp)	0.00819	0.10235	0.04048
(pds)	0.00418	0.10954	0.15105
(pdp)	−0.00325	0.05284	0.05528
(dds)	−0.00093	−0.01956	−0.02986
(ddp)	−0.00016	−0.00955	−0.01246
(ddd)	−0.00002	0.00837	0.01062

Table 5.48 Ruthenium hcp

	Orthogonal			Non-orthogonal		
Band	RMS error mRy	Maximum k	Deviation mRy	RMS error mRy	Maximum k	Deviation mRy
1	13.4	(24 0 0)	25.1	3.2	(16 0 12)	7.3
2	7.1	(24 0 6)	14.8	4.0	(8 0 0)	10.1
3	8.4	(24 0 6)	19.9	5.6	(24 0 6)	14.4
4	8.5	(24 0 9)	21.4	5.3	(24 0 9)	13.3
5	8.0	(8 0 0)	16.0	5.0	(16 0 0)	9.7
6	6.9	(0 6 0)	17.6	5.2	(12 0 0)	9.8
7	9.7	(0 12 12)	25.7	4.7	(0 6 0)	14.5
8	11.6	(0 12 12)	25.7	5.8	(8 0 12)	12.5
9	10.0	(24 0 0)	22.3	5.4	(0 0 0)	11.1
10	7.9	(16 0 12)	17.1	5.3	(8 0 0)	14.1
11	14.8	(24 0 0)	26.1	4.3	(24 0 6)	8.9
12	15.8	(0 18 0)	30.3	3.5	(8 0 0)	8.8
1–12	10.58			4.84		

Table 5.49 Energy values in Ry at selected k-points

	Orthogonal	APW	Non–orthogonal
GAMMA 1+	0.17668	0.16121	0.16309
GAMMA 1+	0.57251	0.58148	0.58032
GAMMA 3+	0.83465	0.83221	0.82950
GAMMA 5+	0.61455	0.61571	0.61413
GAMMA 6+	0.66789	0.65736	0.66843
GAMMA 4−	0.30810	0.30218	0.30246
GAMMA 5−	0.84396	0.84266	0.84297
GAMMA 6−	0.55500	0.56254	0.55489
M1+	0.27219	0.29729	0.30174
M1+	0.62050	0.61410	0.61624
M1+	1.08021	1.05409	1.04711
M2+	0.87141	0.89934	0.89730
M3+	0.33940	0.33857	0.33235
M4+	0.89097	0.89373	0.89066
M1−	0.53811	0.53415	0.53923
M2−	0.36597	0.36494	0.37188
M2−	0.63503	0.64639	0.64655
M2−	1.23067	1.20758	1.20864
M3−	0.86361	0.86294	0.86744
M4−	0.59476	0.60549	0.60331
A1	0.30948	0.30358	0.30603
A1	0.55009	0.55772	0.55525
A3	0.55536	0.55193	0.55058
A3	0.78543	0.79833	0.79503
L1	0.30603	0.30060	0.29566
L1	0.40533	0.42504	0.41985
L1	0.78592	0.78898	0.78555
L1	0.99483	1.00239	1.00159
L2	0.69259	0.67912	0.68051
L2	0.88210	0.88118	0.88772

Table 5.50 Fermi level quantities (non-orthogonal fit)

Energy		Densities of states			
	Total	s	p	t_{2g}	e_g
Ry		States/Ry/atom			
0.7739	11.52	0.16	0.47	5.56	5.34

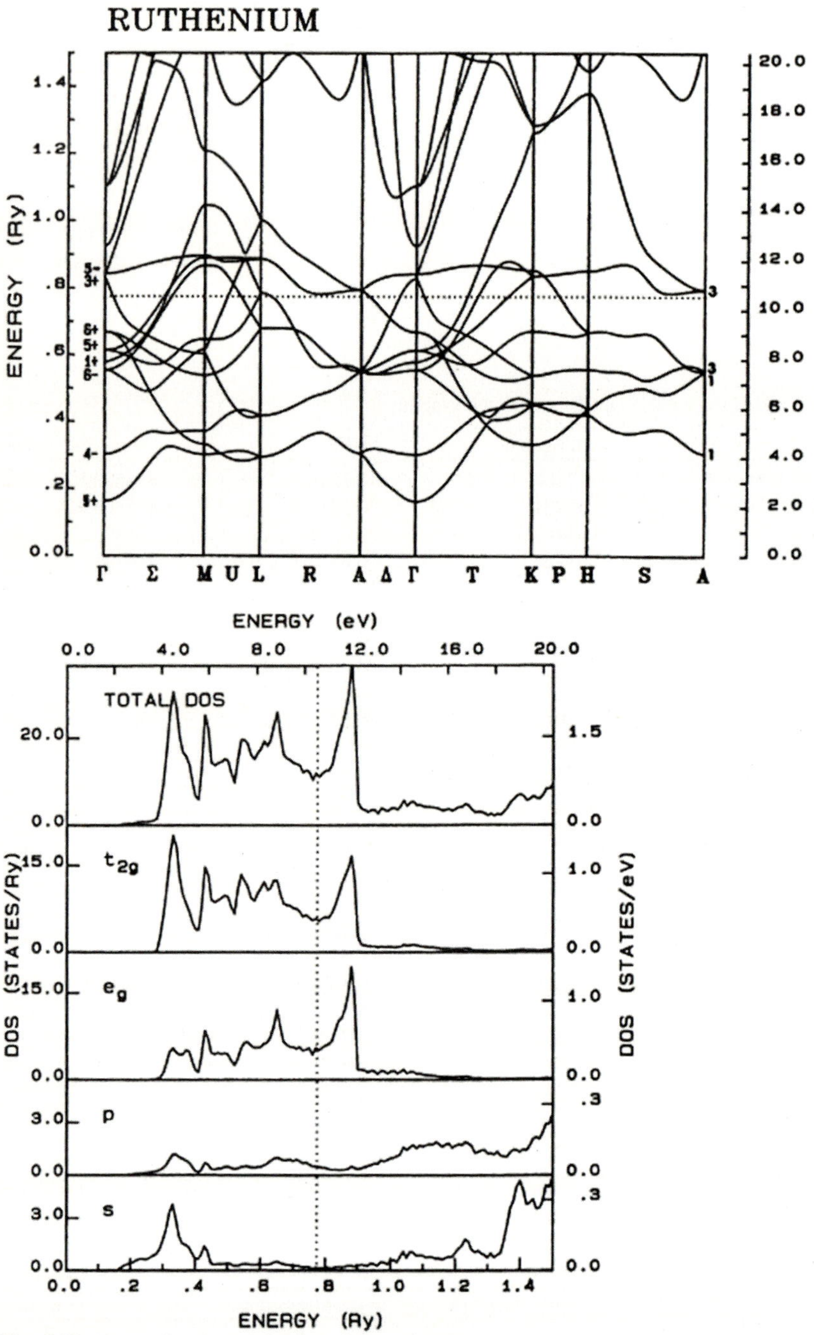

Fig. 5.13 Energy bands and density of states for Ru

5.7 Rhodium

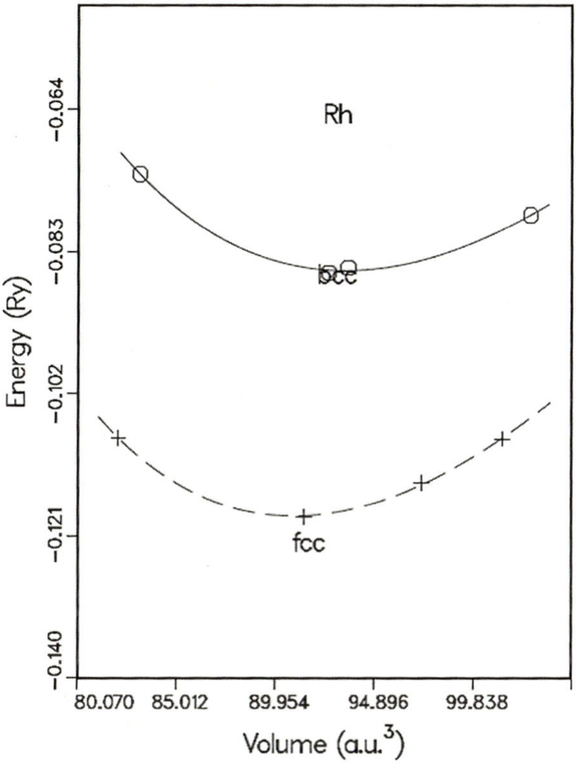

Fig. 5.14 Total energy of Rh

Table 5.51 Lattice constants and bulk modulus

	a (Bohr)	B (mbar)
bcc	5.717	2.792
fcc	7.137	3.073
exp	7.181	2.704
Δ E = 32.8 mRy		

Table 5.52 Birch fit coefficients

	A_1	A_2	A_3
bcc	1.90891	−82.12673	845.44938
fcc	2.01728	−86.32929	872.43591

Table 5.53 Rhodium fcc Z = 45 lattice constant = 7.18000 a.u.

Slater-Koster 3-center parameters	Orthogonal	Non-orthogonal	
	Energy integrals (Ry)	Energy integrals (Ry)	Overlap integrals
On site			
s, s (000)	1.08393	0.96743	
x, x (000)	1.64885	1.43149	
xy, xy (000)	0.51938	0.49693	
d2, d2 (000)	0.48849	0.48121	
First neighbor			
s, s (110)	−0.08691	−0.06618	0.02798
s, x (110)	0.09378	0.05701	−0.04794
s, xy (110)	−0.06236	−0.06467	0.02275
s, d2 (110)	0.03395	0.03893	0.01756
x, x (110)	0.08131	0.05106	−0.03888
x, x (011)	−0.00559	0.07672	0.07017
x, y (110)	0.12739	0.01803	−0.08035
x, xy (110)	−0.06316	−0.05147	0.01655
x, xy (011)	0.00776	−0.01567	−0.04606
z, d2 (011)	0.00830	−0.00404	−0.02928
z, d1 (011)	0.04037	0.02413	−0.01839
xy, xy (110)	−0.04819	−0.01716	0.05726
xy, xy (011)	0.01263	0.00979	−0.01492
xy, xz (011)	0.01146	−0.00055	−0.02898
xy, d2 (110)	0.02533	0.02767	0.00732
d2, d2 (110)	−0.02028	−0.01613	0.00706
d1, d1 (110)	0.03426	0.02768	−0.01436
Second neighbor			
s, s (200)	0.00686	−0.01199	−0.02061
s, x (200)	−0.01674	0.04805	0.03235
s, d2 (002)	0.00650	−0.01987	−0.03920
x, x (200)	−0.03404	0.07581	0.01432
y, y (200)	0.00094	0.01397	0.01673
x, xy (020)	0.00010	−0.01563	−0.00864
z, d2 (002)	0.00616	−0.02734	−0.03296
xy, xy (200)	−0.00246	−0.01542	−0.01357
xy, xy (002)	0.00061	−0.00056	−0.00042
d2, d2 (002)	−0.00410	−0.00828	−0.00642
d1, d1 (002)	0.00016	0.00591	0.01056

Table 5.54 Rhodium fcc

	Orthogonal			Non-orthogonal		
Band	RMS error mRy	Maximum k	Deviation mRy	RMS error mRy	Maximum k	Deviation mRy
1	1.6	(055)	4.9	0.5	(174)	2.0
2	2.9	(007)	5.6	0.9	(008)	2.1
3	2.2	(048)	5.4	1.2	(007)	4.4
4	2.6	(118)	10.8	2.0	(118)	10.3
5	2.1	(044)	4.1	1.0	(000)	2.9
6	4.2	(048)	13.7	1.3	(048)	4.4
1–6	2.7			1.2		

Table 5.55 Energy values in Ry at selected k-points

	Orthogonal	APW	Non-orthogonal
GAMMA 1	0.08219	0.08312	0.08361
GAMMA 12	0.56052	0.55810	0.56105
GAMMA 15	2.21268	2.21018	2.21026
GAMMA 25	0.41901	0.41924	0.42082
X1 (008)	0.17379	0.17476	0.17579
X1 (008)	1.52948	1.52642	1.52639
X2 (008)	0.66687	0.66370	0.66528
X3 (008)	0.21698	0.21998	0.22208
X5 (008)	0.70350	0.70162	0.70234
X4′ (008)	0.91170	0.91041	0.91116
X5′ (008)	1.60688	1.63685	1.63477
L1 (444)	0.21168	0.21231	0.21304
L1 (444)	1.26740	1.27353	1.27349
L3 (444)	0.40766	0.40792	0.40910
L3 (444)	0.66652	0.66956	0.67000
L2′ (444)	0.69403	0.69874	0.69785
L3′ (444)	2.22273	2.22328	2.22302
W1 (048)	0.54510	0.54404	0.54593
W1 (048)	1.47435	1.48020	1.48004
W3 (048)	0.37836	0.37299	0.37458
W3 (048)	1.34725	1.36092	1.36535
W1′ (048)	0.70106	0.70214	0.70381
W2′ (048)	0.26828	0.26655	0.26704
W2′ (048)	1.81379	1.54377	1.54502
EVEN(224)	0.25228	0.25244	0.25252
EVEN (224)	0.40148	0.39778	0.39924
EVEN (224)	0.43375	0.43173	0.43037
EVEN (224)	0.67672	0.67835	0.68007
ODD (224)	0.45747	0.46229	0.46163
ODD (224)	0.65090	0.65006	0.64929

Table 5.56 Fermi level quantities (non-orthogonal fit)

Energy		Densities of states				Velocity	Plasmon energy
	Total	s	p	t_{2g}	e_g		
Ry		States/Ry/atom				cm/s	eV
0.6335	18.68	0.27	0.43	11.56	6.42	0.59 × 10E8	9.57

Integrated densities of states				
Total Electrons	s	p	t_{2g}	e_g
9.00	0.60	0.41	4.54	3.45

Table 5.57 Rhodium fcc

Slater-Koster 2-center parameters	Orthogonal	Non-orthogonal	
	Energy integrals (Ry)	Energy integrals (Ry)	Overlap integrals
On site			
s	1.12862	0.68690	
p	1.50833	1.15107	
d1	0.49824	0.46569	
d2	0.48265	0.46465	
First neighbor			
(sss)	−0.08700	−0.02991	0.11488
(pps)	0.17058	−0.18690	−0.29592
(ppp)	−0.00928	−0.05110	0.02816
(dds)	−0.06303	−0.03417	0.05660
(ddp)	0.03409	0.02417	−0.03176
(ddd)	−0.00610	−0.00416	0.00325
(sps)	0.11907	−0.03829	−0.18905
(sds)	−0.05607	−0.02648	0.06965
(pds)	−0.07627	0.00062	0.10527
(pdp)	0.02397	0.02031	−0.03555
Second neighbor			
(sss)	−0.00094	−0.02099	0.00389
(pps)	0.04716	0.08250	−0.00343
(ppp)	−0.00153	−0.01952	0.00055
(dds)	−0.00405	−0.01655	−0.01262
(ddp)	0.00206	0.00127	−0.00128
(ddd)	−0.00013	0.00073	0.00173
(sps)	0.00790	0.04169	0.00138
(sds)	−0.01022	−0.01584	−0.00128
(pds)	−0.01194	−0.03575	−0.00872
(pdp)	0.00529	0.00430	−0.00095

	Orthogonal			Non-orthogonal		
Band	RMS error mRy	Maximum k	Deviation mRy	RMS error mRy	Maximum k	Deviation mRy
1	7.5	(055)	14.5	1.0	(055)	3.2
2	9.4	(007)	22.2	1.2	(264)	2.3
3	5.1	(033)	10.3	1.4	(007)	4.5
4	7.1	(084)	18.0	2.2	(118)	11.0
5	7.6	(044)	19.0	1.4	(000)	3.6
6	5.9	(044)	13.7	1.2	(000)	3.6
1–6	7.2			1.4		

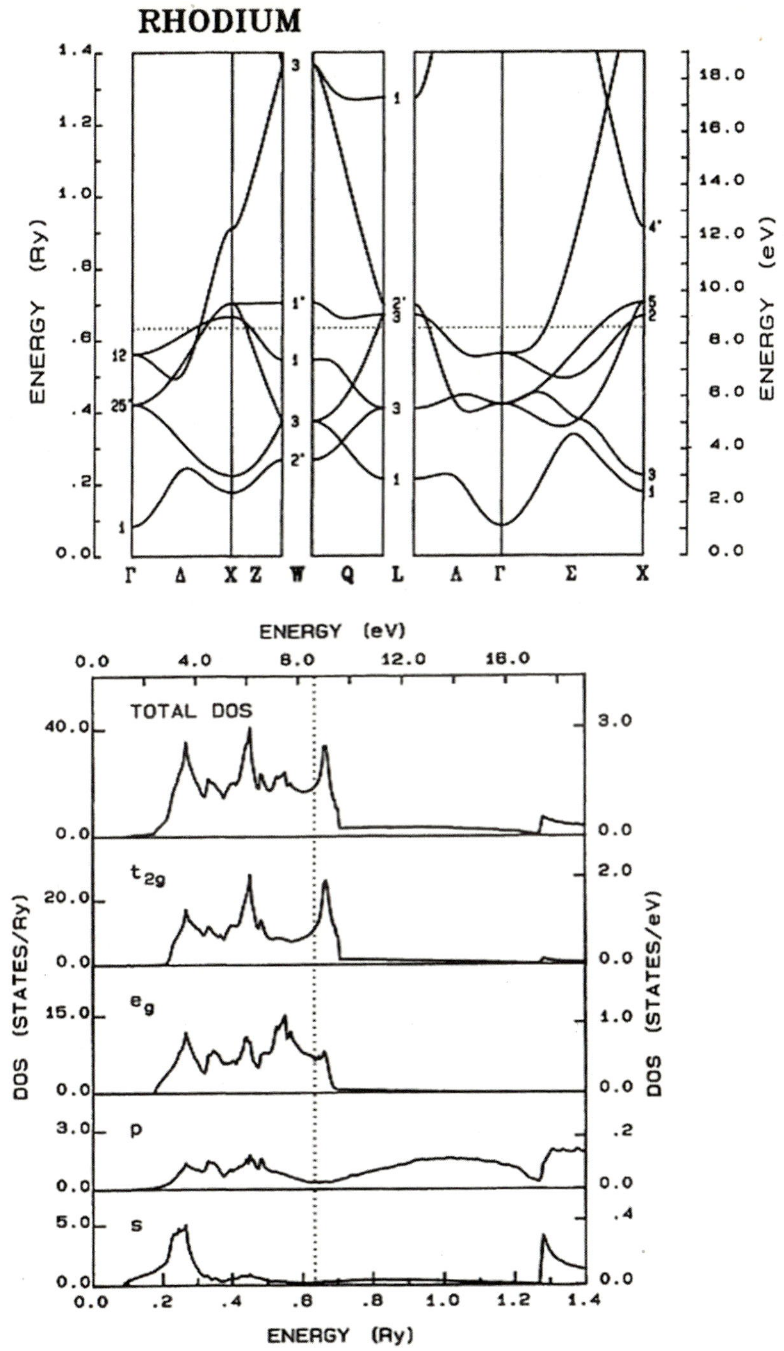

Fig. 5.15 Energy bands and density of states for Rh

5.8 Palladium

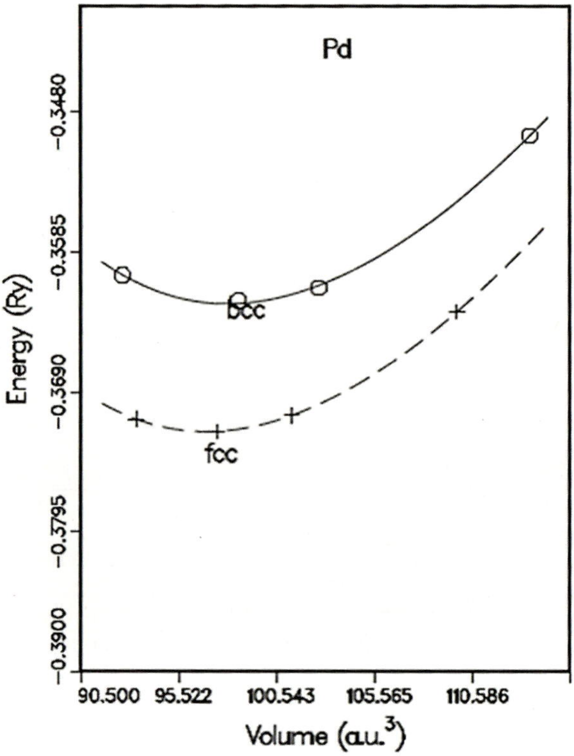

Fig. 5.16 Total energy of Pd

Table 5.58 Lattice constants and bulk modulus

	a (Bohr)	B (mbar)
fcc	7.291	1.915
bcc	5.809	1.936
exp	7.351	1.808
$\Delta E = 9.6$ mRy		

Table 5.59 Birch fit coefficients

	A_1	A_2	A_3
bcc	1.08847	−61.67797	655.51418
fcc	1.04715	−59.87868	631.63453

Table 5.60 Palladium fcc Z = 46 lattice constant = 7.35120 a.u.

Slater-Koster 3-center parameters	Orthogonal	Non-orthogonal	
	Energy integrals (Ry)	Energy integrals (Ry)	Overlap integrals
On site			
s, s (000)	0.90734	0.82970	
x, x (000)	1.49562	1.31863	
xy, xy (000)	0.38611	0.38275	
d2, d2 (000)	0.36787	0.37337	
First neighbor			
s, s (110)	−0.07962	−0.06399	0.02513
s, x (110)	0.08913	0.06102	−0.04073
s, xy (110)	−0.05156	−0.06675	0.00443
s, d2 (110)	0.02878	0.03595	0.02715
x, x (110)	0.08299	0.06305	−0.03671
x, x (011)	−0.00343	0.05460	0.05921
x, y (110)	0.12335	0.03797	−0.06220
x, xy (110)	−0.05202	−0.06246	−0.00783
x, xy (011)	0.00842	−0.01197	−0.04444
z, d2 (011)	0.00733	0.00009	−0.02302
z, d1 (011)	0.03509	0.02787	−0.00836
xy, xy (110)	−0.03997	−0.02191	0.04355
xy, xy (011)	0.01062	0.00842	−0.01449
xy, xz (011)	0.01145	−0.00172	−0.02976
xy, d2 (110)	0.02105	0.02458	0.01199
d2, d2 (110)	−0.01688	−0.01563	−0.00049
d1, d1 (110)	0.02876	0.02740	0.00684
Second neighbor			
s, s (200)	0.00496	−0.01364	−0.02057
s, x (200)	−0.01408	0.05060	0.03489
s, d2 (002)	0.00460	−0.01301	−0.04009
x, x (200)	−0.03385	0.10813	0.04144
y, y (200)	0.00267	0.00812	0.01788
x, xy (020)	0.00054	−0.01078	−0.01037
z, d2 (002)	0.00601	−0.02946	−0.05650
xy, xy (200)	−0.00065	−0.01208	−0.01256
xy, xy (002)	0.00031	−0.00110	−0.00150
d2, d2 (002)	−0.00398	−0.00590	−0.01335
d1, d1 (002)	−0.00027	0.00528	0.01270

Table 5.61 Palladium fcc

	Orthogonal			Non-orthogonal		
Band	RMS error mRy	Maximum k	Deviation mRy	RMS error mRy	Maximum k	Deviation mRy
1	1.2	(055)	2.8	0.5	(174)	1.6
2	2.2	(007)	4.6	0.8	(008)	2.1
3	1.8	(048)	4.1	1.1	(007)	4.1
4	2.3	(118)	9.2	1.8	(118)	9.2
5	1.8	(044)	3.9	0.8	(000)	2.3
6	5.0	(174)	21.3	1.2	(174)	3.3
1–6	2.7			1.1		

Table 5.62 Energy values in Ry at selected k-points

	Orthogonal	APW	Non-orthogonal
GAMMA 1	−0.01838	−0.01723	−0.01701
GAMMA 12	0.42644	0.42442	0.42671
GAMMA 15	2.08884	2.08678	2.08671
GAMMA 25′	0.30924	0.30899	0.31037
X1 (008)	0.10284	0.10350	0.10438
X1 (008)	1.30155	1.29345	1.29356
X2 (008)	0.51391	0.51137	0.51259
X3 (008)	0.13927	0.14119	0.14331
X5 (008)	0.54403	0.54284	0.54347
X4′ (008)	0.76094	0.76019	0.76051
X5′ (008)	1.45232	1.47064	1.47018
L1 (444)	0.12653	0.12646	0.12708
L1 (444)	1.04751	1.05062	1.05082
L3 (444)	0.30072	0.30055	0.30169
L3 (444)	0.51379	0.51655	0.51715
L2′ (444)	0.56582	0.57056	0.57039
L3′ (444)	2.04605	2.04706	2.04713
W1 (048)	0.41679	0.41526	0.41702
W1 (048)	1.25750	1.26607	1.26598
W3 (048)	0.27320	0.26911	0.27058
W3 (048)	1.16699	1.16953	1.17076
W1′ (048)	0.54279	0.54317	0.54494
W2′ (048)	0.18499	0.18334	0.18438
W2′ (048)	1.64388	1.34909	1.34939
EVEN (224)	0.15734	0.15799	0.15814
EVEN (224)	0.29638	0.29298	0.29398
EVEN (224)	0.31889	0.31720	0.31680
EVEN (224)	0.52738	0.52913	0.53007
ODD (224)	0.34226	0.34606	0.34526
ODD (224)	0.50108	0.50086	0.50065

Table 5.63 Fermi level quantities (non-orthogonal fit)

Energy		Densities of states				Velocity	Plasmon energy
	Total	s	p	t_{2g}	e_g		
Ry		States /Ry/atom				cm/s	eV
0.5190	32.16	0.33	0.45	25.83	5.55	0.35 × 10E8	7.18

Integrated densities of states				
Total	s	p	t_{2g}	e_g
Electrons				
10.00	0.60	0.45	5.29	3.67

Table 5.64 Palladium fcc

Slater-Koster 2-center parameters	Orthogonal	Non-orthogonal	
	Energy integrals (Ry)	Energy integrals (Ry)	Overlap integrals
On site			
s	0.94261	0.52122	
p	1.36110	1.00597	
dl	0.37285	0.35083	
d2	0.36265	0.35127	
First neighbor			
(sss)	−0.07962	−0.03712	0.12155
(pps)	0.17119	−0.14554	−0.29494
(ppp)	−0.00540	−0.05335	0.02818
(dds)	−0.05216	−0.03309	0.04968
(ddp)	0.02878	0.02203	−0.02767
(ddd)	−0.00533	−0.00355	0.00311
(sps)	0.11332	−0.02276	−0.19954
(sds)	−0.04885	−0.02949	0.06623
(pds)	−0.06563	−0.01586	0.08650
(pdp)	0.02124	0.02136	−0.03444
Second neighbor			
(sss)	−0.00105	−0.01966	0.00565
(pps)	0.04282	0.07374	−0.00473
(ppp)	−0.00044	−0.02354	−0.00302
(dds)	−0.00385	−0.01096	−0.00808
(ddp)	0.00212	0.00096	−0.00170
(ddd)	−0.00026	0.00025	0.00132
(sps)	0.01048	0.03559	−0.00818
(sds)	−0.00837	−0.01208	0.00233
(pds)	−0.00738	−0.02728	−0.00441
(pdp)	0.00351	0.00507	0.00013

	Orthogonal			Non-orthogonal		
Band	RMS error mRy	Maximum k	Deviation mRy	RMS error mRy	Maximum k	Deviation mRy
1	4.9	(055)	8.9	0.7	(055)	1.9
2	7.1	(007)	17.6	0.9	(008)	2.0
3	3.4	(007)	7.2	1.1	(007)	4.3
4	5.2	(084)	13.6	1.9	(118)	9.9
5	5.9	(044)	14.2	1.1	(000)	3.0
6	5.4	(055)	14.9	1.1	(000)	3.0
1–6	5.4			1.2		

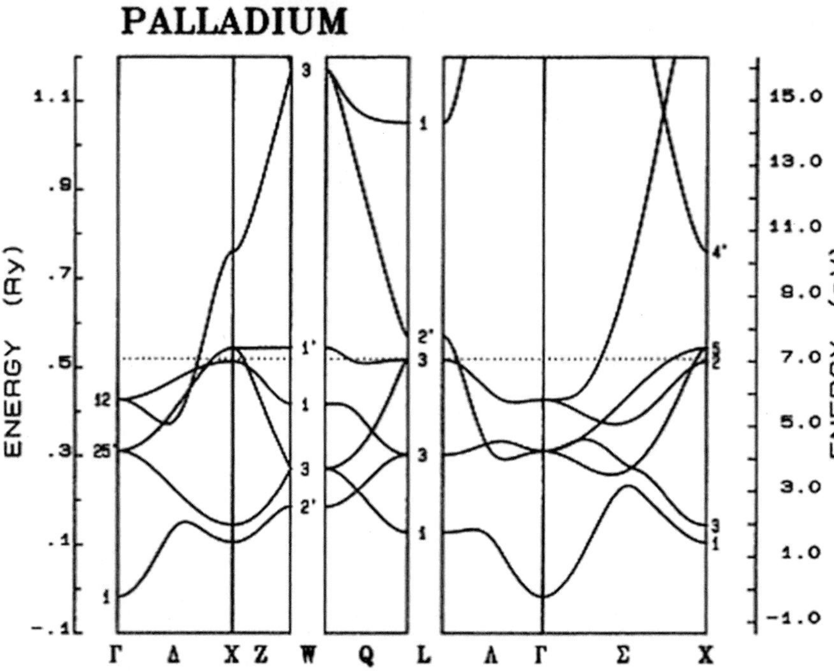

Fig. 5.17 Energy bands for Pd

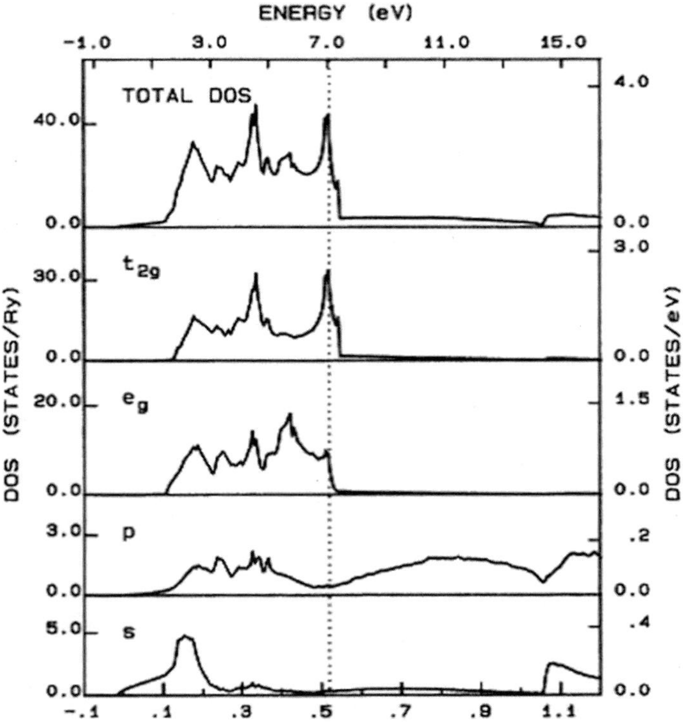

Fig. 5.18 Density of states for Pd

5.9 Silver

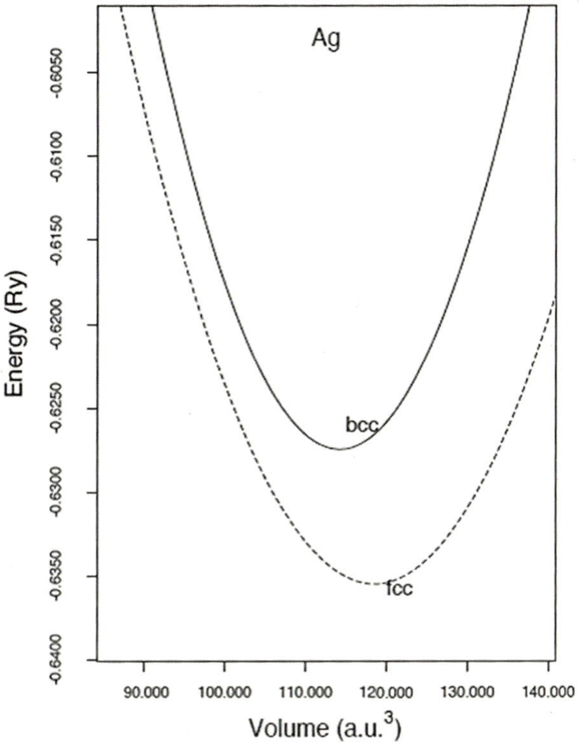

Fig. 5.19 Total energy of Ag

Table 5.65 Lattice constants and bulk modulus

	a (Bohr)	B (mbar)
fcc	7.625	1.319
bcc	6.076	1.301
exp	7.722	1.077
$\Delta E = 9.7$ mRy		

Table 5.66 Birch fit coefficients

	A_1	A_2	A_3	A_4
bcc	0.97182	−85.53899	1385.87362	−6065.19429
fcc	−1.15333	61.61990	−2016.87990	20088.72348

Table 5.67 Silver fcc Z = 47 lattice constant = 7.68000 a.u.

Slater-Koster 3-center parameters	Orthogonal	Non-orthogonal	
	Energy integrals (Ry)	Energy integrals (Ry)	Overlap integrals
On site			
s, s (000)	0.68783	0.59528	
x, x (000)	1.28722	1.06950	
xy, xy (000)	0.12336	0.12267	
d2, d2 (000)	0.12090	0.11864	
First neighbor			
s, s (110)	−0.06646	−0.05741	0.04439
s, x (110)	0.07924	0.05539	−0.04808
s, xy (110)	−0.02804	−0.03759	−0.00040
s, d2 (110)	0.01304	0.02235	0.02544
x, x (110)	0.08198	0.06252	−0.03920
x, x (011)	0.00549	0.03695	0.05067
x, y (110)	0.11338	0.00985	−0.09521
x, xy (110)	−0.02633	−0.02901	0.00905
x, xy (011)	0.00896	0.00341	−0.04268
z, d2 (011)	0.01075	−0.00628	−0.05057
z, d1 (011)	0.01939	0.01788	0.00272
xy, xy (110)	−0.02430	−0.01947	0.02944
xy, xy (011)	0.00664	0.00666	−0.00658
xy, xz (011)	0.00969	0.00658	−0.01553
xy, d2 (110)	0.01261	0.01330	0.00359
d2, d2 (110)	−0.01037	−0.00720	0.01636
d1, d1 (110)	0.01848	0.01429	−0.02554
Second neighbor			
s, s (200)	0.00212	−0.00823	−0.00118
s, x (200)	−0.01177	0.02883	0.01576
s, d2 (002)	0.00177	−0.01001	−0.01771
x, x (200)	−0.04111	0.07959	0.02419
y, y (200)	0.00783	0.03300	0.02886
x, xy (020)	0.00441	−0.00937	−0.02475
z, d2 (002)	0.00667	−0.02003	−0.03984
xy, xy (200)	0.00175	−0.00251	−0.01096
xy, xy (002)	−0.00004	−0.00025	−0.00027
d2, d2 (002)	−0.00343	−0.00563	−0.01031
d1, d1 (002)	−0.00091	0.00065	0.00791

Table 5.68 Silver fcc

	Orthogonal			Non-orthogonal		
Band	RMS error mRy	Maximum k	Deviation mRy	RMS error mRy	Maximum k	Deviation mRy
1	1.6	(222)	4.4	0.4	(174)	1.4
2	1.8	(055)	4.7	0.6	(008)	1.7
3	1.9	(118)	4.9	1.3	(118)	6.0
4	1.8	(222)	6.3	0.5	(048)	1.6
5	1.8	(044)	3.5	0.6	(000)	1.7
6	5.0	(174)	16.6	0.6	(000)	1.7
1–6	2.6			0.7		

Table 5.69 Energy values in Ry at selected k-points

	Orthogonal	APW	Non-orthogonal
GAMMA 1	−0.09705	−0.09449	−0.09372
GAMMA 12	0.15650	0.15426	0.15597
GAMMA 15	1.91410	1.90768	1.90766
GAMMA 25′	0.08616	0.08465	0.08576
X1 (008)	−0.03460	−0.03414	−0.03336
X1 (008)	0.97727	0.95134	0.95147
X2 (008)	0.20706	0.20694	0.20806
X3 (008)	−0.02000	−0.01853	−0.01687
X5 (008)	0.22748	0.22636	0.22677
X4′ (008)	0.60243	0.60048	0.60015
X5′ (008)	1.21435	1.26215	1.26193
L1 (444)	−0.01574	−0.01900	−0.01857
L1 (444)	0.72974	0.72587	0.72640
L3 (444)	0.08188	0.08086	0.08171
L3 (444)	0.20727	0.21061	0.21106
L2′ (444)	0.43108	0.43804	0.43778
L3′ (444)	1.79165	1.79592	1.79588
W1 (048)	0.15871	0.15541	0.15705
W1 (048)	0.96231	0.98668	0.98640
W3 (048)	0.07197	0.06940	0.07045
W3 (048)	0.90183	0.91161	0.91209
W1′ (048)	0.22766	0.22640	0.22764
W2′ (048)	0.01990	0.01887	0.01954
W2′ (048)	1.39652	1.05027	1.05050
EVEN (224)	0.00737	0.00927	0.00952
EVEN (224)	0.08276	0.08267	0.08388
EVEN (224)	0.11938	0.11614	0.11593
EVEN (224)	0.28667	0.28893	0.28942
ODD (224)	0.10775	0.10978	0.10917
ODD (224)	0.19971	0.20194	0.20147

Table 5.70 Fermi level quantities (non-orthogonal fit)

Energy		Densities of states				Velocity	Plasmon energy
	Total	s	p	t_{2g}	e_g		
Ry		States/Ry/atom				cm/s	eV
0.4635	3.59	1.06	1.71	0.64	0.18	1.45 × 10E8	9.28

Integrated densities of states Total Electrons	s	p	t_{2g}	e_g
11.00	0.65	0.35	6.06	3.95

Table 5.71 Silver fcc

Slater-Koster 2-center parameters	Orthogonal	Non-orthogonal	
	Energy integrals (Ry)	Energy integrals (Ry)	Overlap integrals
On site			
s	0.68297	0.39890	
p	1.13432	0.72680	
dl	0.12249	0.11776	
d2	0.12006	0.11834	
First neighbor			
(sss)	−0.06581	−0.04280	0.11070
(pps)	0.15752	−0.14727	−0.35644
(ppp)	0.00649	0.00420	0.07676
(dds)	−0.03151	−0.02825	0.02651
(ddp)	0.01757	0.01580	−0.01514
(ddd)	−0.00336	−0.00235	0.00113
(sps)	0.09781	0.00996	−0.19464
(sds)	−0.03110	−0.03057	0.04683
(pds)	−0.03905	−0.02255	0.09964
(pdp)	0.01519	0.01861	−0.04465
Second neighbor			
(sss)	0.00143	−0.01782	0.00349
(pps)	0.03971	0.04670	−0.04478
(ppp)	0.00434	0.00640	0.01335
(dds)	−0.00282	−0.00466	−0.00776
(ddp)	0.00171	0.00100	0.00005
(ddd)	−0.00038	−0.00007	0.00115
(sps)	0.00545	0.03133	−0.01266
(sds)	−0.00462	−0.00798	−0.00322
(pds)	−0.00065	−0.01648	−0.00199
(pdp)	0.00172	0.00127	−0.00647

	Orthogonal			Non-orthogonal		
	RMS error	Maximum	Deviation	RMS error	Maximum	Deviation
Band	mRy	k	mRy	mRy	k	mRy
1	2.1	(084)	4.0	0.4	(174)	1.0
2	3.4	(007)	9.3	0.7	(008)	1.6
3	1.7	(118)	6.6	1.4	(118)	6.4
4	2.7	(222)	8.6	0.5	(000)	1.4
5	2.9	(044)	6.4	0.8	(000)	2.0
6	3.5	(174)	12.5	0.5	(000)	2.0
1–6	2.8			0.8		

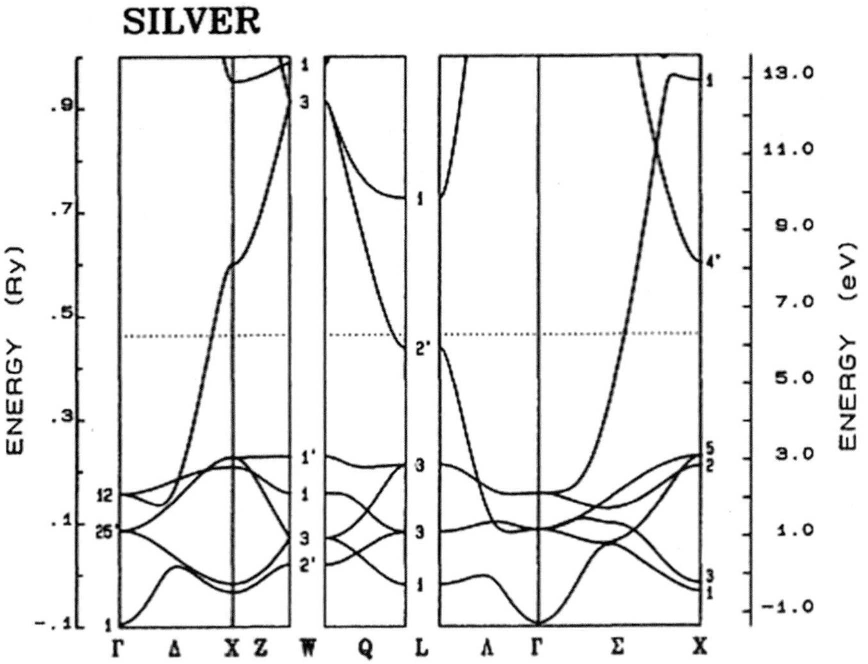

Fig. 5.20 Energy bands for Ag

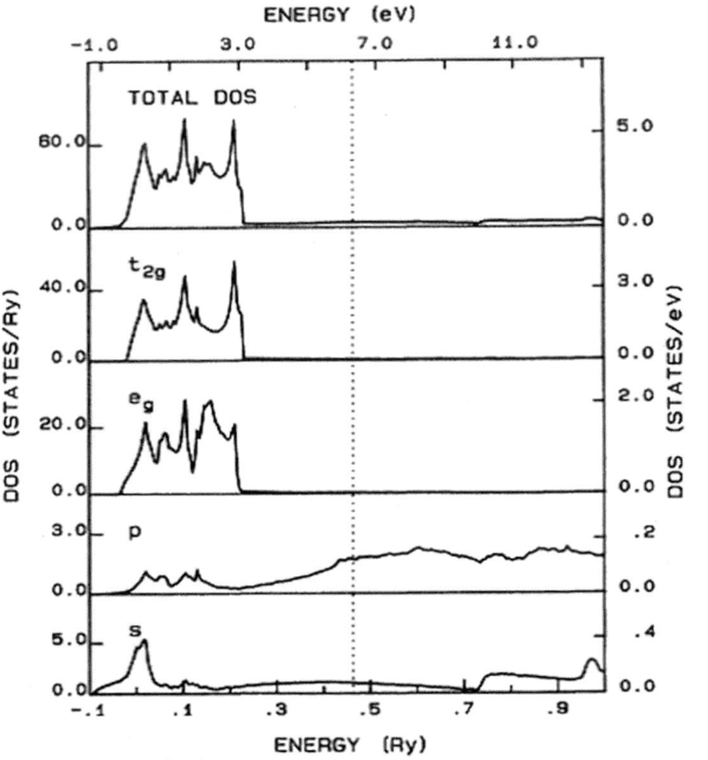

Fig. 5.21 Density of states for Ag

5.10 Cadmium

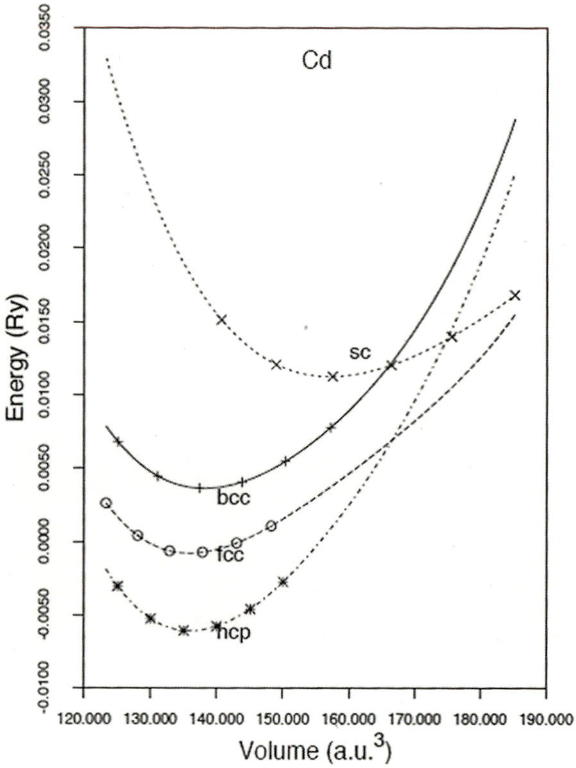

Fig. 5.22 Total energy of Cd

Table 5.72 Lattice constants and bulk modulus

	a (Bohr)	c (Bohr)	B (mbar)
sc	5.397		0.496
bcc	6.514		0.606
fcc	8.161		0.641
hcp	5.527	10.280	0.839
exp	5.631	10.620	0.467
$E_{hcp} - E_{fcc} = \Delta E = 5.3$ mRy			

Table 5.73 Birch fit coefficients

p	A_1	A_2	A_3	A_4
sc	−0.12756	29.50008	−1365.21463	18166.92339
bcc	0.26469	−3.81409	−355.57358	7243.80244
fcc	−0.73575	75.85689	−2469.52833	25849.26121
hcp	0.20170	6.57686	−783.60949	12280.53096

Table 5.74 Cadmium hcp Z = 48 lattice constants = 5.6293 a.u. 10.6159 a.u.

Slater-Koster 3-center parameters	Orthogonal		
	Energy integrals (Ry)		Energy integrals (Ry)
On site			
s, s (0)	0.07673		
x, x (0)	0.64221		
z, z (0)	0.31835		
xy, xy (0)	−0.55403		
yz, yz (0)	−0.55782		
d2, d2 (0)	−0.53604		
s, d2 (0)	0.03036		
x, xy (0)	0.04127		
First neighbor			
s, s (R)	−0.05925	s, s (T)	−0.03885
s, x (R)	0.07893	s, y (T)	−0.03223
s, y (R)	0.01159	s, z (T)	0.04843
s, xy (R)	0.01484	s, d1 (T)	0.00378
s, d1 (R)	−0.00239	s, yz (T)	−0.04077
s, d2 (R)	−0.02393	s, d2 (T)	0.00194
x, x (R)	0.15075	x, x (T)	0.00293
y, y (R)	0.01204	y, y (T)	0.03264
x, y (R)	0.01060	y, z (T)	−0.00806
x, xy (R)	0.04601	x, xy (T)	0.00468
y, d1 (R)	−0.01610	y, d1 (T)	−0.00916
x, d1 (R)	−0.02125	x, xz (T)	0.01054
y, xy (R)	0.05769	y, yz (T)	0.01391
y, d2 (R)	−0.03542	y, d2 (T)	0.02126
x, d2 (R)	−0.05419	z, z (T)	0.03010
z, z (R)	−0.01261	z, d1 (T)	0.01107
z, yz (R)	−0.01663	z, yz (T)	0.04948
z, xz (R)	0.00341	z, d2 (T)	−0.01523
xy, xy (R)	0.00307	xy, xy (T)	0.00150
d1, d1 (R)	−0.01308	d1, d1 (T)	−0.00228
xy, d1 (R)	−0.00117	xy, yz (T)	−0.00021
d1, d2 (R)	−0.00477	yz, d1 (T)	0.00237
xy, d2 (R)	0.00450	d1, d2 (T)	−0.00195
yz, yz (R)	0.00099	yz, yz (T)	−0.00535
xz, xz (R)	0.00099	xz, xz (T)	0.00402
yz, xz (R)	−0.00457	yz, d2 (T)	−0.00198
d2, d2 (R)	0.00055	d2, d2 (T)	0.00718

Table 5.75 Cadmium hcp

	Orthogonal		
Band	RMS error mRy	Maximum k	Deviation mRy
1	6.0	(0 0 12)	13.3
2	5.1	(0 0 12)	13.3
3	6.0	(0 24 12)	14.8
4	5.6	(0 24 12)	14.8
5	2.9	(12 0 0)	6.6
6	6.2	(0 18 0)	17.3
7	5.6	(0 0 3)	10.9
8	4.8	(0 0 3)	10.9
9	2.7	(12 0 0)	4.8
10	3.2	(8 0 0)	8.0
11	10.2	(8 0 12)	22.8
12	16.0	(24 0 9)	34.2
1–12	7.13		

Table 5.76 Energy values in Ry at selected k-points

	Orthogonal	APW
GAMMA 1+	−0.60291	−0.60928
GAMMA 1+	−0.39861	−0.40757
GAMMA 3+	0.06211	0.04604
GAMMA 5+	−0.59495	−0.59825
GAMMA 6+	−0.55590	−0.56040
GAMMA 4−	−0.60374	−0.61687
GAMMA 4−	−0.01776	−0.02476
GAMMA 5−	−0.54791	−0.54506
GAMMA 6−	−0.59309	−0.59246
M1+	−0.63113	−0.63417
M1+	−0.54079	−0.54397
M1+	0.07431	0.07864
M2+	−0.53974	−0.54765
M3+	−0.59448	−0.61092
M4+	−0.52383	−0.52700
M1−	−0.57985	−0.57627
M2−	−0.61751	−0.62008
M2−	−0.55283	−0.56556
M2−	0.07988	0.08700
M3−	−0.52906	−0.52600
M4−	−0.61980	−0.58492
A1	−0.62530	−0.61200
A1	−0.31744	−0.30839
A3	−0.59403	−0.59403
A3	−0.55508	−0.55188
L1	−0.62508	−0.62561
L1	−0.59485	−0.60963
L1	−0.54207	−0.54361
L1	0.16486	0.14428
L2	−0.56101	−0.56224
L2	−0.52560	−0.52600

Table 5.77 Cadmium hcp Z = 48 lattice constants = 5.6293 a.u. 10.6159 a.u.

Slater-Koster 2-center parameters	Orthogonal	Non-orthogonal	
	Energy integrals (Ry)	Energy integrals (Ry)	Overlap integrals
On site			
s	0.10913	−0.18935	
p	0.31785	0.35375	
d0	−0.47518	−0.57454	
d1	−0.57394	−0.57031	
d2	−0.55666	−0.53077	
pd	0.28790	0.00236	0.03991
First neighbor			
(sss)	−0.04693	−0.08844	0.11615
(sps)	0.06335	0.02456	0.02654
(sds)	0.03385	0.07860	−0.08404
(pps)	0.04482	0.08917	0.13172
(ppp)	−0.01905	0.05423	0.09108
(pds)	0.00562	0.06854	−0.07574
(pdp)	0.00812	−0.02678	0.06565
(dds)	−0.01034	0.03181	−0.07448
(ddp)	0.00295	0.04283	−0.06487
(ddd)	0.00004	0.00541	−0.00875
Second neighbor			
(sss)	0.01402	−0.00817	−0.03284
(sps)	0.00103	−0.00927	0.06391
(sds)	0.01785	0.00985	−0.00501
(pps)	0.03797	0.08807	0.19097
(ppp)	−0.00056	0.01851	0.01465
(pds)	0.00347	−0.05604	0.14234
(pdp)	0.00992	0.00569	−0.01219
(dds)	−0.00035	−0.04732	0.07543
(ddp)	−0.00137	0.02536	−0.04386
(ddd)	0.00058	0.04237	−0.07277
Third neighbor			
(sss)	−0.02601	−0.00865	0.05495
(sps)	−0.00177	−0.02266	0.03437
(sds)	0.03078	0.05676	−0.04134
(pps)	0.02188	0.02727	0.12710
(ppp)	0.01437	−0.00259	−0.03681
(pds)	0.03330	0.00545	−0.14944
(pdp)	0.00698	0.00990	−0.02952
(dds)	−0.00017	−0.02622	0.04551
(ddp)	0.00102	−0.03731	0.06446
(ddd)	−0.00077	0.03864	−0.06762

Table 5.78 Cadmium hcp

	Orthogonal			Non-orthogonal		
Band	RMS error mRy	Maximum k	Deviation mRy	RMS error mRy	Maximum k	Deviation mRy
1	14.0	(8 0 0)	34.8	6.8	(0 0 3)	16.2
2	9.5	(24 0 0)	17.2	8.0	(0 0 3)	17.0
3	9.8	(0 0 0)	17.0	4.9	(16 0 0)	8.6
4	7.2	(0 0 0)	17.0	5.2	(0 12 0)	9.0
5	7.6	(0 12 12)	15.5	5.4	(0 6 0)	10.8
6	8.0	(8 0 0)	21.3	7.7	(0 0 0)	17.1
7	5.7	(0 12 0)	9.2	4.5	(0 0 3)	8.7
8	6.6	(16 0 3)	10.4	6.7	(16 0 0)	20.3
9	7.0	(0 18 0)	11.4	10.3	(16 0 12)	25.8
10	5.8	(0 24 12)	10.8	7.5	(16 0 12)	25.8
11	22.3	(16 0 0)	49.7	28.4	(0 0 6)	78.2
12	29.7	(8 0 0)	61.5	15.6	(8 0 12)	41.7
1–12	13.21			11.27		

Table 5.79 Energy values in Ry at selected k-points

	Orthogonal	APW	Non-orthogonal
GAMMA 1+	−0.60143	−0.60928	−0.60932
GAMMA 1+	−0.40088	−0.40757	−0.36420
GAMMA 3+	0.03243	0.04604	0.02817
GAMMA 5+	−0.58127	−0.59825	−0.59616
GAMMA 6+	−0.57764	−0.56040	−0.56056
GAMMA 4−	−0.59932	−0.61687	−0.57534
GAMMA 4−	0.00801	−0.02476	−0.01474
GAMMA 5−	−0.54822	−0.54506	−0.54729
GAMMA 6−	−0.56737	−0.59246	−0.59220
M1+	−0.64679	−0.63417	−0.63475
M1+	−0.55542	−0.54397	−0.54015
M1+	0.08944	0.07864	0.09124
M2+	−0.54843	−0.54765	−0.55175
M3+	−0.59432	−0.61092	−0.60586
M4+	−0.53669	−0.52700	−0.54412
M1−	−0.58973	−0.57627	−0.58334
M2−	−0.63729	−0.62008	−0.61518
M2−	−0.56243	−0.56556	−0.55125
M2−	0.03680	0.08700	0.08519
M3−	−0.53307	−0.52600	−0.55413
M4−	−0.56732	−0.58492	−0.56815
A1	−0.62539	−0.61200	−0.62771
A1	−0.29507	−0.30839	−0.29134
A3	−0.60009	−0.59403	−0.59164
A3	−0.54502	−0.55188	−0.54928
L1	−0.62807	−0.62561	−0.62359
L1	−0.60299	−0.60963	−0.60201
L1	−0.54024	−0.54361	−0.52029
L1	0.15762	0.14428	0.14568
L2	−0.57327	−0.56224	−0.56804
L2	−0.53676	−0.52600	−0.54508

Table 5.80 Fermi level quantities (non-orthogonal fit)

Energy		Densities of states			
	Total	s	p	t_{2g}	e_g
Ry		States/Ry/atom			
0.2249	6.03	1.83	3.69	0.31	0.20

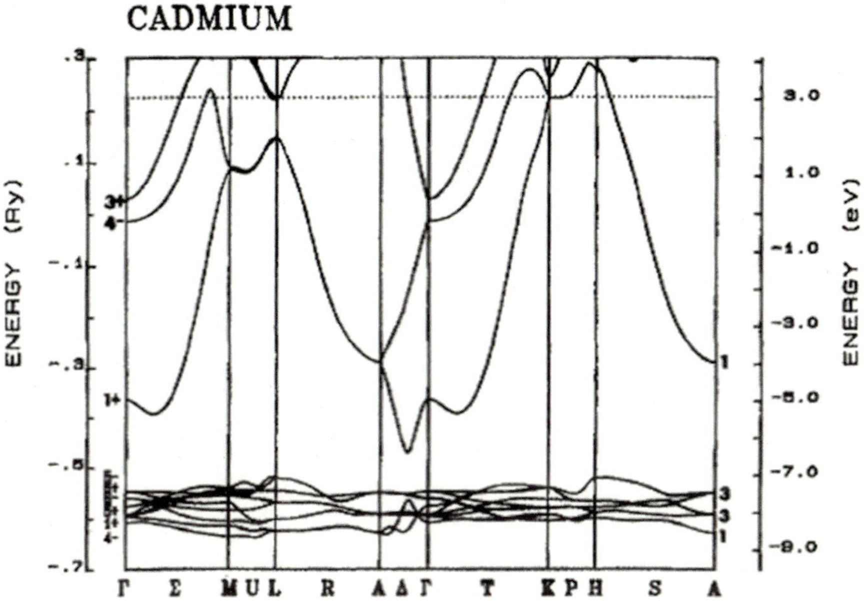

Fig. 5.23 Energy bands for Cd

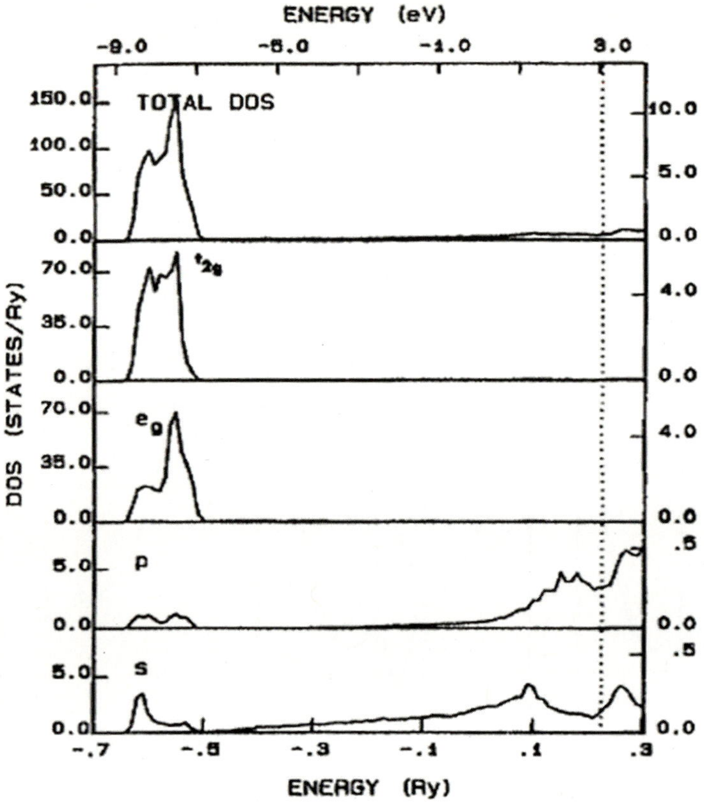

Fig. 5.24 Density of states for Cd

Chapter 6
The 5d Transition Metals

As shown in the total energy plots the ground state of the 5d elements follow the same trend as in the 3d and 4d elements of the same columns of the periodic table. There are two exceptions La (which we have in the section of f-electron metals) that has the dhcp structure and Hg which crystallizes in a rhombohedral structure. It is interesting to note that while in La the f-states are just above the Fermi level in the other 5d elements the f-states are fully occupied and lie deep below the valence bands. Hf is an exception having its f-states close to the valence bands but still below them. The reader should also keep in mind that in this series, as in other heavy elements, the spin-orbit coupling, which is not included in these calculations, may introduce significant changes to the band structure. The main characteristics of the energy bands and DOS found in the 3d and 4d elements are present in the 5d-series when crystal structure and number of valence electrons are taken into account. Again, as in the 3d-4d comparison, a close comparison of the 4d and 5d elements show that a further widening of the d bands occurs in the 5d metals. A comparison of Ir with Rh and Pt with Pd reveals a change in the ordering of levels at L. Also, Au appears to have a band structure much closer to that of Cu than that of Ag.

D.A. Papaconstantopoulos, *Handbook of the Band Structure of Elemental Solids*,
DOI 10.1007/978-1-4419-8264-3_6

6.1 Hafnium

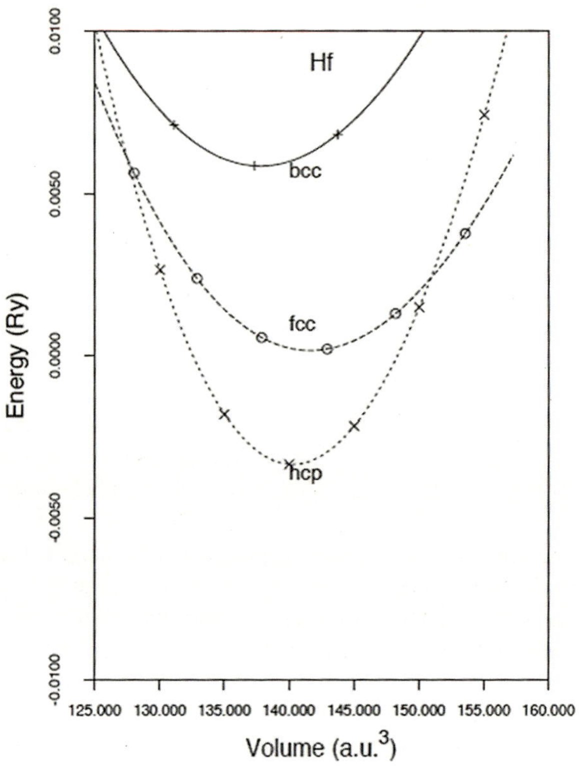

Fig. 6.1 Total energy of Hf

Table 6.1 Lattice constants of bulk modulus

	a (Bohr)	c (Bohr)	B (Mbar)
bcc	6.507		1.119
fcc	8.275		1.138
hcp	5.909	9.277	2.224
exp	6.028	9.543	1.09
$\Delta E = 3.5$ mRy			

Table 6.2 Birch fit coefficients

	A_1	A_2	A_3	A_4
bcc	3.16794	−221.55125	5070.12559	−37618.31312
fcc	2.82450	−196.72527	4435.03874	−31923.76295
hcp	5.63387	−392.15107	8848.05911	−63976.78690

Table 6.3 Hafnium hcp Z = 72 lattice constants 6.0409 a.u. 9.5579 a.u

Slater-Koster 3-center parameters	Orthogonal		
	Energy integrals (Ry)		Energy integrals (Ry)
On site			
s, s (0)	0.74947		
x, x (0)	1.24833		
z, z (0)	1.09764		
xy, xy (0)	0.83209		
yz, yz (0)	0.80941		
d2, d2 (0)	0.73246		
s, d2 (0)	0.06073		
x, xy (0)	−0.01229		
First neighbor			
s, s (R)	−0.04654	s, s (T)	−0.06405
s, x (R)	0.07301	s, y (T)	−0.05084
s, y (R)	0.02961	s, z (T)	0.07887
s, xy (R)	0.02653	s, d1 (T)	−0.02081
s, d1 (R)	0.06421	s, yz (T)	−0.06943
s, d2 (R)	−0.03015	s, d2 (T)	0.05377
x, x (R)	0.11449	x, x (T)	−0.01693
y, y (R)	−0.00863	y, y (T)	0.04728
x, y (R)	0.03410	y, z (T)	−0.07806
x, xy (R)	0.02444	x, xy (T)	0.01966
y, d1 (R)	0.02017	y, d1 (T)	0.01680
x, d1 (R)	0.10353	x, xz (T)	−0.01949
y, xy (R)	−0.00212	y, yz (T)	0.05855
y, d2 (R)	0.00543	y, d2 (T)	−0.03611
x, d2 (R)	−0.05597	z, z (T)	0.04717
z, z (R)	−0.03309	z, d1 (T)	−0.05478
z, yz (R)	−0.01053	z, yz (T)	−0.09443
z, xz (R)	−0.01696	z, d2 (T)	0.02082
xy, xy (R)	0.01734	xy, xy (T)	0.02398
d1, d1 (R)	−0.05554	d1, d1 (T)	−0.00204
xy, d1 (R)	−0.00860	xy, yz (T)	−0.01253
d1, d2 (R)	0.03978	yz, d1 (T)	−0.02018
xy, d2 (R)	0.00295	d1, d2 (T)	0.03387
yz, yz (R)	−0.00105	yz, yz (T)	−0.06137
xz, xz (R)	0.01848	xz, xz (T)	0.02947
yz, xz (R)	−0.01734	yz, d2 (T)	0.03995
d2, d2 (R)	−0.01717	d2, d2 (T)	−0.00312

Table 6.4 Hafnium hcp

Band	Orthogonal RMS error mRy	Maximum k	Deviation mRy
1	6.3	(16 0 0)	17.9
2	8.6	(16 0 0)	27.2
3	4.4	(12 0 0)	12.1
4	5.1	(0 18 0)	9.4
5	4.1	(0 12 0)	10.9
6	2.9	(4 0 0)	9.5
7	4.4	(0 12 12)	12.6
8	5.6	(0 12 12)	12.6
9	5.6	(0 12 12)	15.0
10	6.0	(0 12 12)	15.0
11	6.6	(16 0 12)	22.4
12	7.1	(16 0 12)	22.4
1–12	5.74		

Table 6.5 Energy values in Ry at selected k-points

	Orthogonal	APW
GAMMA 1+	0.01698	0.01564
GAMMA 1+	0.67970	0.67802
GAMMA 3+	0.61606	0.62094
GAMMA 5+	0.71327	0.71200
GAMMA 6+	0.76601	0.76179
GAMMA 4–	0.29669	0.29446
GAMMA 4–	1.20600	1.20482
GAMMA 5–	0.95738	0.95242
GAMMA 6–	0.65155	0.64947
M1+	0.35189	0.35756
M1+	0.52130	0.51762
M1+	1.09015	1.09089
M2+	1.04461	1.03625
M3+	0.39819	0.39760
M4+	1.03397	1.02928
M1–	0.61747	0.62457
M2–	0.32768	0.32539
M2–	0.68627	0.69225
M2–	1.07964	1.08020
M3–	0.98467	0.99214
M4–	0.64032	0.64052
A1	0.16261	0.16305
A1	0.58531	0.58662
A3	0.63945	0.64296
A3	0.90816	0.91585
L1	0.31440	0.31091
L1	0.49974	0.50184
L1	0.66689	0.66285
L1	1.01221	1.01344
L2	0.76556	0.76183
L2	1.01172	1.01250

Table 6.6 Hafnium hcp Z = 72 lattice constants = 6.0409 a.u. 9.5579 a.u.

Slater-Koster 2-center parameters	Orthogonal	Non-orthogonal	
	Energy integrals (Ry)	Energy integrals (Ry)	Overlap integrals
On site			
s	0.83098	0.63766	
p	1.12917	1.05884	
d0	0.76323	0.65758	
d1	0.75470	0.66587	
d2	0.77105	0.67334	
pd	−0.03681	−0.01440	0.00797
First neighbor			
(sss)	−0.06267	−0.04874	0.07561
(sps)	0.08890	0.05939	−0.05645
(sds)	0.06710	0.02743	−0.07160
(pps)	0.08796	0.11724	0.02188
(ppp)	−0.03857	0.05823	0.06873
(pds)	0.08893	−0.01523	−0.07959
(pdp)	−0.03274	0.02012	0.07038
(dds)	−0.09046	0.01836	0.13421
(ddp)	0.04094	0.00971	−0.07486
(ddd)	−0.00406	0.00018	0.01254
Second neighbor			
(sss)	−0.00898	−0.00510	0.01534
(sps)	−0.00499	−0.01001	−0.02787
(sds)	−0.01100	0.00502	−0.00705
(pps)	0.02642	0.03868	−0.00490
(ppp)	−0.00919	0.02201	0.01730
(pds)	0.01285	0.04494	0.01245
(pdp)	−0.01151	0.01741	0.02822
(dds)	−0.00766	−0.03630	−0.01233
(ddp)	0.00157	0.00197	−0.00171
(ddd)	0.00251	0.00106	−0.00323
Third neighbor			
(sss)	−0.00444	0.00506	0.03848
(sps)	0.00660	−0.03111	−0.06452
(sds)	−0.00805	0.00619	0.01851
(pps)	0.00567	0.01776	−0.00421
(ppp)	0.00139	−0.01751	−0.00732
(pds)	−0.00571	−0.00740	0.00228
(pdp)	0.00362	−0.00255	0.00176
(dds)	0.00885	−0.01641	−0.02891
(ddp)	0.00021	−0.00391	−0.00278
(ddd)	−0.00120	0.00916	0.00899

Table 6.7 Hafnium hcp

	Orthogonal			Non-orthogonal		
Band	RMS error mRy	Maximum k	Deviation mRy	RMS error mRy	Maximum k	Deviation mRy
1	13.5	(16 0 0)	31.0	3.1	(16 0 9)	7.1
2	14.5	(16 0 12)	26.5	3.6	(0 18 0)	8.3
3	10.9	(24 0 9)	28.2	3.9	(24 0 0)	7.9
4	10.8	(0 24 12)	28.0	6.1	(0 12 0)	12.3
5	5.9	(0 0 0)	11.1	5.6	(16 0 0)	12.3
6	7.3	(8 0 0)	20.5	6.0	(8 0 0)	12.4
7	15.9	(0 12 12)	35.9	6.1	(8 0 12)	17.2
8	17.7	(0 12 12)	35.9	6.4	(8 0 12)	17.2
9	14.5	(24 0 0)	31.6	7.5	(0 12 12)	20.2
10	14.4	(16 0 12)	24.6	7.7	(0 12 12)	20.2
11	17.9	(0 18 0)	36.2	8.3	(16 0 12)	30.0
12	17.0	(0 12 0)	38.0	8.3	(16 0 12)	30.0
1–12	13.86			6.27		

Table 6.8 Energy values in Ry at selected k-points

	Orthogonal	APW	Non-orthogonal
GAMMA 1+	0.01581	0.01564	0.01545
GAMMA 1+	0.67252	0.67802	0.68092
GAMMA 3+	0.61832	0.62094	0.61888
GAMMA 5+	0.70831	0.71200	0.70721
GAMMA 6+	0.76705	0.76179	0.76729
GAMMA 4–	0.29548	0.29446	0.29139
GAMMA 4–	1.17794	1.20482	1.21744
GAMMA 5–	0.96444	0.95242	0.95441
GAMMA 6–	0.63840	0.64947	0.64449
M1+	0.33237	0.35756	0.35865
M1+	0.51394	0.51762	0.51178
M1+	1.08741	1.09089	1.09048
M2+	0.99762	1.03625	1.03215
M3+	0.40936	0.39760	0.38973
M4+	1.01476	1.02928	1.03272
M1–	0.61884	0.62457	0.63199
M2–	0.33613	0.32539	0.32874
M2–	0.67638	0.69225	0.68308
M2–	1.08005	1.08020	1.08446
M3–	0.99076	0.99214	0.99610
M4–	0.63384	0.64052	0.64099
A1	0.17269	0.16305	0.16385
A1	0.58640	0.58662	0.58564
A3	0.64361	0.64296	0.64472
A3	0.89753	0.91585	0.91447
L1	0.32004	0.31091	0.30840
L1	0.47381	0.50184	0.49552
L1	0.66582	0.66285	0.66165
L1	1.00106	1.01344	1.00964
L2	0.78153	0.76183	0.76053
L2	1.01029	1.01250	1.01700

Table 6.9 Fermi level quantities (non-orthogonal fit)

Energy		Densities of states			
	Total	s	p	t_{2g}	e_g
Ry		States/Ry/atom			
0.5790	6.66	0.14	0.93	3.62	1.97

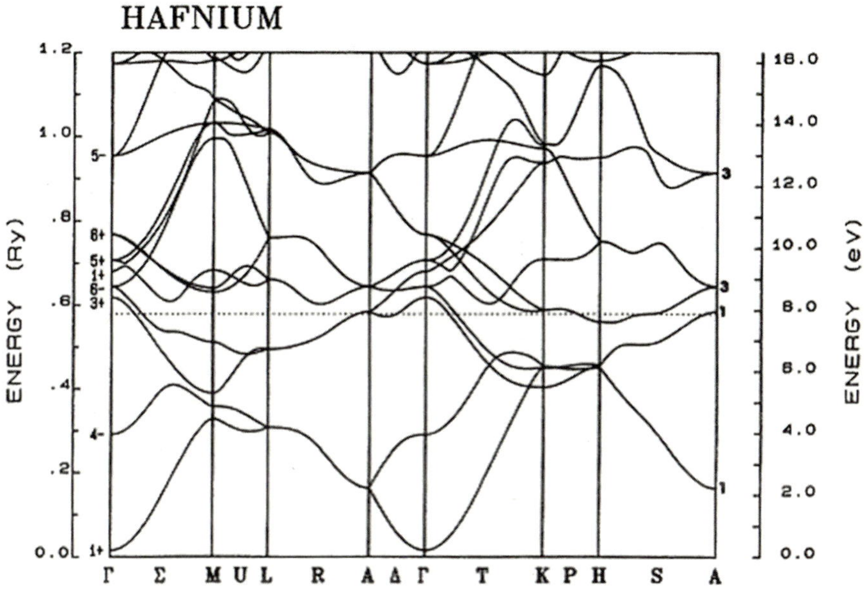

Fig. 6.2 Energy bands for Hf

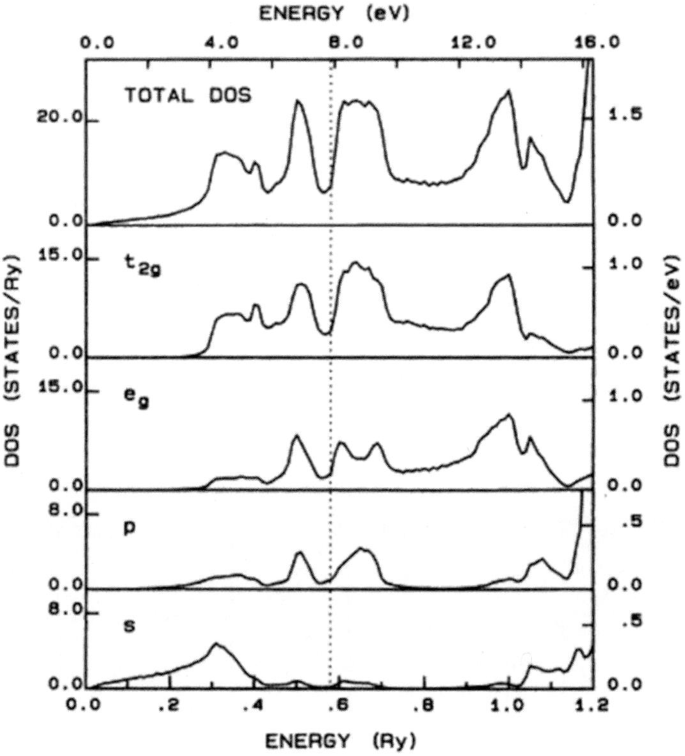

Fig. 6.3 Density of states for Hf

6.2 Tantalum

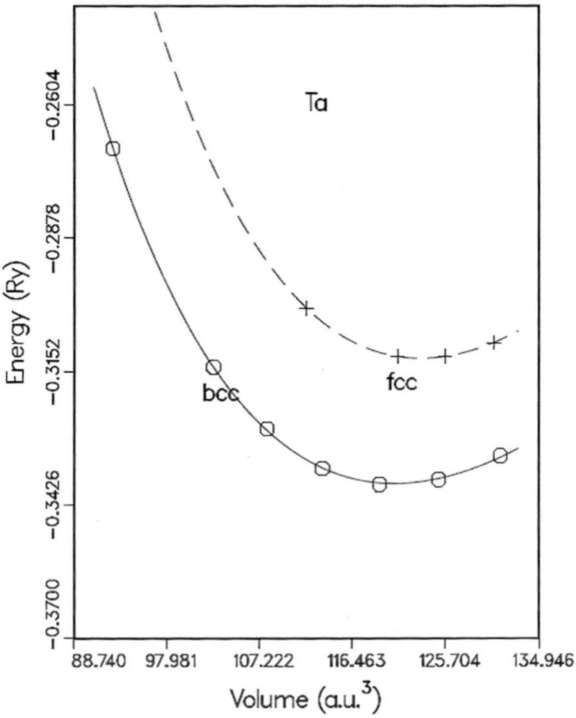

Fig. 6.4 Total energy of Ta

Table 6.10 Lattice constants and bulk modulus

	a (Bohr)	B (MBar)
bcc	6.226	2.008
fcc	7.899	2.483
exp	6.236	2.000
$\Delta E = 25.7$ mRy		

Table 6.11 Birch fit coefficients

	A_1	A_2	A_3
bcc	1.51486	−90.48954	1104.70659
fcc	2.02724	−115.88166	1434.85790

Table 6.12 Tantalum bcc z = 73 lattice constant = 6.24706 a.u.

Slater-Koster 3-center parameters	Orthogonal	Non-orthogonal	
	Energy integrals (Ry)	Energy integrals (Ry)	Overlap integrals
On site			
s, s (000)	0.85026	0.67677	
x, x (000)	1.56880	1.11962	
xy, xy (000)	1.00672	0.81665	
d2, d2 (000)	0.85361	0.74655	
First neighbor			
s, s (111)	−0.10395	−0.03338	0.10961
s, x (111)	0.06093	−0.02178	−0.10938
s, x y (111)	0.05091	0.01370	−0.04996
x, x (111)	0.08781	−0.06771	−0.05806
x, y (111)	0.04643	−0.08342	−0.12856
x, x y (111)	0.05439	−0.01507	−0.04045
x, yz (111)	0.06161	−0.01883	−0.08300
x, d1 (111)	0.01109	0.00097	−0.04741
xy, xy (111)	−0.02697	−0.01279	0.01393
xy, xz (111)	−0.04223	−0.00726	0.03904
xy, d2 (111)	0.00171	0.01073	−0.03187
d2, d2 (111)	0.03455	0.01121	−0.05112
Second neighbor			
s, s (200)	−0.02145	−0.02658	0.05222
s, x (200)	0.05854	−0.00602	−0.11483
s, d2 (002)	−0.06376	0.00322	0.08331
x, x (200)	0.18537	−0.09678	−0.22518
y, y (200)	0.03609	−0.02111	0.00612
x, xy (020)	0.02857	−0.02059	0.00541
z, d2 (002)	−0.12278	0.06352	0.15825
xy, xy (200)	−0.03389	0.02552	−0.00359
xy, xy (002)	0.00920	−0.00724	−0.01111
d2, d2 (002)	−0.06991	0.01023	0.10265
d1, d1 (002)	0.00838	−0.00799	−0.00544
Third neighbor			
s, s (220)	0.01920		
s, x (220)	−0.01956		
s, xy (220)	−0.01889		
s, d2 (220)	−0.00948		
x, x (220)	−0.01083		
x, x (022)	0.00860		
x, y (220)	−0.00644		
x, xy (220)	−0.01294		
x, xy (022)	0.00378		
z, d2 (022)	0.00251		
z, d1 (022)	−0.00757		
xy, xy (220)	0.01853		
xy, xy (022)	−0.00022		
xy, xz (022)	−0.00178		
xy, d2 (220)	0.01273		
d2, d2 (220)	0.00472		
d1, d1 (220)	−0.00417		

Table 6.13 Tantalum bcc

	Orthogonal			Non-orthogonal		
Band	RMS error mRy	Maximum k	Deviation mRy	RMS error mRy	Maximum k	Deviation mRy
1	3.6	(260)	9.4	0.8	(000)	2.9
2	3.0	(030)	9.5	0.8	(004)	1.7
3	2.4	(226)	7.9	0.7	(161)	3.2
4	2.5	(170)	6.1	0.9	(140)	3.3
5	4.0	(131)	15.5	2.0	(131)	12.5
6	4.2	(226)	12.8	2.1	(140)	7.8
1–6	3.4			1.4		

Table 6.14 Energy values in Ry at selected k-points

	Orthogonal	APW	Non-orthogonal
GAMMA 1	0.12032	0.11141	0.11427
GAMMA 12	0.94875	0.95597	0.95500
GAMMA 15	2.73415	2.73605	2.73604
GAMMA 25′	0.74619	0.74671	0.74606
H1 (008)	1.78358	1.79782	1.79720
H12 (008)	0.39590	0.38913	0.39021
H15 (008)	1.32924	1.33196	1.33186
H25′(008)	1.17771	1.17957	1.18144
N1 (044)	0.36235	0.36442	0.36412
N1 (044)	0.95880	0.96036	0.96151
N1 (044)	1.52476	1.53407	1.53542
N2 (044)	0.57638	0.57925	0.57923
N3 (044)	1.25201	1.25135	1.25269
N4 (044)	1.03506	1.03106	1.03097
N1′ (044)	0.79223	0.79890	0.79898
N3′ (044)	1.91617	1.92414	1.92414
N4′ (044)	1.53509	1.53074	1.53077
P1 (444)	1.20936	1.21458	1.21781
P3 (444)	1.04152	1.04349	1.04441
P4 (444)	0.59643	0.59761	0.59684
P4 (444)	1.60131	1.60939	1.60947
(341)	0.40093	0.40082	0.40014
(341)	0.58666	0.58883	0.58903
(341)	0.71616	0.71565	0.71565
(341)	0.95009	0.95019	0.95013
(341)	1.04277	1.04332	1.04355
(341)	1.26636	1.26324	1.26446

Table 6.15 Fermi level quantities (non-orthogonal fit)

Energy		Densities of states				Velocity	Plasmon energy
	Total	s	p	t_{2g}	e_g		
Ry		States/Ry/atom				cm/s	eV
0.6855	17.08	0.54	2.46	11.25	2.83	0.65 × 10E8	8.77

Integrated densities of states Total Electrons	s	p	t_{2g}	e_g
5.00	0.85	0.36	2.28	1.50

Table 6.16 Tantalum bcc

Slater-Koster 2-center parameters	Orthogonal	Non-orthogonal	
	Energy integrals (Ry)	Energy integrals (Ry)	Overlap integrals
On site			
s	0.93239	0.68141	
p	1.55402	1.12204	
d1	0.96397	0.79667	
d2	0.85308	0.75787	
First neighbor			
(sss)	−0.10325	−0.04163	0.10633
(pps)	0.18033	−0.16179	−0.26748
(ppp)	0.04078	0.01014	0.06312
(dds)	−0.11301	−0.00811	0.10891
(ddp)	0.04803	0.02506	−0.07183
(ddd)	0.00459	−0.01105	0.00544
(sps)	0.10680	0.01613	0.18054
(sds)	−0.10683	−0.01624	0.09669
(pds)	−0.16929	−0.08475	−0.18739
(pdp)	0.01934	−0.01074	0.07397
Second neighbor			
(sss)	−0.02365	−0.01548	0.06061
(pps)	0.19286	−0.18370	−0.27748
(ppp)	0.00802	0.02677	0.02715
(dds)	−0.07121	−0.00100	0.08867
(ddp)	−0.01479	0.00849	−0.02525
(ddd)	0.00916	−0.00648	−0.00503
(sps)	0.06068	0.03657	0.13920
(sds)	−0.04758	0.00194	0.07960
(pds)	−0.10435	−0.05725	−0.14851
(pdp)	−0.01250	0.00367	0.02229
Third neighbor			
(sss)	0.01204		
(pps)	0.01606		
(ppp)	−0.00092		
(dds)	0.01356		
(ddp)	−0.00366		
(ddd)	0.00108		
(sps)	−0.01014		
(sds)	0.00786		
(pds)	−0.00025		
(pdp)	−0.00148		

	Orthogonal			Non-orthogonal		
Band	RMS error mRy	Maximum k	Deviation mRy	RMS error mRy	Maximum k	Deviation mRy
1	4.7	(008)	12.7	1.0	(000)	4.0
2	9.9	(044)	27.6	1.0	(121)	2.8
3	7.0	(008)	25.3	1.5	(226)	5.5
4	10.2	(008)	25.3	2.1	(131)	5.1
5	9.0	(008)	25.3	2.6	(131)	12.2
6	10.9	(226)	26.1	2.6	(140)	7.4
1–6	8.9			1.9		

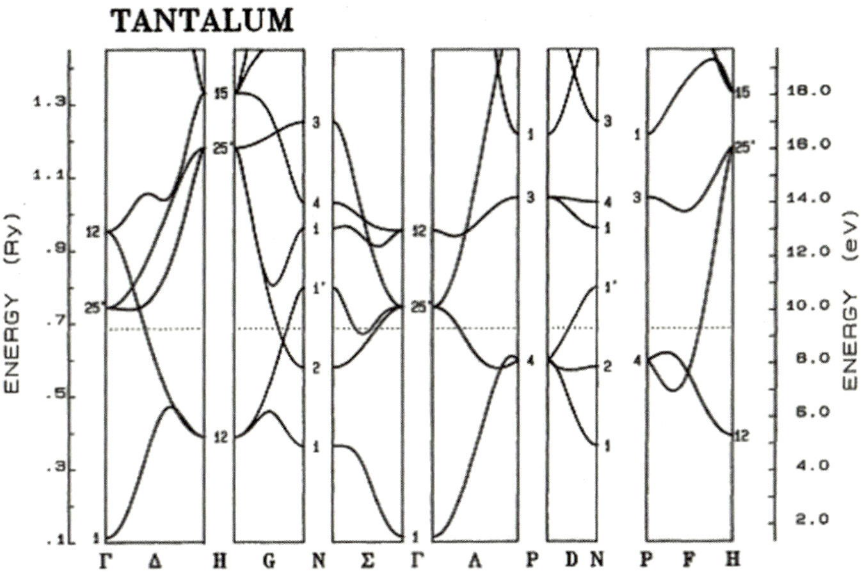

Fig. 6.5 Energy bands for Ta

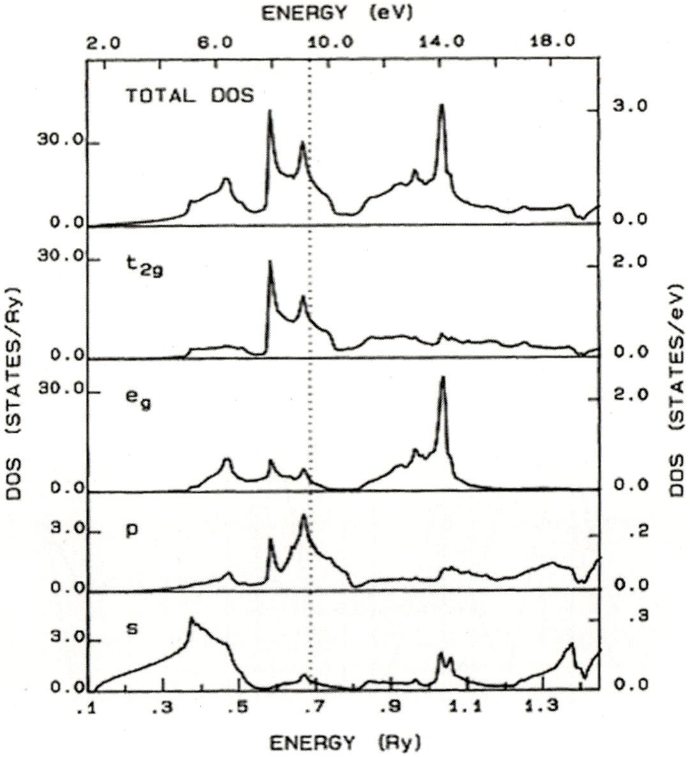

Fig. 6.6 Density of states for Ta

6.3 Tungsten

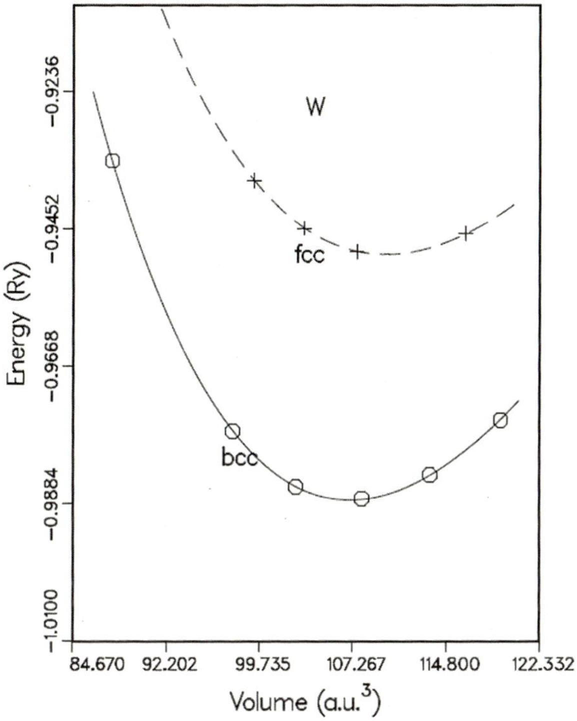

Fig. 6.7 Total energy of W

Table 6.17 Lattice constants and bulk modulus

	a (Bohr)	B (MBar)
bcc	5.983	3.256
fcc	7.605	2.794
exp	5.972	3.232
$\Delta E = 38.6$ mRy		

Table 6.18 Birch fit coefficients

	A_1	A_2	A_3
bcc	1.67804	−120.21872	1355.34568
fcc	1.39980	−107.81907	1237.20774

Table 6.19 Tungsten bcc Z = 74 lattice constant = 5.98042 a.u.

Slater-Koster 3-center parameters	Orthogonal	Non-orthogonal	
	Energy integrals (Ry)	Energy integrals (Ry)	Overlap integrals
On site			
s, s (000)	0.89872	0.71096	
x, x (000)	1.70786	1.22927	
xy, xy (000)	1.01749	0.82432	
d2, d2 (000)	0.86319	0.75774	
First neighbor			
s, s (111)	−0.11385	−0.04258	0.10672
s, x (111)	0.06711	−0.02049	−0.10807
s, xy (111)	0.05646	0.01795	−0.04834
x, x (111)	0.09529	−0.07293	−0.05899
x, y (111)	0.04897	−0.09503	−0.12972
x, x y (111)	0.05790	−0.01541	−0.04079
x, yz (111)	0.06734	−0.01945	−0.08237
x, d1 (111)	0.01159	0.00305	−0.04492
xy, xy (111)	−0.02844	−0.01399	0.01412
xy, xz (111)	−0.04435	−0.01038	0.03753
xy, d2 (111)	0.00240	0.01315	−0.02972
d2, d2 (111)	0.03618	0.01470	−0.04787
Second neighbor			
s, s (200)	−0.02258	−0.03021	0.05088
s, x (200)	0.06290	−0.00364	−0.11167
s, d2 (002)	−0.06800	−0.00019	0.08032
x, x (200)	0.19928	−0.09103	−0.21585
y, y (200)	0.03905	−0.01523	0.00972
x, xy (020)	0.02810	−0.02106	0.00604
z, d2 (002)	−0.13066	0.06360	0.15487
xy, xy (200)	−0.03408	0.02648	−0.00293
xy, xy (002)	0.01043	−0.00704	−0.01064
d2, d2 (002)	−0.07191	0.00370	0.09597
d1, d1 (002)	0.00838	−0.00726	−0.00460
Third neighbor			
s, s (220)	0.02001		
s, x (220)	−0.02078		
s, xy (220)	−0.01918		
s, d2 (220)	−0.00972		
x, x (220)	−0.01172		
x, x (022)	0.01127		
x, y (220)	−0.00617		
x, xy (220)	−0.01274		
x, xy (022)	0.00457		
z, d2 (022)	0.00238		
z, d1 (022)	−0.00757		
xy, xy (220)	0.01849		
xy, xy (022)	−0.00050		
xy, xz (022)	−0.00177		
xy, d2 (220)	0.01280		
d2, d2 (220)	0.00478		
d1, d1 (220)	−0.00437		

Table 6.20 Tungsten bcc

	Orthogonal			Non-orthogonal		
Band	RMS error mRy	Maximum k	Deviation mRy	RMS error mRy	Maximum k	Deviation mRy
1	3.7	(260)	9.3	1.0	(000)	4.2
2	3.0	(030)	9.3	0.8	(260)	1.5
3	2.6	(226)	7.3	0.8	(161)	3.8
4	2.3	(330)	6.0	1.0	(140)	3.7
5	4.0	(131)	16.3	2.1	(131)	13.5
6	4.3	(226)	16.2	2.1	(140)	7.5
1–6	3.4			1.4		

Table 6.21 Energy values in Ry at selected k-points

	Orthogonal	APW	Non-orthogonal
GAMMA 1	0.09253	0.08338	0.08756
GAMMA 12	0.96455	0.97131	0.97032
GAMMA 15	2.97629	2.97853	2.97852
GAMMA 25′	0.74448	0.74512	0.74471
H1 (008)	1.91418	1.92845	1.92755
H12 (008)	0.38566	0.37863	0.37985
H15 (008)	1.45162	1.45497	1.45473
H25′(008)	1.19956	1.20182	1.20378
N1 (044)	0.35096	0.35194	0.35190
N1 (044)	0.97326	0.97367	0.97408
N1 (044)	1.59412	1.60609	1.60739
N2 (044)	0.56790	0.57186	0.57160
N3 (044)	1.27743	1.27628	1.27791
N4 (044)	1.04986	1.04519	1.04507
N1′ (044)	0.87249	0.87915	0.87934
N3′ (044)	2.08901	2.09299	2.09306
N4′ (044)	1.65598	1.65291	1.65296
P1 (444)	1.27436	1.28089	1.28432
P3 (444)	1.05627	1.05745	1.05859
P4 (444)	0.61272	0.61360	0.61272
P4 (444)	1.69462	1.71345	1.71272
(341)	0.39387	0.39355	0.39286
(341)	0.58240	0.58487	0.58507
(341)	0.76376	0.76280	0.76277
(341)	0.96595	0.96592	0.96544
(341)	1.06393	1.06426	1.06453
(341)	1.30685	1.30446	1.30535

Table 6.22 Fermi level quantities (non-orthogonal fit)

Energy		Densities of states				Velocity	Plasmon energy
	Total	s	p	t_{2g}	e_g		
Ry		States/Ry/atom				cm/s	eV
0.8320	5.73	0.18	0.68	3.68	1.18	0.89 × 10E8	7.39

Integrated densities of states Total Electrons	s	p	t_{2g}	e_g
6.00	0.86	0.40	3.06	1.67

Table 6.23 Tungsten bcc

Slater-Koster 2–center parameters	Orthogonal	Non-orthogonal	
	Energy integrals (Ry)	Energy integrals (Ry)	Overlap integrals
On site			
s	0.99861	0.71845	
p	1.68521	1.22942	
d1	0.96813	0.80748	
d2	0.85720	0.76907	
First neighbor			
(sss)	−0.11258	−0.05239	0.10274
(pps)	0.19331	−0.17401	−0.26232
(ppp)	0.04787	0.00752	0.05889
(dds)	−0.11837	−0.01838	0.10257
(ddp)	0.05194	0.02974	−0.06762
(ddd)	0.00248	−0.01064	0.00580
(sps)	0.09945	0.01020	0.17674
(sds)	−0.11301	−0.02597	0.09176
(pds)	−0.18136	−0.08138	−0.18189
(pdp)	0.02424	−0.01365	0.07086
Second neighbor			
(sss)	−0.02498	−0.01762	0.05958
(pps)	0.22704	−0.20384	−0.27698
(ppp)	−0.00848	0.03427	0.02975
(dds)	−0.07313	−0.00755	0.08213
(ddp)	−0.01264	0.00868	−0.02455
(ddd)	0.00908	−0.00595	−0.00450
(sps)	0.08001	0.04064	0.13837
(sds)	−0.05039	−0.00267	0.07543
(pds)	−0.10550	−0.05473	−0.14320
(pdp)	−0.00750	0.00659	0.02438
Third neighbor			
(sss)	0.01143		
(pps)	0.04228		
(ppp)	−0.00822		
(dds)	0.01282		
(ddp)	−0.00313		
(ddd)	0.00118		
(sps)	−0.00285		
(sds)	0.00997		
(pds)	−0.00414		
(pdp)	0.00146		

	Orthogonal			Non-orthogonal		
Band	RMS error mRy	Maximum k	Deviation mRy	RMS error mRy	Maximum k	Deviation mRy
1	4.9	(050)	15.2	1.3	(000)	5.4
2	9.6	(044)	27.8	1.0	(226)	2.2
3	7.0	(008)	19.8	1.6	(226)	5.7
4	9.5	(250)	20.8	2.2	(131)	5.2
5	8.3	(170)	21.1	2.7	(131)	13.3
6	13.1	(342)	37.1	2.6	(140)	7.1
1–6	9.1			2.0		

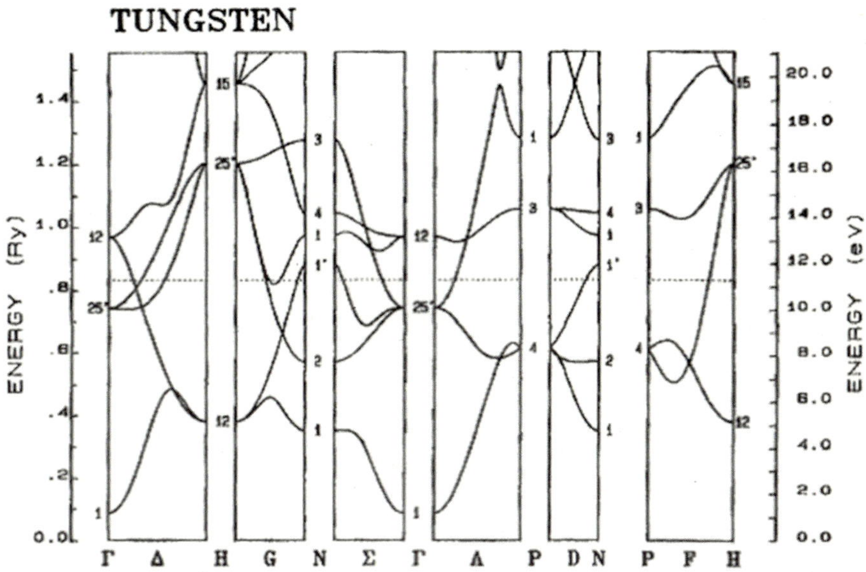

Fig. 6.8 Energy bands for W

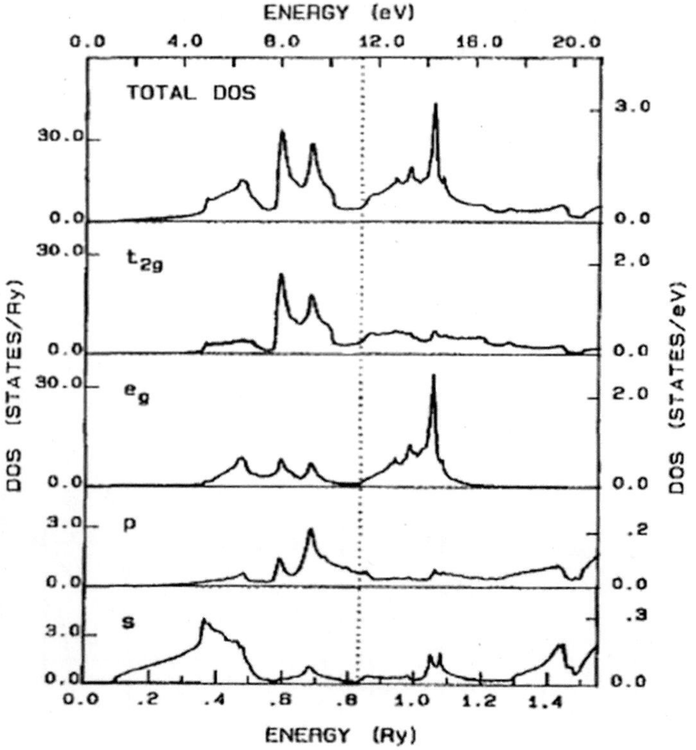

Fig. 6.9 Density of states for W

6.4 Rhenium

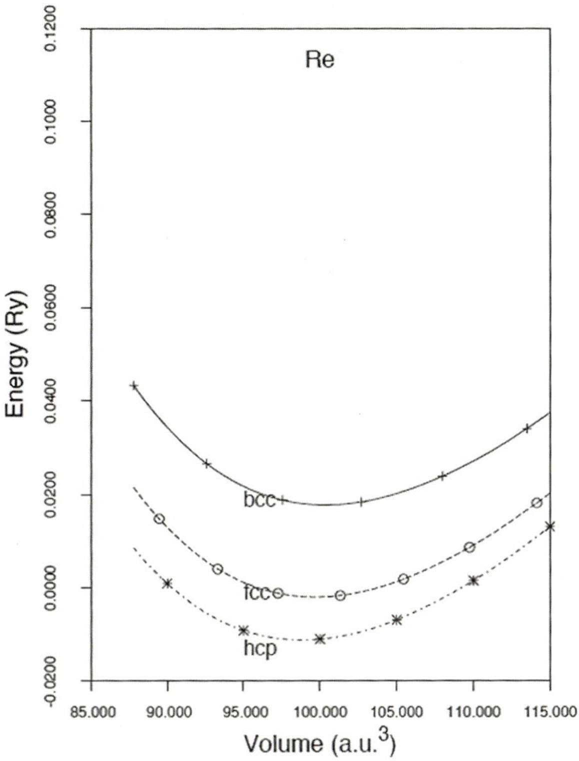

Fig. 6.10 Total energy of Re

Table 6.24 Lattice constants and bulk modulus

	a (Bohr)	c (Bohr)	B (Mbar)
fcc	7.361		3.749
bcc	5.856		3.606
hcp	5.248	8.287	3.690
exp	5.216	8.428	3.72
$\Delta E = 11.1$ mRy			

Table 6.25 Birch fit coefficients

	A_1	A_2	A_3	A_4
bcc	1.15649	−13.98751	−989.68787	16427.74556
fcc	1.03193	−5.23259	−1208.83003	18133.46748
hcp	0.98166	−4.05415	−1187.54847	17540.52111

Table 6.26 Rhenium hcp Z = 75 lattice constants = 5.2172 a.u. 8.4248 a.u.

Slater-Koster 3-center parameters	Orthogonal		
	Energy integrals (Ry)		Energy integrals (Ry)
On site			
s, s (0)	1.04023		
x, x (0)	1.56358		
z, z (0)	1.55123		
xy, xy (0)	0.84287		
yz, yz (0)	0.83850		
d2, d2 (0)	0.76864		
s, d2 (0)	0.00270		
x, xy (0)	−0.02717		
First neighbor			
s, s (R)	−0.05159	s, s (T)	−0.10863
s, x (R)	0.09168	s, y (T)	−0.07462
s, y (R)	0.03559	s, z (T)	0.10426
s, xy (R)	0.00835	s, d1 (T)	−0.03117
s, d1 (R)	0.08149	s, yz (T)	−0.09767
s, d2 (R)	−0.05068	s, d2 (T)	0.02063
x, x (R)	0.09238	x, x (T)	−0.03082
y, y (R)	−0.07462	y, y (T)	0.00225
x, y (R)	0.08188	y, z (T)	−0.11456
x, xy (R)	0.01791	x, xy (T)	0.01858
y, d1 (R)	−0.00705	y, d1 (T)	−0.00061
x, d1 (R)	0.12432	x, xz (T)	−0.03234
y, xy (R)	−0.01318	y, yz (T)	0.05622
y, d2 (R)	0.00880	y, d2 (T)	−0.04250
x, d2(R)	−0.05294	z, z (T)	0.08393
z, z (R)	−0.03867	z, d1 (T)	−0.04640
z, yz (R)	−0.01259	z. yz (T)	−0.12008
z, xz (R)	−0.01862	z, d2 (T)	0.01811
xy, xy (R)	0.02861	xy, xy (T)	0.01441
d1, d1 (R)	−0.07410	d1, d1 (T)	−0.00338
xy, d1 (R)	−0.01462	xy, yz (T)	−0.01478
d1, d2 (R)	0.05075	yz, d1 (T)	−0.02273
xy, d2 (R)	0.00180	d1, d2 (T)	0.03646
yz, yz(R)	−0.00086	yz, yz (T)	−0.07116
xz, xz (R)	0.02867	xz, xz (T)	0.03370
yz, xz (R)	−0.01762	yz, d2 (T)	0.04628
d2, d2 (R)	−0.03877	d2, d2 (T)	0.01395

Table 6.27 Rhenium hcp

Band	Orthogonal RMS error mRy	Maximum k	Deviation mRy
1	4.7	(16 0 0)	7.5
2	4.5	(0 12 0)	10.9
3	4.9	(8 0 12)	10.4
4	6.1	(24 0 0)	19.2
5	7.0	(24 0 3)	17.4
6	5.1	(0 12 12)	13.8
7	5.6	(0 12 12)	13.8
8	6.5	(0 12 12)	13.8
9	6.3	(24 0 3)	13.0
10	6.3	(16 0 0)	20.4
11	6.7	(0 0 9)	13.6
12	6.8	(0 0 9)	13.6
1–12	5.95		

Table 6.28 Energy values in Ry at selected k-points

	Orthogonal	APW
GAMMA 1+	0.02578	0.01942
GAMMA 1+	0.67285	0.67487
GAMMA 3+	0.81561	0.82253
GAMMA 5+	0.73585	0.73386
GAMMA 6+	0.80956	0.80188
GAMMA 4–	0.28730	0.29170
GAMMA 5–	1.03430	1.02801
GAMMA 6–	0.67058	0.66831
M1+	0.33505	0.33926
M1+	0.65781	0.65113
M1+	1.23801	1.23572
M2+	1.11691	1.10792
M3+	0.38049	0.38551
M4+	1.11702	1.11867
M1–	0.62255	0.64177
M2–	0.35149	0.35092
M2–	0.73892	0.74475
M2–	1.29759	1.30568
M3–	1.06915	1.07894
M4–	0.68545	0.69013
A1	0.20173	0.20297
A1	0.61479	0.61424
A3	0.65720	0.66031
A3	0.96379	0.97423
L1	0.30135	0.30654
L1	0.49098	0.49048
L1	0.85366	0.85437
L1	1.14285	1.14027
L2	0.82026	0.81355
L2	1.10472	1.10135

Table 6.29 Rhenium hcp Z = 75 lattice constants = 5.2172 a.u. 8.4248 a.u.

Slater-Koster 2-center parameters	Orthogonal	Non-orthogonal	
	Energy integrals (Ry)	Energy integrals (Ry)	Overlap integrals
On site			
s	1.07617	0.65983	
p	1.51212	1.30399	
d0	0.80316	0.71266	
d1	0.78555	0.71522	
d2	0.80697	0.72152	
pd	−0.05100	−0.05949	−0.06893
First neighbor			
(sss)	−0.08439	−0.04764	0.08499
(sps)	0.11274	0.00195	−0.13535
(sds)	0.08037	0.02221	−0.10266
(pps)	0.10453	0.03476	−0.06824
(ppp)	−0.07072	0.07973	0.09387
(pds)	0.10892	−0.00902	−0.09506
(pdp)	−0.03949	0.03037	0.07831
(dds)	−0.10598	−0.02209	0.09790
(ddp)	0.04907	0.02461	−0.05612
(ddd)	−0.00544	0.00255	0.01431
Second neighbor			
(sss)	−0.00689	−0.00818	−0.00230
(sps)	−0.01395	0.02340	−0.00426
(sds)	−0.01050	0.02821	0.01390
(pps)	0.02055	0.08931	0.01650
(ppp)	−0.01144	0.03380	0.04093
(pds)	0.01744	0.09546	0.06353
(pdp)	−0.01135	0.01320	0.01657
(dds)	−0.00675	−0.03670	−0.01798
(ddp)	0.00049	−0.00033	−0.00291
(ddd)	0.00162	0.00234	0.00145
Third neighbor			
(sss)	−0.00176	0.00249	0.00184
(sps)	−0.00376	−0.04879	−0.05689
(sds)	−0.00266	0.00163	0.00378
(pps)	−0.00015	−0.02435	−0.04159
(ppp)	0.01301	0.01400	0.02707
(pds)	−0.00088	−0.00228	−0.01879
(pdp)	0.00661	0.01241	0.01851
(dds)	0.00114	−0.01088	−0.01297
(ddp)	0.00159	−0.00419	−0.00426
(ddd)	−0.00210	0.00629	0.00608

Table 6.30 Rhenium hcp

Band	RMS error mRy	Maximum k	Deviation mRy	RMS error mRy	Maximum k	Deviation mRy
1	14.7	(24 0 0)	33.6	2.5	(16 0 12)	5.9
2	11.4	(0 18 0)	23.8	2.8	(16 0 12)	5.9
3	10.3	(24 0 9)	26.0	3.1	(24 0 6)	8.3
4	9.7	(24 0 9)	23.9	2.8	(24 0 9)	7.9
5	5.8	(0 24 12)	14.4	3.4	(8 0 0)	7.7
6	5.7	(0 24 12)	14.4	3.0	(0 12 12)	5.3
7	13.1	(16 0 12)	31.3	3.1	(8 0 12)	8.7
8	16.6	(16 0 9)	34.7	3.9	(8 0 12)	8.7
9	12.7	(24 0 0)	32.2	2.9	(0 18 0)	6.9
10	11.7	(0 12 0)	20.8	3.6	(16 0 0)	11.8
11	17.6	(0 18 0)	30.6	1.8	(8 0 12)	4.7
12	17.2	(8 0 0)	46.2	2.3	(4 0 0)	4.8
1–12	12.79			2.99		

Table 6.31 Energy values in Ry at selected k-points

	Orthogonal	APW	Non-orthogonal
GAMMA 1+	0.01860	0.01942	0.02192
GAMMA 1+	0.66560	0.67487	0.67625
GAMMA 3+	0.81547	0.82253	0.81935
GAMMA 5+	0.73477	0.73386	0.73265
GAMMA 6+	0.80451	0.80188	0.80341
GAMMA 4–	0.29556	0.29170	0.29390
GAMMA 5–	1.03467	1.02801	1.03086
GAMMA 6–	0.66225	0.66831	0.66550
M1+	0.30569	0.33926	0.33632
M1+	0.64399	0.65113	0.65462
M1+	1.25586	1.23572	1.23467
M2+	1.07571	1.10792	1.11070
M3+	0.39260	0.38551	0.38289
M4+	1.09865	1.11867	1.11927
M1–	0.63253	0.64177	0.64467
M2–	0.36274	0.35092	0.35187
M2–	0.73560	0.74475	0.73871
M2–	1.29536	1.30568	1.30476
M3–	1.07375	1.07894	1.08034
M4–	0.68550	0.69013	0.68845
A1	0.20929	0.20297	0.20300
A1	0.61918	0.61424	0.61484
A3	0.66068	0.66031	0.66025
A3	0.95736	0.97423	0.97362
L1	0.31738	0.30654	0.30569
L1	0.46698	0.49048	0.48733
L1	0.85715	0.85437	0.85531
L1	1.12772	1.14027	1.13725
L2	0.82795	0.81355	0.81072
L2	1.09671	1.10135	1.10369

Table 6.32 Fermi level quantities 3 (non-orthogonal fit)

Energy	Densities of states				
	Total	s	p	t_{2g}	e_g
Ry		States/Ry/atom			
0.8016	11.04	0.23	1.28	6.01	3.52

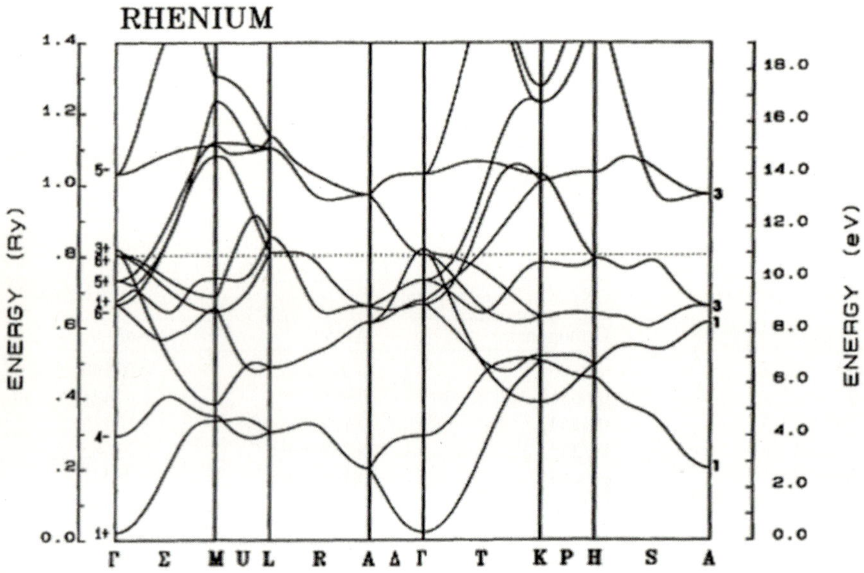

Fig. 6.11 Energy bands for Re

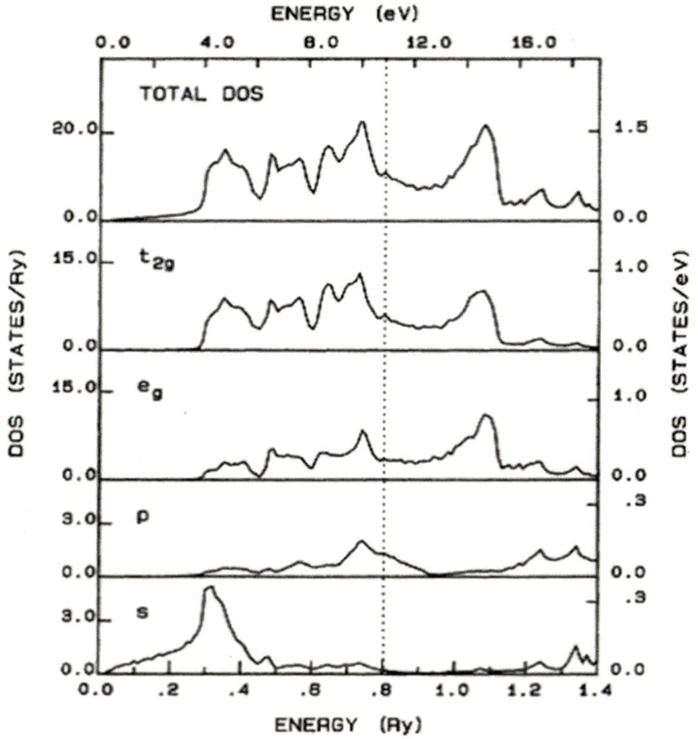

Fig. 6.12 Density of states for Re

6.5 Osmium

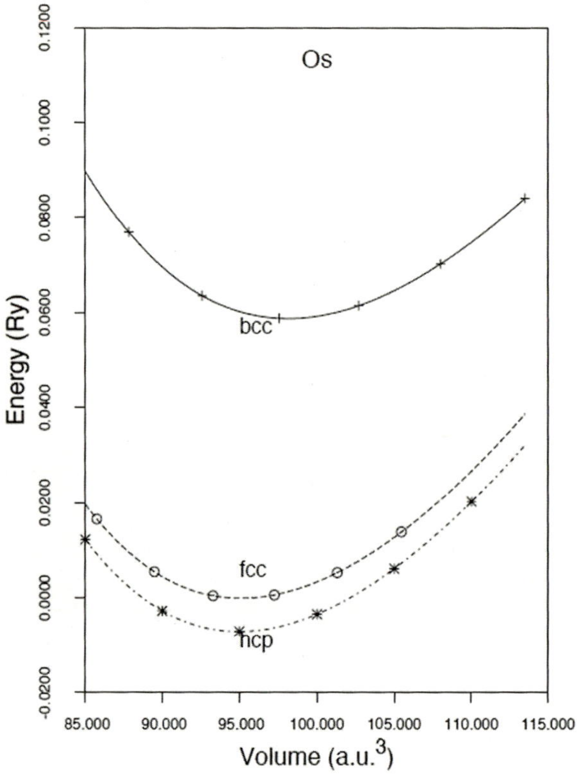

Fig. 6.13 Total energy of Os

Table 6.33 Lattice constants and bulk modulus

	a (Bohr)	c (Bohr)	B (Mbar)
fcc	7.242		4.393
bcc	5.811		4.028
hcp	5.192	8.136	4.420
exp	5.178	8.164	(4.18)
$\Delta E = 7.0$ mRy			

Table 6.34 Birch fit coefficients

	A_1	A_2	A_3	A_4
bcc	2.38587	−84.21189	423.53435	6696.74054
fcc	2.25311	−74.26462	160.45069	8516.11112
hcp	2.40483	−83.78837	352.92331	7204.59861

Table 6.35 Osmium hcp Z = 76 lattice constants = 5.1683 a.u. 8.1617 a.u.

Slater–Koster 3–center parameters	Orthogonal		
	Energy integrals (Ry)		Energy integrals (Ry)
On site			
s, s (0)	1.11951		
x, x (0)	1.78792		
z, z (0)	1.44110		
xy, xy (0)	0.77327		
yz, yz (0)	0.76097		
d2, d2 (0)	0.68812		
s, d2 (0)	−0.04343		
x, xy (0)	0.03811		
First neighbor			
s, s (R)	−0.02594	s, s (T)	−0.14173
s, x (R)	0.06910	s, y (T)	−0.11031
s, y (R)	0.07192	s, z (T)	0.11105
s, xy (R)	0.02369	s, d1 (T)	−0.02832
s, d1 (R)	0.08351	s, yz (T)	−0.08739
s, d2 (R)	−0.04930	s, d2 (T)	0.01452
x, x (R)	0.01928	x, x (T)	0.01604
y, y (R)	−0.17280	y, y (T)	0.04905
x, y (R)	0.17620	y, z (T)	−0.16707
x, xy (R)	0.03013	x, xy (T)	0.01261
y, d1 (R)	−0.01948	y, d1 (T)	−0.00397
x, d1 (R)	0.12554	x, xz (T)	−0.03461
y, xy (R)	−0.03400	y, yz (T)	0.05056
y, d2 (R)	0.01023	y, d2 (T)	−0.05598
x, d2 (R)	−0.04276	z, z (T)	0.04894
z, z (R)	−0.05485	z, d1 (T)	−0.02982
z, yz (R)	−0.00787	z, yz (T)	−0.10533
z, xz (R)	−0.02600	z, d2 (T)	0.01273
xy, xy (R)	0.02808	xy, xy (T)	0.01738
d1, d1 (R)	−0.06851	d1, d1 (T)	−0.00449
xy, d1 (R)	−0.01276	xy, yz (T)	−0.01642
d1, d2 (R)	0.04187	yz, d1 (T)	−0.01986
xy, d2 (R)	−0.00742	d1, d2 (T)	0.03653
yz, yz (R)	−0.00124	yz, yz (T)	−0.07071
xz, xz (R)	0.02935	xz, xz (T)	0.03237
yz, xz (R)	−0.01236	yz, d2 (T)	0.05228
d2, d2 (R)	−0.04036	d2, d2 (T)	0.01289

Table 6.36 Osmium hcp

Band	Orthogonal RMS error mRy	Maximum k	Deviation mRy
1	5.6	(0 18 0)	13.9
2	8.2	(0 18 0)	24.8
3	5.7	(0 12 12)	11.9
4	7.7	(24 0 0)	23.9
5	9.2	(24 0 3)	20.5
6	5.7	(0 12 12)	16.7
7	5.5	(24 0 9)	14.6
8	8.0	(24 0 3)	18.4
9	6.5	(24 0 3)	11.0
10	8.0	(16 0 0)	20.4
11	13.0	(16 0 12)	42.7
12	13.5	(16 0 12)	42.7
1–12	8.45		

Table 6.37 Energy values in Ry at selected k-points

	Orthogonal	APW
GAMMA 1+	−0.00647	−0.01042
GAMMA 1+	0.64334	0.63410
GAMMA 3+	0.81835	0.83038
GAMMA 5+	0.67070	0.67306
GAMMA 6+	0.73028	0.72307
GAMMA 4–	0.25226	0.25406
GAMMA 5–	0.96031	0.95431
GAMMA 6–	0.61116	0.60725
M1+	0.29201	0.29561
M1+	0.62103	0.61301
M1+	1.15866	1.15988
M2+	1.03852	1.03537
M3+	0.34643	0.34811
M4+	1.02179	1.02556
MI–	0.54954	0.57346
M2–	0.31227	0.31314
M2–	0.69469	0.69934
M2–	1.23867	1.24623
M3–	0.96753	0.98407
M4–	0.65016	0.65206
A1	0.17834	0.17778
Al	0.57332	0.57660
A3	0.59356	0.59501
A3	0.89371	0.90155
L1	0.26760	0.26959
L1	0.44558	0.44755
L1	0.82475	0.82460
L1	1.07369	1.07894
L2	0.75771	0.74595
L2	1.01008	1.00818

Table 6.38 Osmium hcp Z = 76 lattice constants = 5.1688 a.u. 8.1617 a.u.

Slater-Koster 2-center parameters	Orthogonal	Non-orthogonal	
	Energy integrals (Ry)	Energy integrals (Ry)	Overlap integrals
On site			
s	1.04541	0.61619	
p	1.51468	1.31857	
d0	0.73187	0.65416	
dl	0.72480	0.66426	
d2	0.74935	0.68179	
pd	−0.04509	−0.06024	−0.09230
First neighbor			
(sss)	−0.08605	−0.04672	0.08278
(sps)	0.11762	−0.00026	−0.15065
(sds)	0.07972	0.02767	−0.10137
(pps)	0.11915	0.03182	−0.08240
(ppp)	−0.04613	0.07394	0.09243
(pds)	0.10441	0.00201	−0.08689
(pdp)	−0.03664	0.02599	0.07701
(dds)	−0.10118	−0.02692	0.09389
(ddp)	0.04527	0.02462	−0.05158
(ddd)	−0.00426	0.00510	0.01632
Second neighbor			
(sss)	−0.00548	−0.01177	−0.01043
(sps)	−0.01346	0.02529	−0.00627
(sds)	−0.00502	0.02784	0.01282
(pps)	0.03197	0.08320	0.00793
(ppp)	0.00158	0.03847	0.04857
(pds)	0.01928	0.10208	0.07944
(pdp)	−0.00915	0.01363	0.01346
(dds)	−0.00920	−0.03661	−0.01935
(ddp)	0.00090	0.00064	−0.00027
(ddd)	0.00308	0.00146	−0.00227
Third neighbor			
(sss)	0.00366	−0.00492	−0.01909
(sps)	0.00488	−0.05507	−0.06894
(sds)	−0.00852	0.00245	0.01092
(pps)	0.01867	−0.01511	−0.05375
(ppp)	0.00255	0.00923	0.03603
(pds)	−0.00546	0.01188	−0.01667
(pdp)	0.00613	0.01743	0.03411
(dds)	0.00552	−0.01720	−0.02607
(ddp)	0.00013	−0.00776	−0.00791
(ddd)	−0.00082	0.00820	0.00809

Table 6.39 Osmium hcp

	Orthogonal			Non-orthogonal		
Band	RMS error mRy	Maximum k	Deviation mRy	RMS error mRy	Maximum k	Deviation mRy
1	13.1	(24 0 0)	25.9	3.4	(16 0 12)	7.1
2	8.7	(8 0 12)	15.9	5.1	(24 0 9)	10.4
3	12.9	(24 0 6)	32.3	6.1	(24 0 6)	16.4
4	12.2	(24 0 9)	35.9	6.7	(24 0 9)	16.3
5	7.2	(0 24 12)	18.7	6.2	(0 12 0)	13.3
6	8.0	(0 24 12)	18.7	6.0	(0 12 0)	14.8
7	14.9	(0 12 12)	36.5	6.1	(8 0 12)	14.2
8	17.0	(0 12 12)	36.5	8.2	(16 0 3)	15.0
9	13.2	(24 0 0)	32.6	6.8	(0 12 12)	15.5
10	12.5	(16 0 0)	21.2	8.4	(16 0 0)	18.9
11	22.0	(16 0 12)	53.7	9.8	(16 0 12)	30.2
12	23.5	(8 0 0)	63.4	9.6	(16 0 12)	30.2
1–12	14.59			7.07		

Table 6.40 Energy values in Ry at selected k-points

	Orthogonal	APW	Non-Orthogonal
GAMMA 1+	−0.01324	−0.01042	−0.01320
GAMMA 1+	0.62686	0.63410	0.63329
GAMMA 3+	0.82067	0.83038	0.82259
GAMMA 5+	0.66817	0.67306	0.66827
GAMMA 6+	0.73526	0.72307	0.73446
GAMMA 4–	0.25655	0.25406	0.25585
GAMMA 5–	0.96088	0.95431	0.95826
GAMMA 6–	0.59677	0.60725	0.59992
M1+	0.26969	0.29561	0.29541
M1+	0.60399	0.61301	0.61893
M1+	1.17732	1.15988	1.15669
M2+	0.99298	1.03537	1.03479
M3+	0.35650	0.34811	0.34468
M4+	1.01542	1.02556	1.02739
M1–	0.57322	0.57346	0.58480
M2–	0.31991	0.31314	0.31053
M2–	0.68842	0.69934	0.69028
M2–	1.23686	1.24623	1.24638
M3–	0.98946	0.98407	0.99415
M4–	0.64387	0.65206	0.64302
A1	0.17766	0.17778	0.18231
A1	0.57654	0.57660	0.57488
A3	0.59660	0.59501	0.59633
A3	0.88345	0.90155	0.89893
L1	0.27393	0.26959	0.26533
L1	0.41703	0.44755	0.44004
L1	0.82969	0.82460	0.82676
L1	1.07625	1.07894	1.07623
L2	0.76467	0.74595	0.74801
L2	1.01011	1.00818	1.01498

Table 6.41 Fermi level quantities (non-orthogonal fit)

Energy		Densities of states			
	Total	s	p	t_{2g}	e_g
Ry		States/Ry/atom			
0.8528	9.49	0.10	0.49	4.54	4.36

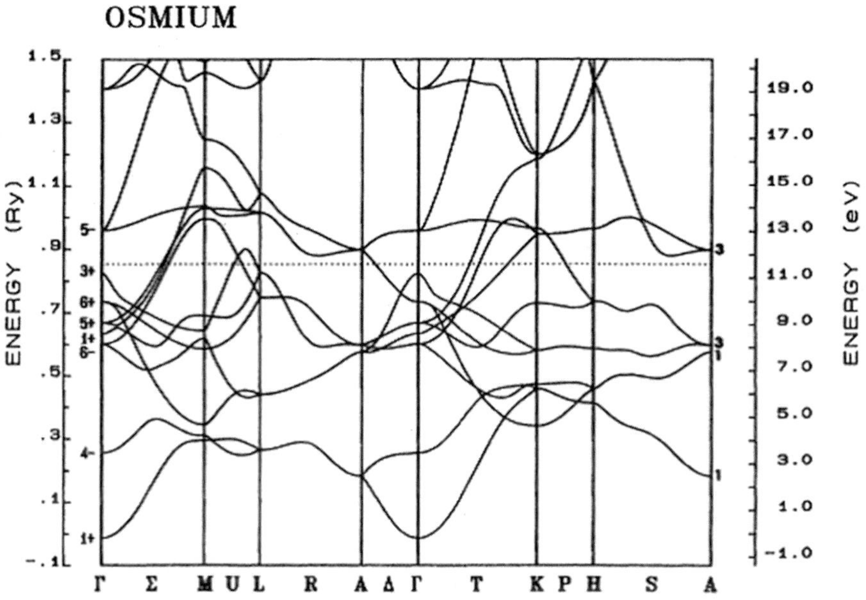

Fig. 6.14 Energy bands for Os

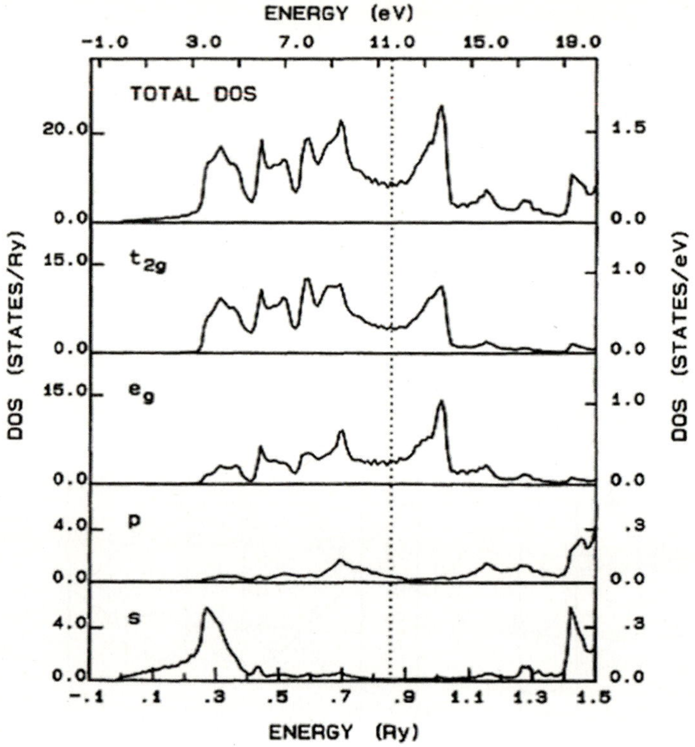

Fig. 6.15 Density of states for Os

6.6 Iridium

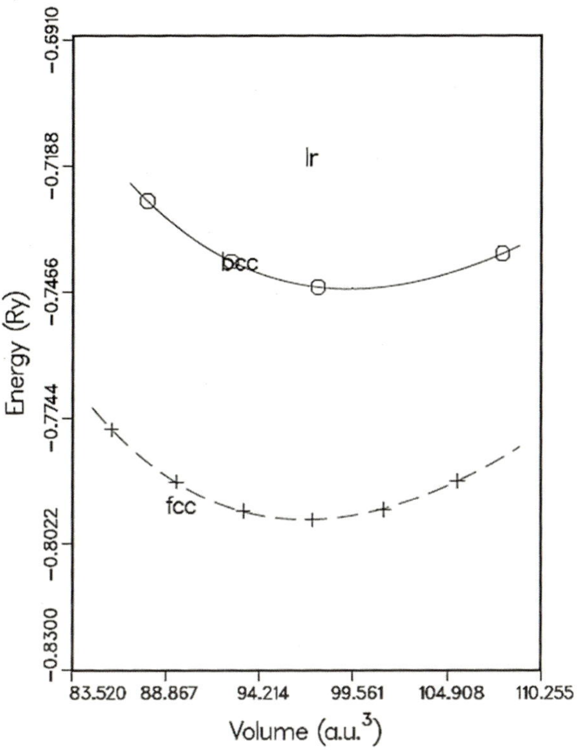

Fig. 6.16 Total energy of Ir

Table 6.42 Lattice constants and bulk modulus

	a (Bohr)	B (MBar)
fcc	7.290	3.763
bcc	5.835	3.481
exp	7.257	3.550
$\Delta E = 50.9$ mRy		

Table 6.43 Birch fit coefficients

	A_1	A_2	A_3
bcc	1.89846	−113.44760	1216.72883
fcc	1.99036	−117.56892	1239.81701

Table 6.44 Iridium fcc Z = 77 lattice constant = 7.25542 a.u.

Slater-Koster 3-center parameters	Orthogonal	Non-orthogonal	
	Energy integrals (Ry)	Energy integrals (Ry)	Overlap integrals
On site			
s, s (000)	0.88951	0.76986	
x, x (000)	1.62418	1.42980	
xy, xy (000)	0.66053	0.58049	
d2, d2 (000)	0.57396	0.57091	
First neighbor			
s, s (110)	−0.08953	−0.06469	0.02370
s, x (110)	0.08652	0.04409	−0.05191
s, xy (110)	−0.07778	−0.05695	0.05195
s, d2 (110)	0.04184	0.03314	0.00631
x, x (110)	0.07847	0.02062	−0.05367
x, x (011)	−0.00483	0.02772	0.04594
x, y (110)	0.12200	0.02397	−0.06761
x, xy (110)	−0.09183	−0.05052	0.03506
x, xy (011)	0.00433	−0.00775	−0.03262
z, d2 (011)	0.00833	0.01725	0.00000
z, d1 (011)	0.04602	0.02195	−0.02667
xy, xy (110)	−0.06408	−0.02822	0.06149
xy, xy (011)	0.01565	0.01115	−0.02210
xy, xz(011)	0.00349	0.00551	−0.02423
xy, d2 (110)	0.03325	0.02913	−0.00477
d2, d2 (110)	−0.02662	−0.01613	0.01365
d1, d1 (110)	0.04344	0.04593	0.00407
Second neighbor			
s, s (200)	0.02162	−0.00841	−0.03995
s, x (200)	−0.02635	0.06216	0.05732
s, d2 (002)	0.01040	−0.02043	−0.04571
x, x (200)	−0.01954	0.13720	0.05804
y, y (200)	−0.00575	−0.00630	0.00823
x, xy (020)	−0.00622	−0.00465	−0.00195
z, d2 (002)	0.00670	−0.04547	−0.05346
xy, xy (200)	−0.01192	−0.00852	0.00064
xy, xy (002)	0.00210	−0.00055	0.00000
d2, d2 (002)	−0.00236	−0.01469	−0.02044
d1, d1 (002)	0.00230	0.00782	0.01282

Table 6.45 Iridium fcc

	Orthogonal			Non-orthogonal		
Band	RMS error Ry	Maximum k	Deviation mRy	RMS error mRy	Maximum k	Deviation mRy
1	3.0	(055)	11.1	0.6	(174)	1.6
2	4.0	(264)	7.6	1.0	(008)	2.8
3	3.5	(044)	7.8	1.3	(007)	5.7
4	3.7	(118)	16.7	2.6	(118)	13.7
5	2.9	(380)	7.0	1.1	(000)	3.1
6	4.1	(380)	7.7	1.1	(000)	3.1
1–6	3.6			1.4		

Table 6.46 Energy values in Ry at selected k-points

	Orthogonal	APW	Non-orthogonal
GAMMA 1	−0.05510	−0.05398	−0.05444
GAMMA 12	0.67470	0.66983	0.67293
GAMMA 15	2.17058	2.16383	2.16388
GAMMA 25′	0.48588	0.48494	0.48672
X1 (008)	0.16729	0.16863	0.16975
X1 (008)	1.46990	1.51284	1.51293
X2 (008)	0.82037	0.81615	0.81794
X3 (008)	0.23550	0.23799	0.24083
X5 (008)	0.87336	0.86928	0.86995
X4′ (008)	0.91499	0.92157	0.92210
X5′ (008)	1.58144	1.62515	1.62462
L1 (444)	0.17743	0.18081	0.18155
L1 (444)	1.25845	1.28221	1.28054
L3 (444)	0.46801	0.46935	0.47043
L3 (444)	0.82414	0.82343	0.82405
L2′ (444)	0.71024	0.71017	0.71045
L3′ (444)	2.17426	2.19880	2.19869
W1 (048)	0.62420	0.62381	0.62578
W1 (048)	1.35635	1.34975	1.34934
W3 (048)	0.42689	0.41987	0.42100
W3 (048)	1.44645	1.44579	1.44750
W1′ (048)	0.86497	0.87023	0.87287
W2′ (048)	0.29173	0.29041	0.29146
W2′ (048)	1.75866	1.71668	1.71718
EVEN (224)	0.20635	0.20653	0.20697
EVEN (224)	0.44097	0.43834	0.43771
EVEN (224)	0.47434	0.47238	0.47251
EVEN (224)	0.77213	0.77333	0.77402
ODD (224)	0.53268	0.53806	0.53757
ODD (224)	0.79667	0.79461	0.79493

Table 6.47 Fermi level quantities (non-orthogonal fit)

Energy		Densities of states				Velocity	Plasmon energy
	Total	s	p	t_{2g}	e_g		
Ry		States/Ry/atom				cm/s	eV
0.7620	12.71	0.12	0.33	7.35	4.91	0.79 × 10E8	10.33

Integrated densities of states				
Total Electrons	s	p	t_{2g}	e_g
9.00	0.77	0.59	4.28	3.37

Table 6.48 Iridium fcc

Slater-Koster 2-center parameters	Orthogonal	Non-orthogonal	
	Energy integrals (Ry)	Energy integrals (Ry)	Overlap integrals
On site			
s	0.97100	0.59582	
p	1.62396	1.12557	
d1	0.59951	0.53710	
d2	0.57185	0.53547	
First neighbor			
(sss)	−0.08547	−0.05395	0.08810
(pps)	0.18728	−0.20949	−0.31121
(ppp)	−0.03000	−0.04106	0.03464
(dds)	−0.08408	−0.03645	0.07780
(ddp)	0.04211	0.02941	−0.04304
(ddd)	−0.00673	−0.00550	0.00410
(sps)	0.12129	−0.00261	−0.17025
(sds)	−0.07255	−0.03194	0.08363
(pds)	−0.10110	0.02779	0.15064
(pdp)	0.02813	0.02005	−0.04338
Second neighbor			
(sss)	−0.00084	−0.00986	0.00184
(pps)	0.02919	0.04795	−0.02402
(ppp)	−0.00647	0.00170	0.01108
(dds)	−0.00245	−0.02346	−0.01468
(ddp)	0.00180	0.00141	−0.00183
(ddd)	0.00007	0.00158	0.00263
(sps)	−0.00120	0.02136	−0.00557
(sds)	−0.00756	−0.01436	−0.00081
(pds)	−0.01703	−0.04351	−0.01031
(pdp)	0.00559	−0.00039	−0.00670

	Orthogonal			Non-orthogonal		
Band	RMS error mRy	Maximum k	Deviation mRy	RMS error mRy	Maximum k	Deviation mRy
1	12.8	(084)	25.9	1.1	(055)	3.0
2	13.7	(007)	32.6	1.6	(264)	3.3
3	9.5	(044)	23.6	2.1	(044)	5.7
4	12.6	(084)	34.8	3.0	(118)	14.6
5	12.7	(044)	29.1	1.6	(000)	4.4
6	15.9	(066)	55.2	1.3	(000)	4.4
1–6	13.0			1.9		

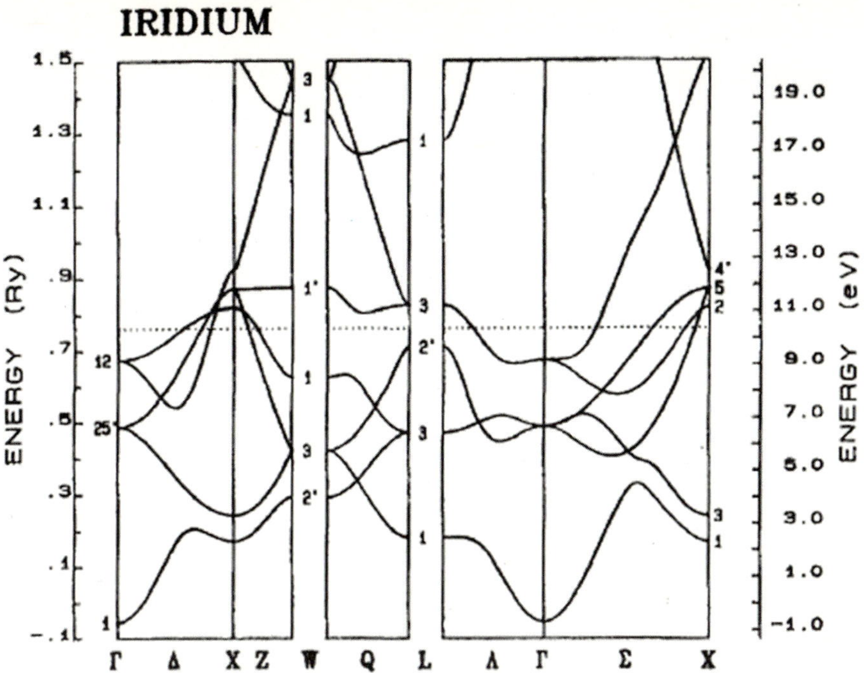

Fig. 6.17 Energy bands for Ir

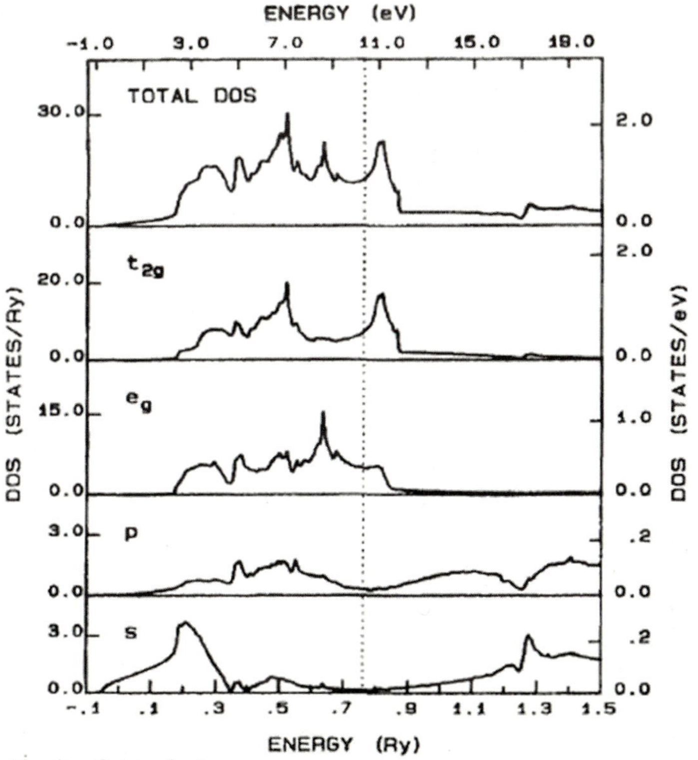

Fig. 6.18 Density of states for Ir

6.7 Platinum

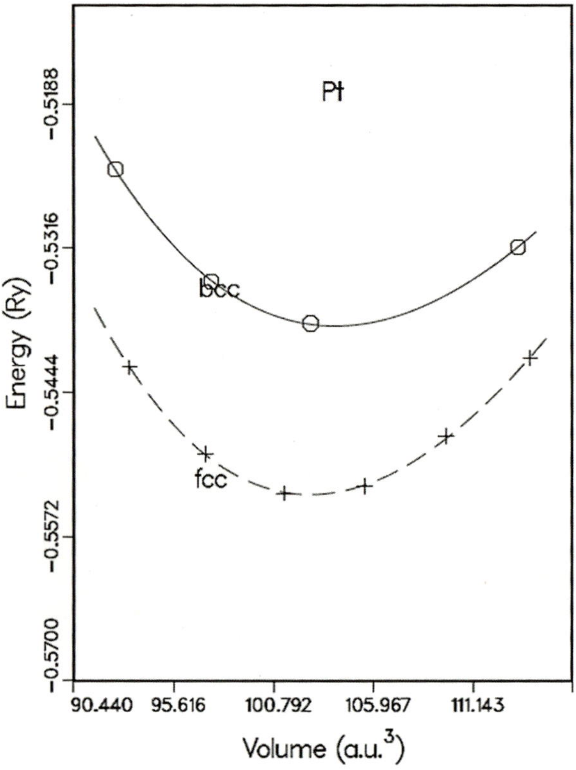

Fig. 6.19 Total energy of Pt

Table 6.49 Lattice constants and bulk modulus

	a (Bohr)	B (MBar)
fcc	7.432	3.335
bcc	5.924	2.721
exp	7.408	2.783
$\Delta E = 15.0$ mRy		

Table 6.50 Birch fit coefficients

	A_1	A_2	A_3
bcc	1.62403	−95.61101	1056.80634
fcc	2.06413	−114.76893	1258.02258

Table 6.51 Platinum fcc Z = 78 lattice constant = 7.41360 a.u.

Slater-Koster 3-center parameters	Orthogonal	Non-orthogonal	
	Energy integrals (Ry)	Energy integrals (Ry)	Overlap integrals
On site			
s, s (000)	0.72116	0.64163	
x, x (000)	1.48718	1.25892	
xy, xy (000)	0.49488	0.44308	
d2, d2 (000)	0.44066	0.43207	
First neighbor			
s, s (110)	−0.08002	−0.06392	0.01518
s, x (110)	0.08105	0.04925	−0.05578
s, xy (110)	−0.06555	−0.05551	0.04250
s, d2 (110)	0.03612	0.02894	0.00335
x, x (110)	0.08029	0.01541	−0.06253
x, x (011)	−0.00130	0.01175	0.04110
x, y (110)	0.11863	0.01312	−0.08233
x, xy (110)	−0.07628	−0.04742	0.02734
x, xy (011)	0.00532	−0.00160	−0.03146
z, d2 (011)	0.00572	0.01409	0.00074
z, d1(011)	0.03962	0.01673	−0.03396
xy, xy (110)	−0.05223	−0.02891	0.05079
xy, xy (011)	0.01300	0.01061	−0.01763
xy, xz (011)	0.00674	0.00623	−0.02344
xy, d2 (110)	0.02765	0.02330	−0.00676
d2, d2 (110)	−0.02166	−0.01394	0.01414
d1, d1 (110)	0.03559	0.03590	−0.00343
Second neighbor			
s, s (200)	0.01424	−0.00378	−0.03439
s, x (200)	−0.01648	0.05371	0.05046
s, d2 (002)	0.00600	−0.01496	−0.04009
x, x (200)	−0.00794	0.12693	0.04623
y, y (200)	−0.01080	0.00037	0.01555
x, xy (020)	−0.00385	−0.00509	−0.00656
z, d2 (002)	0.00441	−0.03906	−0.05062
xy, xy (200)	−0.00684	−0.00673	−0.00320
xy, xy (002)	0.00138	−0.00055	−0.00010
d2, d2 (002)	−0.00278	−0.01311	−0.01867
d1, d1 (002)	0.00098	0.00627	0.01348

Table 6.52 Platinum fcc

	Orthogonal			Non-orthogonal		
Band	RMS error mRy	Maximum k	Deviation mRy	RMS error mRy	Maximum k	Deviation mRy
1	2.5	(055)	6.3	0.6	(174)	1.7
2	3.1	(264)	6.0	0.9	(008)	2.4
3	3.0	(044)	5.8	1.2	(007)	5.1
4	2.9	(118)	13.5	2.3	(118)	11.9
5	2.2	(380)	5.5	1.0	(000)	2.8
6	4.2	(280)	12.4	1.1	(000)	2.8
1–6	3.0			1.3		

Table 6.53 Energy values in Ry at selected k-points

	Orthogonal	APW	Non-orthogonal
GAMMA 1	−0.15363	−0.15189	−0.15177
GAMMA 12	0.51888	0.51529	0.51812
GAMMA 15	2.06519	2.05739	2.05740
GAMMA 25′	0.36537	0.36460	0.36630
X1 (008)	0.09729	0.09847	0.09963
X1 (008)	1.20782	1.24033	1.24145
X2 (008)	0.63641	0.63343	0.63510
X3 (008)	0.15739	0.15871	0.16111
X5 (008)	0.67921	0.67659	0.67707
X4′ (008)	0.78055	0.78967	0.79038
X5′ (008)	1.43330	1.48221	1.48166
L1 (444)	0.08858	0.09366	0.09394
L1 (444)	1.01266	1.03248	1.03142
L3 (444)	0.35176	0.35282	0.35412
L3 (444)	0.64069	0.63919	0.64042
L2′ (444)	0.59726	0.59753	0.59749
L3′ (444)	2.02079	2.04669	2.04663
W1 (048)	0.48450	0.48190	0.48404
W1 (048)	1.11824	1.11467	1.11489
W3 (048)	0.31767	0.31194	0.31339
W3 (048)	1.27811	1.25966	1.26134
W1′ (048)	0.67368	0.67711	0.67953
W2′ (048)	0.20901	0.20739	0.20863
W2′ (048)	1.54505	1.50142	1.50173
EVEN(224)	0.10784	0.10727	0.10793
EVEN(224)	0.33153	0.32957	0.32920
EVEN(224)	0.35666	0.35551	0.35597
EVEN(224)	0.60495	0.60721	0.60790
ODD (224)	0.40645	0.41045	0.40961
ODD (224)	0.61933	0.61765	0.61768

Table 6.54 Fermi level quantities (non-orthogonal fit)

Energy		Densities of states				Velocity	Plasmon energy
	Total	s	p	t_{2g}	e_g		
Ry		States/Ry/atom				cm/s	eV
0.6380	29.90	0.23	0.36	22.41	6.91	$0.44 \times 10E8$	8.56

Integrated densities of states				
Total	s	p	t_{2g}	e_g
Electrons				
10.00	0.75	0.50	5.06	3.68

Table 6.55 Platinum fcc

Slater-Koster 2-center parameters	Orthogonal	Non-orthogonal	
	Energy integrals (Ry)	Energy integrals (Ry)	Overlap integrals
On site			
s	0.77393	0.45649	
P	1.48490	0.98826	
d1	0.45468	0.41348	
d2	0.43701	0.41334	
First neighbor			
(sss)	−0.07835	−0.05907	0.08585
(pps)	0.18677	−0.17826	−0.31622
(ppp)	−0.02465	−0.04410	0.03518
(dds)	−0.06856	−0.03753	0.06647
(ddp)	0.03528	0.02719	−0.03588
(ddd)	−0.00588	−0.00488	0.00315
(sps)	0.11192	0.01642	−0.16881
(sds)	−0.06197	−0.03709	0.07376
(pds)	−0.08564	0.00562	0.13563
(pdp)	0.02446	0.02296	−0.03910
Second neighbor			
(sss)	0.00157	−0.00958	0.00247
(pps)	0.03487	0.04464	−0.02729
(ppp)	−0.00958	−0.00209	0.00991
(dds)	−0.00300	−0.01668	−0.01229
(ddp)	0.00226	0.00159	−0.00121
(ddd)	−0.00029	0.00081	0.00211
(sps)	0.00074	0.01934	−0.00730
(sds)	−0.00770	−0.01086	0.00099
(pds)	−0.01321	−0.03342	−0.00673
(pdp)	0.00522	0.00256	−0.00307

	Orthogonal			Non-orthogonal		
Band	RMS error mRy	Maximum k	Deviation mRy	RMS error mRy	Maximum k	Deviation mRy
1	9.2	(084)	17.6	0.7	(444)	1.7
2	10.6	(007)	26.1	1.2	(000)	2.5
3	6.5	(044)	15.1	1.5	(007)	5.1
4	9.3	(084)	28.5	2.5	(118)	12.8
5	9.0	(044)	20.7	1.3	(000)	3.8
6	13.1	(066)	39.2	1.2	(000)	3.8
1–6	9.8			1.5		

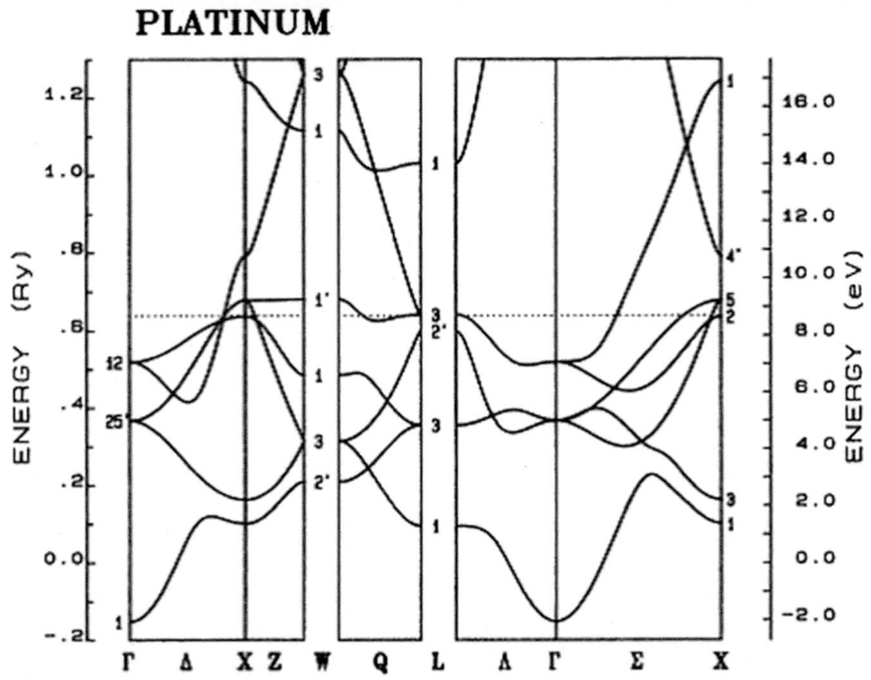

Fig. 6.20 Energy bands for Pt

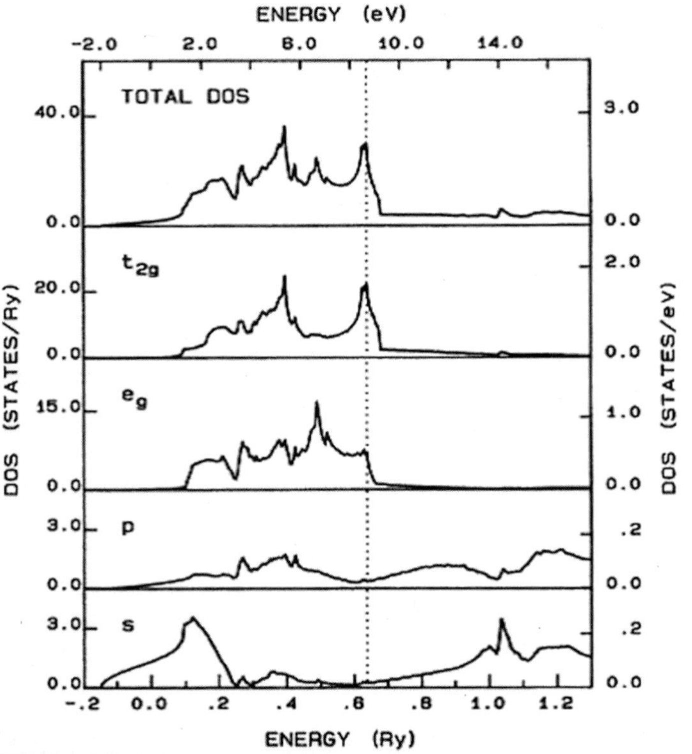

Fig. 6.21 Density of states for Pt

6.8 Gold

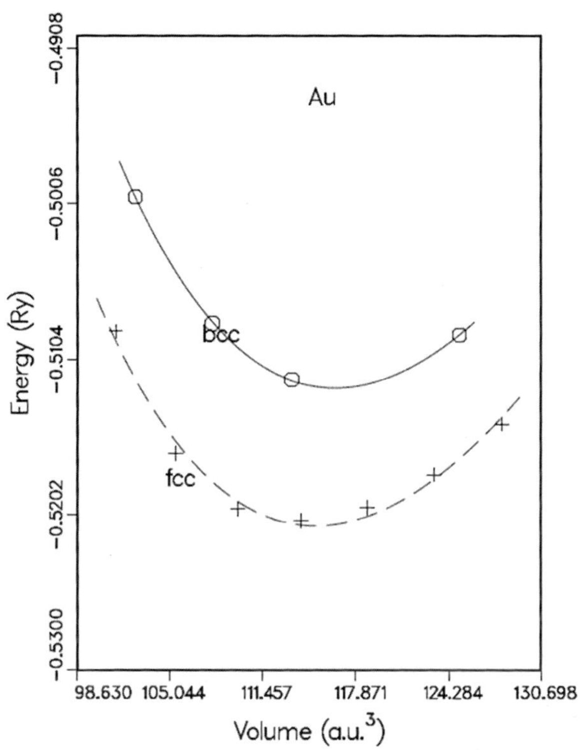

Fig. 6.22 Total energy of Au

Table 6.56 Lattice constants and bulk modulus

	a (Bohr)	B (MBar)
fcc	7.722	1.676
bcc	6.153	1.763
exp	7.710	1.732
$\Delta E = 8.7$ mRy		

Table 6.57 Birch fit coefficients

	A_1	A_2	A_3
bcc	1.05815	−74.89527	892.993592
fcc	0.95475	−69.83923	826.33084

Table 6.58 GOLD fcc Z = 79 lattice constant = 7.70687 a.u.

Slater-Koster 3-center parameters	Orthogonal	Non-orthogonal	
	Energy integrals (Ry)	Energy integrals (Ry)	Overlap integrals
On site			
s, s (000)	0.54761	0.29456	
x, x (000)	1.28585	0.73919	
xy, xy (000)	0.27657	0.24197	
d2, d2 (000)	0.25897	0.24122	
First neighbor			
s, s (110)	−0.06668	−0.05882	0.08647
s, x (110)	0.07250	0.01879	−0.13183
s, xy (110)	−0.04821	−0.03214	0.06048
s, d2 (110)	0.02863	0.01963	−0.02599
x, x (110)	0.07979	−0.08546	−0.14716
x, x (011)	0.00401	−0.00591	0.07081
x, y (110)	0.11001	−0.07940	−0.21454
x, xy (110)	−0.05205	−0.01031	0.07221
x, xy (011)	0.00737	0.01322	−0.03533
z, d2 (011)	0.00139	0.00967	−0.00906
z, d1 (011)	0.03173	0.00806	−0.06391
xy, xy (110)	−0.03782	−0.02576	0.04543
xy, xy (011)	0.00963	0.00963	−0.01226
xy, xz (011)	0.00972	0.01279	−0.01475
xy, d2 (110)	0.02063	0.01546	−0.01529
d2, d2 (110)	−0.01550	−0.00995	0.02050
d1, d1 (110)	0.02544	0.02318	−0.02731
Second neighbor			
s, s (200)	0.00529	−0.00573	0.00068
s, x (200)	−0.00545	0.02475	−0.00654
s, d2 (002)	0.00049	−0.01043	−0.00446
x, x (200)	0.00281	0.05739	−0.04206
y, y (200)	−0.01425	0.00080	0.01447
x, xy (020)	−0.00069	−0.00097	−0.00755
z, d2 (002)	0.00091	−0.02774	−0.01025
xy, xy (200)	−0.00112	−0.00015	−0.00203
xy, xy (002)	0.00056	−0.00020	0.00041
d2, d2 (002)	−0.00266	−0.00918	−0.00714
d1, d1 (002)	−0.00052	0.00108	0.00628

Table 6.59 Gold fcc

	Orthogonal			Non-orthogonal		
Band	RMS error mRy	Maximum k	Deviation mRy	RMS error mRy	Maximum k	Deviation mRy
1	2.1	(444)	4.5	0.5	(174)	1.7
2	2.2	(048)	4.3	0.8	(008)	2.3
3	2.3	(007)	4.9	1.0	(007)	4.1
4	2.1	(118)	9.2	1.8	(118)	9.4
5	1.7	(380)	3.8	0.9	(000)	2.4
6	4.5	(280)	14.6	0.7	(000)	2.4
1–6	2.6			1.0		

Table 6.60 Energy values in Ry at selected k-points

	Orthogonal	APW	Non-orthogonal
GAMMA 1	−0.22085	−0.21851	−0.21826
GAMMA 12	0.30907	0.30734	0.30973
GAMMA 15	1.88889	1.87979	1.87979
GAMMA 25′	0.19894	0.19823	0.19970
X1 (008)	0.00353	0.00444	0.00585
X1 (008)	0.90836	0.92867	0.92868
X2 (008)	0.39331	0.39240	0.39394
X3 (008)	0.04492	0.04559	0.04784
X5 (008)	0.42451	0.42382	0.42415
X4′ (008)	0.61218	0.62311	0.62327
X5′ (008)	1.21844	1.27579	1.27560
L1 (444)	−0.00999	−0.00553	−0.00472
L1 (444)	0.72802	0.74227	0.74202
L3 (444)	0.18953	0.19108	0.19231
L3 (444)	0.39779	0.39722	0.39813
L2′ (444)	0.45712	0.45810	0.45812
L3′ (444)	1.77726	1.80609	1.80607
W1 (048)	0.29642	0.29194	0.29372
W1 (048)	0.85060	0.84982	0.84983
W3 (048)	0.16888	0.16458	0.16585
W3 (048)	1.04261	1.01116	1.01154
W1′ (048)	0.42226	0.42409	0.42601
W2′ (048)	0.09195	0.09009	0.09098
W2′ (048)	1.26498	1.21732	1.21749
EVEN (224)	0.00636	0.00629	0.00650
EVEN (224)	0.18633	0.18377	0.18380
EVEN (224)	0.19566	0.19560	0.19582
EVEN (224)	0.38727	0.38929	0.38952
ODD (224)	0.23203	0.23424	0.23365
ODD (224)	0.38347	0.38257	0.38233

Table 6.61 Fermi level quantities (non-orthogonal fit)

Energy		Densities of states				Velocity	Plasmon energy
	Total	s	p	t_{2g}	e_g		
Ry		States/Ry/atom				cm/s	eV
0.5380	3.99	0.98	0.58	1.75	0.67	1.32 × 10E8	8.84

Integrated densities of states Total Electrons	s	p	t_{2g}	e_g
11.00	0.86	0.25	5.93	3.96

Table 6.62 Gold fcc

Slater-Koster 2-center parameters	Orthogonal	Non-orthogonal	
	Energy integrals (Ry)	Energy integrals (Ry)	Overlap integrals
On site			
s	0.56220	0.29240	
p	1.27897	0.73634	
d1	0.26097	0.24145	
d2	0.25309	0.24249	
First neighbor			
(sss)	−0.06680	−0.05747	0.09205
(pps)	0.17866	−0.16644	−0.36368
(ppp)	−0.01645	−0.00709	0.06975
(dds)	−0.04971	−0.03600	0.05031
(ddp)	0.02624	0.02259	−0.02668
(ddd)	−0.00457	−0.00357	0.00256
(sps)	0.09721	0.02800	−0.18632
(sds)	−0.04722	−0.03739	0.06401
(pds)	−0.06399	−0.01303	0.12648
(pdp)	0.01896	0.01950	−0.05120
Second neighbor			
(sss)	0.00277	−0.01099	0.00279
(pps)	0.03707	0.04491	−0.04205
(ppp)	−0.01025	0.00304	0.01061
(dds)	−0.00305	−0.00903	−0.00832
(ddp)	0.00240	0.00133	−0.00082
(ddd)	−0.00057	0.00010	0.00135
(sps)	0.00261	0.02431	−0.01054
(sds)	−0.00784	−0.00901	0.00047
(pds)	−0.00762	−0.02429	−0.00506
(pdp)	0.00470	0.00132	−0.00495

	Orthogonal			Non-orthogonal		
Band	RMS error mRy	Maximum k	Deviation mRy	RMS error mRy	Maximum k	Deviation mRy
1	5.6	(333)	10.6	0.5	(174)	1.2
2	7.1	(007)	18.2	0.9	(000)	2.0
3	3.2	(007)	8.4	1.2	(007)	3.9
4	5.8	(084)	18.3	1.9	(118)	10.0
5	5.5	(044)	12.1	1.1	(000)	3.0
6	9.6	(008)	25.7	0.8	(000)	3.0
1–6	6.4			1.2		

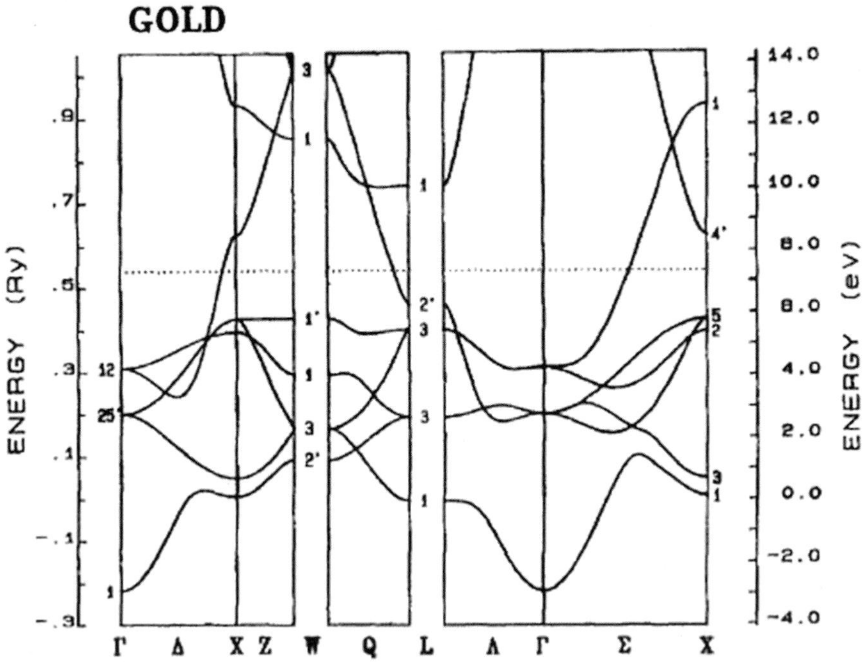

Fig. 6.23 Energy bands for Au

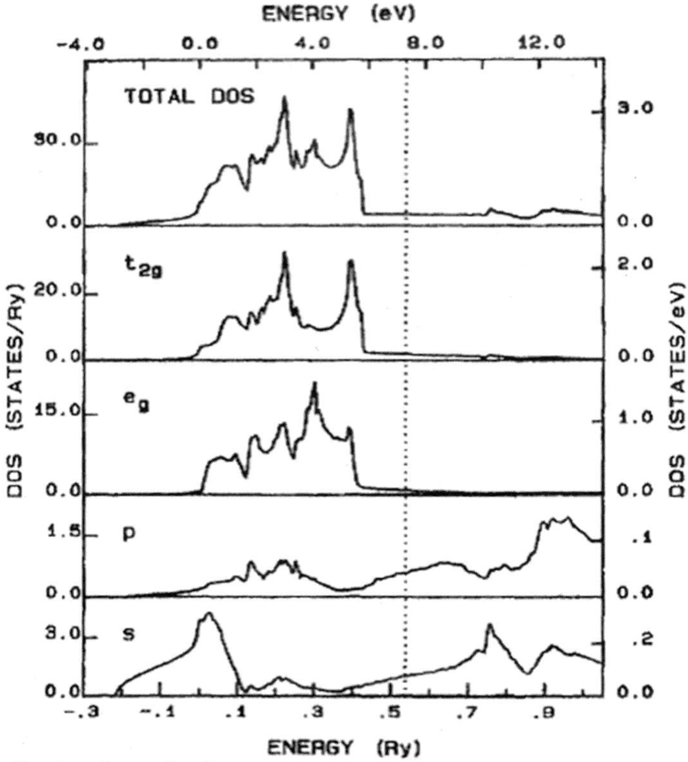

Fig. 6.24 Density of states for Au

6.9 Mercury

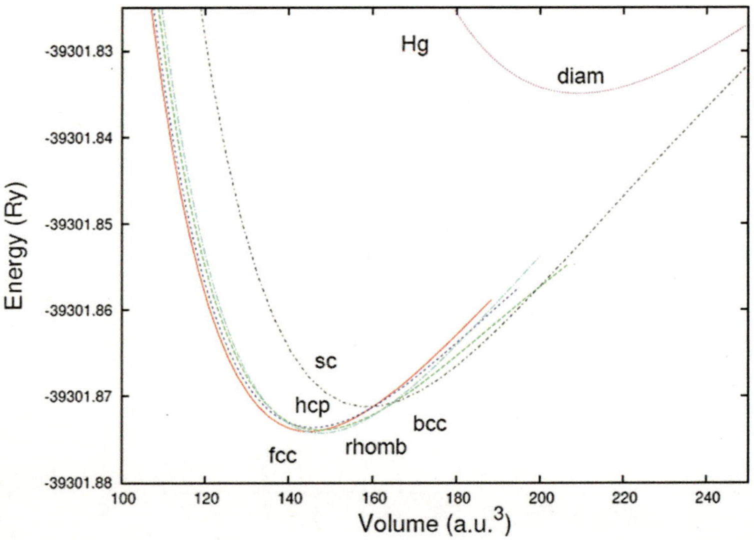

Fig. 6.25 Total energy of Hg

Table 6.63 Lattice constants and bulk modulus

	a (Bohr)	c (Bohr)	B (MBar)
sc	5.41573		0.67545
bcc	6.65822		0.62106
fcc	8.33801		0.68032
hcp	3.17463	33.42170	0.64164
rhomb ($\gamma = 70°$)	8.39267		0.70185
diam	11.87253		0.46193
$\Delta E_{fcc-rhomb} = 0.187$ mRy			

Table 6.64 Birch fit coefficients

	A_1	A_2	A_3	A_4
sc	−39301.96618	32.41715	−1656.65460	23097.56639
bcc	−39302.29800	55.10365	−2085.61573	24504.68413
fcc	−39302.12639	41.68555	−1724.04767	21133.84665
hcp	−78604.51131	163.78158	−9973.00646	186818.76107
ang70	−39302.06185	37.90350	−1679.32596	21422.52937
diam	−78604.00625	139.13167	−12407.69788	317593.33863

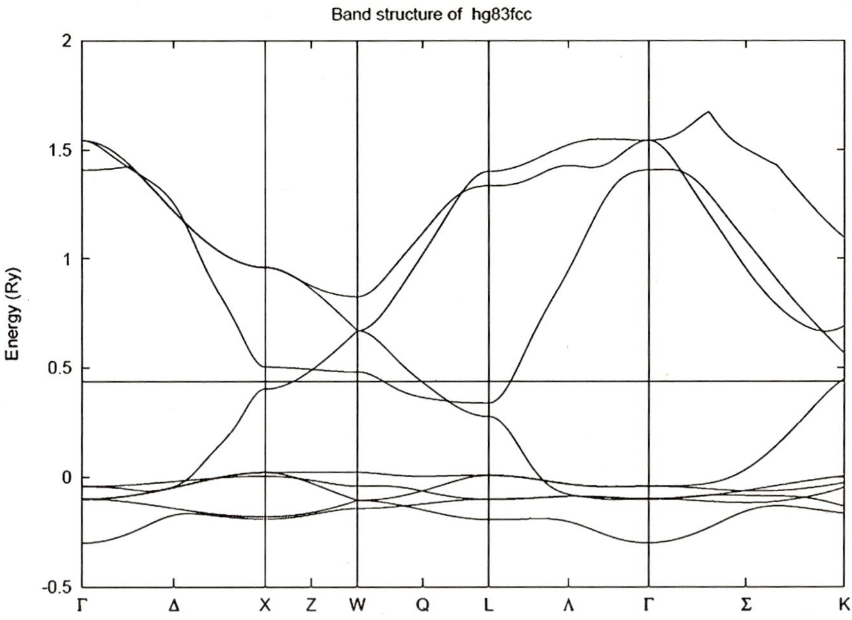

Fig. 6.26 Energy bands for Hg

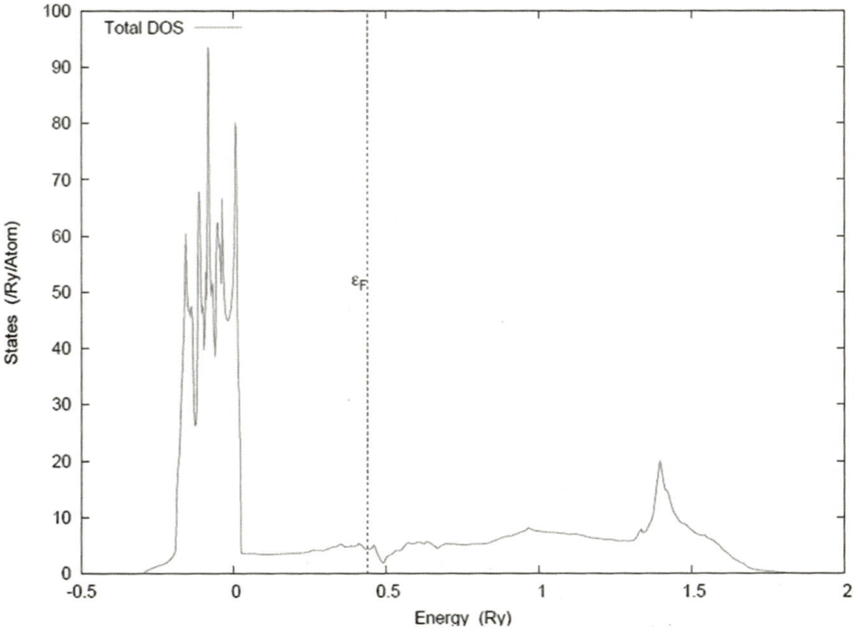

Fig. 6.27 Density of States for Hg

Table 6.65 DOS at the Fermi level (APW results)

Energy Ry	Total	s States/Ry/atom	p	e_g	t_{2g}	f	Velocity cm/s	Plasmon energy eV
0.440	4.325	1.761	1.182	0.140	0.329	0.106	1.222 × 10E8	7.650

Table 6.66 Mercury fcc structure a = 8.30 Bohr
Tight-binding parameters in Ry
Non-orthogonal

On-site									
s	p	d							
0.39809	0.93225	0.12041							
sss	sps	pps	ppp	sds	pds	pdp	dds	ddp	ddd
Hopping dist = 5.86899									
−0.04456	0.04616	−0.03144	−0.05042	−0.02915	−0.03411	0.02289	−0.02557	0.01291	−0.00192
Overlap dist = 5.86899									
0.02942	−0.08461	−0.19735	0.00554	0.01915	0.06572	0.02772	0.00957	0.00792	−0.00232
Hopping dist = 8.30000									
−0.00036	0.00160	0.00024	−0.00033	−0.00026	−0.00104	0.00036	−0.00019	0.00000	0.00000
Overlap dist = 8.30000									
0.00026	−0.00125	−0.00426	0.00003	−0.00009	−0.00022	0.00009	0.00001	0.00001	0.00000

Table 6.67 Fitting rms error for 7 bands 4 mRy
Orthogonal

On-site									
s	p	d							
0.046559	1.07529	0.11892							
sss	sps	pps	ppp	sds	pds	pdp	dds	ddp	ddd
Hopping dist = 5.86899									
−0.05000	0.08018	0.13572	−0.00591	−0.02951	−0.03862	0.01198	−0.02679	0.01271	−0.00224
Hopping dist = 8.30000									
0.00041	−0.00859	−0.00076	0.01926	−0.00060	−0.00187	0.00209	−0.00046	0.00015	−0.00022

Fitting rms error for 7 bands 8 mRy

Chapter 7
Free-Electron-Like Metals of Groups III and IV

In this category the fcc materials Al and Pb are included together with Ga and In in an assumed fcc structure and Tl in the hcp structure. The DOS of Al has strong free-electron character, similar to that of Na. In the unoccupied states the p-like levels mix with the d-like levels. For example, the p state Γ_{15} lies between the d states $\Gamma_{25'}$ and Γ_{12}. Pb has unique features: an energy gap appears between the s-like and p-like states and the d levels lie much higher than the p levels. Consequently, in Pb $N(E_F)$ is almost exclusively p-like while in Al it has a strong s-like contribution. It should be noted here that for Pb the SK fit is good only for four bands throughout the whole Brillouin zone. Ga and In, which are calculated in a hypothetical fcc structure, are similar to Zn and Cd in that they have a set of narrow d bands far below E_F, and free-electron characteristics near E_F. The hcp T1 is also characterized by narrow d bands, as in Cd, but near E_F the details of the band structure are different.

D.A. Papaconstantopoulos, *Handbook of the Band Structure of Elemental Solids*,
DOI 10.1007/978-1-4419-8264-3_7

7.1 Boron

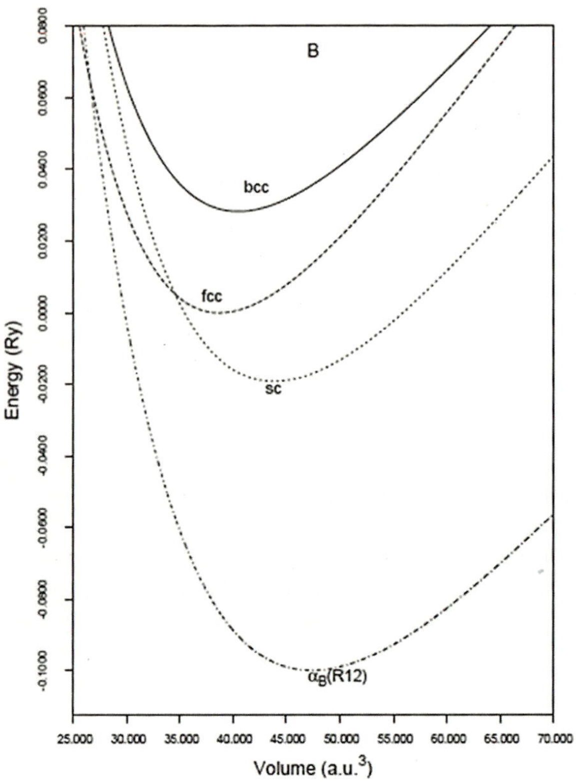

Fig. 7.1 Total energy of B

Table 7.1 Lattice constants and bulk modulus

	a (Bohr)	Volume (Bohr^3)	B (MBar)
sc	3.526	43.820	2.507
bcc	4.326	40.478	2.353
fcc	5.369	38.683	2.831
α_B ($R12$)		47.381	2.220
exp			1.780
$\Delta E_{\alpha B\text{–}fcc} = 100$ mRy			

Table 7.2 Birch fit coefficients

	A_1	A_2	A_3	A_4
sc	0.95858	−26.01732	193.56913	−264.16971
bcc	0.81712	−19.30927	126.38593	−98.76595
fcc	0.96730	−23.61361	160.51597	−194.22596
α_B ($R12$)	9.96856	−1637.05182	66891.37695	−490245.51874

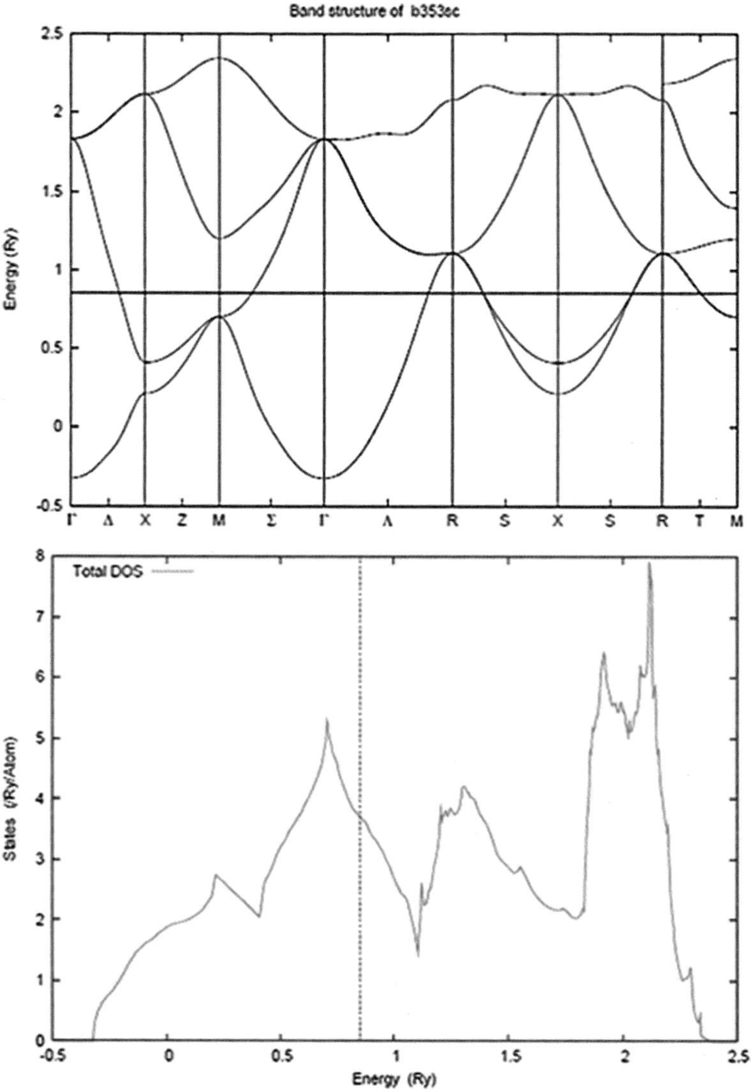

Fig. 7.2 Energy bands and density of states for B

Table 7.3 DOS at the Fermi level (APW results)

Energy Ry	Total	s States/Ry/atom	p	e_g	t_{2g}	f	Velocity cm/s	Plasmon energy eV
0.852	3.704	0.304	2.197	0.097	0.039	0.021	$1.607 \times 10E8$	16.789

Boron sc structure a = 3.53 Bohr
Tight-binding parameters in Ry

Non-orthogonal On-site	
s	p
0.53334	0.96831

sss	sps	pps	ppp
Hopping dist = 3.53000			
−0.15764	−0.08876	−0.04266	0.04186
Overlap dist = 3.53000			
0.13396	0.20001	−0.30272	0.11999
Hopping dist = 4.99217			
−0.00840	−0.05077	0.08438	−0.01170
Overlap dist = 4.99217			
−0.00437	0.00344	−0.02400	−0.00525
Hopping dist = 6.11414			
−0.00024	−0.00483	0.01018	−0.00065
Overlap dist = 6.11414			
−0.00094	−0.00027	−0.00149	−0.00121

Fitting rms error for 3 bands 12 mRy

Orthogonal On-site	
s	p
0.77566	1.29272

sss	sps	pps	ppp
Hopping dist = 3.53000			
−0.19320	−0.23358	0.34552	−0.09519
Hopping dist = 4.99217			
−0.00100	0.00355	0.00877	0.00465
Hopping dist = 6.11414			
0.00021	0.00296	−0.00029	0.00015

Fitting rms error for 3 bands 52 mRy

7.2 Aluminum

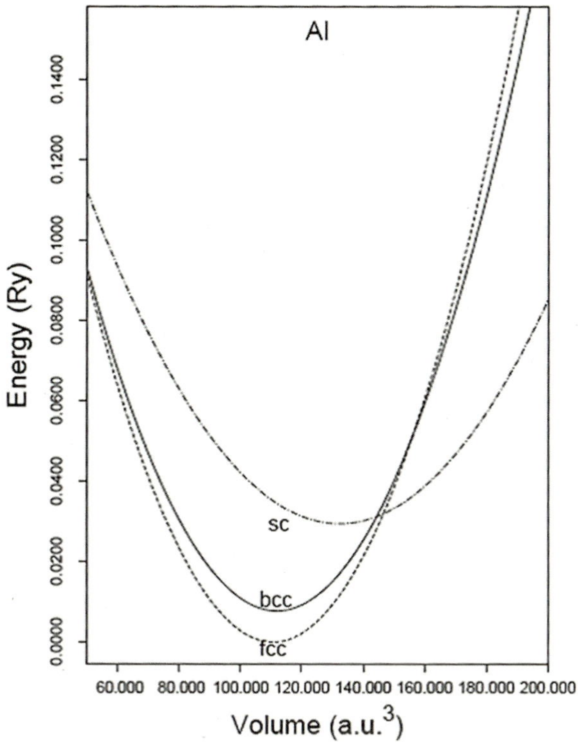

Fig. 7.4 Total energy of Al

Table 7.4 Lattice constants and bulk modulus

	a (Bohr)	B (MBar)
sc	5.077	0.573
bcc	6.053	0.717
fcc	7.550	0.849
exp	7.653	0.722
$\Delta E = E_{fcc} - E_{bcc} = 7.9$ mRy		

Table 7.5 Birch fit coefficients

	A_1	A_2	A_3	A_4
sc	0.60439	−29.69720	384.66639	−33.36742
bcc	0.86128	−45.05986	716.15593	−3018.14457
fcc	0.41491	−12.35314	−78.00743	3283.03362

Table 7.6 Aluminum fcc Z = 13 lattice constant = 7.60000 a.u.

Slater-Koster 3-center parameters	Orthogonal	Non-orthogonal	
	Energy integrals (Ry)	Energy integrals (Ry)	Overlap integrals
On site			
s, s (000)	0.46185	0.25757	
x, x (000)	1.03995	0.65840	
xy, xy (000)	1.81664	1.25923	
d2, d2 (000)	1.69063	1.39913	
First neighbor			
s, s (110)	−0.07167	−0.05210	−0.10041
s, x (110)	0.06667	0.03040	−0.11194
s, xy (110)	−0.02211	−0.00720	0.11386
s, d2 (110)	0.04913	0.01586	−0.04296
x, x (110)	0.07781	−0.01813	−0.10675
x, x (011)	−0.00049	−0.00643	0.05784
x, y (110)	0.10132	−0.00606	−0.16176
x, xy (110)	0.08799	0.08003	0.15545
x, xy (011)	−0.00445	−0.02237	−0.09092
z, d2 (011)	−0.00899	−0.02164	−0.03715
z, d1 (011)	0.06868	−0.01543	−0.08474
xy, xy (110)	−0.12735	0.16895	0.19034
xy, xy (011)	0.03369	−0.11841	−0.13887
xy, xz (011)	0.02566	−0.12587	−0.14614
xy, d2 (110)	−0.08717	−0.04335	−0.05024
d2, d2 (110)	−0.07258	0.02266	0.05904
d1, d1 (110)	0.11980	−0.17231	−0.21080
Second neighbor			
s, s (200)	0.02904	−0.01483	0.00530
s, x (200)	−0.04909	0.02173	−0.00654
s, d2 (002)	0.03595	−0.02307	0.00157
x, x (200)	−0.01894	0.04038	−0.01176
y, y (200)	0.00575	−0.00385	−0.00397
x, xy (020)	0.00631	0.01552	0.00991
z, d2 (002)	0.01734	−0.02844	0.02727
xy, xy (200)	−0.01804	0.05085	0.02270
xy, xy (002)	0.01156	−0.04850	−0.01910
d2, d2 (002)	0.00737	−0.03697	0.03010
d1, d1 (002)	−0.01851	0.02826	0.01806

Table 7.7 Aluminum fcc

	Orthogonal			Non-orthogonal		
Band	RMS error mRy	Maximum k	Deviation mRy	RMS error mRy	Maximum k	Deviation mRy
1	11.5	(444)	29.2	1.2	(008)	3.2
2	14.9	(444)	28.4	1.7	(380)	3.8
3	16.8	(444)	52.5	2.6	(354)	7.6
4	19.7	(444)	52.5	2.5	(003)	6.4
5	29.8	(044)	89.1	4.2	(354)	10.5
6	33.4	(055)	101.8	14.3	(055)	77.5
1–6	22.4			6.3		

Table 7.8 Energy values in Ry at selected k-points

	Orthogonal	APW	Non-orthogonal
GAMMA 1	−0.22398	−0.20557	−0.20416
GAMMA 12	1.94056	2.03530	2.03571
GAMMA 15	1.64560	1.64099	1.63610
GAMMA 25′	1.52769	1.55469	1.55639
X1 (008)	0.45962	0.48773	0.48614
X1 (008)	1.25637	1.26416	1.26678
X2 (008)	2.33231	2.45166	2.42160
X3 (008)	0.98869	1.02023	1.02180
X5 (008)	2.27700	2.57312	2.36872
X4′ (008)	0.40062	0.40889	0.41209
X5′ (008)	1.02706	1.02820	1.02782
L1 (444)	0.27073	0.29912	0.30168
L1 (444)	1.67729	1.74785	1.73282
L3 (444)	1.40200	1.45829	1.45782
L3 (444)	2.29038	2.41847	2.41715
L2′ (444)	0.24426	0.27347	0.27493
L3′ (444)	1.46008	1.45453	1.44956
W1 (048)	0.71068	0.69873	0.70224
W1 (048)	2.01397	1.95962	1.96171
W3 (048)	0.57032	0.56230	0.56001
W3 (048)	1.82552	1.78875	1.78839
W1′ (048)	2.23075	2.58907	2.40414
W2′ (048)	0.62293	0.61179	0.61422
W2′ (048)	1.73194	1.80196	1.79144
EVEN(224)	0.04535	0.04313	0.04238
EVEN(224)	0.68668	0.68026	0.68187
EVEN(224)	1.30339	1.29275	1.29731
EVEN(224)	1.36586	1.36945	1.37965
ODD (224)	1.23521	1.24103	1.24352
ODD (224)	1.80976	1.80900	1.81123

Table 7.9 Fermi level quantities (Non-orthogonal fit)

Energy		Densities of states				Velocity	Plasmon energy
	Total	s	p	t_{2g}	e_g		
Ry		States/Ry/atom				cm/s	eV
0.6195	5.46	1.91	3.20	0.20	0.15	1.55 × 10E8	12.44

Integrated densities of states Total Electrons	s	p	t_{2g}	e_g
3.00	1.55	1.38	0.04	0.03

Table 7.10 Aluminum fcc

Slater-Koster 2-center parameters	Orthogonal	Non-orthogonal	
	Energy integrals (Ry)	Energy integrals (Ry)	Overlap integrals
On site			
s	0.48220	0.20639	
p	1.04964	0.58673	
d1	1.73912	1.11415	
d2	1.62982	1.15157	
First neighbor			
(sss)	−0.05831	−0.05137	0.11591
(pps)	0.17149	−0.06630	−0.31915
(ppp)	−0.01020	−0.02184	0.05724
(dds)	−0.17004	0.30226	0.26750
(ddp)	0.07394	−0.12743	−0.23221
(ddd)	−0.00490	−0.01282	0.03033
(sps)	−0.09315	0.02672	−0.19400
(sds)	0.08296	0.05157	0.19477
(pds)	−0.16259	0.20498	0.30813
(pdp)	0.03088	−0.00371	−0.11467
Second neighbor			
(sss)	−0.00163	−0.01440	0.00522
(pps)	0.00026	0.03906	−0.02899
(ppp)	−0.00316	−0.00337	0.00145
(dds)	−0.01163	−0.02273	0.06680
(ddp)	0.00315	0.02855	−0.00588
(ddd)	−0.00283	−0.00559	−0.00071
(sps)	0.00519	0.02296	−0.01417
(sds)	−0.03032	−0.02465	0.02140
(pds)	0.02447	−0.03543	0.04495
(pdp)	−0.00266	0.01481	0.00008

	Orthogonal			Non-orthogonal		
Band	RMS error mRy	Maximum k	Deviation mRy	RMS error mRy	Maximum k	Deviation mRy
1	27.4	(008)	57.3	1.4	(008)	3.2
2	25.7	(004)	54.0	3.2	(111)	10.0
3	31.4	(008)	65.3	3.7	(354)	12.3
4	36.3	(003)	71.4	4.3	(002)	12.9
5	43.9	(084)	100.7	7.7	(111)	21.3
6	51.4	(055)	164.3	8.8	(224)	24.8
1–6	37.1			5.5		

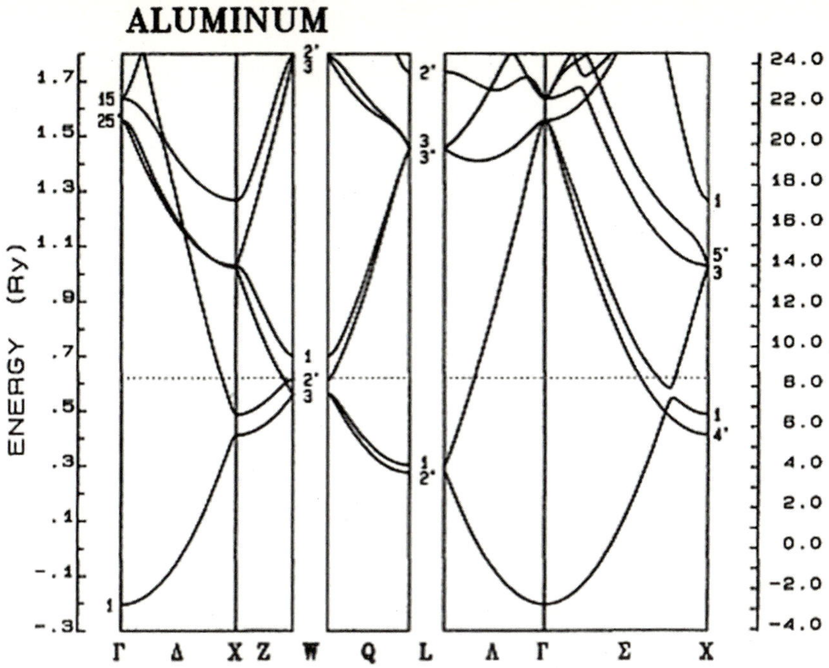

Fig. 7.5 Energy bands for Al

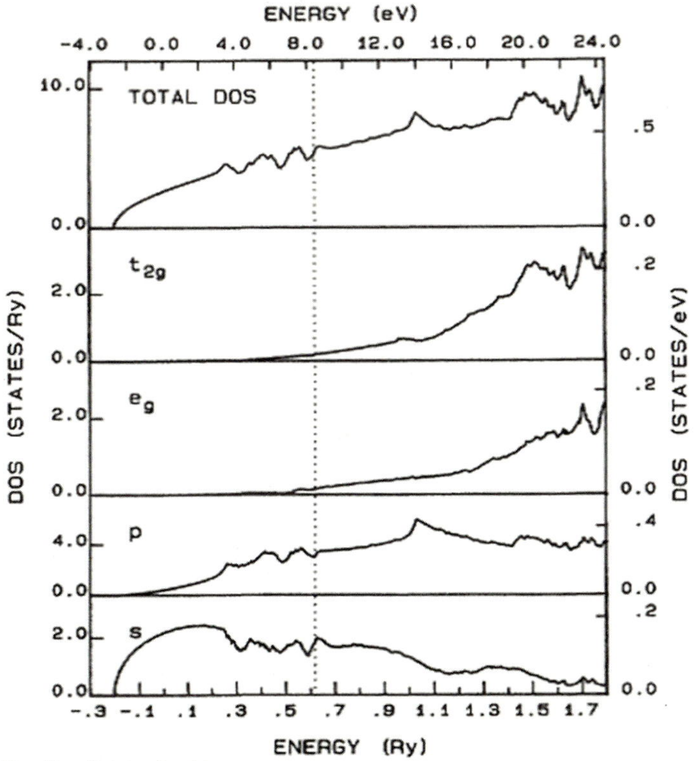

Fig. 7.6 Density of states for Al

7.3 Gallium

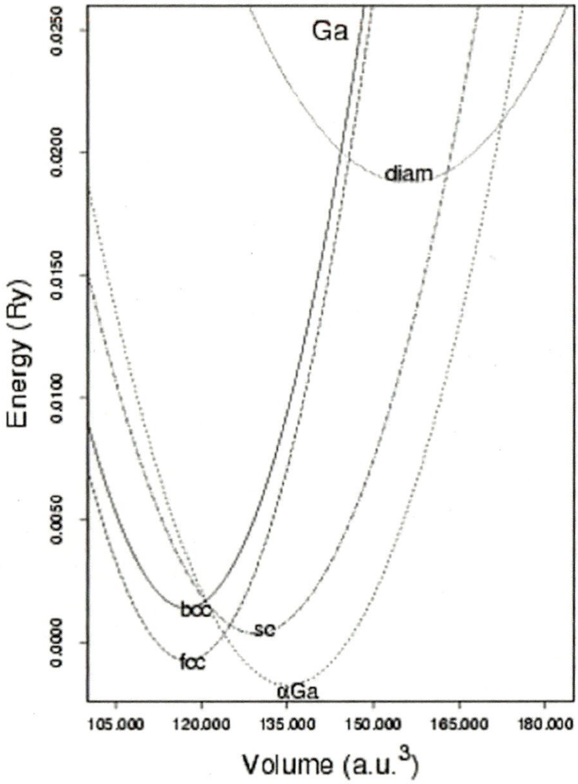

Fig. 7.7 Total energy of Ga

Table 7.11 Lattice constants and bulk modulus

	a (Bohr)	b (Bohr)	c (Bohr)	B (MBar)
sc	5.032			0.661
bcc	6.142			0.872
fcc	7.751			0.861
diam	10.760			0.426
αGa	8.602	14.580	8.613	0.687
exp	8.523	14.446	8.534	0.569
$\Delta E = E_{fcc} - E_{\alpha Ga} = 0.9$ mRy				

Table 7.12 Birch fit coefficients

	A_1	A_2	A_3	A_4
sc	0.46554	−19.02491	68.73208	2905.12252
bcc	0.30745	−3.45186	−354.34268	6262.99825
fcc	−0.16893	30.31121	−1158.40171	12668.29641

Table 7.13 Gallium fcc z = 31 lattice constant = 7.83000 a.u.

Slater-Koster 3-center parameters	Orthogonal	Non-orthogonal	
	Energy integrals (Ry)	Energy integrals (Ry)	Overlap integrals
On site			
s, s (000)	0.13673	−0.01606	
x, x (000)	0.77883	0.32458	
xy, xy (000)	−0.65473	−0.65397	
d2, d2(000)	−0.64463	−0.65444	
First neighbor			
s, s (110)	−0.04294	−0.06280	0.11798
s, x (110)	−0.04938	−0.03553	0.15354
s, xy (110)	−0.00070	0.01257	−0.02763
s, d2 (110)	0.01621	−0.01760	0.02885
x, x (110)	0.06925	−0.02491	−0.12485
x, x (011)	0.01765	0.01543	0.09901
x, y (110)	0.08447	−0.05240	−0.29528
x, xy (110)	−0.00030	−0.03947	0.05120
x, xy (011)	0.00164	0.01775	−0.02673
z, d2 (011)	0.00340	0.01620	−0.02330
z, d1 (011)	0.00198	0.03852	−0.05195
xy, xy (110)	−0.00308	−0.02106	0.02750
xy, xy(011)	0.00072	−0.00156	0.00365
xy, xz (011)	0.00102	−0.03528	0.05517
xy, d2 (110)	0.00153	−0.02478	0.03540
d2, d2 (110)	−0.00117	0.02012	−0.03304
dl, dl (110)	0.00028	0.00132	0.00091
Second neighbor			
s, s (200)	0.00177	−0.01646	0.00921
s, x (200)	0.00464	−0.02532	0.04079
s, d2 (002)	0.02646	0.01112	−0.00570
x, x (200)	−0.00653	0.03191	−0.10233
y, y (200)	0.00206	0.00390	−0.00124
x, xy (020)	−0.00077	−0.00045	−0.00080
z, d2 (002)	−0.05914	−0.01606	0.01997
xy, xy (200)	0.00009	−0.00722	0.01056
xy, xy (002)	0.00048	0.02633	−0.04046
d2, d2(002)	−0.00071	0.00425	−0.00685
dl, dl (002)	0.00039	−0.02528	0.03927

Table 7.14 Gallium fcc

	Orthogonal			Non-orthogonal		
Band	RMS error mRy	Maximum k	Deviation mRy	RMS error mRy	Maximum k	Deviation mRy
1	3.7	(055)	9.6	0.9	(354)	1.8
2	2.2	(004)	6.7	0.7	(008)	1.9
3	1.9	(444)	4.4	0.5	(444)	1.2
4	2.4	(174)	5. 1	0.6	(226)	1.2
5	2.5	(044)	7. 1	0.7	(055)	1.9
6	11.7	(048)	35.6	1.7	(000)	4.9
7	35.8	(111)	99.8	7.0	(002)	26.9
1–7	14.4			2.8		

Table 7.15 Energy values in Ry at selected k-points

	Orthogonal	APW	Non-orthogonal
GAMMA 1	−0.36795	−0.35639	−0.35145
GAMMA 12	−0.65095	−0.65164	−0.65104
GAMMA 15	1.39858	1.41982	1.41978
GAMMA 25′	−0.66006	−0.66087	−0.66049
X1 (008)	−0.66668	−0.67029	−0.66839
X1 (008)	0.33616	0.30904	0. 31099
X2 (008)	−0.63801	−0.64467	−0.64412
X3 (008)	−0.67150	−0.67353	−0.67365
X5 (008)	−0.64110	−0.64188	−0.64133
X4′ (008)	0.29060	0.30553	0.30433
X5' (008)	0.70341	0.86988	0.86802
L1 (444)	−0.66422	−0.66652	−0.66596
L1 (444)	0.12616	0.11403	0.11464
L3 (444)	−0.65639	−0.66083	−0.65963
L3 (444)	−0.63923	−0.64419	−0.64306
L2′ (444)	0.10791	0.18695	0.18600
L3′ (444)	1.12155	1.24718	1.24710
W1 (048)	−0.66780	−0.64986	−0.65017
W1 (048)	0.34176	0.37733	0.37529
W3 (048)	−0.65669	−0.65839	−0.65841
W3 (048)	0.48882	0.48462	0.48488
W1′ (048)	−0.64300	−0.64191	−0.64284
W2′ (048)	−0.64844	−0.66353	−0.66525
W2′ (048)	0.72972	0.59085	0.59207
EVEN(224)	−0.66403	−0.66417	−0.66537
EVEN(224)	−0.65734	−0.65919	−0.65890
EVEN(224)	−0.64990	−0.65030	−0.65023
EVEN(224)	−0.10292	−0.10518	−0.10476
ODD (224)	−0.65918	−0.65663	−0.65646
ODD (224)	−0.64741	−0.64489	−0.64469

Table 7.16 Fermi level quantities (Non-orthogonal fit)

Energy		Densities of states				Velocity	Plasmon energy
	Total	s	p	t_{2g}	e_g		
Ry		States/Ry/atom				cm/s	eV
0.4960	5.39	2.20	2.80	0.20	0.18	1.64 × 10E8	12.82

Integrated densities of states Total Electrons	s	p	t_{2g}	e_g
3.00	2.19	0.65	0.08	0.07

Table 7.17 Gallium fcc

Slater-Koster 2-center parameters	Orthogonal	Non-orthogonal	
	Energy integrals (Ry)	Energy integrals (Ry)	Overlap integrals
On site			
s	0.14657	0.00375	
p	0.76463	0.38883	
d1	−0.65490	−0.65472	
d2	−0.65244	−0.65364	
First neighbor			
(sss)	−0.04540	−0.06134	0.10346
(pps)	0.14327	−0.06287	−0.37749
(ppp)	0.00345	0.00647	0.09803
(dds)	−0.00408	−0.02134	0.02607
(ddp)	0.00188	−0.01172	0.02044
(ddd)	−0.00028	−0.00974	0.01470
(sps)	0.07024	0.04979	−0.19950
(sds)	−0.00072	−0.01886	0.03594
(pds)	0.00744	−0.05676	0.07406
(pdp)	0.00550	0.02678	−0.03332
Second neighbor			
(sss)	0.00502	−0.01237	0.00548
(pps)	0.00818	0.05702	−0.03142
(ppp)	0.00386	−0.00146	−0.00051
(dds)	−0.00091	0.02848	−0.04389
(ddp)	0.00017	−0.02372	0.03622
(ddd)	0.00019	−0.01782	0.02725
(sps)	−0.00945	0.02381	−0.01245
(sds)	0.01822	−0.00301	0.01444
(pds)	−0.01213	−0.00594	−0.00222
(pdp)	−0.00464	0.00001	−0.00818

	Orthogonal			Non-orthogonal		
Band	RMS error mRy	Maximum k	Deviation mRy	RMS error mRy	Maximum k	Deviation mRy
1	1.0	(044)	3.8	0.9	(008)	2.2
2	1.7	(003)	5.0	0.6	(000)	1.5
3	0.9	(044)	2.5	0.7	(118)	2.0
4	1.2	(033)	3.5	0.7	(444)	2.5
5	0.7	(033)	1.9	0.8	(444)	2.5
6	25.9	(008)	82.2	1.4	(055)	4.1
7	45.2	(222)	113.2	7.6	(002)	25.6
1–7	19.7			3.0		

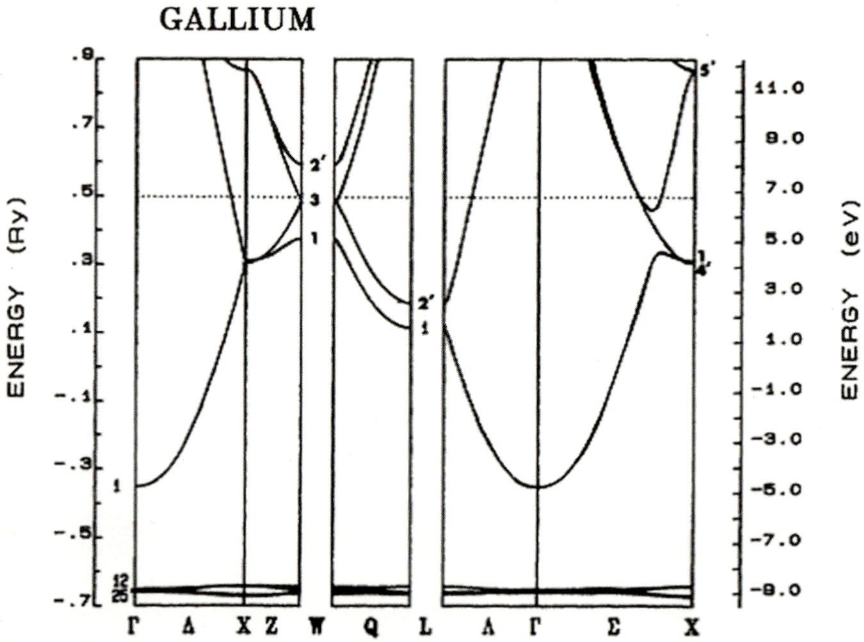

Fig. 7.8 Energy bands for Ga

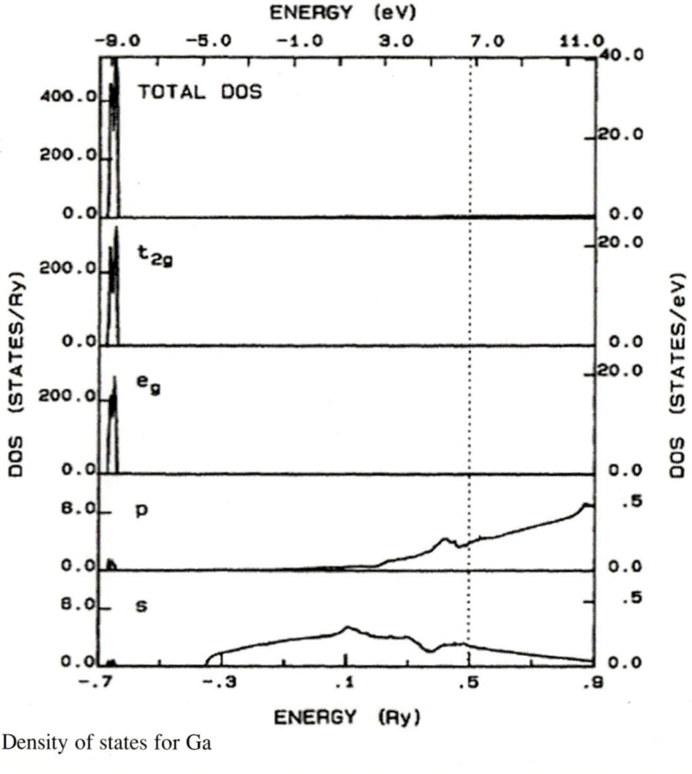

Fig. 7.9 Density of states for Ga

7.4 Indium

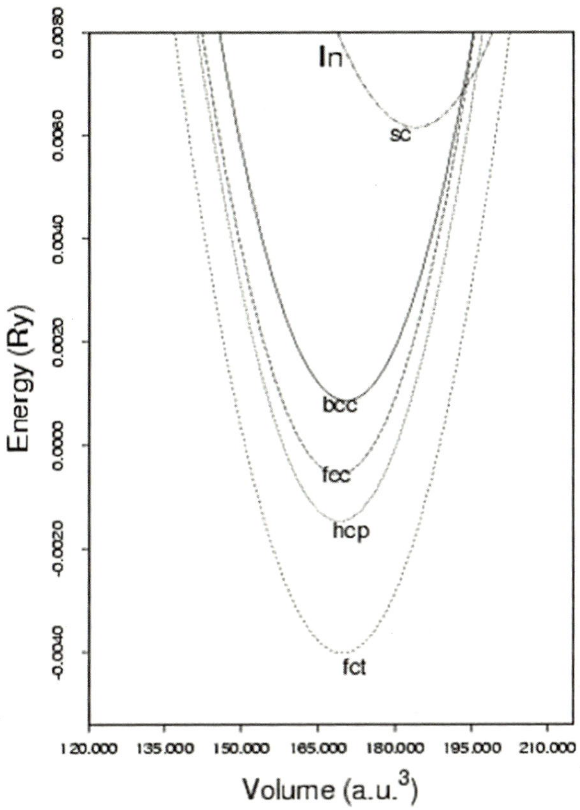

Fig. 7.10 Total energy of In

Table 7.18 Lattice constants and bulk modulus

	a (Bohr)	c (Bohr)	B (MBar)
sc	5.672		0.416
bcc	6.970		0.551
fcc	8.760		0.582
hcp	6.190	10.164	0.577
fct	8.394	9.611	0.599
exp	8.690	9.350	0.411
$\Delta E = E_{fct} - E_{hcp} = 1.2$ mRy			

Table 7.19 Birch fit coefficients

	A_1	A_2	A_3	A_4
sc	0.96536	−73.91363	1777.70917	−12628.12287
bcc	0.80519	−52.01612	923.74703	−2605.56415
fcc	0.54126	−26.73560	120.88389	5811.39178
fct	0.96649	−65.36870	1274.17012	5589.79971

Table 7.20 Indium fcc z = 49 lattice constant = 8.95000 a.u.

Slater-Koster 3-center parameters	Orthogonal	Non-orthogonal	
	Energy integrals (Ry)	Energy integrals (Ry)	Overlap integrals
On site			
s, s (000)	0.04772	0.01359	
x, x (000)	0.57493	0.37059	
xy, xy(000)	−0.80665	−0.80677	
d2, d2(000)	−0.79856	−0.80266	
First neighbor			
s, s (110)	−0.03277	−0.04411	0.04116
s, x (110)	−0.03833	−0.04312	0.08586
s, xy (110)	−0.00534	0.00600	−0.00098
s, d2(110)	0.00631	−0.00750	−0.01085
x, x (110)	0.05268	0.00353	−0.08655
x, x (011)	0.00718	−0.00134	0.05586
x, y (110)	0.06162	0.00202	−0.19412
x, xy(110)	−0.00379	−0.00885	0.00743
x, xy(011)	−0.00797	0.02136	−0.02387
z, d2 (011)	−0.02575	0.01024	−0.01276
z, d1(011)	0.00139	0.03112	−0.02240
xy, xy (110)	−0.00346	0.00162	−0.00631
xy, xy (011)	0.00076	0.00460	−0.00470
xy, xz (011)	0.00065	−0.00002	0.00109
xy, d2 (110)	−0.00163	0.00238	−0.00078
d 2, d 2(110)	−0.00050	−0.02149	0.02581
d1, d1 (110)	0.00137	0.00825	−0.00988
Second neighbor			
s, s(200)	0.00180	−0.00270	0.02291
s, x (200)	0.00505	−0.00759	0.01718
s, d2 (002)	0.02628	0.01338	−0.01301
x, x (200)	0.00265	0.04082	−0.03053
y, y (200)	0.00431	−0.01183	−0.01938
x, xy (020)	0.01099	0.00072	0.00481
z, d2 (002)	−0.00478	−0.02308	0.02313
xy, xy (200)	0.00005	−0.00178	0.00252
xy, xy (002)	0.00070	−0.00335	0.00410
d2, d2 (002)	−0.00280	0.01463	−0.01777
dl, d1 (002)	0.00137	−0.02496	0.03116

Table 7.21 Indium fcc

	Orthogonal			Non-orthogonal		
Band	RMS error mRy	Maximum k	Deviation mRy	RMS error mRy	Maximum k	Deviation mRy
1	2.2	(044)	7.7	0.5	(174)	1.4
2	4.1	(004)	10.7	0.5	(008)	1.2
3	1.8	(444)	4.3	0.5	(118)	1.8
4	2.7	(000)	6.7	0.5	(380)	1.2
5	2.9	(000)	6.7	0.4	(033)	1.0
6	6.6	(066)	17.8	2.3	(444)	7.8
7	24.3	(444)	60.9	2.7	(000)	8.3
1–7	9.8			1.4		

Table 7.22 Energy values in Ry at selected k-points

	Orthogonal	APW	Non-orthogonal
GAMMA 1	−0.33471	−0.32859	−0.32609
GAMMA 12	−0.79767	−0.80442	−0.80399
GAMMA 15	1.04763	1.09833	1.09000
GAMMA 25′	−0.81286	−0.81431	−0.81360
X1 (008)	−0.81460	−0.82121	−0.81998
X1 (008)	0.19210	0.17801	0.18048
X2 (008)	−0.79709	−0.79654	−0.79515
X3 (008)	−0.82496	−0.82644	−0.82556
X5 (008)	−0.79124	−0.79269	−0.79257
X4′ (008)	0.20469	0.21051	0.20892
X5′ (008)	0.56875	0.64943	0.64309
L1 (444)	−0.81505	−0.81666	−0.81660
L1 (444)	0.03854	0.02914	0.03698
L3 (444)	−0.80974	−0.81403	−0.81288
L3 (444)	−0.79016	−0.79582	−0.79523
L2′ (444)	0.05948	0.12037	0.11892
L3′ (444)	0.79888	0.95438	0.95101
W1 (048)	−0.80681	−0.80183	−0.80295
W1 (048)	0.19956	0.21358	0.21166
W3 (048)	−0.80852	−0.81035	−0.81022
W3 (048)	0.36973	0.35873	0.35870
W1′ (048)	−0.79403	−0.79276	−0.79220
W2′ (048)	−0.81390	−0.81486	−0.81565
W2′ (048)	0.55825	0.45479	0.44423
EVEN (224)	−0.81714	−0.81610	−0.81660
EVEN (224)	−0.81030	−0.81174	−0.81089
EVEN (224)	−0.80604	−0.80196	−0.80251
EVEN (224)	−0.13339	−0.13223	−0.13423
ODD (224)	−0.81053	−0.80935	−0.80949
ODD (224)	−0.79597	−0.79671	−0.79656

Table 7.23 Fermi level quantities (Non-orthogonal fit)

Energy		Densities of states				Velocity	Plasmon energy
	Total	s	p	t_{2g}	e_g		
Ry		States/Ry/atom				cm/s	eV
0.3540	6.22	2.36	3.73	0.04	0.10	1.44 × 10E8	9.88

Integrated densities of states				
Total	s	p	t_{2g}	e_g
Electrons				
3.00	2.03	0.92	0.01	0.04

Table 7.24 Indium fcc

Slater-Koster 2-center parameters	Orthogonal	Non-orthogonal	
	Energy integrals (Ry)	Energy integrals (Ry)	Overlap integrals
On site			
s	0.05099	−0.05892	
p	0.55930	0.25103	
dl	−0.80391	−0.80642	
d2	−0.80237	−0.80580	
First neighbor			
(sss)	−0.03356	−0.05085	0.09739
(pps)	0.10719	−0.05222	−0.40364
(ppp)	0.00285	−0.00320	0.09784
(dds)	−0.00481	−0.02095	0.02034
(ddp)	0.00166	−0.02384	0.03145
(ddd)	−0.00008	−0.00485	0.00576
(sps)	0.05536	0.04693	−0.20532
(sds)	0.00527	−0.02569	0.02516
(pds)	−0.02565	−0.07034	0.07707
(pdp)	−0.01799	0.02816	−0.02477
Second neighbor			
(sss)	0.00298	−0.00789	0.00444
(pps)	0.00796	0.03874	−0.04633
(ppp)	0.00756	−0.00081	0.00180
(dds)	−0.00140	0.01513	−0.01857
(ddp)	−0.00039	−0.01461	0.01802
(ddd)	0.00077	−0.02541	0.03153
(sps)	−0.00806	0.01912	−0.01306
(sds)	−0.02545	−0.00039	0.00505
(pds)	−0.00827	−0.00051	−0.00812
(pdp)	0.02420	−0.00141	0.00132

Band	Orthogonal RMS error mRy	Maximum k	Deviation mRy	Non-orthogonal RMS error mRy	Maximum k	Deviation mRy
1	3.0	(044)	9.1	0.6	(008)	1.6
2	3.6	(004)	9.7	0.6	(008)	1.8
3	3.0	(005)	8.6	0.6	(118)	2.2
4	3.3	(005)	8.6	0.6	(444)	1.9
5	2.3	(444)	5.4	0.7	(444)	1.9
6	7. 1	(066)	15.3	1.1	(000)	2.3
7	28.5	(222)	75.1	4.2	(011)	16.8
1–6	11.4			1.7		

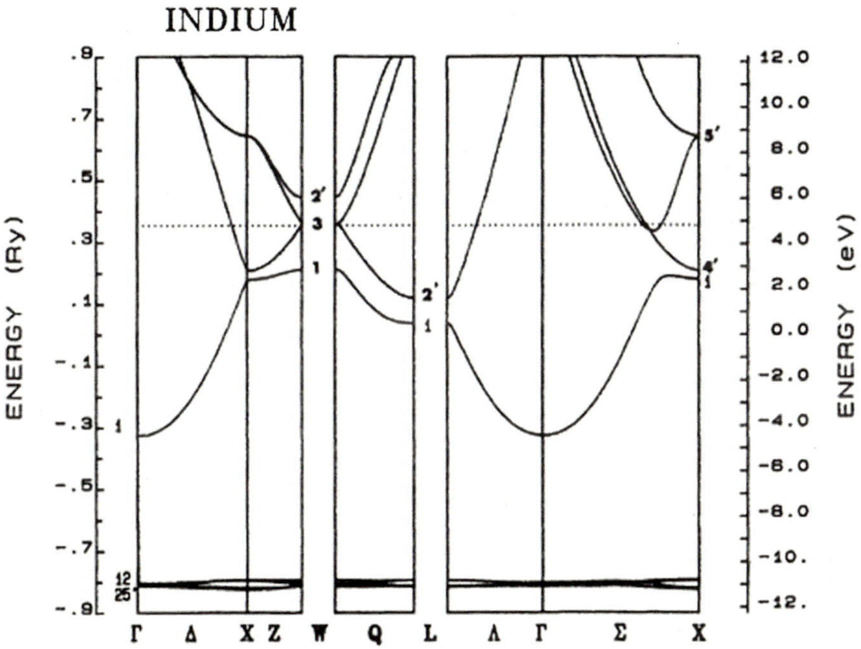

Fig. 7.11 Energy bands for In

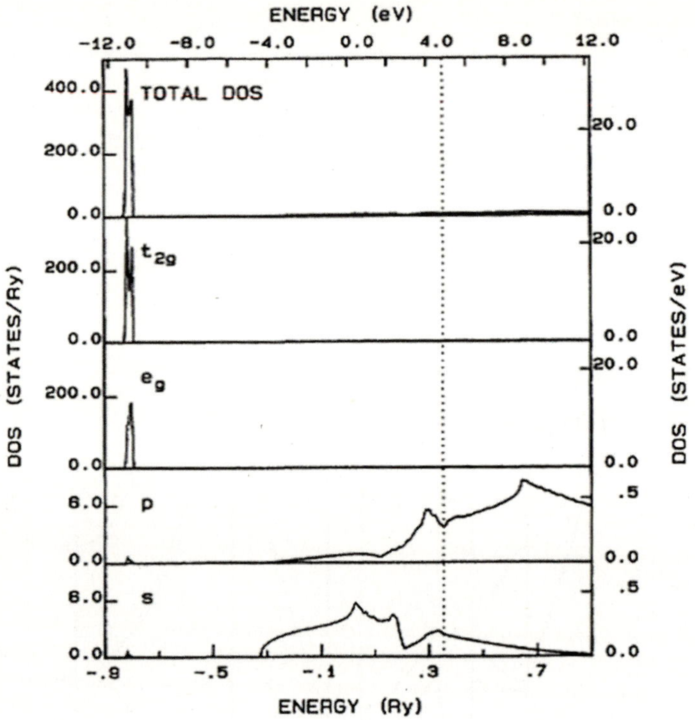

Fig. 7.12 Density of states for In

7.5 Thallium

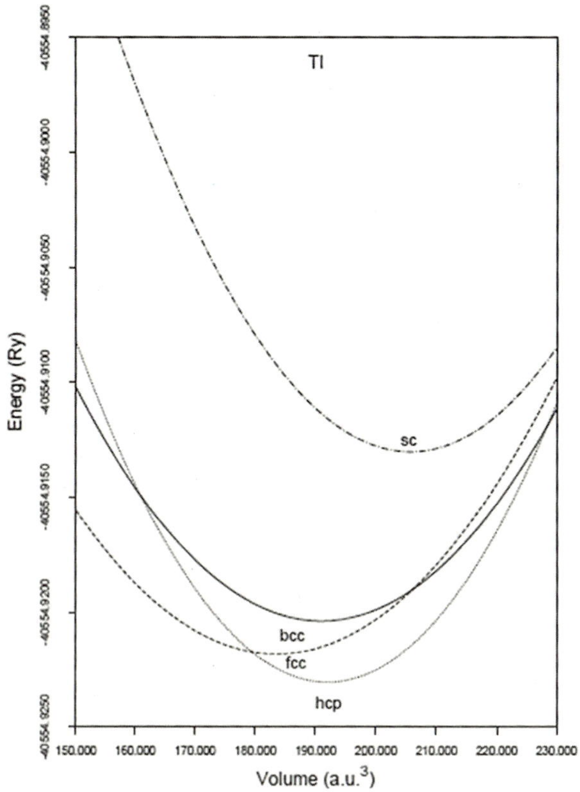

Fig. 7.13 Total energy of Tl

Table 7.25 Lattice constants and bulk modulus

	a (Bohr)	c (Bohr)	B (MBar)
sc	5.837		0.372
bcc	7.179		0.424
fcc	9.008		0.441
hcp	6.470	10.190	0.434
exp	6.538	10.431	0.359
$\Delta E = E_{fcc} – E_{hcp} = 0.7$ mRy			

Table 7.26 Birch fit coefficients

	A_1	A_2	A_3	A_4
sc	−40554.74546	2.24804	−734.04511	15802.11673
bcc	−40554.66358	−5.57375	−451.80597	11735.69251
fcc	−40554.73952	2.22180	−711.36777	14504.96080

Table 7.27 Thallium hcp z = 81 lattice constants = 6.5320 a.u. 10.4404 a.u.

Slater-Koster 3-center parameters	Orthogonal		
	Energy integrals (Ry)		Energy integrals (Ry)
On site			
s, s(0)	−0.04447		
x, x(0)	0.19825		
z, z(0)	0.46468		
xy, xy(0)	−0.38075		
yz, yz(0)	−0.61949		
d2, d2(0)	−0.54783		
s, d2(0)	−0.03671		
x, xy(0)	0.44523		
First neighbor			
s, s(R)	−0.08282	s, s(T)	−0.00704
s, x(R)	0.06145	s, y(T)	−0.06062
s, y(R)	−0.00895	s, z(T)	0.05291
s, xy(R)	0.01944	s, d1(T)	0.02462
s, d1(R)	−0.00536	s, yz(T)	0.04005
s, d2(R)	−0.00569	s, d2(T)	0.02911
x, x(R)	0.13816	x, x(T)	0.00071
y, y(R)	0.01734	y, y(T)	0.05918
x, y(R)	−0.01493	y, z(T)	−0.04349
x, xy(R)	0.04167	x, xy(T)	−0.00447
y, d1(R)	0.00440	y, d1(T)	−0.01281
x, d1(R)	−0.01722	x, xz(T)	−0.01246
y, xy(R)	0.00382	y, yz(T)	−0.02203
y, d2(R)	−0.01787	y, d2(T)	−0.02029
x, d2(R)	−0.06359	z, z(T)	0.06246
z, z(R)	−0.00077	z, d1(T)	0.07043
z, yz(R)	−0.00343	z, yz(T)	−0.04257
z, xz(R)	−0.00949	z, d2(T)	0.00476
xy, xy(R)	0.00604	xy, xy(T)	−0.00668
d1, d1(R)	−0.00584	d1, d1(T)	0.00337
xy, d1(R)	−0.00689	xy, yz(T)	−0.00310
d1, d2(R)	−0.01187	yz, d1(T)	−0.00907
xy, d2(R)	−0.03389	d1, d2(T)	−0.01652
yz, yz(R)	0.00124	yz, yz(T)	−0.00557
xz, xz(R)	−0.00132	xz, xz(T)	0.00157
yz, xz(R)	−0.00317	yz, d2(T)	−0.00076
d2, d2(R)	0.02278	d2, d2(T)	−0.01986

Table 7.28 Thallium hcp

Band	Orthogonal RMS error mRy	Maximum k	Deviation mRy
1	8.4	(0 0 6)	21.5
2	6.0	(0 0 0)	20.1
3	3.2	(8 0 0)	10.0
4	3.9	(24 0 9)	10.1
5	3.9	(16 0 12)	8.7
6	4.4	(0 0 9)	13.9
7	4.6	(4 0 0)	14.9
8	5.5	(4 0 0)	16.9
9	5.8	(0 12 12)	17.7
10	4.9	(0 12 12)	17.7
11	14.4	(0 0 9)	33.2
12	19.0	(0 18 0)	47.5
1–12	8.39		

Table 7.29 Energy values in Ry at selected k-points

	Orthogonal	APW
GAMMA 1+	−0.66409	−0.64193
GAMMA 1+	−0.44979	−0.48244
GAMMA 3+	0.08534	0.10223
GAMMA 5+	−0.64103	−0.63744
GAMMA 6+	−0.63174	−0.63364
GAMMA 4-	−0.66202	−0.65389
GAMMA 4-	−0.12909	−0.14856
GAMMA 5-	−0.60775	−0.61188
GAMMA 6-	−0.64179	−0.64074
M1+	−0.65804	−0.66322
M1+	−0.61096	−0.61319
M1+	−0.11908	−0.09917
M2+	−0.60625	−0.60867
M3+	−0.65182	−0.65355
M4+	−0.61373	−0.60871
M1-	−0.64281	−0.64431
M2-	−0.64769	−0.64767
M2-	−0.63216	−0.62824
M2-	−0.20509	−0.18872
M3-	−0.60989	−0.61172
M4-	−0.62927	−0.62907
A1	−0.64885	−0.64639
A1	−0.39813	−0.37816
A3	−0.64528	−0.64283
A3	−0.62304	−0.61894
L1	−0.65987	−0.65399
L1	−0.64970	−0.65319
L1	−0.60637	−0.61027
L1	−0.12302	−0.11947
L2	−0.62824	−0.62764
L2	−0.60774	−0.60962

Table 7.30 Thallium hcp z = 81 lattice constants = 6.5320 a.u. 10.4404 a.u.

Slater-Koster 2-center parameters	Orthogonal	Non-orthogonal	
	Energy integrals (Ry)	Energy integrals (Ry)	Overlap integrals
On site			
s	−0.00218	−0.21445	
p	0.41325	0.41586	
d0	−0.56873	−0.62926	
dl	−0.62975	−0.62588	
d2	−0.62789	−0.62622	
pd	0.23507	0.00358	−0.01151
First neighbor			
(sss)	−0.03118	−0.10250	0.15418
(sps)	0.08473	0.07436	−0.07351
(sds)	−0.01266	0.04976	−0.05387
(pps)	0.12807	0.12073	−0.03569
(ppp)	−0.00159	0.00096	0.06847
(pds)	−0.02716	−0.03671	0.05103
(pdp)	−0.01048	−0.03177	0.04749
(dds)	−0.00642	0.04314	−0.07817
(ddp)	−0.00040	0.01124	−0.01587
(ddd)	−0.00014	0.00987	−0.01439
Second neighbor			
(sss)	−0.00013	−0.02183	0.07625
(sps)	−0.00039	−0.00626	−0.11030
(sds)	−0.02855	0.02264	−0.01462
(pps)	−0.01104	−0.03421	−0.13855
(ppp)	0.00197	0.01310	0.03573
(pds)	−0.00210	0.02790	−0.08098
(pdp)	0.00387	0.01746	−0.04233
(dds)	−0.00102	0.00588	−0.01013
(ddp)	0.00144	0.02375	−0.03525
(ddd)	0.00080	0.06380	−0.10275
Third neighbor			
(sss)	−0.05817	−0.00981	0.00438
(sps)	0.04273	0.01494	0.01684
(sds)	−0.01359	0.06234	−0.09615
(pps)	0.08816	0.09474	0.20750
(ppp)	0.02234	−0.06093	−0.08768
(pds)	0.03585	−0.05123	0.02888
(pdp)	0.00976	0.06505	−0.10805
(dds)	0.00019	−0.03167	0.04972
(ddp)	0.00080	0.10973	−0.17483
(ddd)	−0.00203	0.09932	−0.15485

Table 7.31 Thallium hcp

	Orthogonal			Non-orthogonal		
Band	RMS error mRy	Maximum k	Deviation mRy	RMS error mRy	Maximum k	Deviation mRy
1	17.1	(8 0 0)	47.3	5.7	(24 0 0)	15.6
2	6.5	(8 0 12)	15.7	4.2	(8 0 12)	11.5
3	5.7	(0 24 12)	15.7	3.4	(0 6 0)	6.6
4	6.1	(0 24 12)	15.7	3.5	(4 0 0)	5.8
5	4.6	(0 0 0)	9.5	3.0	(24 0 6)	5.1
6	5.3	(0 0 0)	9.5	3.5	(24 0 9)	8.5
7	6.1	(16 0 12)	11.5	4.0	(0 24 12)	8.7
8	6.1	(16 0 12)	11.5	5.1	(0 6 0)	11.7
9	4.6	(0 24 12)	9.0	4.5	(24 0 9)	8.1
10	5.6	(24 0 6)	9.0	5.1	(24 0 0)	13.4
11	24.7	(12 0 0)	60.1	10.0	(16 0 12)	18.3
12	25.3	(16 0 9)	43.4	17.7	(24 0 9)	55.4
1–12	12.35			7.03		

Table 7.32 Energy values in Ry at selected k-points

	Orthogonal	APW	Non-orthogonal
GAMMA 1+	−0.64651	−0.64193	−0.63520
GAMMA 1+	−0.48860	−0.48244	−0.47550
GAMMA 3+	0.08516	0.10223	0.07237
GAMMA 5+	−0.62794	−0.63744	−0.63910
GAMMA 6+	−0.64175	−0.63364	−0.63288
GAMMA 4–	−0.63835	−0.65389	−0.65298
GAMMA 4–	−0.11316	−0.14856	−0.12579
GAMMA 5–	−0.61776	−0.61188	−0.61313
GAMMA 6–	−0.62507	−0.64074	−0.64180
M1+	−0.65536	−0.66322	−0.67882
M1+	−0.61301	−0.61319	−0.60671
M1+	−0.11679	−0.09917	−0.10309
M2+	−0.61340	−0.60867	−0.60355
M3+	−0.63145	−0.65355	−0.65454
M4+	−0.60893	−0.60871	−0.59525
M1–	−0.64288	−0.64431	−0.64017
M2–	−0.66711	−0.64767	−0.65093
M2–	−0.63803	−0.62824	−0.63363
M2–	−0.21470	−0.18872	−0.19693
M3–	−0.60107	−0.61172	−0.61635
M4–	−0.62965	−0.62907	−0.62154
A1	−0.64786	−0.64639	−0.64115
A1	−0.36273	−0.37816	−0.36564
A3	−0.64287	−0.64283	−0.64152
A3	−0.61444	−0.61894	−0.62342
L1	−0.66773	−0.65399	−0.65556
L1	−0.63752	−0.65319	−0.64786
L1	−0.61707	−0.61027	−0.61515
L1	−0.12194	−0.11947	−0.12164
L2	−0.63267	−0.62764	−0.63229
L2	−0.60067	−0.60962	−0.61898

Table 7.33 Fermi level quantities (Non-orthogonal fit)

Energy		Densities of states			
	Total	s	p	t_{2g}	e_g
Ry		States/Ry/atom			
0.3090	3.74	1.49	2.08	0.13	0.04

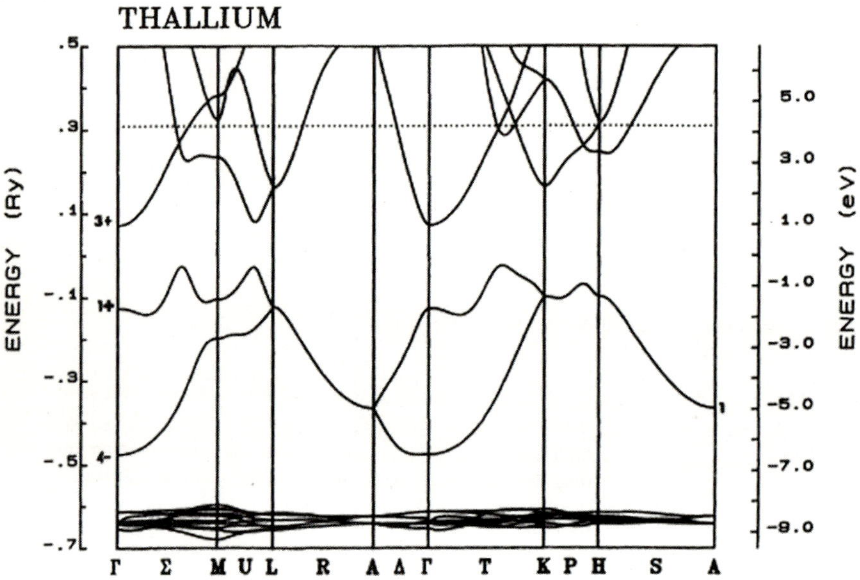

Fig. 7.14 Energy bands for Tl

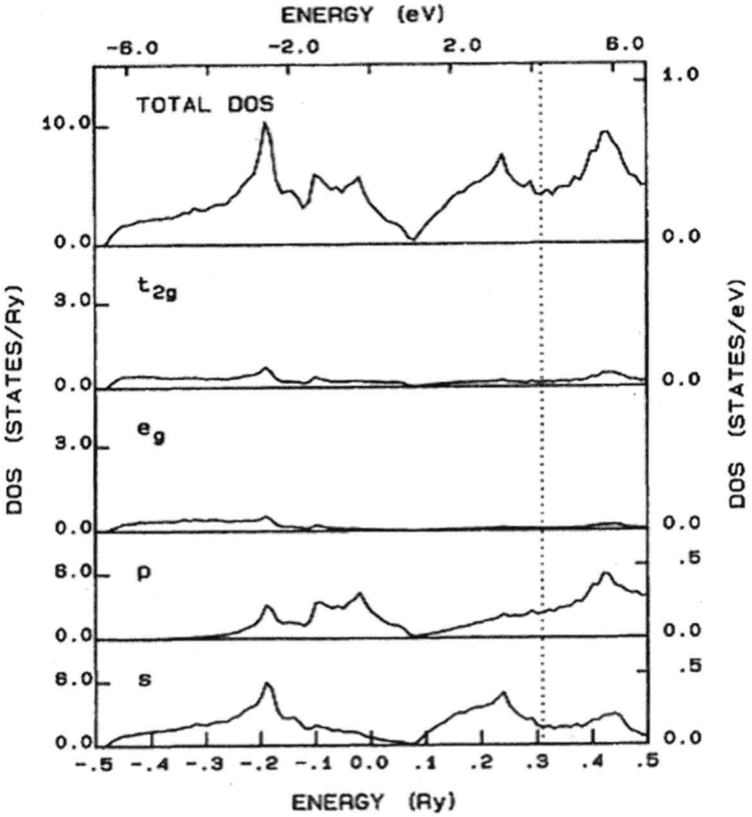

Fig. 7.15 Density of states for Tl

7.6 Lead

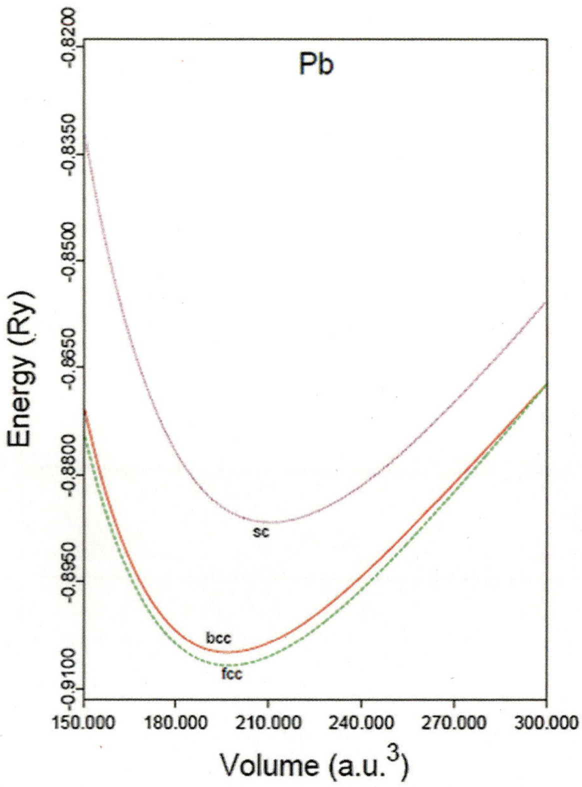

Fig. 7.16 Total energy of Pb

Table 7.34 Lattice constants and bulk modulus

	a (Bohr)	B (MBar)
bcc	7.328	0.513
fcc	9.238	0.500
sc	5.948	0.483
exp	9.354	(0.430)
$\Delta E_{fcc-bcc} = 1.85$ mRy		

Table 7.35 Birch fit coefficients

	A_1	A_2	A_3	A_4
bcc	−0.71973	7.29742	−1129.50887	22690.13470
fcc	−0.52695	−13.04321	−423.33056	14544.02520
sc	−0.46343	−17.39746	−358.22706	15707.50355

Table 7.36 Lead fcc z = 82 lattice constant = 9.35550 a.u.

Slater-Koster 3-center parameters	Orthogonal	Non-orthogonal	
	Energy integrals (Ry)	Energy integrals (Ry)	Overlap integrals
On site			
s, s (000)	−0.16616	−0.29270	
x, x (000)	0.53625	0.35378	
xy, xy (000)	1.37006	0.87801	
d2, d2 (000)	1.28414	0.84525	
First neighbor			
s, s (110)	−0.04257	−0.03226	0.02104
s, x (110)	0.03882	0.04489	−0.05533
s, xy (110)	−0.02402	−0.04282	0.12083
s, d2 (110)	0.05617	0.02273	−0.09353
x, x (110)	0.04878	0.01058	−0.07737
x, x (011)	−0.00058	−0.01767	0.02317
x, y (110)	0.06335	0.02222	−0.12653
x, xy (110)	0.07910	0.05846	0.16859
x, xy (011)	−0.00183	0.00896	−0.06216
z, d2 (011)	0.00223	0.00975	−0.02557
z,d1 (011)	0.02283	−0.04098	−0.13580
xy, xy (110)	−0.12750	0.21709	0.22548
xy, xy (011)	0.02719	−0.08620	−0.12381
xy, xz (011)	0.01432	0.01421	−0.08004
xy, d2 (110)	0.00032	−0.08800	−0.07554
d2, d2 (110)	−0.08359	0.04994	0.09402
d1, d1 (110)	0.14087	−0.13712	−0.26990
Second neighbor			
s, s (200)	0.01999	−0.00317	0.00606
s, x (200)	−0.04712	0.00355	0.00290
s, d2 (002)	0.08720	−0.01215	−0.01355
x, x (200)	0.01967	0.02445	−0.00833
y, y (200)	−0.01380	−0.00261	−0.00202
x, xy (020)	−0.01296	0.00714	−0.00061
z, d2 (002)	0.01876	−0.01465	0.04942
xy, xy (200)	0.07448	0.02218	−0.01956
xy, xy (002)	−0.03823	−0.03842	−0.00487
d2, d2 (002)	0.10776	−0.04257	0.04155
d1, d1 (002)	−0.07270	−0.03608	−0.00658

Table 7.37 Lead fcc

	Orthogonal			Non-orthogonal		
Band	RMS error mRy	Maximum k	Deviation mRy	RMS error mRy	Maximum k	Deviation mRy
1	11.4	(008)	24.3	1.1	(000)	2.9
2	15.3	(062)	46.0	2.7	(001)	7.3
3	21.1	(005)	102.9	1.2	(008)	3.5
4	16.9	(022)	45.5	1.8	(011)	4.3
1–4	16.5			1.8		

Table 7.38 Energy values in Ry at selected k-points

	Orthogonal	APW	Non-orthogonal
GAMMA 1	−0.55706	−0.54511	−0.54222
GAMMA 12	1.73302	1.73698	1.73696
GAMMA 15	0.90832	0.90411	0.90476
GAMMA 25′	1.29901	1.29779	1.29778
X1 (008)	−0.21637	−0.19211	−0.19254
X1 (008)	0.71727	0.73430	0.73423
X3 (008)	0.86403	0.87301	0.87287
X5 (008)	2.10154	2.13224	2.13224
X4′ (008)	0.12788	0.13793	0.14044
X5′ (008)	0.52272	0.52097	0.52447
L1 (444)	−0.30672	−0.28778	−0.28670
L1 (444)	1.05465	1.06294	1.06290
L3 (444)	1.17888	1.19240	1.19237
L2′ (444)	0.04529	0.06188	0.06754
L3′ (444)	0.80551	0.77804	0.77942
W1 (048)	−0.17935	−0.18548	−0.18535
W3 (048)	0.26638	0.27601	0.27541
W3 (048)	1.29895	1.26994	1.26994
W2′ (048)	0.39766	0.38080	0.38101
W2′ (048)	1.16300	1.14323	1.14324
EVEN (224)	−0.36081	−0.37308	−0.37361
EVEN (224)	0.25804	0.26298	0.26250
EVEN (224)	0.69230	0.69832	0.69736
ODD (224)	0.70190	0.70917	0.70748

Table 7.39 Fermi level quantities (Non-orthogonal fit)

Energy		Densities of states				Velocity	Plasmon energy
	Total	s	p	t_{2g}	e_g		
Ry		States/Ry/atom				cm/s	eV
0.3700	6.86	0.73	6.03	0.06	0.04	$1.38 \times 10E8$	9.12

Integrated densities of states				
Total	s	p	t_{2g}	e_g
Electrons				
4.00	2.18	1.81	0.01	0.00

Table 7.40 Lead fcc

Slater-Koster 2-center parameters	Orthogonal	Non-orthogonal	
	Energy integrals (Ry)	Energy integrals (Ry)	Overlap integrals
On site			
s	−0.23417	−0.30195	
P	0.52776	0.36526	
dl	1.46854	0.87728	
d2	1.36143	0.84284	
First neighbor			
(sss)	−0.02719	−0.03743	0.03874
(pps)	0.10763	0.03634	−0.19279
(ppp)	−0.00651	−0.01486	0.03069
(dds)	−0.13575	0.30007	0.29044
(ddp)	0.09885	−0.13415	−0.26929
(ddd)	−0.02685	−0.03011	0.02939
(sps)	−0.04094	0.06009	−0.07980
(sds)	0.05329	−0.04732	0.15983
(pds)	−0.10170	0.09429	0.27482
(pdp)	0.00988	0.00102	−0.10057
Second neighbor			
(sss)	−0.00141	0.00179	−0.01605
(pps)	0.00486	0.02273	0.00472
(ppp)	−0.00565	−0.00375	−0.00431
(dds)	0.02686	−0.04124	0.04314
(ddp)	0.01689	0.02294	−0.01957
(ddd)	−0.01195	−0.03765	−0.00496
(sps)	0.01936	0.01035	0.01621
(sds)	−0.05739	−0.01930	0.00484
(pds)	0.01481	−0.01412	0.04030
(pdp)	−0.00707	0.01081	0.00289

	Orthogonal			Non-orthogonal		
Band	RMS error mRy	Maximum k	Deviation mRy	RMS error mRy	Maximum k	Deviation mRy
1	13.5	(222)	29.1	1.5	(000)	4.1
2	25.1	(008)	53.6	3.4	(001)	8.7
3	15.0	(005)	33.4	1.8	(111)	4.1
4	26.6	(033)	70.0	2.4	(022)	5.8
1–4	20.9			2.4		

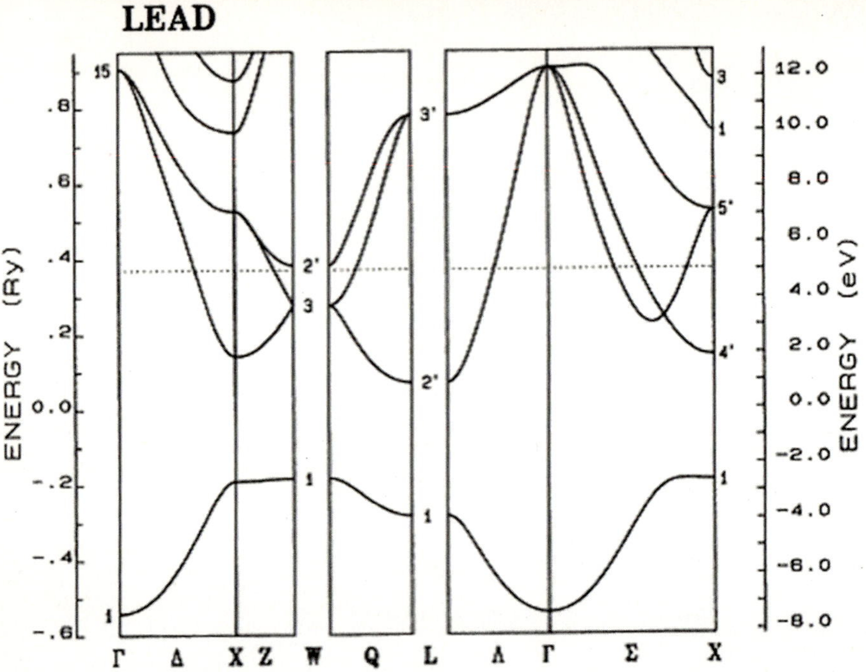

Fig. 7.17 Energy bands for Pb

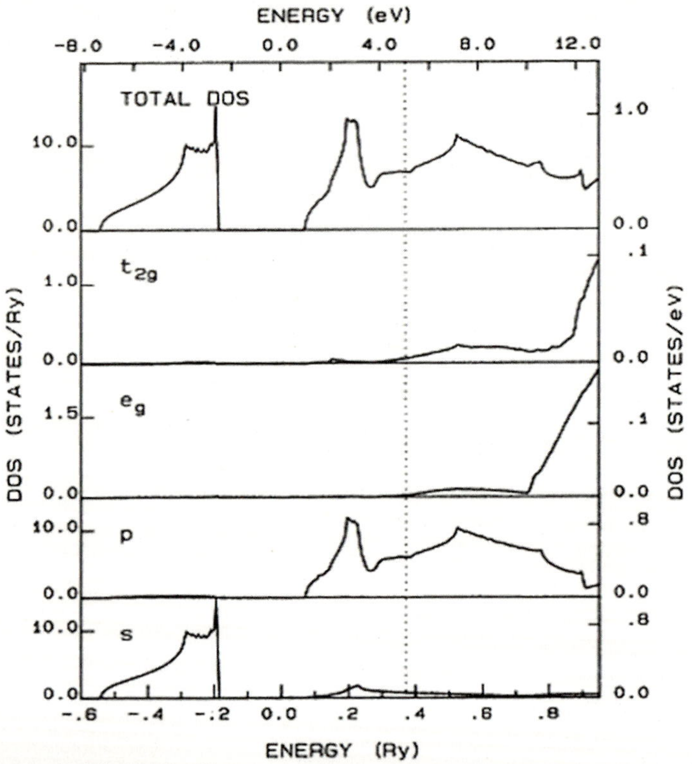

Fig. 7.18 Density of states for Pb

Chapter 8
The Diamond Structure

The total energy plots generated by the LAPW method show for carbon the diamond and graphite structures as almost indistinguishable ground states. For silicon and germanium the diamond structure is clearly the ground state. However, for tin our total energy calculations show the diamond, beta-tin and simple hexagonal (shex) structures very close in energy. The LDA gives the shex as the ground state and the GGA gives the diamond as the lowest energy. In the diamond structure, unlike the other structures, the d states were omitted in the tight-binding Hamiltonian. For this reason the conduction bands should not be considered very reliable. Also the reader is reminded that units of eV are used for the SK parameters and not Ry as in the rest of this handbook. The SK parameters of Si, Ge, and Sn were derived from fitting to empirical pseudopotential calculations and, for C, to an APW calculation using overlapping spheres, unlike the first-principles APW touching muffin tin calculations used in the rest of this book. These calculations reproduce band gaps that are close to experiment and should be considered accurate for the four valence bands and first two conduction bands.

D.A. Papaconstantopoulos, *Handbook of the Band Structure of Elemental Solids*,
DOI 10.1007/978-1-4419-8264-3_8

8.1 Carbon

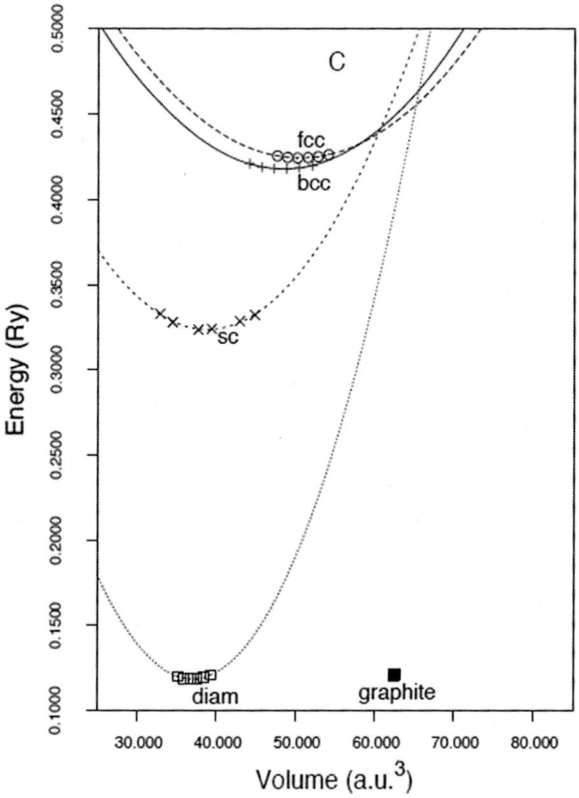

Fig. 8.1 Total energy of C

Table 8.1 Lattice constants and bulk modulus

	a (Bohr)	c (Bohr)	B (MBar)
sc	3.364		2.859
bcc	4.584		2.210
fcc	5.845		2.270
diam	6.658		4.817
exp diam	6.741		4.430
graphite	4.646	13.369	
exp graphite	4.658	12.689	
$\Delta E_{\text{sc-diam}} = 2.0$ mRy			

Table 8.2 Birch fit coefficients

	A_1	A_2	A_3	A_4
sc	0.82300	−7.49500	−21.91418	484.13797
bcc	0.69499	−0.46925	−132.74156	1196.83503
fcc	1.88427	−48.10706	497.93727	−1546.42007
diam	1.34375	−25.68758	117.93250	180.23944

Table 8.3 Carbon diamond Z = 6 lattice constant = 6.74030 a.u.

Slater-Koster 3-center parameters	Orthogonal	Non-orthogonal	
	Energy integrals (eV)	Energy integrals (eV)	Overlap integrals
On site			
s, s (000)	−8.51517	−11.81840	
p, p (000)	3.88980	0.86088	
First neighbor			
s, s (111)	−3.82441	−4.81108	0.17460
s, x (111)	1.75801	2.77047	−0.16131
x, x (111)	0.64149	0.25413	−0.18839
x, y (111)	2.54011	2.12793	0.02679
Second neighbor			
s, s (220)	0.41003	−0.52661	0.00863
s, x (022)	0.48691	−0.37355	0.02780
s, x (220)	−0.26779	0.31437	0.01115
x, x (220)	0.01505	0.46662	−0.00630
x, x (022)	−0.31135	−0.90778	0.05739
x, y (220)	−0.16693	0.15269	−0.01081
x, y (022)	−0.50501	0.25849	−0.01834
Third neighbor			
s, s (311)	−0.20371	−0.24534	0.00851
s, x (311)	0.22956	−0.01232	−0.01442
s, x (113)	−0.08436	−0.02559	0.01052
x, x (311)	0.06576	−0.20717	−0.00895
x, x (113)	0.00675	0.09259	−0.00769
x, y (311)	0.07728	0.02348	−0.00416
x, y (113)	−0.10111	−0.03102	0.00265

Table 8.4 Carbon diamond

	Orthogonal			Non-orthogonal		
Band	RMS error eV	Maximum k	Deviation eV	RMS error eV	Maximum k	Deviation eV
1	0.22	(660)	0.42	0.05	(333)	0.19
2	0.19	(400)	0.38	0.06	(480)	0.10
3	0.20	(460)	0.37	0.07	(462)	0.25
4	0.16	(400)	0.33	0.05	(620)	0.12
5	0.32	(550)	0.76	0.09	(462)	0.29
6	0.41	(262)	1.50	0.11	(462)	0.25
1–6	0.27			0.07		

Table 8.5 Energy values in eV at selected k-points

	Orthogonal	Pseudopotential	Non-orthogonal
GAMMA 1	−21.33696	−21.17955	−21.17780
GAMMA 15	5.64773	5.69786	5.76192
GAMMA 2′	14.14730	13.15828	13.36727
GAMMA 25′	−0.11819	0.00000	0.01680
X1 (080)	−12.16867	−12.49527	−12.44633
X1 (080)	4.53743	4.62836	4.59875
X3 (080)	16.31829	16.52768	15.97179
X4 (080)	−6.04790	−6.12258	−6.12528
L1 (444)	−13.28700	−13.22491	−13.30503
L1 (444)	8.41033	8.81022	8.70221
L3 (444)	−2.47132	−2.70871	−2.62996
L3 (444)	8.91546	8.48532	8.69575
L2′ (444)	−15.52507	−15.37806	−15.41541
W (480)	−12.03332	−11.67124	−11.77387
W (480)	−8.43536	−8.21671	−8.16639
W (480)	9.74771	10.31071	10.16049
W (480)	13.36012	12.28189	12.80551
(260)	−15.45848	−15.61824	−15.62850
(260)	−8.83553	−8.76248	−8.85563
(260)	−7.02504	−6.85018	−6.81627
(260)	−5.77224	−5.81209	−5.69693
(260)	6.48237	6.23329	6.26221
(260)	8.21345	8.39025	8.53773

Table 8.6 Energy GAP in eV

Orthogonal	Pseudopotential	Non-orthogonal
4.279	4.052	4.038

Table 8.7 Carbon diamond

Slater-Koster 2-center parameters	Orthogonal	Non-orthogonal	
	Energy integrals (eV)	Energy integrals (eV)	Overlap integrals
On site			
s	−5.16331	−14.61534	
p	2.28887	−3.25210	
First neighbor			
(sss)	−4.43338	−6.31297	0.33491
(sps)	3.78614	4.51067	−0.40587
(pps)	5.65984	3.88932	−0.38891
(ppp)	−1.82861	−2.28558	0.06779
Second neighbor			
(sss)	0.13579	−1.21543	0.03198
(sps)	−0.01536	1.30055	−0.05404
(pps)	0.62602	1.35152	−0.07324
(ppp)	−0.24384	−0.04971	0.01874

(continued)

Table 8.7 (continued)

Slater-Koster 2-center parameters	Orthogonal	Non-orthogonal	
	Energy integrals (eV)	Energy integrals (eV)	Overlap integrals
Third neighbor			
(sss)	−0.01124	−0.37034	0.00032
(sps)	−0.16336	0.53454	−0.00078
(pps)	−0.06670	0.70763	0.01863
(ppp)	0.06517	−0.01395	0.00552

	Orthogonal			Non-orthogonal		
Band	RMS error (eV)	Maximum (k)	Deviation (eV)	RMS error (eV)	Maximum (k)	Deviation (eV)
1	0.34	(444)	0.90	0.53	(480)	1.28
2	0.28	(500)	0.60	0.77	(550)	1.70
3	0.45	(800)	0.84	0.39	(462)	1.75
4	0.51	(800)	0.84	0.14	(800)	0.28
5	0.94	(480)	2.39	0.72	(444)	2.05
6	1.34	(550)	3.25	0.59	(262)	1.17
1–6	0.75			0.56		

Table 8.8 Energy GAP in eV

Orthogonal	Pseudopotential	Non-orthogonal
5.241	4.052	4.444

Carbon

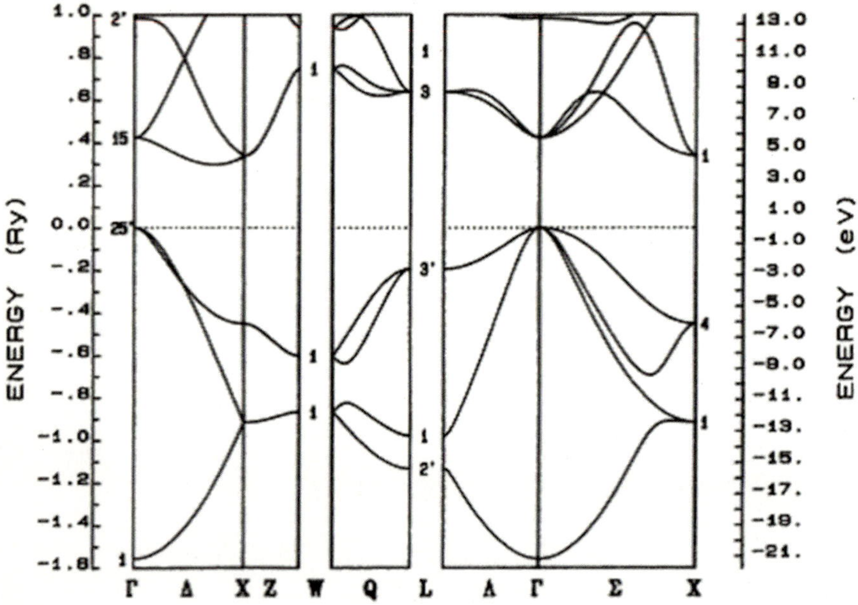

Fig. 8.2 Energy bands for C

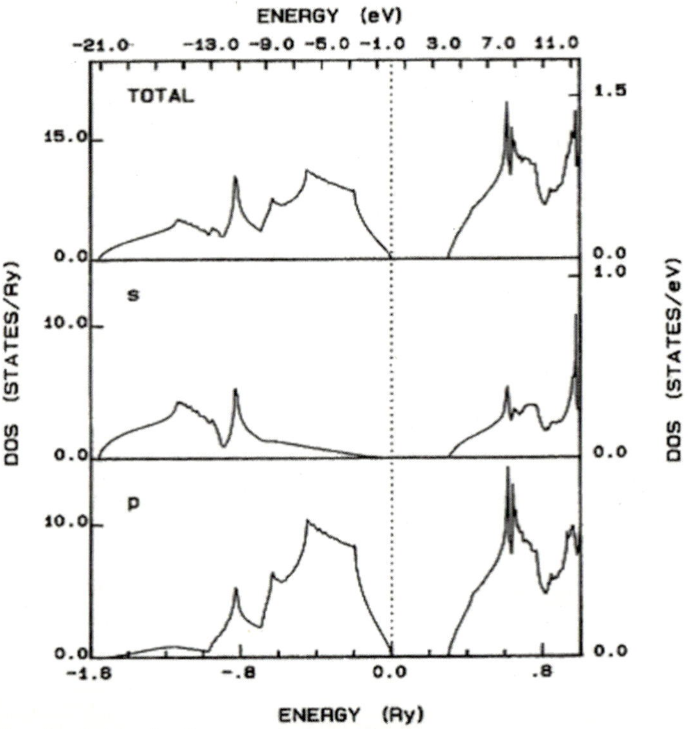

Fig. 8.3 Density of states for C

8.2 Silicon

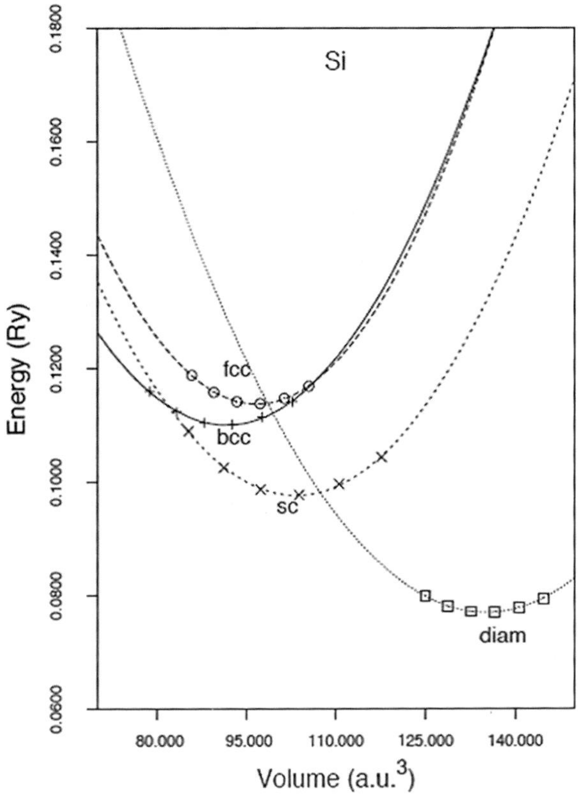

Fig. 8.4 Total energy of Si

Table 8.8 Lattice constants and bulk modulus

	a (Bohr)	B (MBar)
sc	4.682	1.000
bcc	5.669	0.924
fcc	7.278	1.159
diam	10.256	1.081
exp	10.265	0.998
$\Delta E = 20.7$ mRy		

Table 8.9 Birch fit coefficients

	A_1	A_2	A_3	A_4
sc	2.08930	−113.63993	2110.62283	−12637.92504
bcc	1.43270	−67.38469	1102.35954	−5673.21217
fcc	2.01363	−101.86959	1764.31262	−9720.24748
diam	1.74046	−101.89049	1908.44382	−9975.18410

Table 8.10 Silicon diamond Z = 14 lattice constant = 10.26250 a. u.

Slater-Koster 3-center parameters	Orthogonal	Non-orthogonal	
	Energy integrals (eV)	Energy integrals (eV)	Overlap integrals
On site			
s, s (000)	−6.47049	−7.87943	
p, p (000)	2.14983	0.75909	
First neighbor			
s, s (111)	−1.84407	−1.77153	0.05619
s, x (111)	1.01852	1.63062	−0.16106
x, x (111)	0.29947	0.33744	−0.13370
x, y (111)	1.44623	0.93686	0.04495
Second neighbor			
s, s (220)	0.21149	0.09522	−0.00845
s, x (022)	0.04306	−0.26778	0.04765
s, x (220)	−0.14350	−0.15005	0.04396
x, x (220)	0.09858	0.17560	−0.05872
x, x (022)	−0.32894	−0.43609	0.08565
x, y (220)	−0.12292	0.15207	−0.02249
x, y (022)	−0.18240	0.06200	−0.03543
Third neighbor			
s, s (311)	−0.11562	−0.16764	0.01177
s, x (311)	0.12484	−0.18440	−0.00520
s, x (113)	−0.07116	−0.15335	0.01760
x, x (311)	0.07920	−0.31399	0.01883
x, x (113)	0.02620	0.03630	−0.02252
x, y (311)	0.08797	0.09070	−0.02645
x, y (113)	−0.07875	−0.06077	0.00470

Table 8.11 Silicon diamond

	Orthogonal			Non-orthogonal		
Band	RMS error (eV)	Maximum (k)	Deviation (eV)	RMS error (eV)	Maximum (k)	Deviation (eV)
1	0.10	(480)	0.21	0.07	(480)	0.13
2	0.14	(500)	0.25	0.13	(444)	0.29
3	0.23	(660)	0.44	0.08	(462)	0.22
4	0.19	(480)	0.41	0.07	(480)	0.15
5	0.25	(480)	0.59	0.11	(440)	0.25
6	0.47	(440)	1.10	0.17	(440)	0.45
1–6	0.26			0.11		

Table 8.12 Energy values in eV at selected k-points

	Orthogonal	Pseudopotential	Non-orthogonal
GAMHA 1	−12.69632	−12.59938	−12.52180
GAMMA 15	3.34696	3.41326	3.44736
GAMMA 2′	4.83115	4.46350	4.43237
GAMMA 25′	−0.10150	0.00000	0.02333
X1 (080)	−8.38757	−8.36346	−8.30383
X1 (080)	1.11656	1.22086	1.22270
X3 (080)	10.26921	12.23020	11.89757
X4 (080)	−3.33806	−3.07943	−3.12532
L1 (444)	−7.26198	−7.35174	−7.06290
L1 (444)	2.48864	2.21386	2.38631
L3 (444)	−1.06071	−1.29369	−1.28233
L3 (444)	4.37700	3.98364	4.34846
L2′ (444)	−10.42481	−10.26009	−10.35616
L2′ (444)	8.52358	8.75096	8.78873
W (480)	−8.41517	−8.20127	−8.32897
W (480)	−4.42182	−4.00909	−3.85531
W (480)	4.31404	4.90314	4.98978
W (480)	8.18309	5.93171	6.29060
(260)	−9.93051	−9.96684	−9.90532
(260)	−6.59705	−6.40348	−6.60957
(260)	−3.74057	−3.52905	−3.45211
(260)	−3.14979	−2.91295	−2.81020
(260)	2.46955	2.52442	2.43139
(260)	3.28259	3.20291	3.11387

Table 8.13 Energy GAP in eV

Orthogonal	Pseudopotential	Non-orthogonal
1.011	1.099	1.114

Table 8.14 Silicon diamond

Slater-Koster 2-center parameters	Orthogonal	Non-orthogonal	
	Energy integrals (eV)	Energy integrals (eV)	Overlap integrals
On site			
s	−5.19278	−9.43024	
p	1.05825	−0.98539	
First neighbor			
(sss)	−2.36233	−3.64630	0.30363
(sps)	1.86401	2.71033	−0.37426
(pps)	2.85882	2.14624	−0.24550
(ppp)	−0.94687	−1.33192	0.13067
Second neighbor			
(sss)	0.13595	−0.48994	0.01846
(sps)	−0.26377	0.69041	−0.04226
(pps)	0.41007	0.68720	−0.14030
(ppp)	−0.13704	−0.13386	0.02574

(continued)

Table 8.14 (continued)

Slater-Koster 2-center parameters	Orthogonal	Non-orthogonal	
	Energy integrals (eV)	Energy integrals (eV)	Overlap integrals
Third neighbor			
(sss)	0.04603	−0.07037	−0.00161
(sps)	−0.19434	0.19543	−0.00657
(pps)	−0.05261	0.38545	−0.01572
(ppp)	0.08081	−0.01848	−0.02499

	Orthogonal			Non-orthogonal		
Band	RMS error (eV)	Maximum (k)	Deviation (eV)	RMS error (eV)	Maximum (k)	Deviation (eV)
1	0.32	(444)	0.80	0.11	(060)	0.25
2	0.18	(300)	0.51	0.18	(444)	0.54
3	0.32	(800)	0.63	0.10	(660)	0.21
4	0.36	(800)	0.63	0.08	(330)	0.16
5	0.53	(480)	1.61	0.17	(480)	0.69
6	0.86	(440)	1.86	0.32	(462)	0.79
1–6	0.48			0.18		

Table 8.15 Energy GAP in eV

Orthogonal	Pseudopotential	Non-orthogonal
1.401	1.099	1.195

Silicon

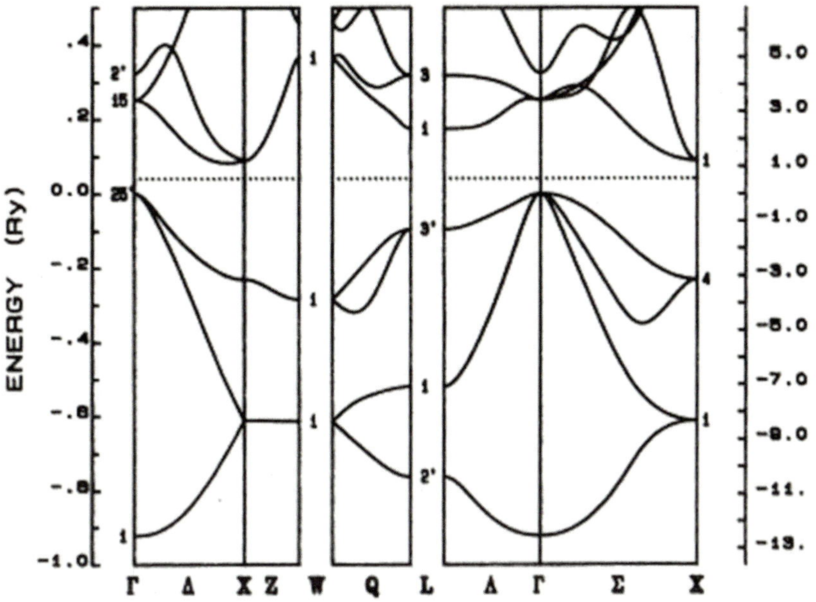

Fig. 8.5 Energy bands for Si

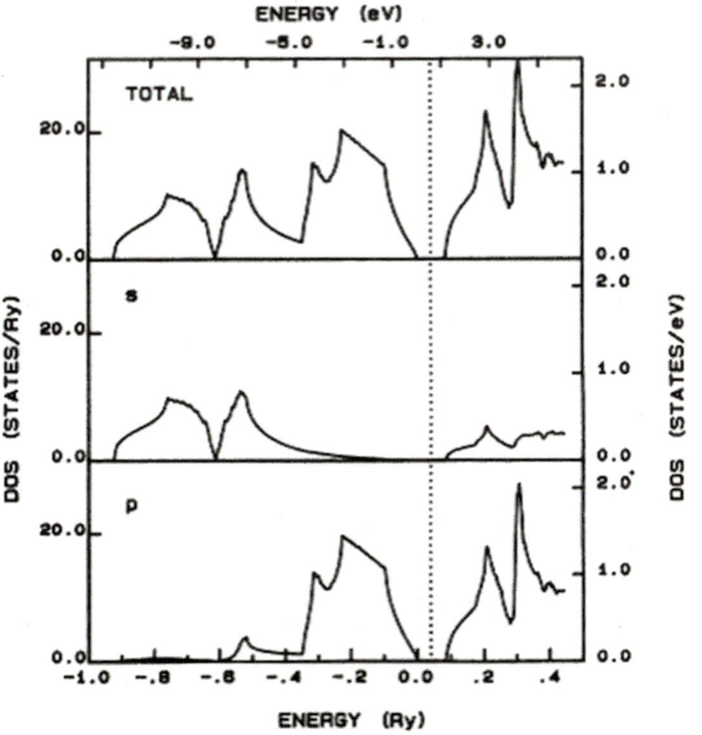

Fig. 8.6 Density of states for Si

8.3 Germanium

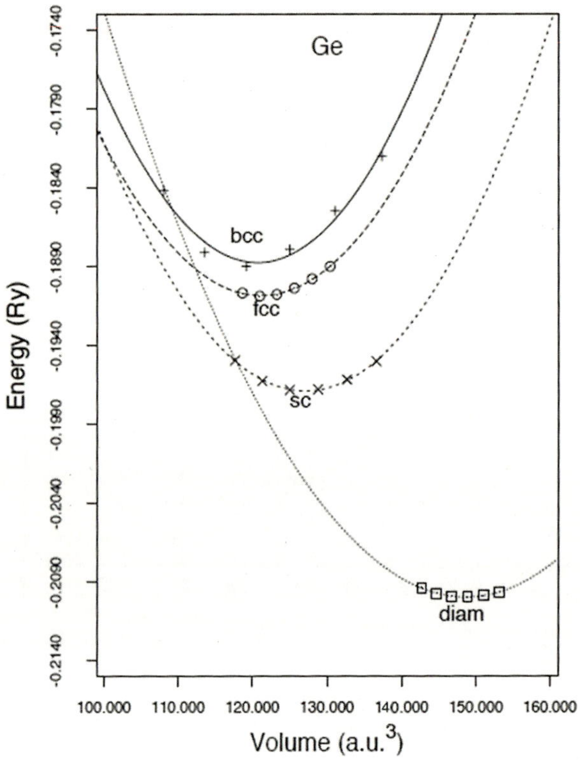

Fig. 8.7 Total energy of Ge

Table 8.16 Lattice constants and bulk modulus

	a (Bohr)	B (MBar)
sc	5.020	0.809
bcc	6.200	1.123
fcc	7.855	0.921
diam	10.591	0.667
exp	10.692	0.772
$\Delta E = 13.1$ mRy		

Table 8.17 Birch fit coefficients

	A_1	A_2	A_3	A_4
sc	0.38560	−24.31541	115.49029	3209.17889
bcc	−1.35980	109.45721	−3238.31087	30857.35716
fcc	−1.35950	108.40383	−3206.85158	30684.58161
diam	−0.13702	15.09027	−1018.33177	15081.23076

Table 8.18 Germanium diamond Z = 32 lattice constant = 10.69080 a. u.

Slater-Koster 3-center parameters	Orthogonal	Non-orthogonal	
	Energy integrals (eV)	Energy integrals (eV)	Overlap integrals
On site			
s, s (000)	−7.21995	−7.42226	
p, p (000)	1.76430	−0.40474	
First neighbor			
s, s (111)	−1.46162	−0.02536	−0.14129
s, x (111)	0.89350	0.85766	−0.12571
x, x (111)	0.33939	0.03624	−0.04407
x, y (111)	1.22493	1.04047	−0.18257
Second neighbor			
s, s (220)	0.13417	0.26057	−0.00457
s, x (022)	−0.06443	0.51217	−0.03815
s, x (220)	−0.14005	−0.21928	0.02901
x, x (220)	0.07884	−0.01160	−0.02567
x, x (022)	−0.19580	0.29422	−0.05343
x, y (220)	−0.07246	0.14058	−0.00945
x, y (022)	−0.21980	−0.33270	0.00851
Third neighbor			
s, s (311)	−0.09273	−0.43201	0.03319
s, x (311)	0.10312	0.06926	0.00902
s, x (113)	−0.07145	−0.21534	0.02057
x, x (311)	0.12053	0.06330	0.00400
x, x (113)	−0.03262	0.03199	0.00053
x, y (311)	0.05259	−0.04946	0.00116
x, y (113)	−0.03056	−0.11989	−0.00842

Table 8.19 Germanium diamond

	Orthogonal			Non-orthogonal		
Band	RMS error eV	Maximum k	Deviation eV	RMS error eV	Maximum k	Deviation eV
1	0.15	(444)	0.39	0.07	(333)	0.14
2	0.15	(500)	0.32	0.07	(040)	0.14
3	0.25	(460)	0.54	0.08	(660)	0.22
4	0.11	(480)	0.30	0.10	(330)	0.28
5	0.24	(480)	0.90	0.15	(480)	0.41
6	0.46	(440)	1.21	0.21	(480)	0.41
1–6	0.26			0.12		

Table 8.20 Energy values in eV at selected k-points

	Orthogonal	Pseudopotential	Non-orthogonal
GAMMA 1	−12.56909	−12.41689	−12.30962
GAMMA 15	3.19051	3.22020	3.13805
GAMMA 2′	1.34927	0.94300	0.89039
GAMMA 25′	0.03311	0.00000	0.03428
X1 (080)	−8.51341	−8.57072	−8.65260
X1 (080)	1.10717	1.07078	1.24514
X3 (080)	7.99021	11.83362	5.06083
X4 (080)	−2.89521	−2.96659	−3.06535
L1 (444)	−7.04258	−7.35471	−7.26587
L1 (444)	0.74497	0.76066	0.73936
L3 (444)	−1.27214	−1.30418	−1.17434
L3 (444)	4.22103	4.16687	4.22067
L2′ (444)	−10.73159	−10.34541	−10.27904
L2′ (444)	7.27730	7.68636	11.15426
W (480)	−8.60295	−8.46839	−8.49201
W (480)	−4.10218	−3.80574	−3.75594
W (480)	4.05368	4.95215	4.54369
W (480)	6.34021	5.00457	4.77017
(260)	−9.88775	−10.03757	−10.01521
(260)	−7.03714	−6.80287	−6.85932
(260)	−3.64361	−3.39240	−3.42262
(260)	−2.74646	−2.79232	−2.78633
(260)	2.29500	2.24532	2.29088
(260)	2.92757	2.94416	3.34081

Table 8.21 Energy GAP in eV

Orthogonal	Pseudopotential	Non-orthogonal
0.712	0.761	0.705

Table 8.22 Germanium diamond

Slater-Koster 2-center parameters	Orthogonal	Non-orthogonal	
	Energy integrals (eV)	Energy integrals (eV)	Overlap integrals
On site			
s	−6.09231	−9.00687	
p	1.17426	−1.28186	
First neighbor			
(sss)	−1.86035	−2.82594	0.21480
(sps)	1.90405	3.11310	−0.35031
(pps)	2.79070	1.69491	−0.23134
(ppp)	−0.92858	−1.53592	0.13770
Second neighbor			
(sss)	0.04434	−0.12370	−0.00186
(sps)	−0.20314	0.72871	−0.05036
(pps)	0.28255	0.61066	−0.24376
(ppp)	−0.07537	−0.07671	0.03664

(continued)

Table 8.22 (continued)

Slater-Koster 2-center parameters	Orthogonal	Non-orthogonal	
	Energy integrals (eV)	Energy integrals (eV)	Overlap integrals
Third neighbor			
(sss)	0.05290	0.04246	−0.00838
(sps)	−0.08846	0.22448	0.03324
(pps)	0.01713	0.77052	−0.01239
(ppp)	0.04430	−0.08876	0.00169

	Orthogonal			Non-orthogonal		
Band	RMS error eV	Maximum k	Deviation eV	RMS error eV	Maximum k	Deviation eV
1	0.24	(444)	0.69	0.16	(060)	0.34
2	0.12	(020)	0.24	0.16	(444)	0.41
3	0.25	(800)	0.57	0.09	(020)	0.19
4	0.26	(800)	0.57	0.10	(020)	0.19
5	0.68	(480)	1.82	0.25	(480)	0.75
6	0.81	(480)	1.82	0.36	(480)	0.75
1–6	0.47			0.21		

Table 8.23 Energy GAP in eV

Orthogonal	Pseudopotential	Non-orthogonal
0.515	0.761	0.645

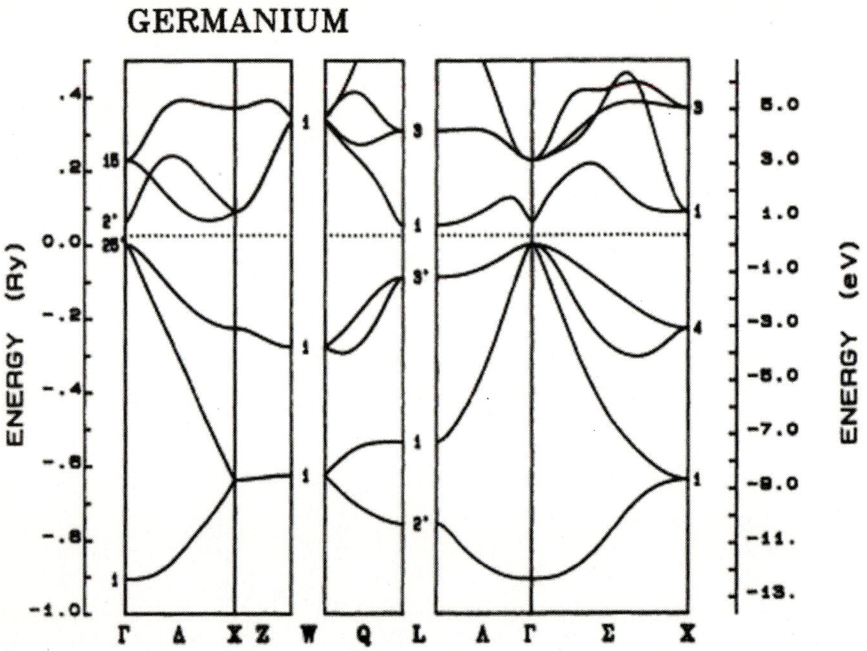

Fig. 8.8 Energy bands for Ge

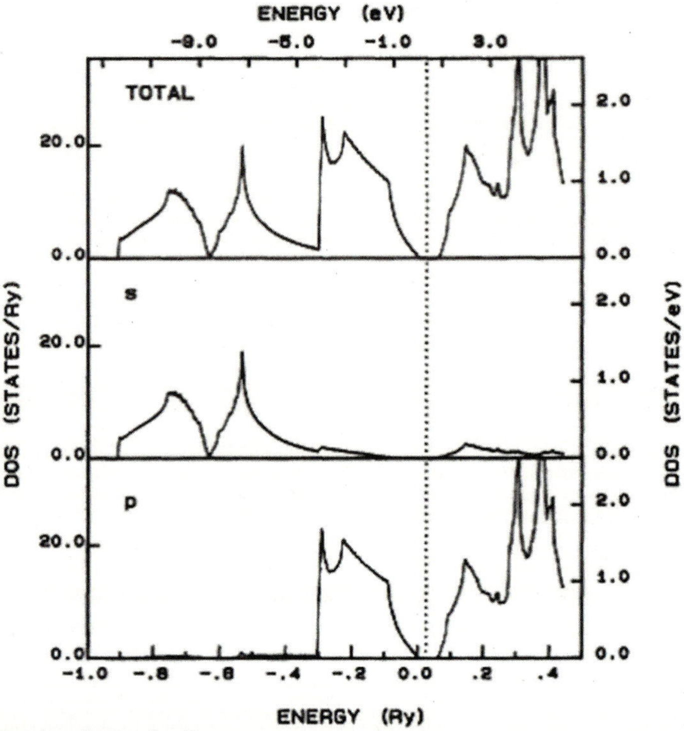

Fig. 8.9 Density of states for Ge

8.4 Tin (LDA)

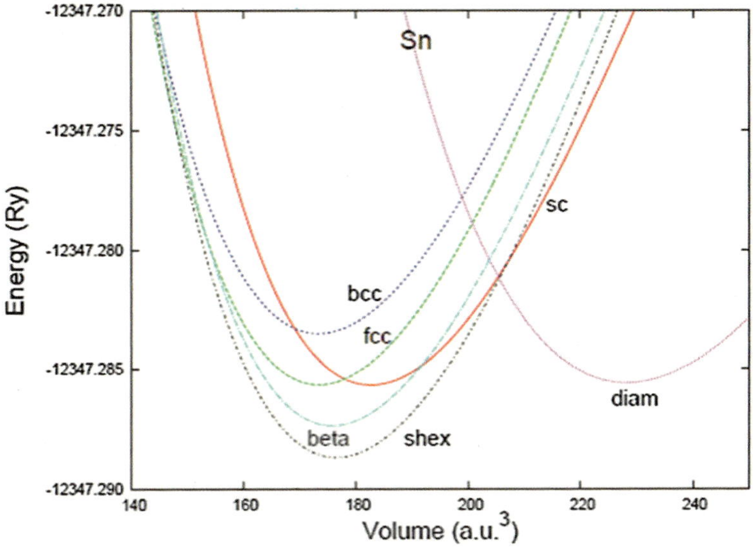

Fig. 8.10 Total energy of Sn (LDA)

Table 8.24 Lattice constants and bulk modulus

	a (Bohr)	c (Bohr)	B (MBar)
sc	5.674		0.598
bcc	7.026		0.578
fcc	8.854		0.618
diam	12.223		0.459
beta	10.855	5.970	0.619
shex	5.987	5.688	0.620
exp diam	12.268		
exp beta	11.021	6.011	
$\Delta E_{shex-diam} = 3.11$ mRy			

Table 8.25 Birch fit coefficients

	A_1	A_2	A_3	A_4
sc	−12346.82467	−17.63220	−298.20332	12493.49209
bcc	−12346.738100	−27.05408	100.24694	6643.89369
fcc	−12346.79624	−20.16951	−165.89411	9946.42952
diam	−24693.65639	−67.70203	−1617.99378	143274.03442
beta	−24693.72327	−44.33801	−1923.88823	100614.24393
shex	−12346.73336	−26.08472	−8.39887	8790.33264

8.5 Tin (GGA)

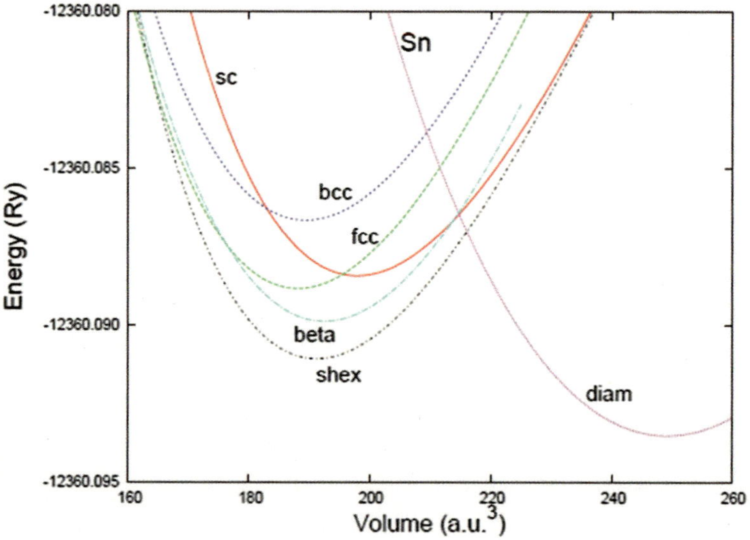

Fig. 8.11 Total energy of Sn (GGA)

Table 8.26 Lattice constants and bulk modulus

	a (Bohr)	c (Bohr)	B (MBar)
sc	5.828		0.473
bcc	7.235		0.459
fcc	9.095		0.487
diam	12.586		0.373
beta	11.186	6.152	0.463
shex	6.148	5.840	0.482
exp diam	12.268		
exp beta	11.021	6.011	
$\Delta E_{diam-shex} = 2.45$ mRy			

Table 8.27 Birch fit coefficients

	A_1	A_2	A_3	A_4
sc	−12359.74788	−10.38254	−473.12268	14703.89444
bcc	−12359.52343	−33.77938	390.44781	3660.87655
fcc	−12359.75351	−10.01447	−426.37508	12926.62874
diam	−24717.61881	−394.84445	19195.70594	−284362.96041
beta	−24718.29599	−226.96203	8195.72979	−77282.96987
shex	−12360.00197	14.52973	−1258.60248	22510.24185

Table 8.28 Tin diamond Z = 50 lattice constant = 12.26660 a. u.

Slater-Koster 3-center parameters	Orthogonal	Non-orthogonal	
	Energy integrals (eV)	Energy integrals (eV)	Overlap integrals
On site			
s, s (000)	−6.76930	−6.62760	
p, p (000)	1.74194	1.13265	
First neighbor			
s, s (111)	−1.14086	−0.41821	−0.10096
s, x (111)	0.63610	0.58014	−0.06141
x, x (111)	0.36542	0.11272	−0.17083
x, y (111)	1.01161	0.74447	0.00915
Second neighbor			
s, s (220)	0.11541	0.26113	−0.00307
s, x (022)	−0.12793	−0.05400	0.05171
s, x (220)	−0.11711	−0.38058	0.03790
x, x (220)	0.05114	−0.06174	−0.02540
x, x (022)	−0.23052	−0.08968	0.12980
x, y (220)	−0.07524	−0.13220	−0.05765
x, y (022)	−0.22961	−0.29888	0.01412
Third neighbor			
s, s (311)	−0.10261	−0.11677	0.00296
s, x (311)	0.11146	0.18404	−0.02364
s, x (113)	−0.03489	−0.06270	0.00432
x, x (311)	0.05966	0.00505	−0.02384
x, x (113)	−0.04058	0.00165	−0.03235
x, y (311)	0.07739	−0.01017	−0.00338
x, y (113)	−0.05947	0.06094	−0.00198

Table 8.29 Tin diamond

	Orthogonal			Non-orthogonal		
Band	RMS error eV	Maximum k	Deviation eV	RMS error eV	Maximum k	Deviation eV
1	0.11	(444)	0.24	0.03	(040)	0.06
2	0.22	(444)	0.56	0.10	(500)	0.23
3	0.23	(660)	0.45	0.09	(354)	0.23
4	0.19	(480)	0.31	0.10	(440)	0.24
5	0.23	(550)	0.44	0.23	(660)	0.54
6	0.32	(462)	0.63	0.24	(660)	0.62
1-6	0.23			0.15		

Table 8.30 Energy values in eV at selected k-points

	Orthogonal	Pseudopotential	Non-orthogonal
GAMMA 1	−11.17918	−11.05832	−11.04249
GAMMA 15	2.60463	2.71865	2.74404
GAMMA 2′	0.41036	−0.17345	−0.31542
GAMMA 25′	−0.14665	0.00000	−0.08680
X1 (080)	−7.64202	−7.60897	−7.58169
X1 (080)	0.82181	1.14101	0.83101
X3 (080)	7.56750	8.65359	8.31596
X4 (080)	−2.23943	−2.47733	−2.50687
L1 (444)	−5.76740	−6.33191	−6.32888
L1 (444)	0.59250	0.39068	0.15274
L3 (444)	−0.92763	−1.17479	−1.38653
L3 (444)	3.80956	3.86522	3.89460
L2′ (444)	−9.41072	−9.16992	−9.15081
L2′ (444)	5.73482	5.63706	6.16034
W (480)	−7.63109	−7.54808	−7.59566
W (480)	−3.35108	−3.04083	−3.10802
W (480)	3.77445	3.95088	4.23249
W (480)	5.71558	4.24760	4.96647
(260)	−9.31052	−8.86463	−9.26796
(260)	−5.74281	−6.11825	−5.97298
(260)	−2.33965	−2.74456	−2.38303
(260)	−0.96569	−2.28318	−1.18922
(260)	1.73328	1.99650	1.49318
(260)	3.84762	2.49374	3.57410

Table 8.31 Tin diamond

Slater-Koster 2-center parameters	Orthogonal	Non-orthogonal	
	Energy integrals (eV)	Energy integrals (eV)	Overlap integrals
On site			
s	−5.88191	−8.22749	
p	1.26821	0.07822	
First neighbor			
(sss)	−1.46728	−2.69973	0.24057
(sps)	1.53221	2.05073	−0.28144
(pps)	2.42232	2.10817	−0.10456
(ppp)	−0.85908	−0.82931	0.13737
Second neighbor			
(sss)	0.03642	−0.18726	0.00202
(sps)	−0.19583	0.33354	−0.00872
(pps)	0.16708	0.47212	−0.06540
(ppp)	−0.06653	−0.12179	0.00335
Third neighbor			
(sss)	0.02531	0.02681	−0.00726
(sps)	−0.09026	0.06710	0.01970
(pps)	0.03846	0.13062	−0.01467
(ppp)	0.03693	−0.02388	−0.02830

	Orthogonal			Non-orthogonal		
Band	RMS error eV	Maximum k	Deviation eV	RMS error eV	Maximum k	Deviation eV
1	0.17	(444)	0.43	0.07	(060)	0.18
2	0.12	(100)	0.23	0.12	(444)	0.33
3	0.22	(444)	0.48	0.10	(442)	0.18
4	0.24	(444)	0.48	0.10	(800)	0.18
5	0.53	(480)	1.18	0.19	(480)	0.55
6	0.61	(550)	1.53	0.24	(550)	0.58
1–6	0.37			0.15		

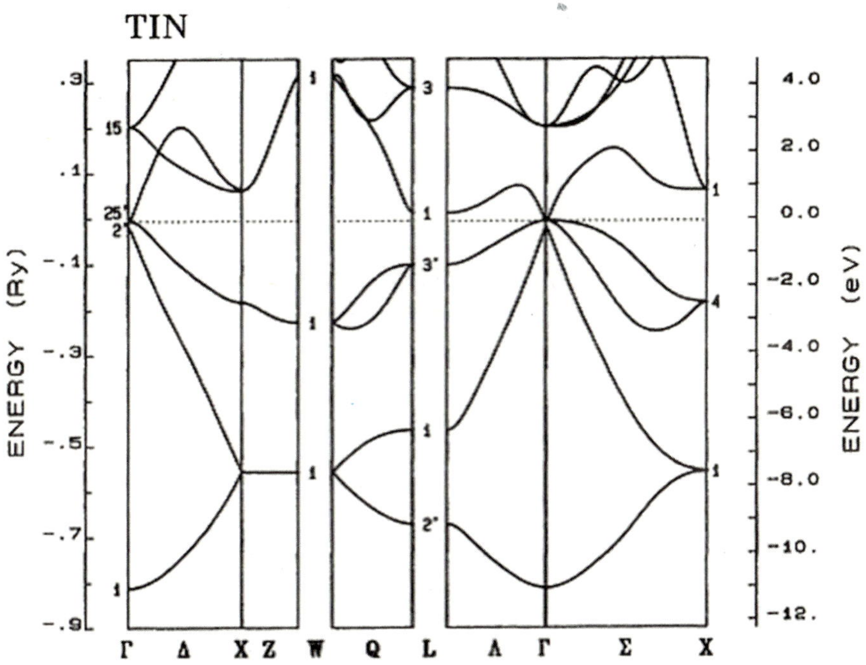

Fig. 8.12 Energy bands for Sn

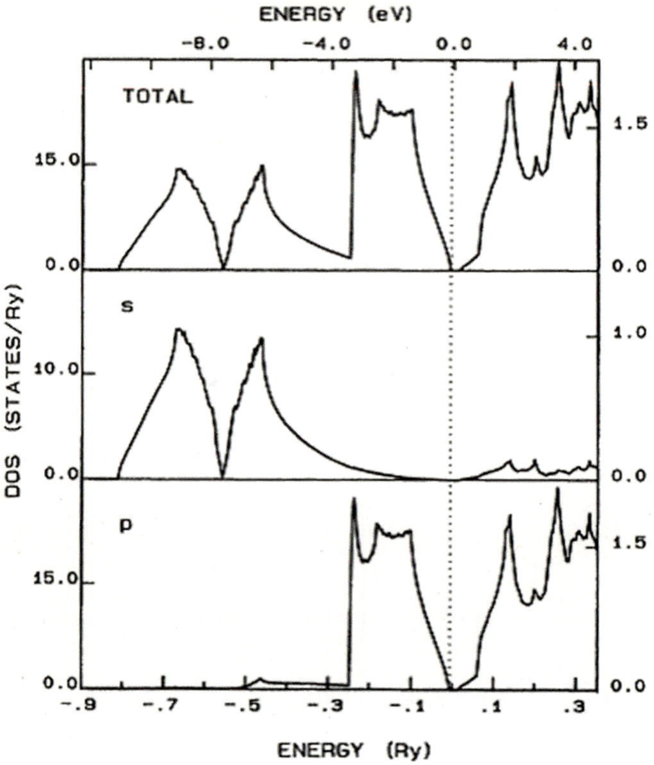

Fig. 8.13 Density of states for Sn

Chapter 9
Group 15 Elements: Pnictogens

These elements have the following complex crystal structures; N: hcp molecular, P: triclinic, As: trigonal, Sb: trigonal and Bi: monoclinic. In this work we performed calculations in only the cubic phases and found that the lowest energy occurs in the simple cubic structure. In addition to the total energy diagrams we present energy bands and densities of states, with values at the Fermi level similar to the free-electron-like metals. We also include a set of two-center tight-binding parameters.

D.A. Papaconstantopoulos, *Handbook of the Band Structure of Elemental Solids*,
DOI 10.1007/978-1-4419-8264-3_9

9.1 Nitrogen

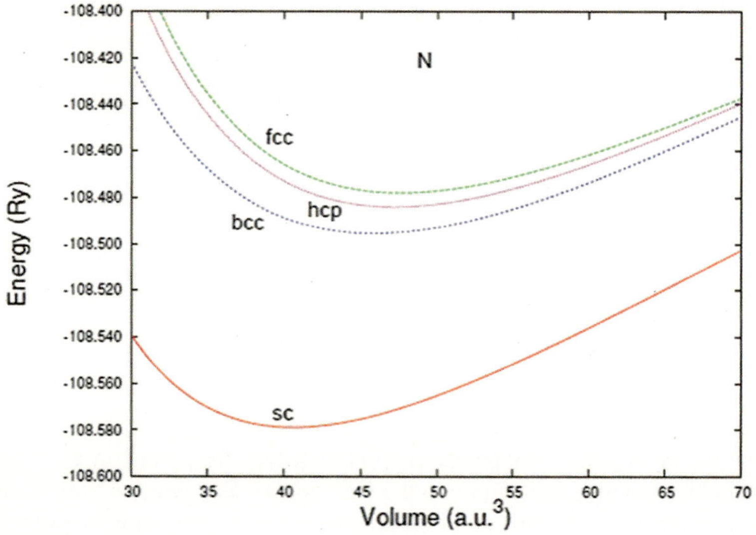

Fig. 9.1 Total energy of N

Table 9.1 Lattice constants and bulk modulus

	a (Bohr)	c (Bohr)	B (MBar)
fcc	5.751		2.231
hcp	4.084	6.534	2.291
bcc	4.509		2.198
sc	3.437		2.598
$\Delta E_{sc-bcc} = 83.5$ mRy			

Table 9.2 Birch fit coefficients

	A_1	A_2	A_3	A_4
fcc	−107.76272	−17.51544	90.12556	217.22961
hcp	−215.26935	−71.34656	768.45397	−402.08317
bcc	−107.48134	−29.09363	246.27312	−512.18041
sc	−107.71783	−20.98560	135.30197	−89.19936

Nitrogen (sc)

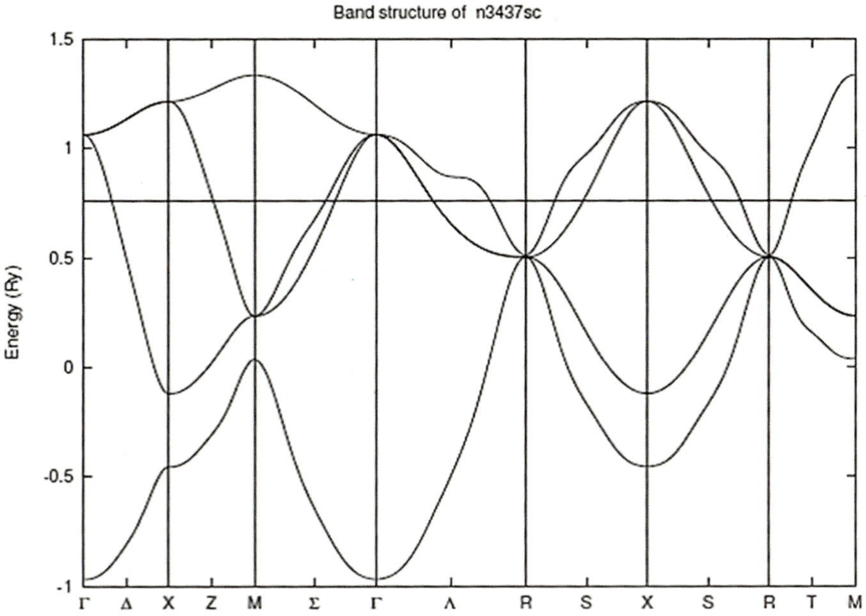

Fig. 9.2 Energy bands for N

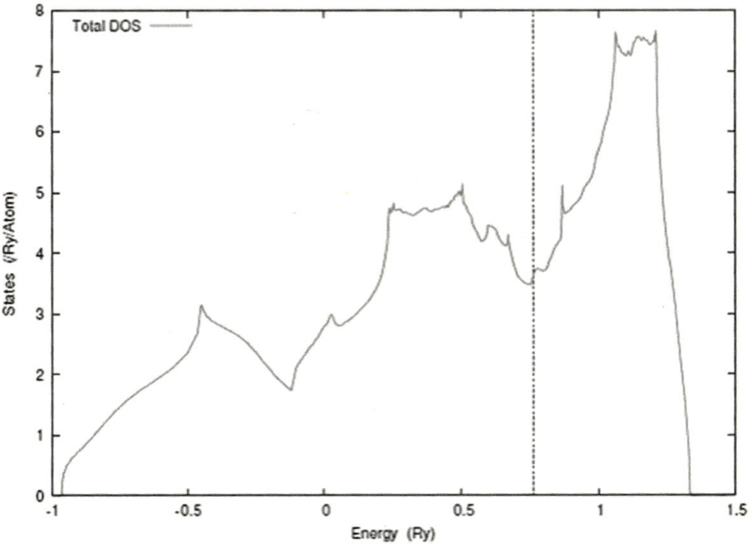

Fig. 9.3 Density of states for N

Table 9.3 DOS at the Fermi level (APW results)

Energy Ry	Total	s States/Ry/atom	p	e_g	t_{2g}	f	Velocity cm/s	Plasmon energy eV
0.763	3.642	0.394	2.685	0.105	0.014	0.011	1.526 × 10E8	16.456

9.2 Phosphorus

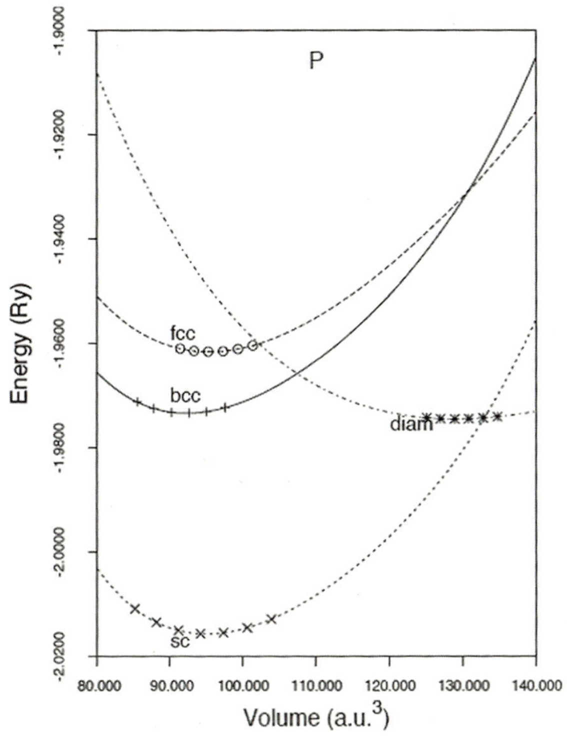

Fig. 9.4 Total energy of P

Table 9.4 Lattice constants and bulk modulus

	a (Bohr)	B (MBar)
fcc	7.251	1.026
bcc	5.693	1.126
diam	10.102	0.528
sc	4.562	1.172
P4	13.549	0.304
$\Delta E = E_{diam} - E_{sc} = 41.1$ mRy		

Table 9.5 Birch fit coefficients

	A_1	A_2	A_3	A_4
bcc	−1.01035	−42.77715	542.31980	−1437.07325
fcc	−7.39247	−60.90487	945.10678	−4307.96556
diam	−1362.97584	−76.08698	1374.09196	4501.09462
sc	−1.16439	−35.43964	368.90896	−1.10175

Phosphorus (sc)

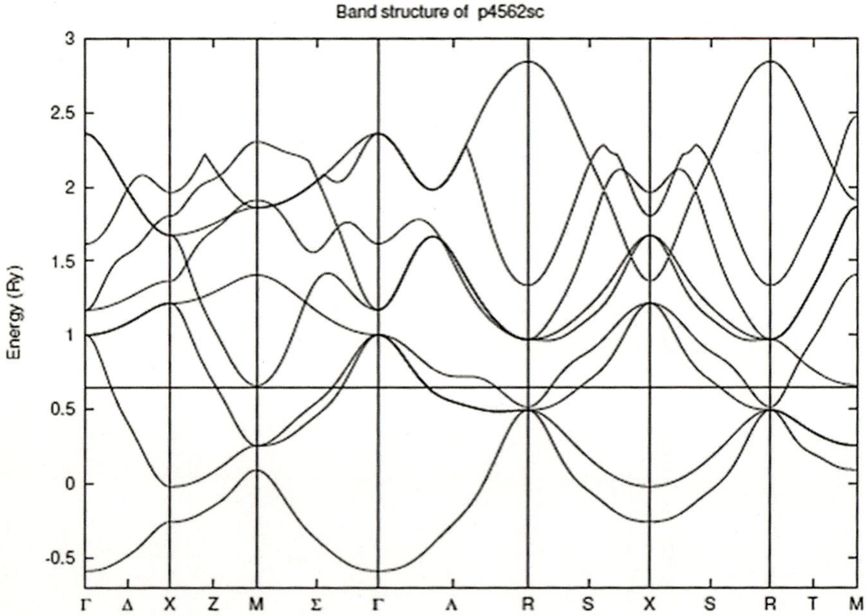

Fig. 9.5 Energy bands for P

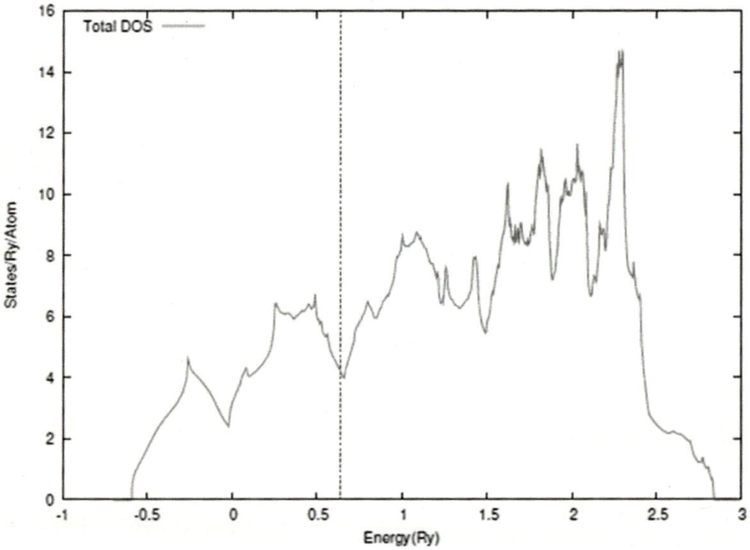

Fig. 9.6 Density of states for P

Table 9.6 DOS at the Fermi level (APW results)

Energy \Ry	Total	s	p	e_g	t_{2g}	f	Velocity cm/s	Plasmon energy eV
		States/Ry/atom						
0.643	4.165	0.376	2.242	0.480	0.069	0.023	1.797 × 10E8	13.549

Phosphorus sc structure a = 4.56 Bohr
Tight-binding parameters in Ry

Non-orthogonal On-site			
s	p		
−0.21951	0.32471		
sss	sps	pps	ppp
Hopping dist = 4.56000			
−0.09753	−0.09646	−0.00910	−0.02660
Overlap dist = 4.56000			
0.13967	0.22527	−0.36682	0.07845
Hopping dist = 6.44881			
−0.02248	−0.04843	0.07378	−0.01223
Overlap dist = 6.44881			
0.00261	0.00832	−0.03855	0.00081
Hopping dist = 7.89815			
−0.00045	−0.00213	0.01210	−0.00021
Overlap dist = 7.89815			
−0.00003	0.00022	−0.00096	0.00000

Fitting rms error for 4 bands 11 mRy

9.3 Arsenic

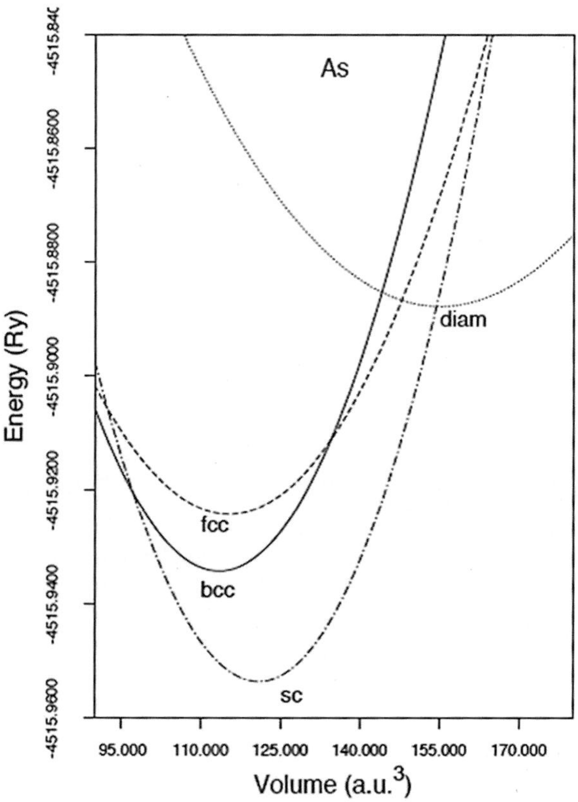

Fig. 9.7 Total energy of As

Table 9.7 Lattice constants and bulk modulus

	a (Bohr)	B (MBar)
fcc	7.644	1.542
bcc	6.045	1.453
diam	10.732	0.946
sc	4.897	1.456
rhl	7.805	0.394
$\Delta E = E_{sc} - E_{bcc} = 16.7$ mRy		

Table 9.8 Birch fit coefficients

	A_1	A_2	A_3	A_4
fcc	−4514.77529	−49.51838	440.44706	2066.31457
bcc	−4514.91472	−42.07013	317.99304	2551.73275
sc	−4514.62586	−63.88074	779.70803	−218.78209

Arsenic (sc)

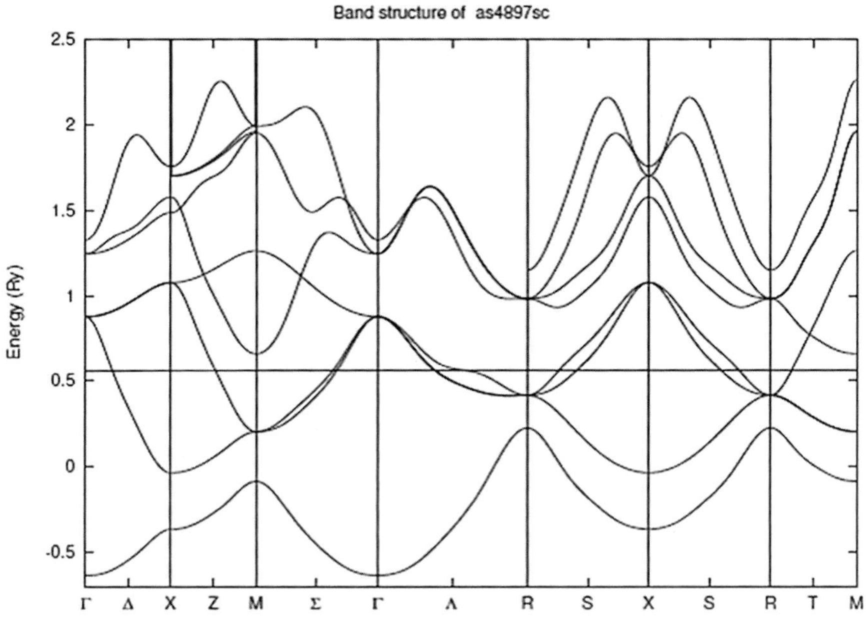

Fig. 9.8 Energy bands for As

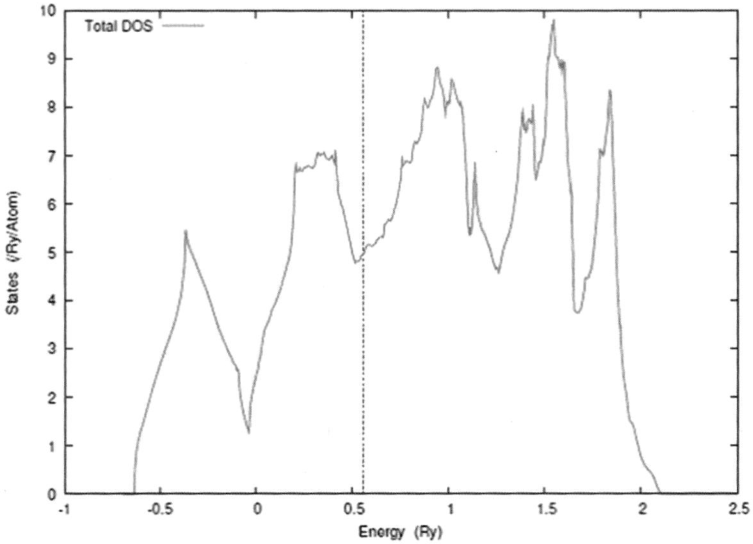

Fig. 9.9 Density of states for As

Table 9.9 DOS at the Fermi level (APW results)

Energy Ry	Total	s	p	e_g	t_{2g}	f	Velocity cm/s	Plasmon energy eV
		States/Ry/atom						
0.558	4.966	0.485	2.813	0.318	0.080	0.033	1.920 × 10E8	14.213

Arsenic sc structure a = 4.90 Bohr
Tight-binding parameters in Ry

Non-orthogonal On-site			
s	p		
−0.17843	0.45104		
sss	sps	pps	ppp
Hopping dist = 4.90000			
−0.08756	−0.04404	−0.06262	0.01014
Overlap dist = 4.90000			
0.16366	0.25675	−0.36018	0.09460
Hopping dist = 6.92965			
−0.02292	−0.03554	0.06219	−0.00011
Overlap dist = 6.92965			
0.01390	0.02177	−0.04136	0.01247
Hopping dist = 8.48705			
−0.00038	−0.00106	0.00656	0.00000
Overlap dist = 8.48705			
0.00026	0.00043	−0.00084	0.00017

Fitting rms error for 4 bands 22 mRy

9.4 Antimony

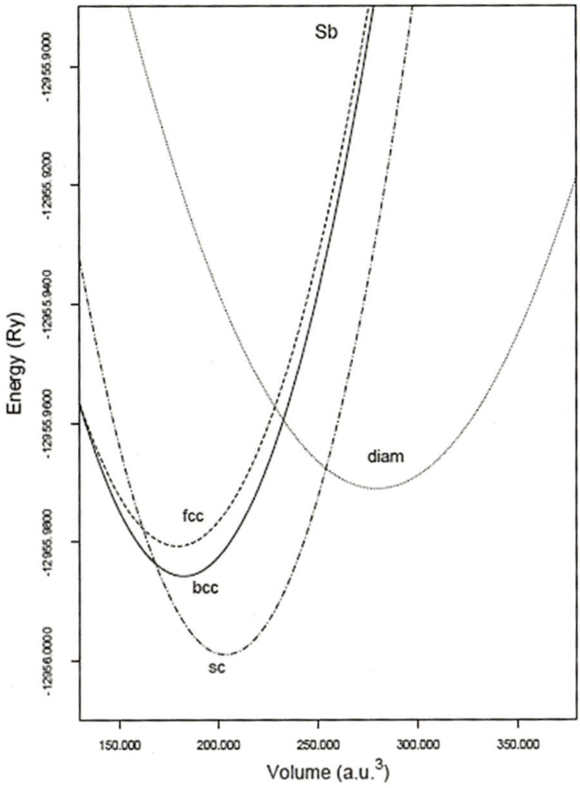

Fig. 9.10 Total energy of Sb

Table 9.10 Lattice constants and bulk modulus

	a (Bohr)	B (MBar)
bcc	7.106	0.657
fcc	8.948	0.735
sc	5.821	0.635
diam	13.058	0.356
exp	8.526	0.383
$\Delta E_{bcc-sc} = 10.9$ mRy		

Table 9.11 Birch fit coefficients

	A_1	A_2	A_3	A_4
bcc	−12955.11688	−54.28006	814.37458	1037.68359
fcc	−12956.13394	46.56175	−2495.99181	37205.05788
sc	−12955.40379	−27.96498	−152.15914	14141.29816
diam	−25910.36282	−217.91380	7804.01449	−19458.00624

Antimony (sc)

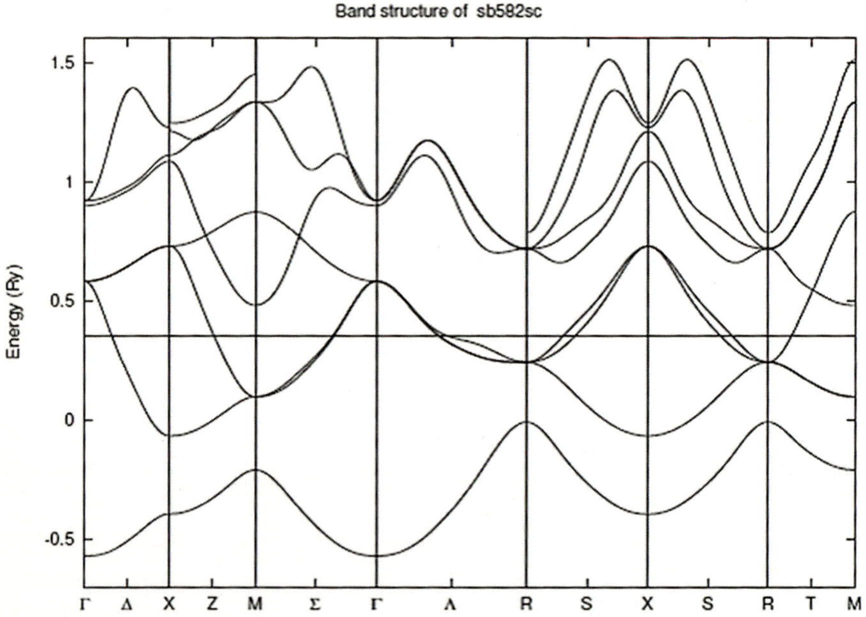

Fig. 9.11 Energy bands for Sb

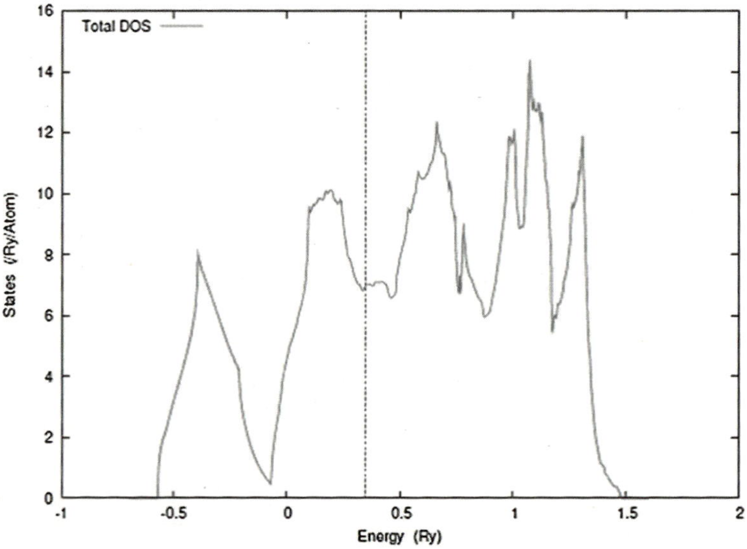

Fig. 9.12 Density of states for Sb

Table 9.12 DOS at the Fermi level (APW results)

Energy	Total	s	p	e_g	t_{2g}	f	Velocity	Plasmon energy
Ry		States/Ry/atom					cm/s	eV
0.350	7.009	0.585	4.111	0.351	0.103	0.056	1.768 × 10E8	12.001

Antimony sc structure a = 5.82 Bohr
Tight-binding parameters in Ry

Non-orthogonal On-site			
s	p		
−0.28402	0.29154		
sss	sps	pps	ppp
Hopping dist = 5.82000			
−0.04549	0.07988	0.09411	−0.03390
Overlap dist = 5.82000			
0.01114	−0.01394	−0.15980	0.00964
Hopping dist = 8.23072			
−0.00014	0.00114	0.01602	−0.00101
Overlap dist = 8.23072			
0.00001	0.01424	−0.00059	−0.00033

Fitting rms error for 4 bands 12 mRy

Orthogonal On-site			
s	p		
−0.35120	0.36932		
sss	sps	pps	ppp
Hopping dist = 5.82000			
−0.04401	0.00018	0.15138	−0.03942
Hopping dist = 8.23072			
−0.00008	0.00000	0.00059	−0.00024

Fitting rms error for 4 bands 28 mRy

9.5 Bismuth

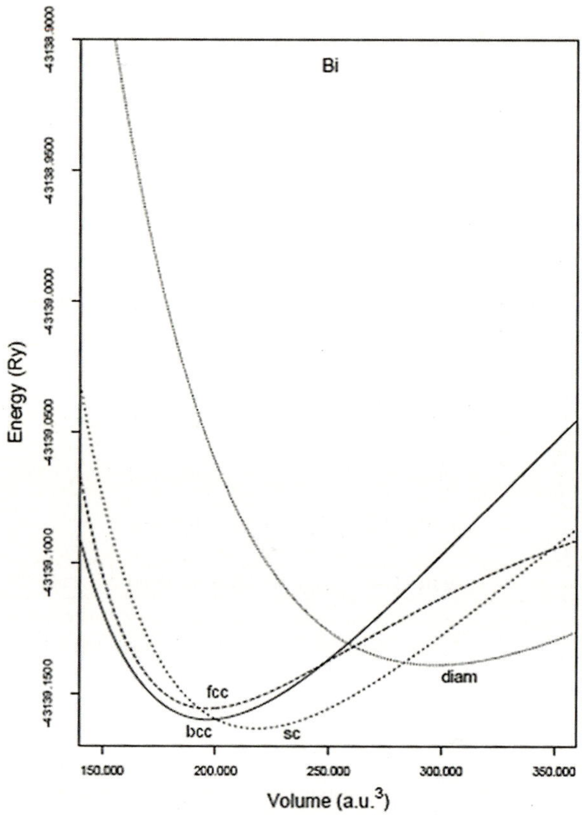

Fig. 9.13 Total energy of Bi

Table 9.13 Lattice constants and bulk modulus

	a (Bohr)	B (MBar)
bcc	7.326	0.702
fcc	9.230	0.690
sc	6.018	0.613
diam	13.348	0.387
exp (monoclinic)	8.976	0.315
$\Delta E_{sc-bcc} = 3.4$ mRy		

Table 9.14 Birch fit coefficients

	A_1	A_2	A_3	A_4
bcc	−43138.14316	−67.40349	1071.84034	1524.63586
fcc	−43139.48041	68.01752	−3485.00004	52634.85505
sc	−43138.40666	−45.15110	294.12511	12638.73360
diam	−86276.24844	−305.84881	12819.08873	−94576.53184

Bismuth (sc)

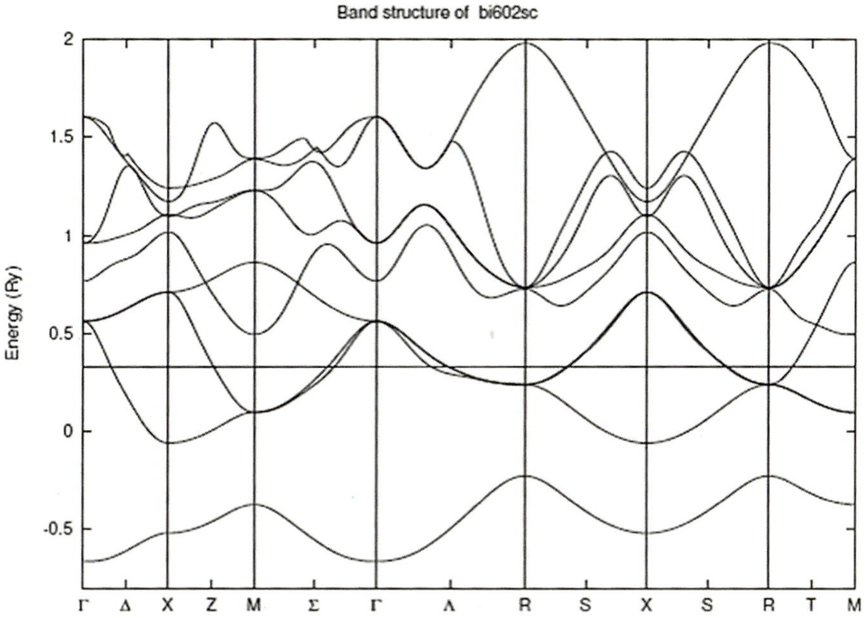

Fig. 9.14 Energy bands for Bi

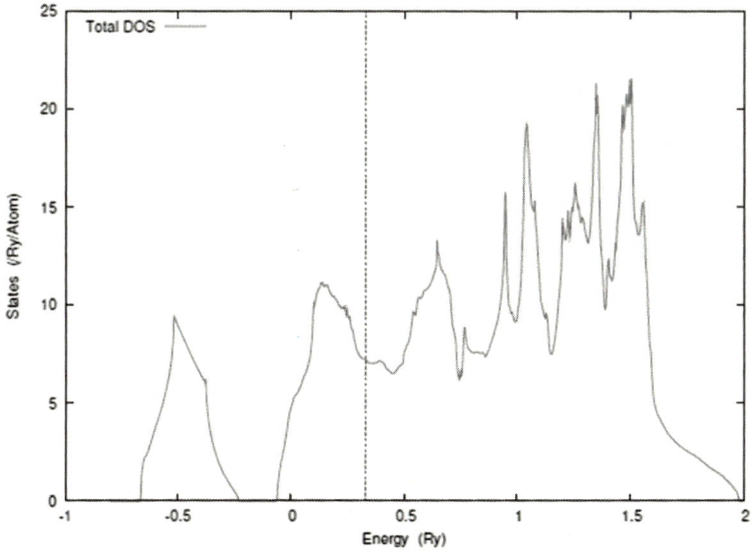

Fig. 9.15 Density of states for Bi

Table 9.15 DOS at the Fermi level (APW results)

Energy	Total	s	p	e_g	t_{2g}	f	Velocity	Plasmon energy
Ry		States/Ry/atom					cm/s	eV
0.330	7.225	0.393	4.368	0.278	0.076	0.077	1.804 × 10E8	11.817

Bismuth sc structure a = 6.02 Bohr
Tight-binding parameters in Ry

Non-orthogonal On-site			
s	p		
−0.44378	0.29793		
sss	sps	pps	ppp
Hopping dist = 6.02000			
−0.03564	0.06752	0.09804	−0.03514
Overlap dist = 6.02000			
0.00122	−0.01628	−0.13646	0.00864
Hopping dist = 8.51357			
−0.00006	0.00054	0.01141	−0.00064
Overlap dist = 8.51357			
0.00000	0.00825	−0.00028	−0.00006

Fitting rms error for 4 bands 14 mRy

Orthogonal On-site			
s	p		
−0.47783	0.34629		
sss	sps	pps	ppp
Hopping dist = 6.02000			
−0.03586	−0.00009	0.13816	−0.03665
Hopping dist = 8.51357			
−0.00006	0.00000	0.00390	−0.00121

Fitting rms error for 4 bands 34 mRy

Chapter 10
Group 16 Elements: Chalcogens

These elements have the following crystal structures: S: orthorhombic, Se: monoclinic, Te: trigonal and Po: simple cubic. In this work we performed calculations in only the cubic structures and found that the lowest energy occurs in the simple cubic lattice. For Se and Te we indicate in the total energy diagram, the lowest energy corresponding to monoclinic and trigonal lattices [1]. Polonium is the only element in the periodic table that crystallizes in the simple cubic structure and we correctly predict it. We also present energy bands and DOS in the simple cubic lattice as well as TB parameters.

D.A. Papaconstantopoulos, *Handbook of the Band Structure of Elemental Solids*,
DOI 10.1007/978-1-4419-8264-3_10

10.1 Oxygen

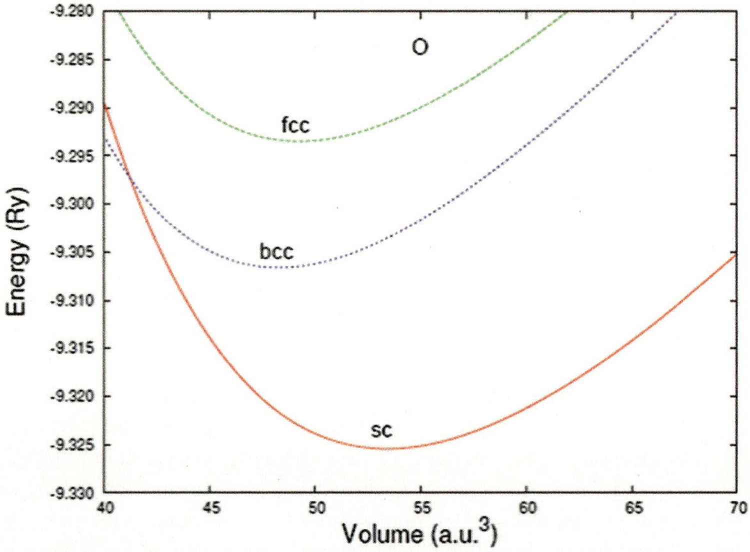

Fig. 10.1 Total energy of O

Table 10.1 Lattice constants and bulk modulus

	a (Bohr)	B (MBar)
fcc	5.819	1.878
bcc	4.588	1.994
sc	3.766	1.885
$\Delta E_{sc-bcc} = 18.81$ mRy		

Table 10.2 Birch fit coefficients

	A_1	A_2	A_3	A_4
fcc	−8.83101	−9.13671	−4.96134	594.33743
bcc	−8.84443	−8.62395	−15.06996	638.61216
sc	−8.70390	−15.52475	65.26348	424.09160

Oxygen (sc)

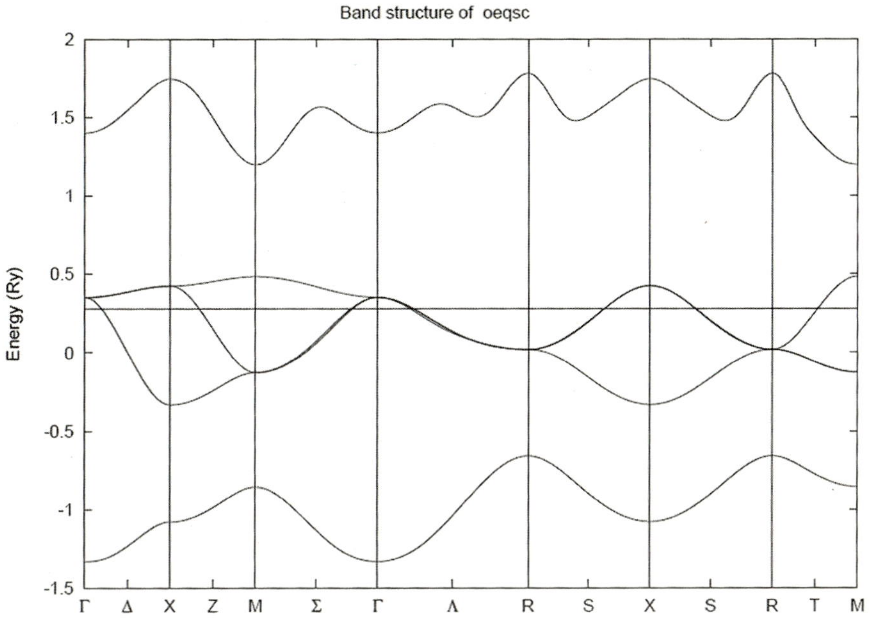

Fig. 10.2 Energy bands for O

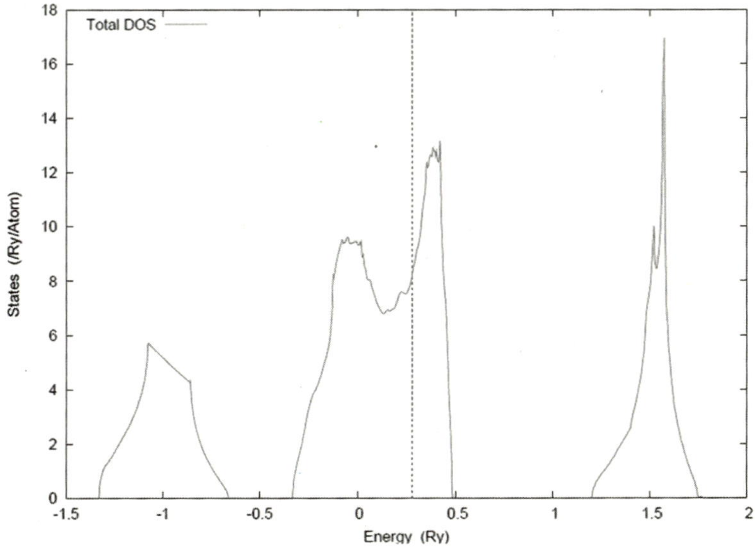

Fig. 10.3 Density of states for O

Table 10.3 DOS at the Fermi level (APW results)

Energy Ry	Total	s States/Ry/atom	p	e_g	t_{2g}	f	Velocity cm/s	Plasmon energy eV
0.277	8.139	0.209	7.328	0.085	0.011	0.019	$1.012 \times 10E8$	14.223

10.2 Sulfur

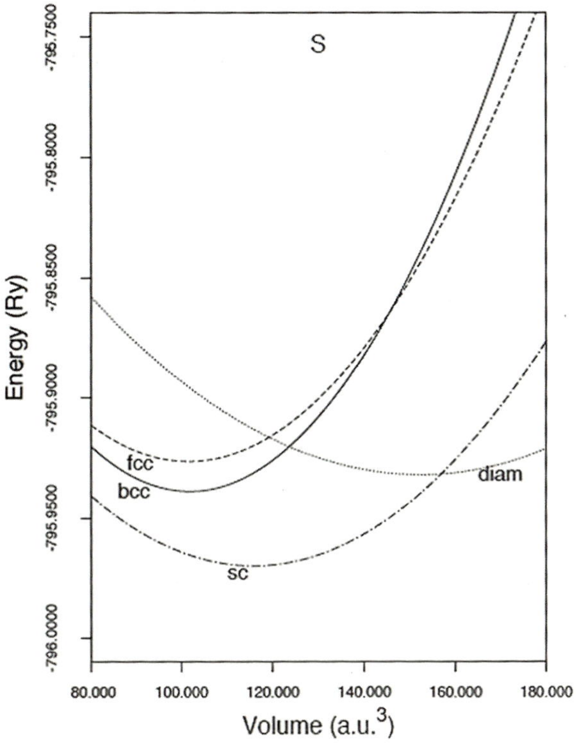

Fig. 10.4 Total energy of S

Table 10.4 Lattice constants and bulk modulus

	a (Bohr)	B (MBar)
fcc	7.363	0.997
bcc	5.834	1.032
diam	10.658	0.520
sc	4.778	1.002
orc	19.785	0.178
$\Delta E = E_{sc} - E_{bcc} = 33.7$ mRy		

Table 10.5 Birch fit coefficients

	A_1	A_2	A_3	A_4
fcc	−795.07246	−38.76533	481.86426	−929.88005
bcc	−794.97089	−45.44570	614.27551	−1815.65186
sc	−795.18042	−35.14550	366.58850	527.50297

Sulfur (sc)

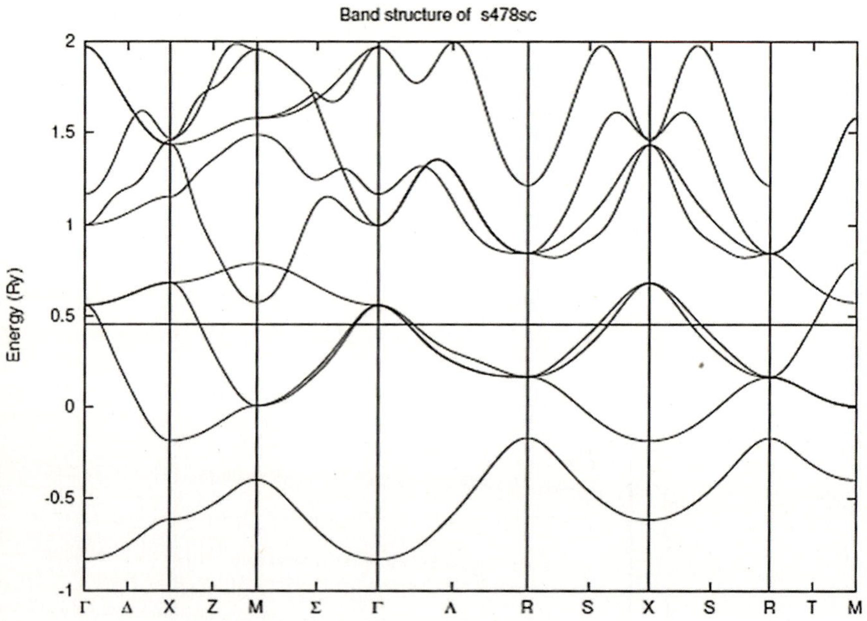

Fig. 10.5 Energy bands for S

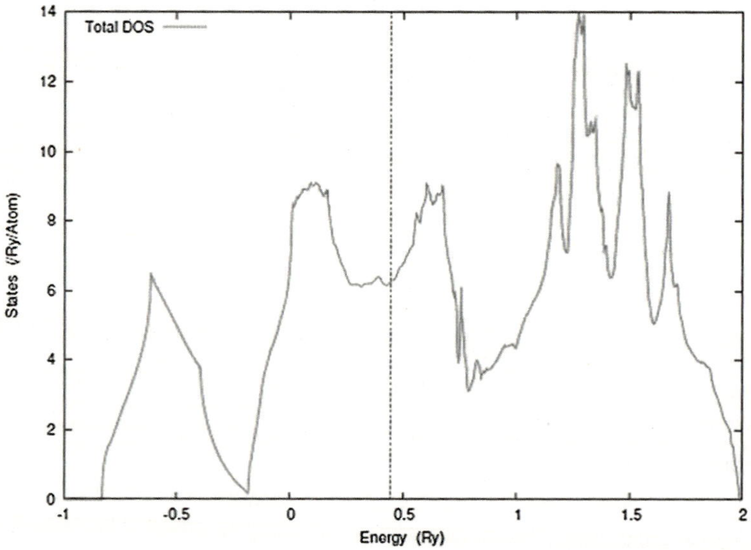

Fig. 10.6 Density of states for S

Table 10.6 DOS at the Fermi level (APW results)

Energy Ry	Total	s States/Ry/atom	p	e_g	t_{2g}	f	Vel cm/s	Plasmon energy eV
0.449	6.293	0.374	4.569	0.335	0.064	0.043	1.579 × 10E8	13.650

Sulfur sc structure a = 4.78 Bohr
Tight-binding parameters in Ry

Non-orthogonal On-site			
s	p		
−0.45800	0.21920		
sss	sps	pps	ppp
Hopping dist = 4.78000			
−0.03701	−0.10450	0.11850	−0.04056
Overlap dist = 4.78000			
−0.03945	0.01490	−0.15675	−0.00137
Hopping dist = 6.75994			
−0.00042	0.00129	0.01805	0.00020
Overlap dist = 6.75994			
0.00929	0.00010	−0.00143	0.00077

Fitting rms error for 4 bands 12 mRy

Orthogonal On-site			
s	p		
−0.55203	0.30320		
sss	sps	pps	ppp
Hopping dist = 4.78000			
−0.05298	−0.00011	0.16453	0.03907
Hopping dist = 6.75994			
−0.00057	0.00001	0.00690	0.00007

Fitting rms error for 4 bands 25 mRy

10.3 Selenium

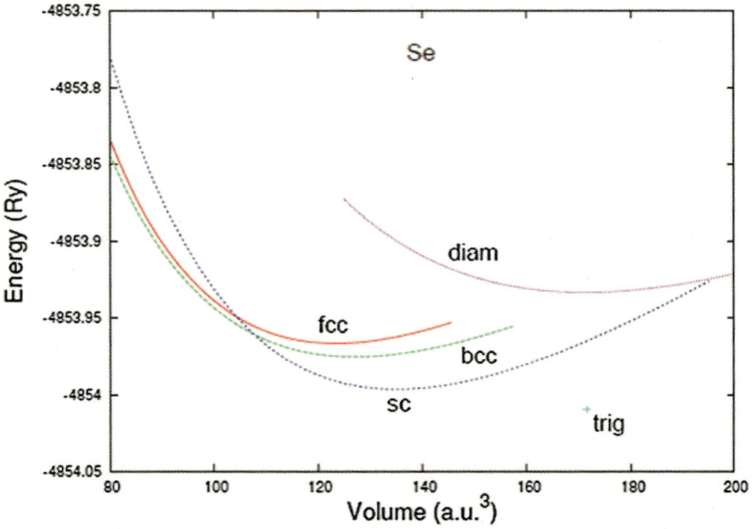

Fig. 10.7 Total energy of Se

Table 10.7 Lattice constants and bulk modulus

	a (Bohr)	B (MBar)
fcc	7.904	1.314
bcc	6.335	1.131
diam	11.102	0.920
sc	5.131	1.360
$\Delta E_{trig-sc} = 13$ mRy		

Table 10.8 Birch fit coefficients

	A_1	A_2	A_3	A_4
fcc	−4852.55349	−74.34241	1080.58104	−2627.33682
bcc	−4852.61317	−75.51488	1206.53131	−4247.40599
diam	−9704.81611	−329.96647	10381.10434	−75357.71651
sc	−4852.21138	−104.01927	1765.03167	−6946.55059

Selenium (sc)

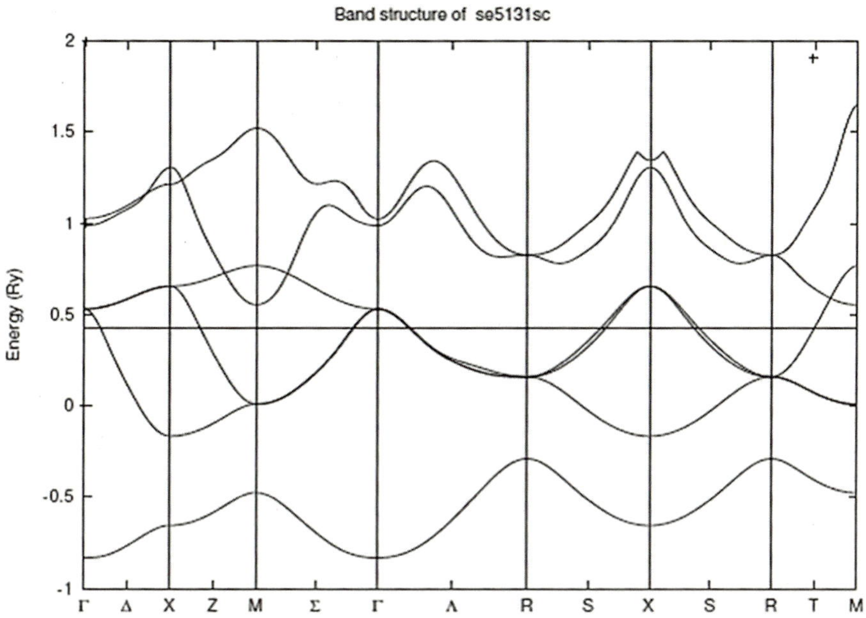

Fig. 10.8 Energy bands for Se

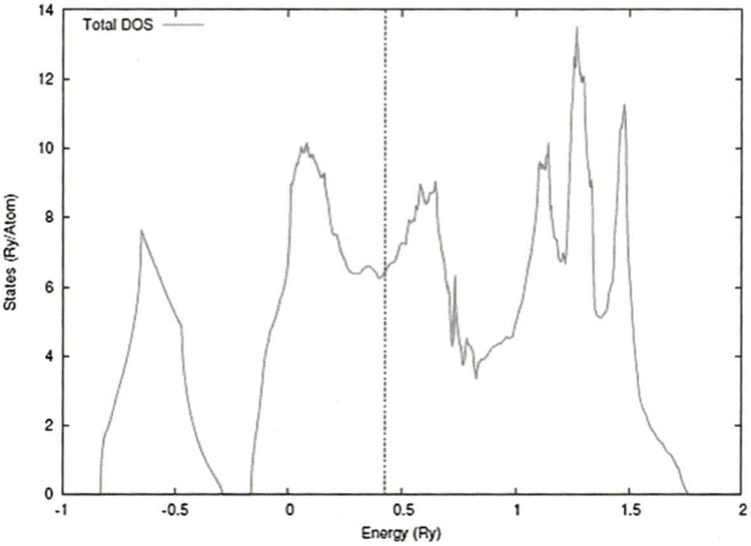

Fig. 10.9 Density of states for Se

Table 10.9 DOS at the Fermi level (APW results)

Energy Ry	Total	s States/Ry/atom	p	e_g	t_{2g}	f	Velocity cm/s	Plasmon energy eV
0.424	6.429	0.298	4.679	0.311	0.046	0.051	1.734 × 10E8	13.617

Selenium sc structure a = 5.131 Bohr
Tight-binding parameters in Ry

Non-orthogonal On-site			
s	p		
−0.56795	0.24931		
sss	sps	pps	ppp
Hopping dist = 5.13100			
−0.05356	−0.06042	0.12888	−0.04502
Overlap dist = 5.13100			
0.01586	−0.01064	−0.10473	−0.01761
Hopping dist = 7.25633			
−0.00064	−0.00048	0.00043	−0.00033
Overlap dist = 7.25633			
−0.00120	−0.00005	−0.01649	−0.00002

Fitting rms error for 4 bands 11 mRy

Orthogonal On-site			
s	p		
−0.59982	0.28963		
sss	sps	pps	ppp
Hopping dist = 5.13100			
−0.04435	0.00003	0.15794	−0.03740
Hopping dist = 7.25633			
−0.00051	−0.00001	0.00053	−0.00028

Fitting rms error for 4 bands 27 mRy

10.4 Tellurium

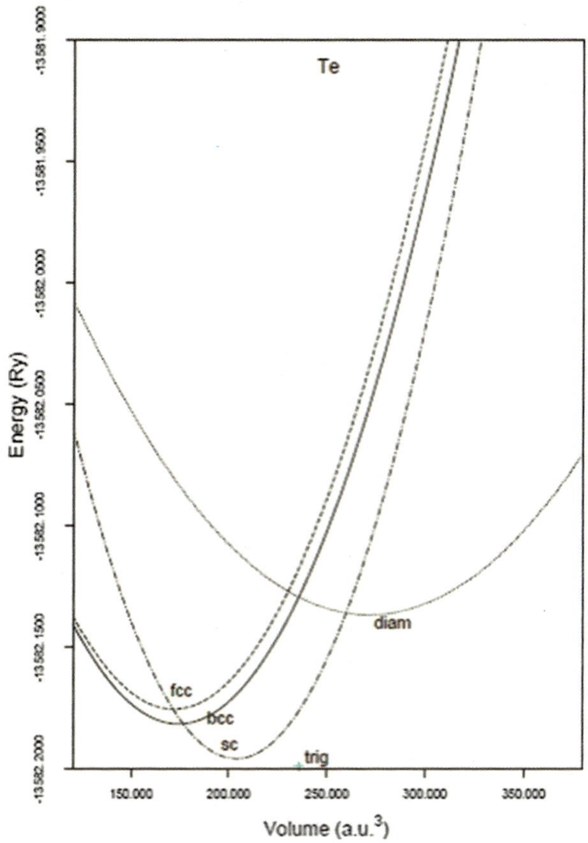

Fig. 10.10 Total energy of Te

Table 10.10 Lattice constants and bulk modulus

	a (Bohr)	c (Bohr)	B (MBar)
bcc	7.010		0.875
fcc	8.805		0.861
sc	5.828		0.459
diam	12.913		0.435
exp	8.409	11.184	0.230
$\Delta E_{trig-sc} = 3$ mRy			

Table 10.11 Birch fit coefficients

	A_1	A_2	A_3	A_4
bcc	−13581.73691	−5.63221	−930.16263	20994.90311
fcc	−13581.67913	−11.23104	−718.35207	18276.87569
sc	−13581.18521	−63.48572	830.30619	5612.87967
diam	−27162.47602	−238.11173	7912.07929	−1489.97173

Tellurium (sc)

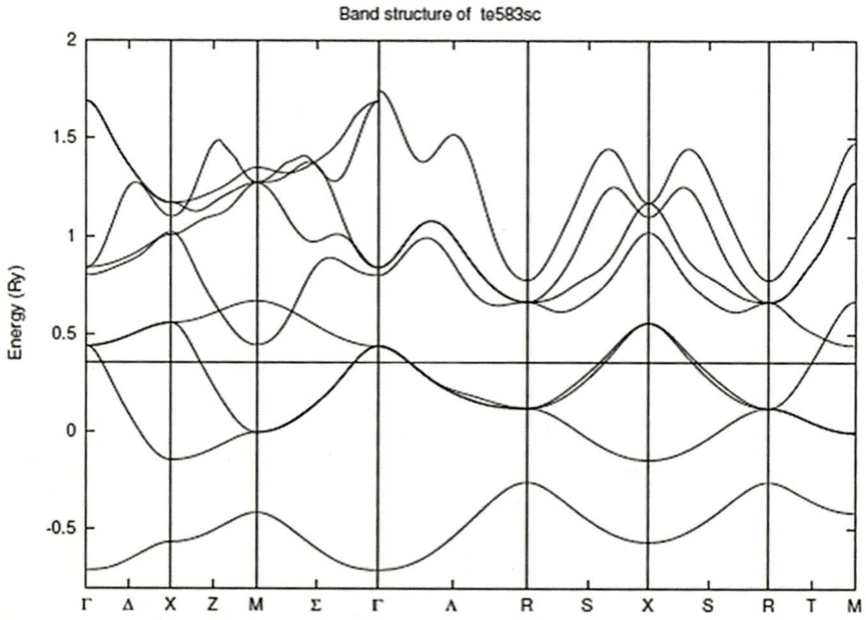

Fig. 10.11 Energy bands for Te

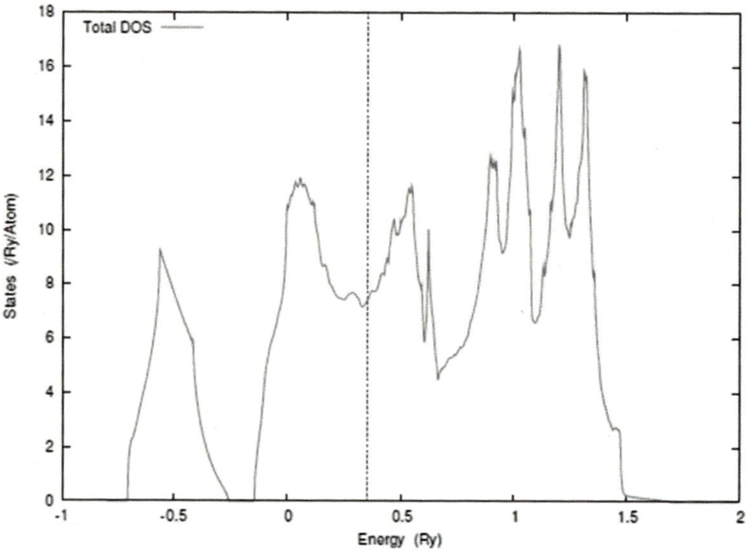

Fig. 10.12 Density of states for Te

Table 10.12 DOS at the Fermi level (APW results)

Energy Ry	Total	s States/Ry/atom	p	e_g	t_{2g}	f	Velocity cm/s	Plasmon energy eV
0.352	7.417	0.327	5.148	0.382	0.068	0.084	1.718 × 10E8	11.964

Tellurium sc structure a = 5.83 Bohr
Tight-binding parameters in Ry

Non-orthogonal On-site			
s	p		
−0.20905	0.46982		
sss	sps	pps	ppp
Hopping dist = 5.83000			
−0.04114	0.05962	0.07420	−0.04625
Overlap dist = 5.83000			
0.01200	−0.01240	−0.12380	−0.03140
Hopping dist = 8.24487			
−0.00006	0.00014	0.00235	−0.00025
Overlap dist = 8.24487			
0.00006	0.00005	−0.00046	−0.00006

Fitting rms error for 4 bands 16 mRy

Orthogonal On-site			
s	p		
−0.23698	0.51368		
sss	sps	pps	ppp
Hopping dist = 5.83000			
−0.03773	−0.00002	0.13812	−0.03041
Hopping dist = 8.24487			
−0.00005	0.00000	0.00165	−0.00016

Fitting rms error for 4 bands 25 mRy

10.5 Polonium

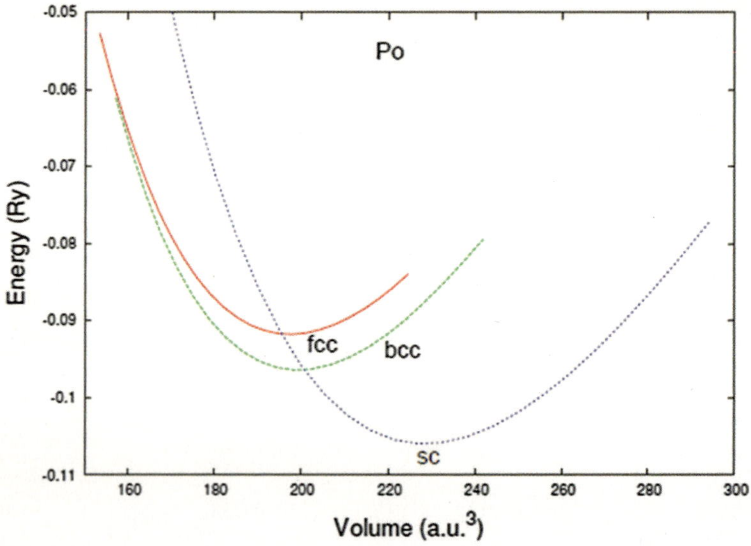

Fig. 10.13 Total energy of Po

Table 10.13 Lattice constants and bulk modulus

	a (Bohr)	B (MBar)
fcc	9.245	0.768
bcc	7.360	0.771
sc	6.113	0.685
$\Delta E_{sc-bcc} = 9.58$ mRy		

Table 10.14 Birch fit coefficients

	A_1	A_2	A_3	A_4
fcc	1.03368	−75.18358	1215.99820	1336.02051
bcc	0.95842	−67.85857	946.21389	4817.92225
sc	1.08325	−88.58489	1638.86035	405.60774

Polonium (sc)

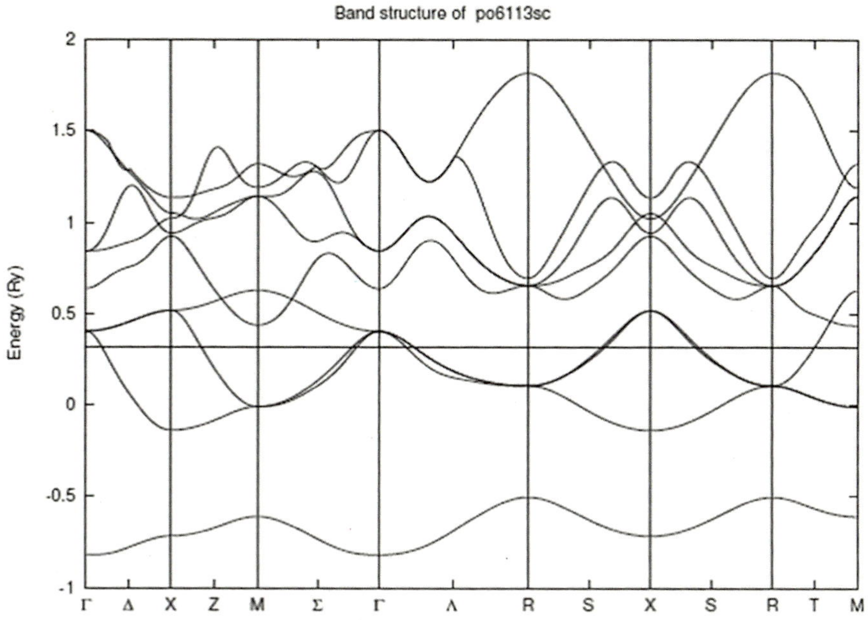

Fig. 10.14 Energy bands for Po

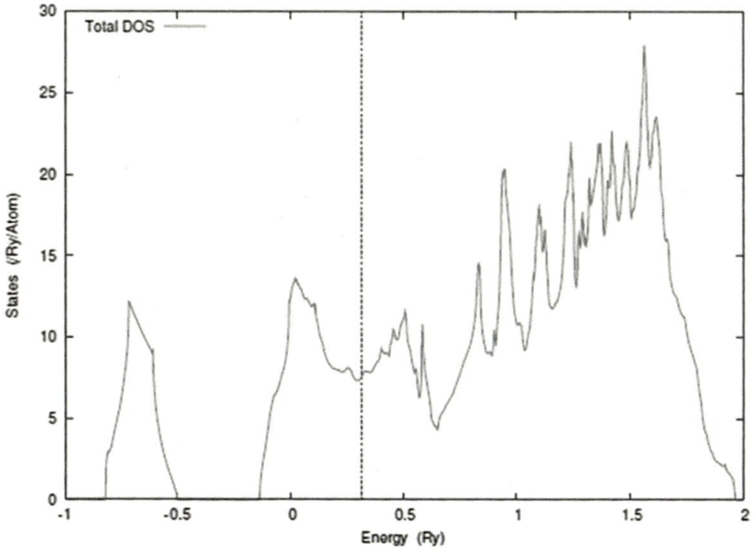

Fig. 10.15 Density of states for Po

Table 10.15 DOS at the Fermi level (APW results)

Energy Ry	Total	s States/Ry/atom	p	e_g	t_{2g}	f	Velocity cm/s	Plasmon energy eV
0.318	7.614	0.231	5.350	0.361	0.050	0.089	1.709 × 10E8	11.232

Polonium sc structure a = 6.20 Bohr
Tight-binding parameters in Ry

Non-orthogonal On-site			
s	p		
−0.51455	0.30897		
sss	sps	pps	ppp
Hopping dist = 6.20000			
−0.03532	0.04840	0.07541	−0.03366
Overlap dist = 6.20000			
0.01839	−0.00908	−0.13271	−0.01704
Hopping dist = 8.76812			
−0.00115	0.00030	0.00257	−0.00033
Overlap dist = 8.76812			
0.00039	−0.00044	−0.00614	−0.00015

Fitting rms error for 4 bands 14 mRy

Orthogonal On-site			
s	p		
−0.53856	0.34327		
sss	sps	pps	ppp
Hopping dist = 6.20000			
−0.02465	0.00006	0.11881	0.02593
Hopping dist = 8.76812			
−0.00045	0.00000	0.00255	−0.00083

Fitting rms error for 4 bands 25 mRy

Reference

1. J. Li, A. Ciani, H. Gayles, D.A. Papconstantopoulos, N. Kioussis, C. Grein, F. Aqariden, Non-orthogonal tight-binding model for tellurium and selenium. Philos. Mag. **93**(23), 3216–3230 (2013)

Chapter 11
Group 17 Elements: Halogens

The halogens, with the exception of Astatine, form diatomic molecules. Fluorine has a monoclinic crystal structure, while Chlorine, Bromine, and Iodine are orthorhombic in nature. In this study we present calculations of hypothetical cubic structures of the Halogens for the purpose of making comparisons with the neighboring 16th and 18th columns of the periodic table. They are all predicted as metals in these hypothetical structures with ground states: sc for F, bcc for Cl and Br, and fcc for I and At.

D.A. Papaconstantopoulos, *Handbook of the Band Structure of Elemental Solids*,
DOI 10.1007/978-1-4419-8264-3_11

11.1 Fluorine

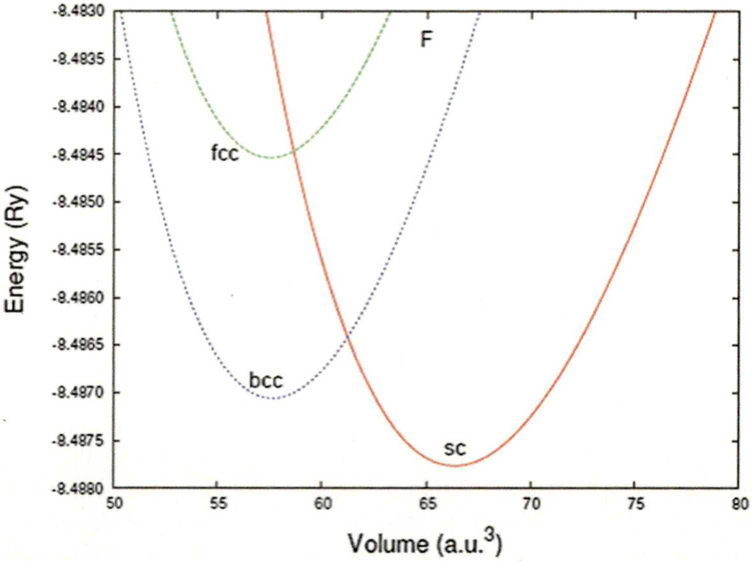

Fig. 11.1 Total energy of F

Table 11.1 Lattice constants and bulk modulus

	a (Bohr)	B (MBar)
fcc	6.128	0.967
bcc	4.868	0.976
sc	4.049	0.846
$\Delta E_{sc-bcc} = 0.704$ mRy		

Table 11.2 Birch fit coefficients

	A_1	A_2	A_3	A_4
fcc	−8.28803	−2.44528	−58.03535	757.59184
bcc	−8.28047	−2.82794	−53.64712	743.80542
sc	−8.27816	−3.26988	−61.75113	967.65531

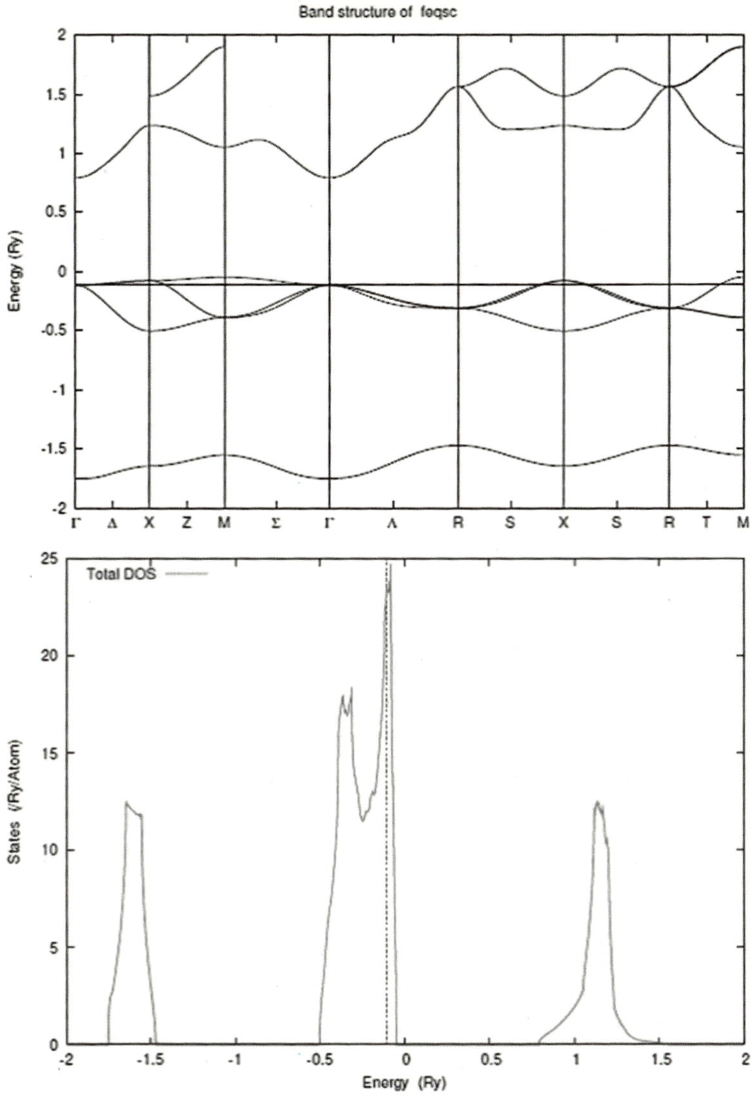

Fig. 11.2 Energy bands and density of states for F

Table 11.3 Dos at the Fermi level (APW results)

Energy	Total	s	p	e_g	t_{2g}	f	Velocity	Plasmon energy
Ry		States/Ry/atom					cm/s	eV
−0.106	23.252	0.047	22.582	0.055	0.009	0.049	0.371 × 10E8	7.896

11.2 Chlorine

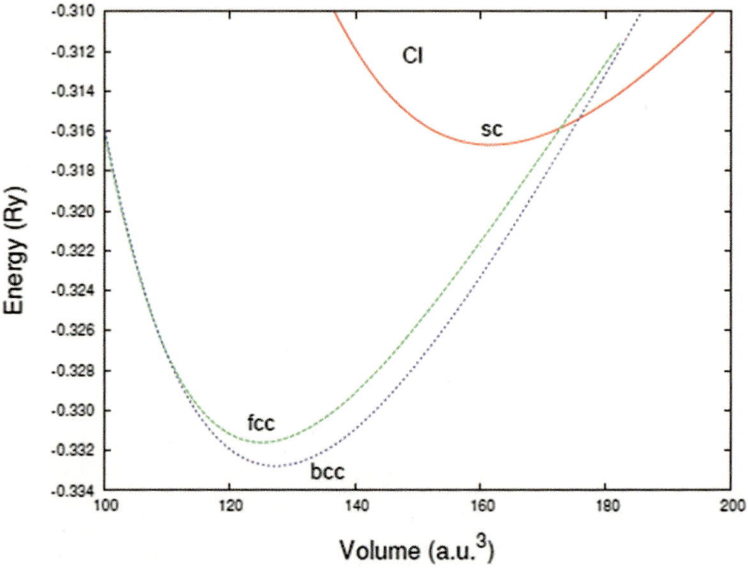

Fig. 11.3 Total energy of Cl

Table 11.4 Lattice constants and bulk modulus

	a (Bohr)	B (MBar)
fcc	7.942	0.541
bcc	6.341	0.527
sc	5.446	0.370
ΔE_{bcc}–f_{cc} = 1.18 mRy		

Table 11.5 Birch fit coefficients

	A_1	A_2	A_3	A_4
fcc	−0.30316	10.84230	−596.32151	7687.21718
bcc	−0.04765	−8.66309	−109.86753	3707.48915
sc	−0.06237	−9.06806	−133.38546	5297.72700

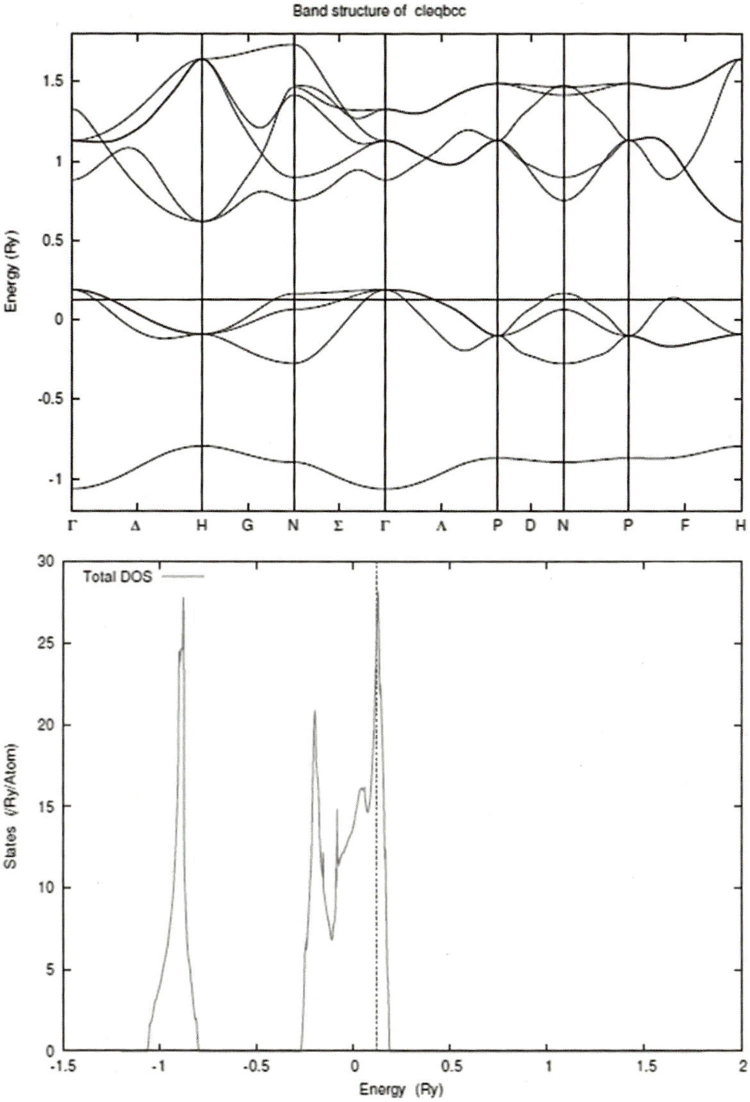

Fig. 11.4 Energy bands and density of states for Cl

Table 11.6 Dos at the Fermi level (APW results)

Energy	Total	s	p	e_g	t_{2g}	f	Velocity	Plasmon energy
Ry		States/Ry/atom					cm/s	eV
0.125	27.224	0.119	25.509	0.031	0.302	0.164	$0.426 \times 10E8$	7.095

11.3 Bromine

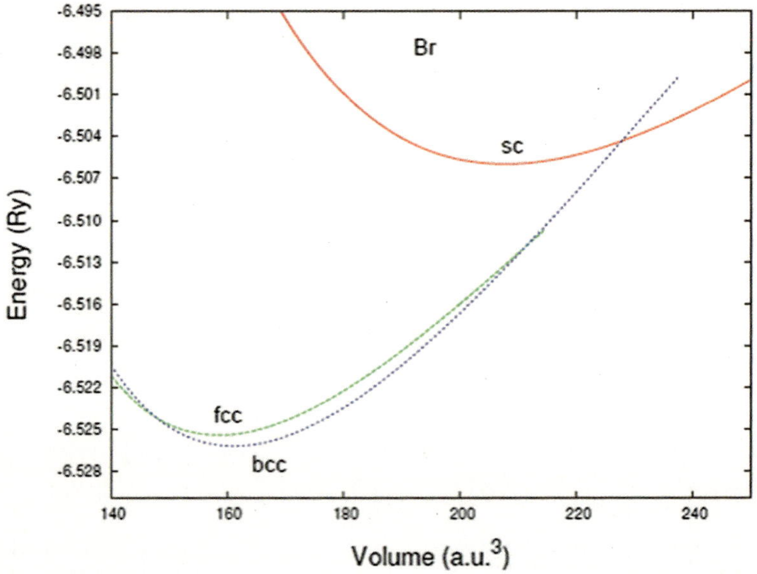

Fig. 11.5 Total energy of Br

Table 11.7 Lattice constants and bulk modulus

	a (Bohr)	B (MBar)
fcc	8.594	0.433
bcc	6.857	0.462
sc	5.922	0.301
$\Delta E_{bcc\text{-}fcc} = 0.785$ mRy		

Table 11.8 Birch fit coefficients

	A_1	A_2	A_3	A_4
fcc	−6.43761	7.67952	−676.50072	11019.29321
bcc	−6.20704	−11.48212	−160.02151	6518.91241
sc	−6.27372	−7.64998	−320.29240	10624.00616

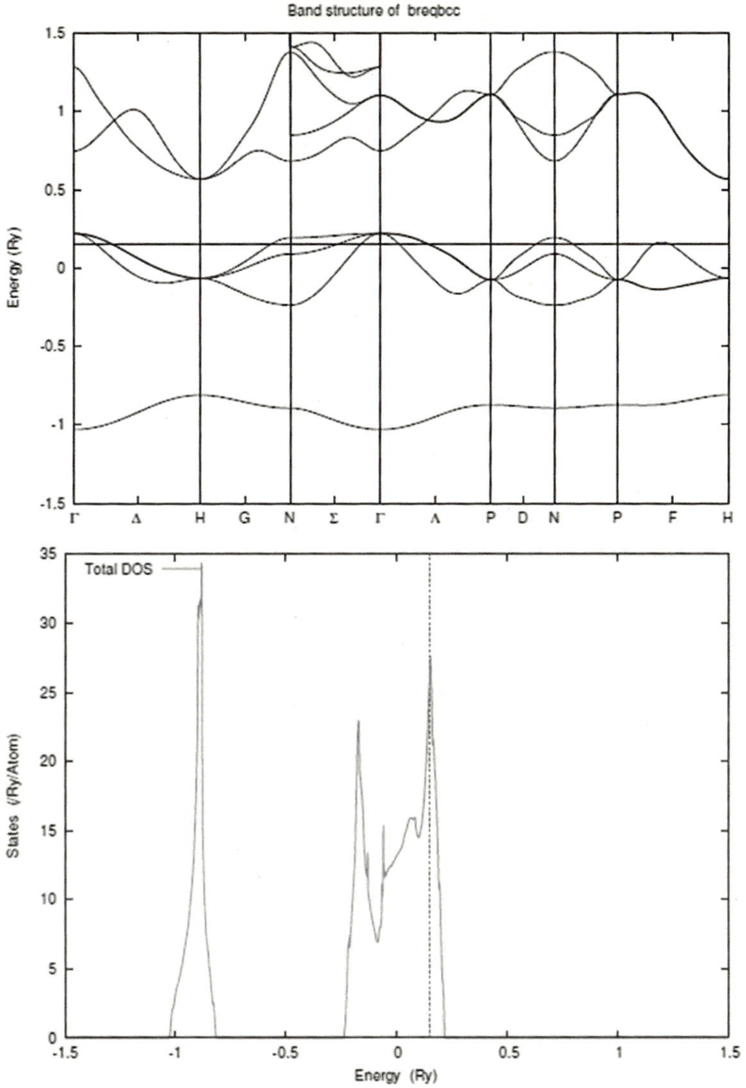

Fig. 11.6 Energy bands and density of states for Br

Table 11.9 Dos at the Fermi level (Apw results)

Energy	Total	s	p	e_g	t_{2g}	f	Velocity	Plasmon energy
Ry		States/Ry/atom					cm/s	eV
0.151	26.257	0.121	24.218	0.038	0.347	0.208	0.480 × 10E8	6.973

11.4 Iodine

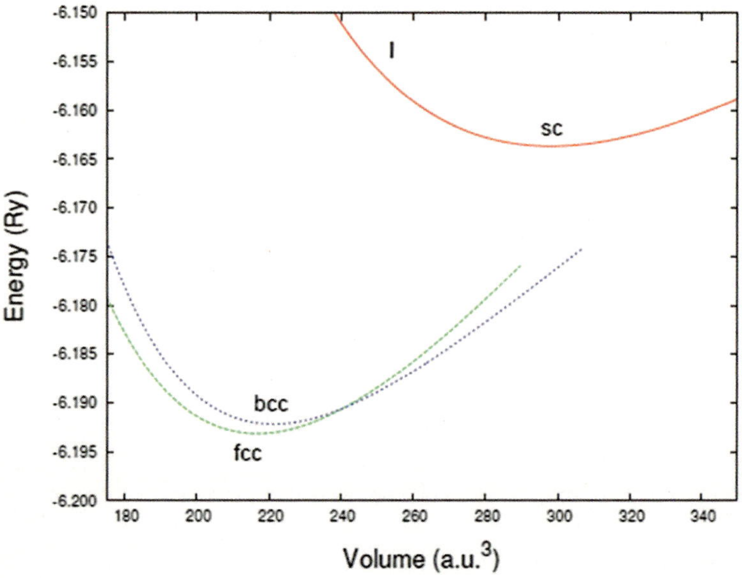

Fig. 11.7 Total energy of I

Table 11.10 Lattice constants and bulk modulus

	a (Bohr)	B (MBar)
fcc	9.535	0.349
bcc	7.619	0.337
sc	6.678	0.216
ΔE_{fcc}–b_{cc} = 0.976 mRy		

Table 11.11 Birch fit coefficients

	A_1	A_2	A_3	A_4
fcc	−5.72981	−29.28105	303.56770	5403.22156
bcc	−6.14246	15.41175	−1326.54443	25470.25374
sc	−5.90326	−12.93496	−399.90315	20462.68360

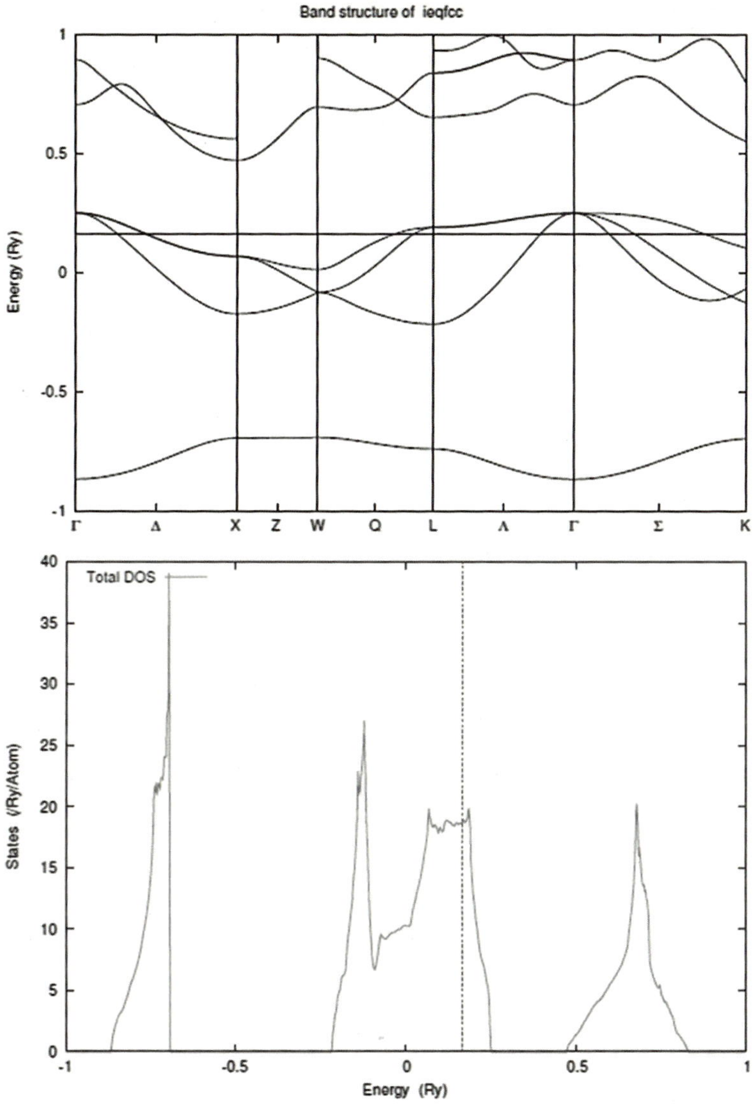

Fig. 11.8 Energy bands and density of states for I

Table 11.12 Dos at the Fermi level (APW results)

Energy	Total	s	p	e_g	t_{2g}	f	Velocity	Plasmon energy
Ry		States/Ry/atom					cm/s	eV
0.165	18.690	0.067	17.080	0.172	0.275	0.228	0.856 × 10E8	9.047

11.5 Astatine

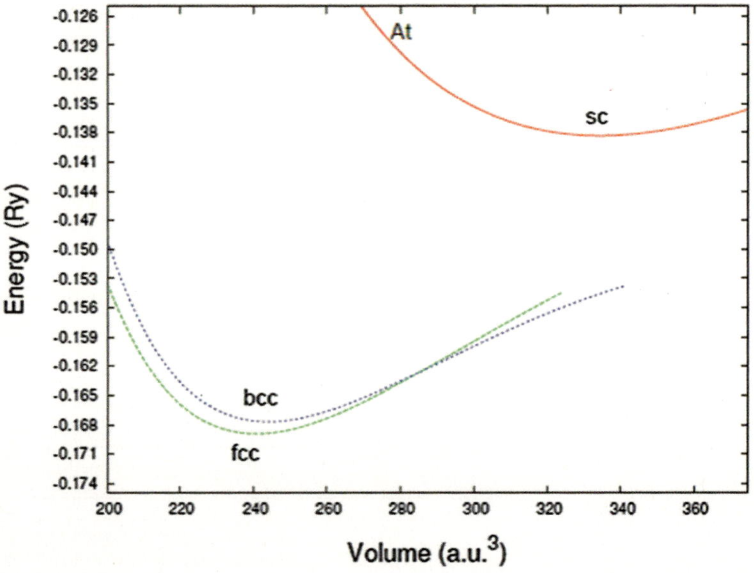

Fig. 11.9 Total energy of At

Table 11.13 Lattice constants and bulk modulus

	a (Bohr)	B (MBar)
fcc	9.876	0.382
bcc	7.873	0.368
sc	6.940	0.201
$\Delta E_{fcc}–b_{cc} = 1.22$ mRy		

Table 11.14 Birch fit coefficients

	A_1	A_2	A_3	A_4
fcc	−0.88095	109.90344	−5307.60533	82073.57350
bcc	−1.17365	144.61519	−6692.06956	100704.36033
sc	0.09669	−9.19821	−749.11771	31162.48725

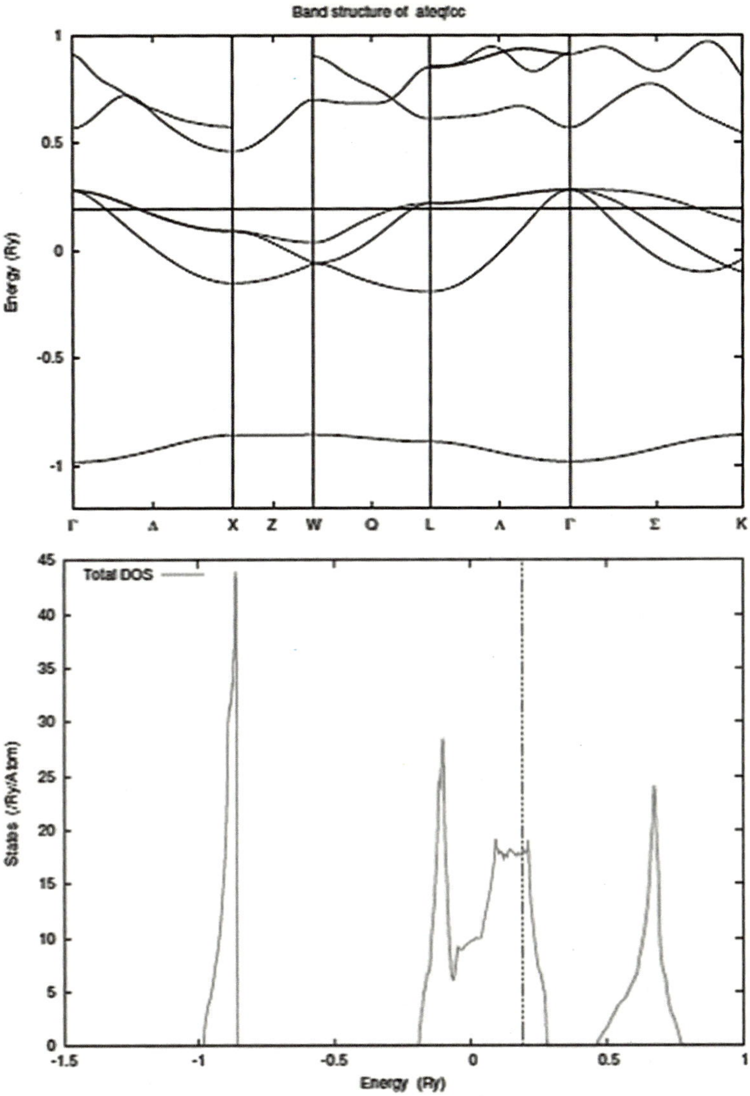

Fig. 11.10 Energy bands and density of states for At

Table 11.15 Dos at the Fermi level (APW results)

Energy	Total	s	p	e_g	t_{2g}	f	Velocity	Plasmon energy
Ry		States/Ry/atom					cm/s	eV
0.189	17.822	0.066	16.068	0.178	0.291	0.266	0.916×10E8	8.969

Chapter 12
The Noble Gases

The last group of elements in the periodic table consists of the noble gases, which have solid phases in the fcc structure at low temperatures. The total energy plots give the fcc as the ground state for all these elements in agreement with experiment. These materials become metallic at large pressures. For these materials the SK fit was done for the empty states only. The dotted line indicating E_F represents the top of a very narrow set of p-like bands and is the beginning of a large gap indicating that they are insulators. These energy bands and DOS, corresponding to the unoccupied states of the noble gases in the solid state, resemble somewhat the partially or fully occupied bands in the transition metals, with the exception of the position of p-like states like Γ_{15}, which now lie close to the d states.

D.A. Papaconstantopoulos, *Handbook of the Band Structure of Elemental Solids*,
DOI 10.1007/978-1-4419-8264-3_12

12.1 Helium

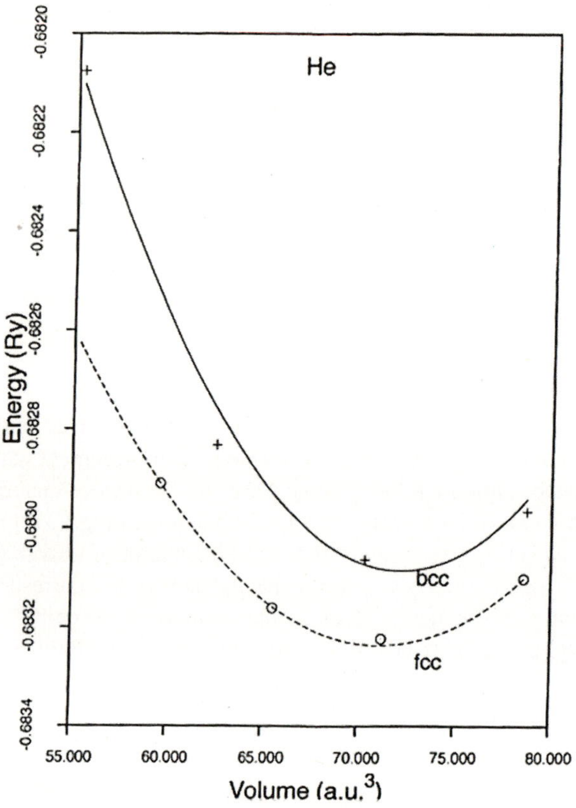

Fig. 12.1 Total energy of He

Table 12.1 Lattice constants and bulk modulus

	a (Bohr)	B (MBar)
fcc	6.575	0.051
bcc	5.224	0.040
exp	6.746	
$\Delta E = .2$ mRy		

Helium (fcc)

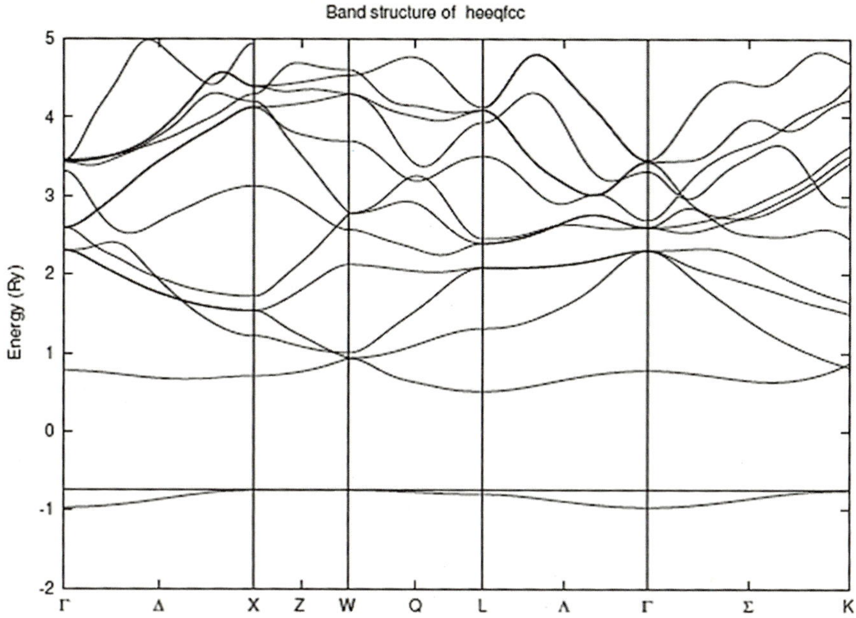

Fig. 12.2 Energy bands for He

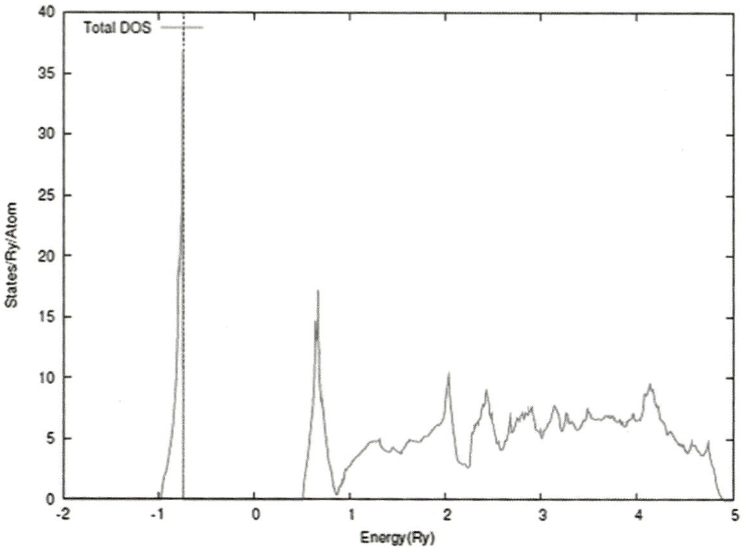

Fig. 12.3 Density of states for He

Table 12.2 Helium fcc structure a = 6.40 Bohr Tight-binding parameters in Ry

Non-orthogonal
On-site
s
−0.78835
sss
Hopping dist = 4.52548
−0.03055
Overlap dist = 4.52548
0.01684
Hopping dist = 6.40000
−0.00219
Overlap dist = 6.40000
0.00069

Fitting rms error for 1 band 0.6 mRy

Orthogonal
On-site
s
−0.78507
sss
Hopping dist = 4.52548
−0.01623
Hopping dist = 6.40000
−0.00060

Fitting rms error for 1 band 4 mRy

12.2 Neon

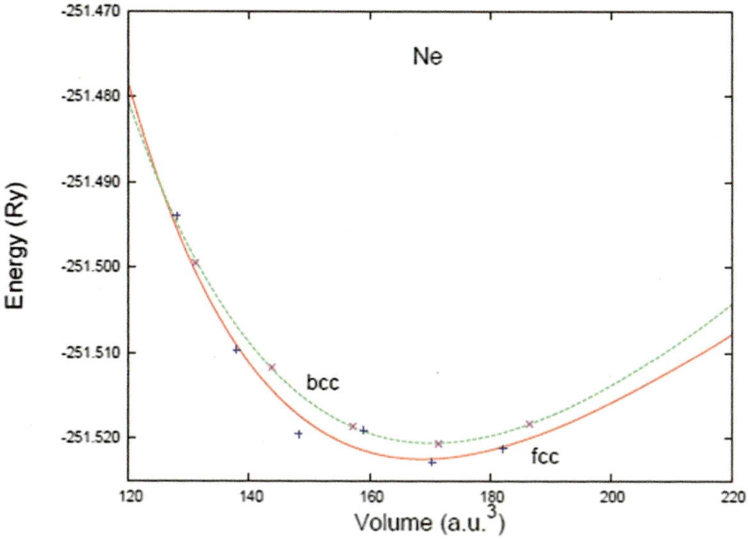

Fig. 12.4 Total energy of Ne

Table 12.3 Lattice constants and bulk modulus

	LDA		GGA	
	a (Bohr)	B (MBar)	a (Bohr)	B (MBar)
bcc	5.796	0.100	6.984	0.483
fcc	7.260	0.100	8.772	0.465
exp	8.428	0.010		
$\Delta E_{fcc-bcc} = 1.87$ mRy				

Table 12.4 Birch fit coefficients

	A_1	A_2	A_3	A_4
bcc	–250.68683	–57.54006	1174.57889	–5951.81721
fcc	–251.20284	–10.93390	–226.19258	8002.77097

Table 12.5 Neon fcc Z = 10 lattice constant = 8.36960 a.u.

Slater-Koster 3-center parameters	Orthogonal	Non-orthogonal	
	Energy integrals (Ry)	Energy integrals (Ry)	Overlap integrals
On site			
s,s(000)	0.92186	0.54996	
x,x(000)	1.26965	0.80266	
xy,xy(000)	1.66458	1.52202	
d2,d2(000)	1.89133	1.56901	
First neighbor			
s,s(110)	−0.05635	−0.01959	0.11279
s,x(110)	0.06940	−0.03044	−0.15013
s,xy(110)	−0.08592	0.04394	0.10464
s,d2(110)	0.05863	−0.02013	−0.07406
x,x(110)	0.06994	−0.07121	−0.11098
x,x(011)	−0.00214	−0.02583	0.02849
x,y(110)	0.09812	−0.10165	−0.23030
x,xy(110)	−0.10328	0.03656	0.08834
x,xy(011)	0.00331	0.01234	−0.03256
z,d2(011)	−0.02559	0.02544	−0.02008
z,d1(011)	0.08696	−0.03280	−0.09940
xy,xy(110)	−0.15911	−0.13216	0.00916
xy,xy(011)	0.03175	−0.07184	−0.08184
xy,xz(011)	0.03445	0.01733	−0.01240
xy,d2(110)	0.09785	0.14391	0.06513
d2,d2(110)	−0.06919	0.02054	0.07468
d1,d1(110)	0.07584	0.10528	−0.08381
Second neighbor			
s,s(200)	−0.02143	−0.00200	0.00574
s,x(200)	0.01556	0.00821	−0.02915
s,d2(002)	0.00046	−0.03525	−0.01284
x,x(200)	0.01804	−0.02372	−0.10758
y,y(200)	−0.01828	0.01181	0.01538
x,xy(020)	0.02415	−0.01545	−0.02637
z,d2(002)	−0.00495	0.01870	0.06425
xy,xy(200)	0.03261	−0.01334	−0.03836
xy,xy(002)	0.01088	−0.02042	−0.00260
d2,d2(002)	0.00733	0.09357	0.04732
d1,d1(002)	−0.04748	0.10753	0.14512

Table 12.6 Neon fcc

	Orthogonal			Non-orthogonal		
Band	RMS error mRy	Maximum k	Deviation mRy	RMS error mRy	Maximum k	Deviation mRy
1	5.8	(444)	13.4	1.8	(444)	4.2
2	24.1	(002)	65.0	2.5	(007)	6.6
3	25.8	(033)	48.0	4.2	(226)	13.4
4	17.5	(444)	34.2	2.9	(442)	9.3
5	35.9	(044)	93.4	18.9	(224)	81.4
6	67.3	(264)	122.8	73.2	(226)	187.9
1–6	35.1			31.0		

Table 12.7 Energy values in Ry at selected k-points

	Orthogonal	APW	Non-orthogonal
GAMMA 1	0.11702	0.12350	0.12684
GAMMA 12	1.81078	1.92080	1.92261
GAMMA 15	1.78352	1.78434	1.78520
GAMMA 25′	1.43434	1.45572	1.45326
X1 (008)	0.62237	0.62529	0.62891
X1 (008)	1.57376	1.58371	1.58908
X2 (008)	2.33772	2.34802	2.34981
X3 (008)	0.92633	0.96275	0.96160
X5 (008)	2.45320	2.44908	2.43120
X4′ (008)	0.66456	0.68462	0.68656
X5′ (008)	1.24118	1.23498	1.23671
L1 (444)	0.54111	0.56721	0.56669
L1 (444)	1.74617	1.89775	1.88964
L3 (444)	1.31863	1.35288	1.34762
L3 (444)	2.34333	2.28856	2.28647
L2′ (444)	0.52178	0.53522	0.53941
L3′ (444)	1.69916	1.70520	1.70479
W1 (048)	1.04689	1.04581	1.04998
W1 (048)	2.07584	1.99277	1.98860
W3 (048)	0.78976	0.76676	0.76591
W3 (048)	1.79556	1.71344	1.71804
W1′ (048)	2.40969	2.48352	2.50041
W2′ (048)	0.67596	0.68771	0.68811
W2′ (048)	2.15051	1.98674	1.98693
EVEN(224)	0.36169	0.35545	0.35437
EVEN(224)	0.82953	0.84087	0.83907
EVEN(224)	1.25674	1.21496	1.22329
EVEN(224)	1.51143	1.51821	1.59961
ODD (224)	1.31259	1.32323	1.32218

Table 12.8 Fermi level quantities (Non-orthogonal fit)

Energy Gap Ry
0.84084

Table 12.9 Neon fcc

Slater-Koster 2-center parameters	Orthogonal	Non-orthogonal	
	Energy integrals (Ry)	Energy integrals (Ry)	Overlap integrals
On site			
s	0.97222	0.69758	
p	1.22025	0.85547	
d1	1.61934	1.08863	
d2	1.90076	1.10415	
First neighbor			
(sss)	−0.04992	−0.02398	0.10141
(pps)	0.15460	−0.12973	−0.30366
(ppp)	−0.00686	0.03146	0.07657
(dds)	−0.20672	0.25905	0.25701
(ddp)	0.07078	−0.12880	−0.23116
(ddd)	−0.00772	0.01370	0.04301
(sps)	0.09241	−0.03208	−0.18647
(sds)	−0.10669	0.04465	0.16415
(pds)	−0.17690	0.21374	0.28882
(pdp)	0.01169	−0.05252	−0.12805
Second neighbor			
(sss)	−0.04251	−0.02031	0.00331
(pps)	0.06895	0.09194	0.01234
(ppp)	−0.03088	−0.00714	−0.00138
(dds)	−0.04581	−0.03844	0.05326
(ddp)	0.04275	0.01677	−0.00946
(ddd)	0.00917	−0.00158	−0.00035
(sps)	0.04228	0.06629	0.04091
(sds)	−0.00193	−0.12001	−0.09173
(pds)	−0.02956	−0.10014	−0.00900
(pdp)	0.02958	0.03444	0.01706

	Orthogonal			Non-orthogonal		
Band	RMS error mRy	Maximum k	Deviation mRy	RMS error mRy	Maximum k	Deviation mRy
1	15.2	(264)	33.3	6.9	(033)	16.4
2	30.1	(002)	67.6	11.4	(002)	29.8
3	27.2	(042)	76.5	10.2	(354)	24.8
4	30.1	(033)	77.4	7.5	(042)	19.4
5	59.6	(444)	176.4	74.8	(064)	264.6
6	76.7	(444)	176.4	94.5	(380)	287.4
1–6	45.1			49.8		

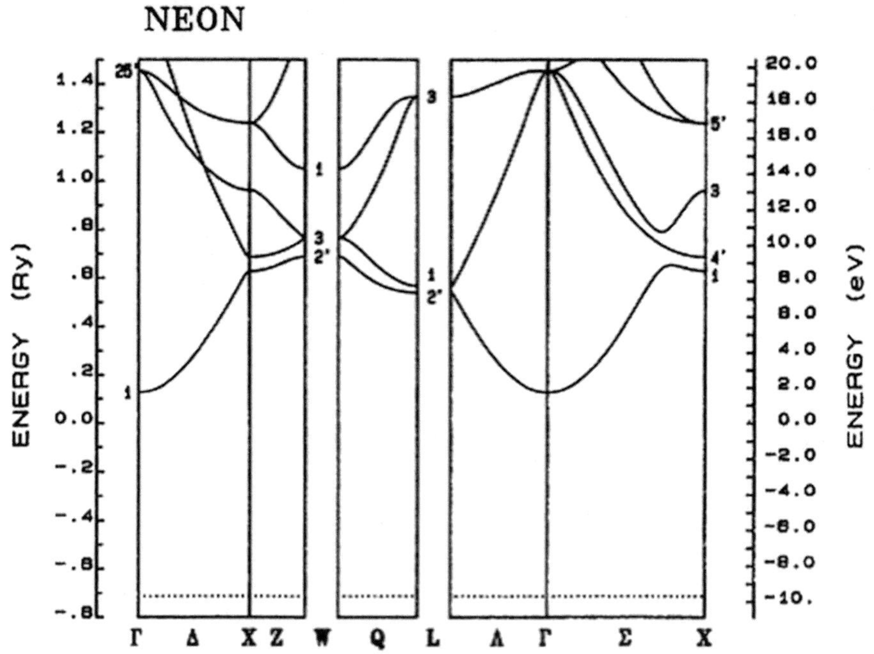

Fig. 12.5 Energy bands for Ne

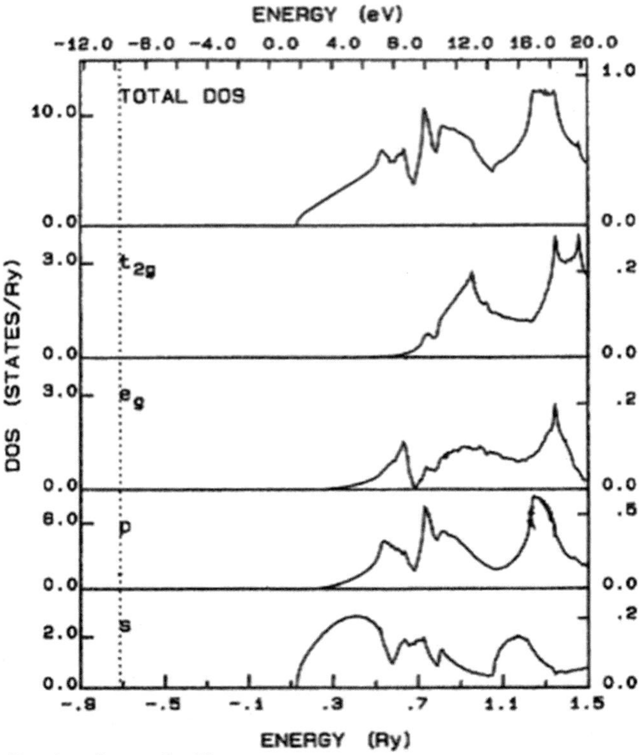

Fig. 12.6 Density of states for Ne

12.3 Argon

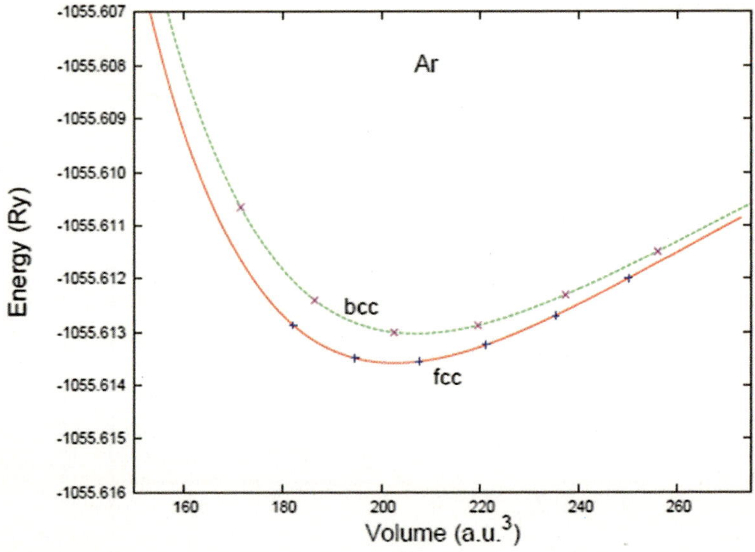

Fig. 12.7 Total energy of Ar

Table 12.10 Lattice constants and bulk modulus

	a (Bohr)	B (MBar)
bcc	7.453	0.070
fcc	9.326	0.076
exp	10.034	0.013
$\Delta E_{fcc-bcc} = 0.55$ mRy		

Table 12.11 Birch fit coefficients

	A_1	A_2	A_3	A_4
bcc	−1055.67042	9.89714	−481.83800	7200.87016
fcc	−1055.68052	10.97253	−518.22657	7567.26682

Table 12.12 Argon fcc Z = 18 lattice constant = 9.93240 a.u.

Slater-Koster 3-center parameters	Orthogonal	Non-orthogonal	
	Energy integrals (Ry)	Energy integrals (Ry)	Overlap integrals
On site			
s,s(000)	0.69509	0.67320	
x,x(000)	1.01635	0.94296	
xy,xy(000)	0.91161	0.72408	
d2,d2(000)	0.75852	0.71599	
First neighbor			
s,s(110)	−0.06064	−0.04528	0.04394
s,x(110)	0.04716	0.00991	−0.06598
s,xy(110)	−0.03557	0.01969	0.10661
s,d2(110)	0.03054	0.04816	0.03210
x,x(110)	0.03773	−0.02153	−0.06870
x,x(011)	−0.00541	0.04786	0.07350
x,y(110)	0.06033	0.03950	−0.03193
x,xy(110)	−0.05958	0.01333	0.07824
x,xy(011)	0.00140	−0.03347	−0.06651
z,d2(011)	0.00962	0.00386	−0.02153
z,d1(011)	0.03581	0.00611	−0.03003
xy,xy(110)	−0.05769	−0.00536	0.07803
xy,xy(011)	0.01507	−0.01883	−0.06477
xy,xz(011)	−0.00945	−0.01027	−0.05225
xy,d2(110)	0.02475	−0.00880	−0.03872
d2,d2(110)	−0.02637	0.03094	0.06757
d1,d1(110)	0.04702	0.01327	−0.06482
Second neighbor			
s,s(200)	0.03506	0.02338	−0.00288
s,x(200)	−0.04651	−0.01157	0.01751
s,d2(002)	0.01915	0.01025	0.00162
x,x(200)	−0.03385	0.03633	0.03604
y,y(200)	0.01343	−0.04451	−0.03729
x,xy(020)	−0.01451	0.02954	0.02459
z,d2(002)	0.01342	−0.00667	0.00023
xy,xy(200)	−0.02169	0.03022	0.03258
xy,xy(002)	0.00347	−0.00487	−0.00377
d2,d2(002)	−0.00008	0.01301	0.03771
d1,d1(002)	0.00530	−0.00473	−0.00287

Table 12.13 Argon fcc

	Orthogonal			Non-orthogonal		
Band	RMS error mRy	Maximum k	Deviation mRy	RMS error mRy	Maximum k	Deviation mRy
1	3.5	(004)	7.9	1.2	(222)	3.1
2	7.7	(055)	18.9	2.2	(044)	4.6
3	6.4	(264)	13.4	2.2	(066)	4.5
4	5.1	(222)	12.4	1.9	(022)	4.2
5	4.8	(048)	11.0	3.4	(226)	8.7
6	5.2	(055)	10.8	4.4	(042)	10.9
1–6	5.6			2.8		

Table 12.14 Energy values in Ry at selected k-points

	Orthogonal	APW	Non-orthogonal
GAMMA 1	0.17772	0.17743	0.17889
GAMMA 12	0.89810	0.89543	0.89746
GAMMA 15	1.28256	1.28090	1.28376
GAMMA 25′	0.72163	0.72305	0.72331
X1 (008)	0.36325	0.36966	0.36964
X1 (008)	1.22408	1.23890	1.23691
X2 (008)	1.02648	1.01920	1.02167
X3 (008)	0.48044	0.49061	0.49382
X5 (008)	1.06255	1.05811	1.05672
X4′ (008)	0.67892	0.68083	0.68001
X5′ (008)	1.02399	1.02582	1.02738
L1 (444)	0.39445	0.39463	0.39485
L1 (444)	1.15728	1.15741	1.15409
L3 (444)	0.68759	0.69192	0.69231
L3 (444)	1.00891	1.01826	1.01039
L2′ (444)	0.54767	0.55272	0.54959
L3′ (444)	1.27166	1.34395	1.27071
W1 (048)	0.73511	0.73827	0.73826
W1 (048)	1.15267	1.13438	1.13783
W3 (048)	0.58218	0.57782	0.57753
W3 (048)	1.07381	1.07483	1.07063
W1′ (048)	1.04866	1.05964	1.06098
W2′ (048)	0.44933	0.44825	0.45034
W2′ (048)	1.27496	1.29143	1.28187
EVEN(224)	0.36027	0.36264	0.36323
EVEN(224)	0.56516	0.56312	0.56272
EVEN(224)	0.67437	0.66628	0.66943
EVEN(224)	0.89570	0.89278	0.89208
ODD (224)	0.72958	1.10038	0.73804
ODD (224)	0.98329	1.27971	0.99196

Table 12.15 Fermi level quantities (Non-orthogonal fit)

Energy Gap Ry
0.59719

Table 12.16 Argon fcc

Slater-Koster 2-center parameters	Orthogonal	Non-orthogonal	
	Energy integrals (Ry)	Energy integrals(Ry)	Overlap integrals
On site			
s	0.84708	0.58911	
p	1.00975	0.78837	
d1	0.82645	0.70294	
d2	0.76903	0.68826	
First neighbor			
(sss)	−0.05442	−0.02651	0.05672
(pps)	0.09070	−0.15278	−0.28728
(ppp)	−0.01339	−0.03026	0.01639
(dds)	−0.07536	0.03639	0.13603
(ddp)	0.03869	−0.00725	−0.09162
(ddd)	−0.00500	−0.00214	0.00972
(sps)	0.07870	−0.03098	−0.16313
(sds)	−0.06412	0.02375	0.12818
(pds)	−0.08275	0.09409	0.20082
(pdp)	0.01695	−0.01473	−0.07296
Second neighbor			
(sss)	−0.00387	0.00409	−0.00166
(pps)	0.00220	0.05311	0.01639
(ppp)	0.00157	−0.00839	−0.00216
(dds)	0.00052	−0.05815	−0.04173
(ddp)	−0.00321	−0.00079	−0.00553
(ddd)	0.00283	0.00056	0.00051
(sps)	−0.01466	−0.00127	−0.01006
(sds)	0.01669	−0.01632	−0.00924
(pds)	−0.00892	−0.04698	−0.01681
(pdp)	0.00250	0.00314	−0.00170

	Orthogonal			Non-orthogonal		
Band	RMS error mRy	Maximum k	Deviation mRy	RMS error mRy	Maximum k	Deviation mRy
1	24.2	(264)	54.8	4.7	(084)	13.3
2	22.1	(264)	38.5	4.1	(264)	15.1
3	23.7	(033)	51.9	5.2	(042)	14.2
4	18.4	(280)	35.0	4.1	(224)	10.0
5	25.5	(066)	58.7	4.0	(444)	9.8
6	22.5	(044)	51.8	4.3	(264)	13.5
1–6	22.8			4.4		

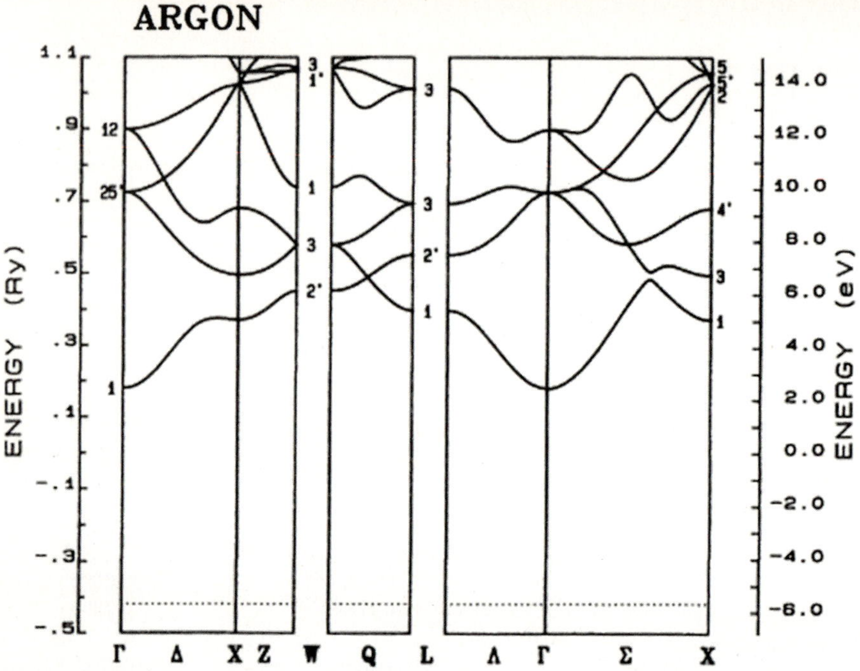

Fig. 12.8 Energy bands for Ar

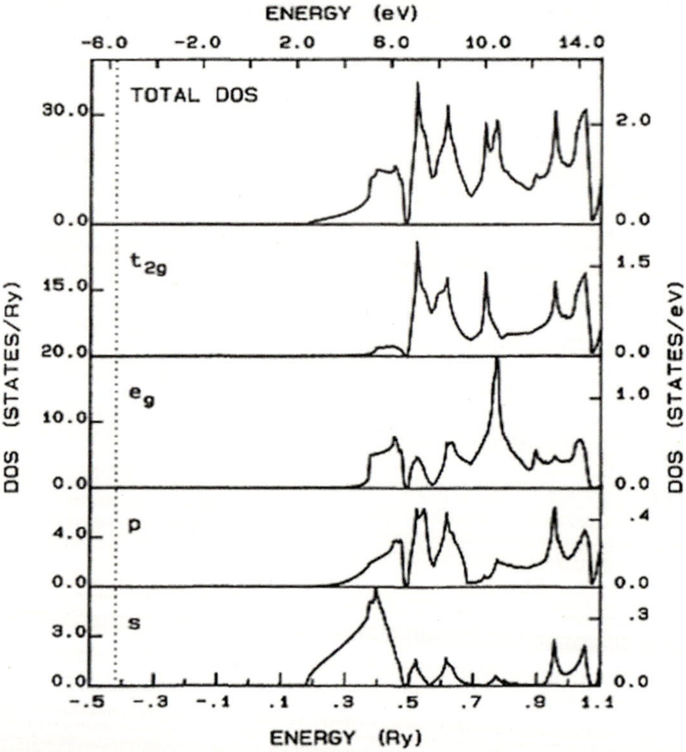

Fig. 12.9 Density of states for Ar

12.4 Krypton

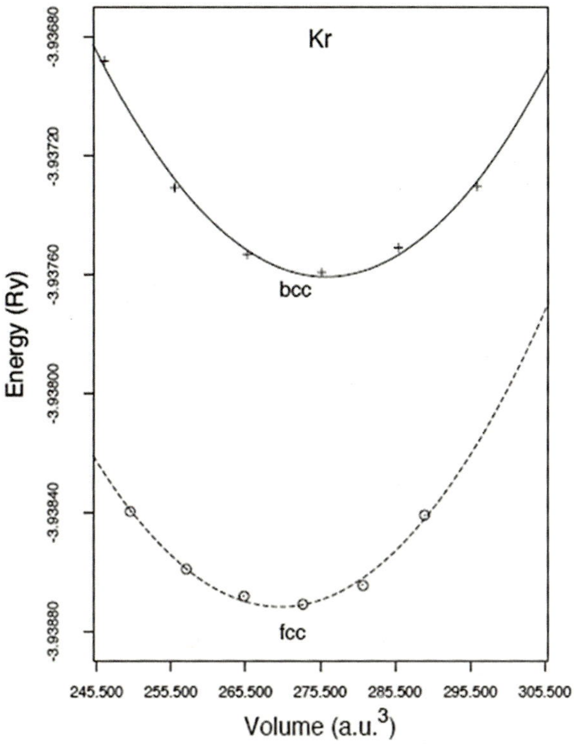

Fig. 12.10 Total energy of Kr

Table 12.17 Lattice constants and bulk modulus

	a (Bohr)	B (MBar)
bcc	8.188	0.058
fcc	10.221	0.058
exp	10.658	0.018
$\Delta E = 1.1$ mRy		

Table 12.18 Birch fit coefficients

	A_1	A_2	A_3	A_4
bcc	−3.90103	0.53019	−240.53916	6458.68107
fcc	−3.10797	−98.50105	3878.94523	−50708.70190

Table 12.19 Krypton fcc Z = 36 lattice constant = 10.81110 a. u.

Slater-Koster 3-center parameters	Orthogonal	Non-orthogonal	
	Energy integrals (Ry)	Energy integrals (Ry)	Overlap integrals
On site			
s,s(000)	0.57589	0.49715	
x,x(000)	0.88264	0.75526	
xy,xy(000)	0.83520	0.61418	
d2,d2(000)	0.68624	0.59364	
First neighbor			
s,s(110)	−0.05259	−0.03476	0.01985
s,x(110)	0.03866	−0.03151	−0.11148
s,xy(110)	−0.02564	0.04094	0.16374
s,d2(110)	0.03159	0.02649	0.01249
x,x(110)	0.02710	−0.09662	−0.17403
x,x(011)	−0.00669	0.04690	0.10713
x,y(110)	0.04942	−0.00640	−0.05077
x,xy(110)	−0.05114	0.04286	0.12596
x,xy(011)	0.00213	−0.05246	−0.11114
z,d2(011)	0.00694	0.01732	0.00599
z,d1(011)	0.03835	−0.00238	−0.00943
xy,xy(110)	−0.06335	0.01580	0.10738
xy,xy(011)	0.01432	−0.01327	−0.06228
xy,xz(011)	−0.01081	−0.01265	−0.06607
xy,d2(110)	0.02674	−0.01064	−0.05102
d2,d2(110)	−0.02835	0.04771	0.10092
d1,d1(110)	0.04409	−0.01116	−0.12899
Second neighbor			
s,s(200)	0.03288	0.00919	−0.04552
s,x(200)	−0.04976	0.04799	0.10041
s,d2(002)	0.02033	−0.04311	−0.07373
x,x(200)	−0.04276	0.21316	0.24104
y,y(200)	0.01842	−0.02788	−0.01560
x,xy(020)	−0.01419	0.00842	−0.00199
z,d2(002)	0.01590	−0.09455	−0.12901
xy,xy(200)	−0.01910	0.00112	−0.00635
xy,xy(002)	0.00390	−0.00308	−0.00234
d2,d2(002)	0.00506	−0.00420	0.04438
d1,d1(002)	0.00441	−0.00634	−0.00280

Table 12.20 Krypton fcc

	Orthogonal			Non-orthogonal		
	RMS error	Maximum	Deviation	RMS error	Maximum	Deviation
Band	mRy	k	mRy	mRy	k	mRy
1	4.4	(264)	11.1	4.3	(003)	9.7
2	8.7	(003)	17.2	5.3	(354)	16.3
3	7.7	(003)	16.0	5.4	(354)	12.8
4	5.9	(222)	13.3	4.7	(066)	12.4
5	10.6	(444)	37.6	4.9	(354)	15.0
6	13.3	(444)	37.6	6.4	(442)	15.0
1–6	8.9			5.2		

Table 12.21 Energy values in Ry at selected k-points

	Orthogonal	APM	Non-orthogonal
GAMMA 1	0.14206	0.14153	0.14011
GAMMA 12	0.80905	0.81614	0.81698
GAMMA 15	1.06084	1.06271	1.06301
GAMMA 25′	0.62774	0.63143	0.63201
X1 (008)	0.29995	0.30428	0.31309
X1 (008)	1.07694	1.07881	1.07857
X2 (008)	0.97295	0.96688	0.97020
X3 (008)	0.39864	0.41085	0.41203
X5 (008)	1.02003	1.01582	1.01648
X4′ (008)	0.62725	0.62691	0.62666
X5′ (008)	0.89759	0.89563	0.89034
L1 (444)	0.33078	0.32500	0.32246
L1 (444)	1.03812	1.05357	1.05160
L3 (444)	0.59355	0.59952	0.60469
L3 (444)	0.92484	0.96248	0.95418
L2′ (444)	0.49914	0.51003	0.49840
L3′ (444)	1.09214	1.13841	1.09970
W1 (048)	0.62104	0.62729	0.61912
W1 (048)	1.03879	1.03126	1.03105
W3 (048)	0.51867	0.51103	0.51195
W3 (048)	0.95577	0.96407	0.96126
W1′ (048)	1.00444	1.01876	1.02160
W2′ (048)	0.38331	0.38497	0.37976
W2′ (048)	1.20598	1.24538	1.25005
EVEN(224)	0.30834	0.31250	0.31305
EVEN(224)	0.49504	0.49218	0.48443
EVEN(224)	0.58331	0.57266	0.57868
EVEN(224)	0.80778	0.79911	0.80011
ODD (224)	0.63674	0.64842	0.64329
ODD (224)	0.91155	0.91140	0.92017

Table 12.22 Fermi level quantities (Non-orthogonal fit)

Energy Gap Ry
0.49511

Table 12.23 Krypton fcc

Slater-Koster 2-center parameters	Orthogonal	Non-orthogonal	
	Energy integrals (Ry)	Energy integrals (Ry)	Overlap integrals
On site			
s	0.73834	0.72588	
p	0.88830	0.70981	
d1	0.74938	0.60041	
d2	0.68640	0.61921	
First neighbor			
(sss)	−0.04626	−0.02756	0.06604
(pps)	0.07024	−0.09139	−0.23527
(ppp)	−0.01443	0.07468	0.10877
(dds)	−0.08194	0.04391	0.16091
(ddp)	0.03817	−0.01715	−0.11850
(ddd)	−0.00633	0.00774	0.02629
(sps)	0.07203	0.03562	0.06371
(sds)	−0.05954	−0.03141	0.04668
(pds)	−0.07675	0.08606	0.19396
(pdp)	0.01553	−0.02055	−0.09333
Second neighbor			
(sss)	−0.00780	−0.01665	0.00219
(pps)	−0.01234	0.13710	0.10039
(ppp)	0.00471	0.02454	0.02662
(dds)	0.00466	−0.04681	−0.01946
(ddp)	−0.00049	−0.00264	−0.01187
(ddd)	0.00413	0.00674	0.00905
(sps)	−0.01879	0.07292	0.12652
(sds)	0.02067	−0.11845	−0.24567
(pds)	−0.00015	−0.07857	−0.04349
(pdp)	0.00557	0.03267	0.02244

	Orthogonal			Non-orthogonal		
Band	RMS error mRy	Maximum k	Deviation mRy	RMS error mRy	Maximum k	Deviation mRy
1	24.8	(264)	61.5	20.5	(033)	50.3
2	25.2	(264)	54.9	15.8	(222)	41.5
3	29.6	(006)	60.7	23.4	(006)	70.2
4	25.0	(118)	45.9	31.0	(062)	80.5
5	33.5	(066)	66.2	21.4	(226)	60.8
6	28.4	(062)	58.4	44.2	(004)	154.9
1–6	27.9			27.6		

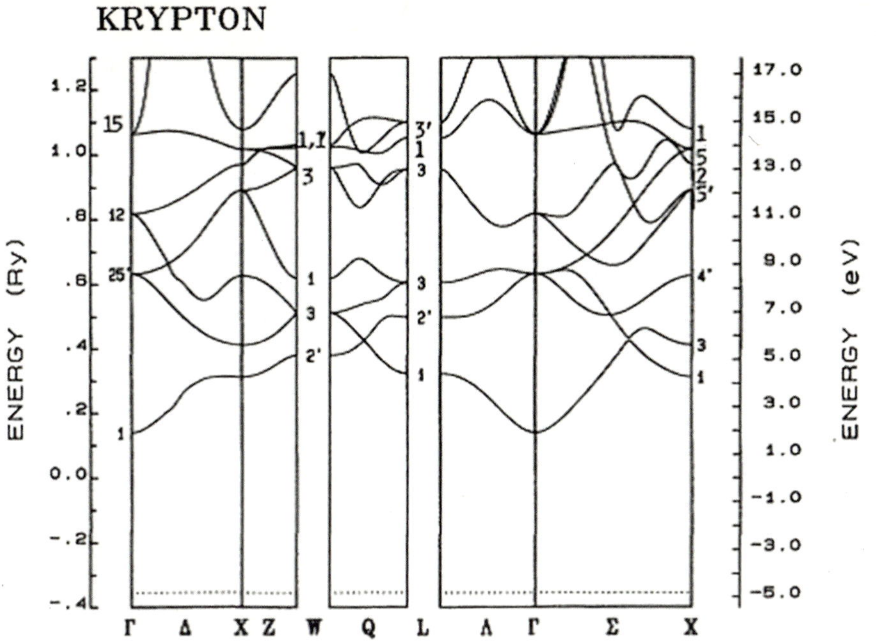

Fig. 12.11 Energy bands for Kr

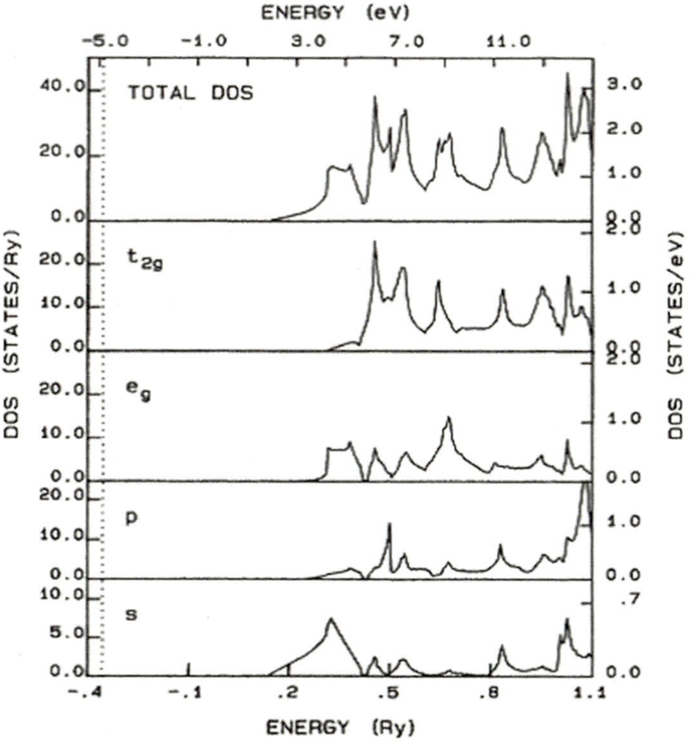

Fig. 12.12 Density of states for Kr

12.5 Xenon

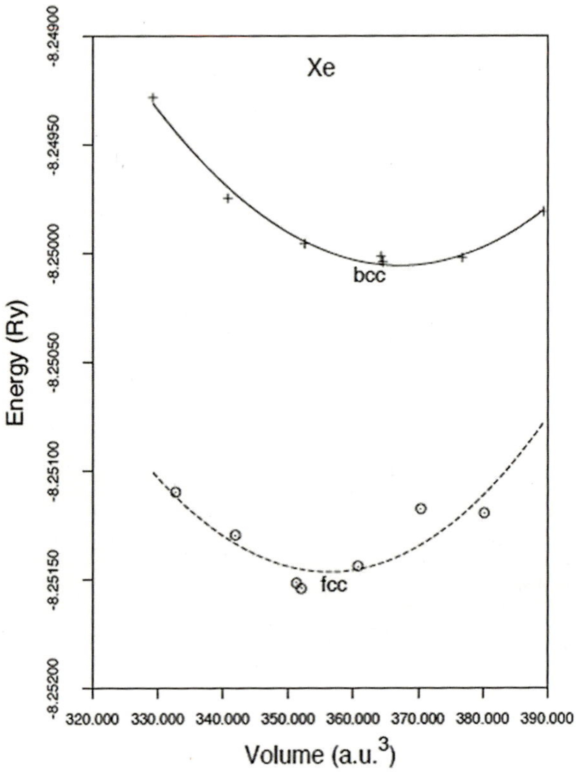

Fig. 12.13 Total energy of Xe

Table 12.24 Lattice constants and bulk modulus

	a (Bohr)	B (MBar)
bcc	8.998	0.059
fcc	11.208	0.055
exp	11.588	
$\Delta E = 1.5$ mRy		

Table 12.25 Birch fit coefficients

	A_1	A_2	A_3	A_4
bcc	−8.30119	14.72011	−1103.49801	24776.31258
fcc	−10.71232	379.54403	−19490.61040	333276.01301

Table 12.26 Xenon fcc Z = 54 lattice constant = 11.71060 a.u.

Slater-Koster 3-center parameters	Orthogonal	Non-orthogonal	
	Energy integrals (Ry)	Energy integrals (Ry)	Overlap integrals
On site			
s,s(000)	0.53404	0.48871	
x,x(000)	0.75177	0.74692	
xy,xy(000)	0.68469	0.52038	
d2,d2(000)	0.58040	0.51231	
First neighbor			
s,s(110)	−0.04590	−0.04172	0.00259
s,x(110)	0.02803	−0.01594	−0.06507
s,xy(110)	−0.02865	0.03597	0.15883
s,d2(110)	0.02981	0.04064	0.05214
x,x(110)	0.01432	−0.06392	−0.11665
x,x(011)	−0.00822	0.02882	0.06734
x,y(110)	0.03221	0.04038	0.02699
x,xy(110)	−0.04341	0.02343	0.09153
x,xy(011)	0.00293	−0.04484	−0.09446
z,d2(011)	0.00852	0.01034	−0.00257
z,d1(011)	0.03352	0.01208	0.02124
xy,xy(110)	−0.05803	0.00315	0.09642
xy,xy(011)	0.01252	−0.01019	−0.06094
xy,xz(011)	−0.00695	−0.00705	−0.05676
xy,d2(110)	0.02439	0.00052	−0.03972
d2,d2(110)	−0.02506	0.05305	0.12059
d1,d1(110)	0.03640	0.01345	−0.09204
Second neighbor			
s,s(200)	0.02753	0.02427	−0.02812
s,x(200)	−0.04075	0.03188	0.10682
s,d2(002)	0.01876	−0.03966	−0.09005
x,x(200)	−0.01636	0.16144	0.21412
y,y(200)	0.01171	−0.03603	−0.03572
x,xy(020)	−0.00436	−0.01037	−0.02982
z,d2(002)	0.01123	−0.06401	−0.13025
xy,xy(200)	−0.01241	−0.00215	−0.01055
xy,xy(002)	0.00292	0.00045	0.00097
d2,d2(002)	0.00615	0.02570	0.08288
d1,d1(002)	0.00173	−0.00403	0.00174

Table 12.27 Xenon fcc

	Orthogonal			Non-orthogonal		
Band	RMS error mRy	Maximum k	Deviation mRy	RMS error mRy	Maximum k	Deviation mRy
1	5.7	(055)	15.2	5.7	(264)	10.9
2	11.3	(222)	29.4	7.3	(442)	22.7
3	6.8	(048)	14.5	5.2	(264)	12.4
4	6.9	(354)	16.9	4.5	(444)	11.7
5	14.1	(444)	52.6	5.2	(380)	11.4
6	18.6	(444)	52.6	6.5	(062)	14.7
1–6	11.5			5.8		

Table 12.28 Energy values in Ry at selected k-points

	Orthogonal	APW	Non-orthogonal
GAMMA 1	0.14837	0.14747	0.15499
GAMMA 12	0.67208	0.68193	0.68506
GAMMA 15	0.84755	0.85165	0.85211
GAMMA 25′	0.50889	0.51541	0.51735
X1 (008)	0.24621	0.24597	0.25491
X1 (008)	0.97213	0.98006	0.98215
X2 (008)	0.82721	0.82581	0.82575
X3 (008)	0.30858	0.32511	0.33104
X5 (008)	0.87301	0.87367	0.87111
X4′ (008)	0.61843	0.62210	0.62395
X5′ (008)	0.79878	0.79410	0.79762
L1 (444)	0.28927	0.27657	0.27761
L1 (444)	0.86370	0.93550	0.93102
L3 (444)	0.48924	0.48855	0.48293
L3 (444)	0.76824	0.82086	0.81379
L2′ (444)	0.47999	0.50319	0.49145
L3′ (444)	0.86648	0.92703	0.88704
W1 (048)	0.52644	0.52966	0.52317
W1 (048)	0.92596	0.90781	0.91062
W3 (048)	0.46542	0.45089	0.44550
W3 (048)	0.83679	0.84263	0.83279
W1′ (048)	0.86133	0.87693	0.87386
W2′ (048)	0.31950	0.32976	0.32472
W2′ (048)	0.99624	1.07018	1.07013
EVEN(224)	0.28699	0.29293	0.29588
EVEN(224)	0.43894	0.43120	0.43198
EVEN(224)	0.48592	0.47256	0.46899
EVEN(224)	0.68429	0.67656	0.68122
ODD (224)	0.52318	0.53661	0.53094
ODD (224)	0.76650	0.76217	0.76454

Table 12.29 Fermi level quantities (Non-orthogonal fit)

Energy Gap Ry
0.41669

Table 12.30 Xenon fcc

Slater-Koster 2-center parameters	Orthogonal	Non-orthogonal	
	Energy integrals (Ry)	Energy integrals (Ry)	Overlap integrals
On site			
s	0.68705	0.59191	
p	0.76732	0.70053	
d1	0.61748	0.52889	
d2	0.56271	0.54744	
First neighbor			
(sss)	−0.04480	−0.00786	0.11805
(pps)	0.04070	−0.09443	−0.23229
(ppp)	−0.01402	0.07303	0.11075
(dds)	−0.07426	0.02701	0.15731
(ddp)	0.03440	0.01796	−0.03975
(ddd)	−0.00604	0.00271	0.00784
(sps)	−0.05550	0.02706	0.07145
(sds)	−0.05019	−0.03055	0.02679
(pds)	0.05646	0.07159	0.16523
(pdp)	−0.01539	0.01046	−0.04045
Second neighbor			
(sss)	−0.00269	−0.01345	0.01542
(pps)	−0.01192	0.13376	0.10335
(ppp)	0.01023	0.02446	0.02674
(dds)	0.00574	−0.05785	−0.07860
(ddp)	0.00239	−0.00657	−0.00904
(ddd)	0.00075	0.01671	0.02735
(sps)	0.02204	0.05070	0.11259
(sds)	0.00852	−0.09633	−0.25398
(pds)	−0.00812	−0.02857	0.01243
(pdp)	−0.00952	0.03859	0.03937

	Orthogonal			Non-orthogonal		
Band	RMS error mRy	Maximum k	Deviation mRy	RMS error mRy	Maximum k	Deviation mRy
1	24.6	(264)	55.6	16.3	(042)	45.9
2	20.8	(222)	41.9	10.0	(224)	31.5
3	25.8	(006)	56.1	19.7	(006)	59.2
4	21.7	(084)	56.2	23.5	(062)	60.1
5	33.3	(066)	67.8	25.0	(226)	49.1
6	27.7	(055)	63.8	29.5	(005)	96.1
1–6	26.0			21.6		

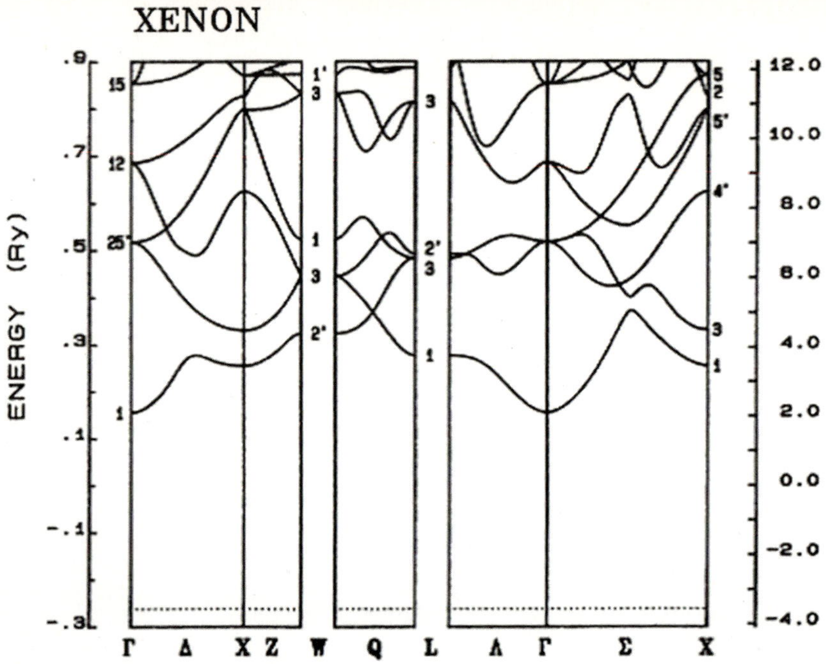

Fig. 12.14 Energy bands for Xe

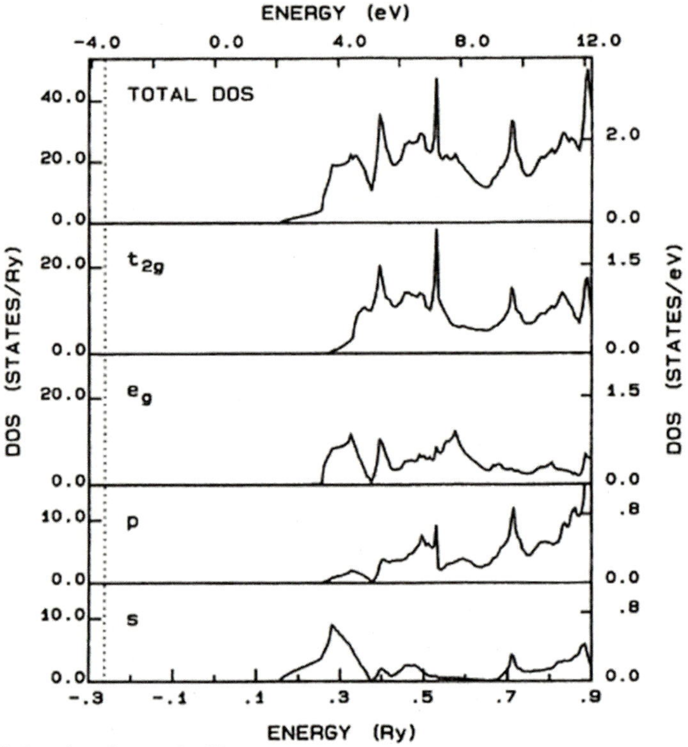

Fig. 12.15 Density of states for Xe

12.6 Radon

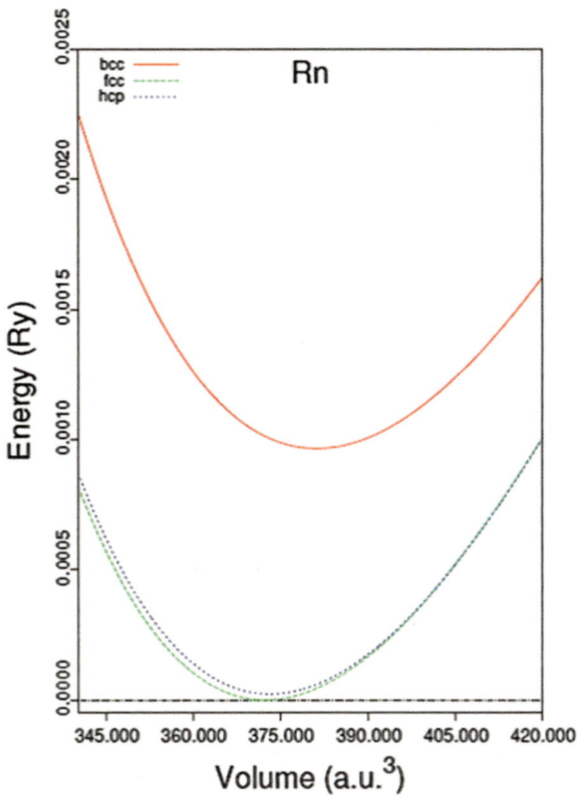

Fig. 12.16 Total energy of Rn

Table 12.31 Lattice constants and bulk modulus

	a (Bohr)	c (Bohr)	B (MBar)
bcc	9.134		0.063
fcc	11.420		0.068
hcp	11.206	18.299	0.067
$\Delta E_{hcp\text{–}fcc} = 0.02$ mRy			

Table 12.32 Birch fit coefficients

	A_1	A_2	A_3	A_4
bcc	−47217.00278	24.38878	−1792.53939	40353.09107
fcc	−47217.01679	26.42563	−1882.59636	41361.18942
hcp	−94439.79492	70.49263	−8387.06461	300962.53066

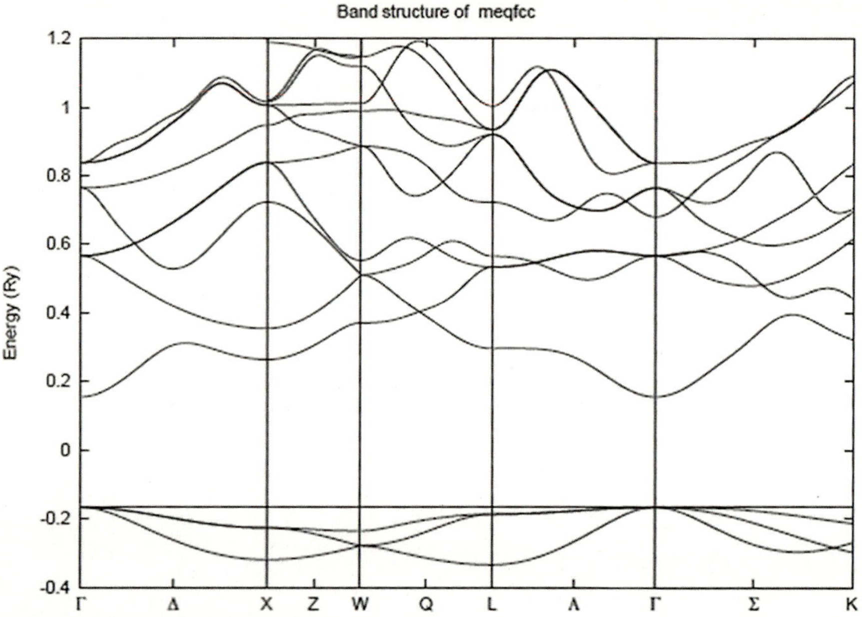

Fig. 12.17 Energy bands for Rn

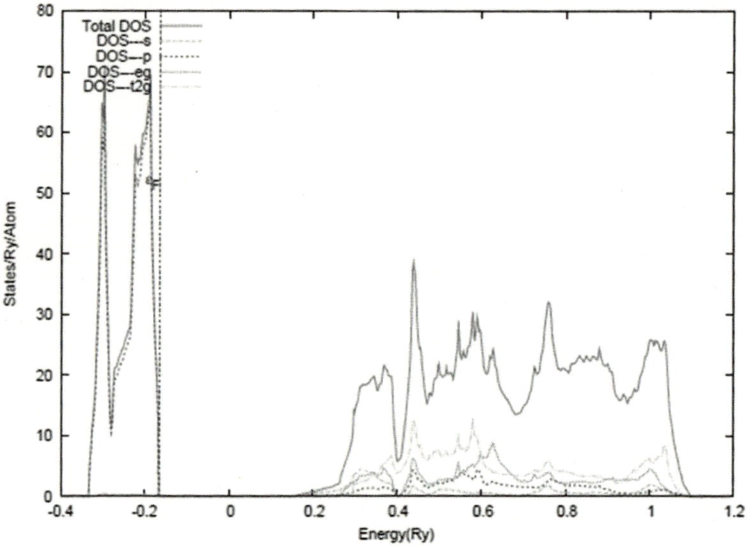

Fig. 12.18 Density of states for Rn

Table 12.33 Energy Gap

Ry
0.36

Chapter 13
Lanthanides

For the Lanthanides we present a total energy versus volume graph for La which shows the correct ground state as dhcp. As it is shown in the energy bands and DOS figures the f-bands are positioned above the Fermi level for La. As we examine the energy bands and DOS for the rest of the elements (except for Lu for which we show total energies for fcc and bcc with a standard method) in this row we note that Ef is located inside the f-bands and finally in Yb and Lu lies above the f-bands. The, at-least, partially occupied f-bands in these materials present a serious difficulty in determining the equilibrium volume and structure using standard density functional theory. We have found that the only way to calculate a reasonable equilibrium volume for these materials is to push the f-electrons into the core and use only s- and d- electrons in the valence [1]. We demonstrate this for Ce, Pr, Nd, Pm, and Sm where the equilibrium fcc, bcc and sc volumes are predicted by forcing the f-bands into the core. For Eu, we show fcc, bcc, as well as hcp equilibrium volumes using the same method. For the rest of the elements in this row, which are mostly existing in hcp lattice, we present energy bands and DOS for the equivalent fcc volume to show trends.

D.A. Papaconstantopoulos, *Handbook of the Band Structure of Elemental Solids*,
DOI 10.1007/978-1-4419-8264-3_13

13.1 Lanthanum Z = 57

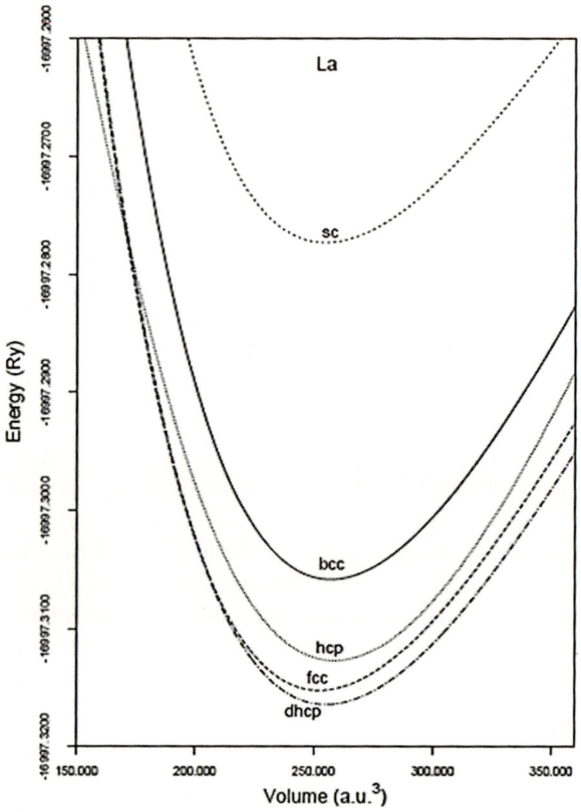

Fig. 13.1 Total energy of La

Table 13.1 Lattice constants and bulk modulus

	a (Bohr)	c (Bohr)	B (MBar)
sc	6.338		0.240
bcc	8.011		0.270
fcc	10.032		0.248
hcp	7.185	11.855	0.264
dchp	6.84	22.07	0.300
exp	7.13	22.83	0.240
$\Delta E_{fcc\text{–}dhcp} = 1.2$ mRy			

Table 13.2 Birch fit coefficients

	A_1	A_2	A_3	A_4
sc	−16996.90361	−26.28179	302.38950	60392.19115
bcc	−16996.55887	−69.12300	1926.50289	−14264.67872
fcc	−16996.67954	−57.01302	1512.13350	−9948.83479
hcp	−33992.51350	−340.89968	17625.09865	−285296.26332
dhcp	−67986.68678	−234.67719	6359.17857	−43927.77984

Lanthanum

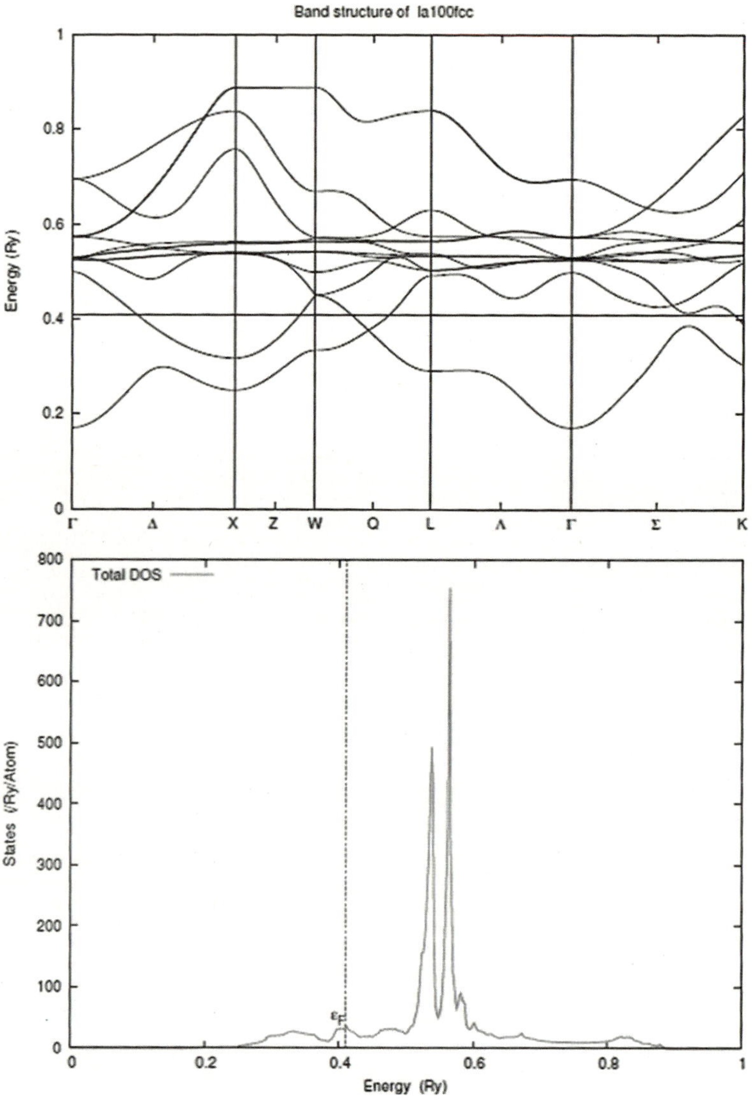

Fig. 13.2 Energy bands and density of states for La

Table 13.3 DOS at the Fermi level (APW results)

Energy	Total	s	p	e_g	t_{2g}	f	Velocity	Plasmon energy
Ry		States/Ry/atom					cm/s	eV
0.410	32.280	0.993	2.898	5.673	11.717	3.809	0.347 × 10E8	4.492

Lanthanum fcc structure a = 10.00 Bohr
Tight-binding parameters in Ry

Non-orthogonal
On-site

s	p	d							
0.57153	0.98716	0.67157							
sss	sps	pps	ppp	sds	pds	pdp	dds	ddp	ddd
Hopping dist = 7.07107									
−0.01268	0.04110	0.24050	0.22087	−0.03252	−0.04721	−0.01670	0.14515	0.00509	−0.10044
Overlap dist = 7.07107									
0.03903	−0.10887	0.18627	0.21817	−0.15152	−0.06313	−0.07895	0.18368	−0.06484	−0.03820
Hopping dist = 10.00000									
−0.03031	0.00282	0.02273	0.01481	−0.00243	−0.00498	0.00000	0.01150	0.00196	−0.00678
Overlap dist = 10.00000									
0.00965	0.00259	0.00734	0.02127	−0.01660	−0.00742	−0.04224	0.01618	−0.01357	−0.00274
Hopping dist = 12.24745									
−0.00156	0.00003	0.00034	0.00017	−0.00003	−0.00008	0.00000	0.00015	0.00004	−0.00008
Overlap dist = 12.24745									
0.00011	0.00013	0.00006	0.00033	−0.00029	−0.00012	−0.00089	0.00022	−0.00032	−0.00003

Fitting rms error for 4 bands 55 mRy

Orthogonal
On-site

s	p	d							
1.19419	1.13992	0.73800							
sss	sps	pps	ppp	sds	pds	pdp	dds	ddp	ddd
Hopping dist = 7.07107									
−0.05940	0.14955	0.23981	0.17580	−0.06973	−0.12695	−0.04101	0.05797	−0.03104	−0.04014
Hopping dist = 10.00000									
−0.04689	0.01435	0.07391	0.00970	0.04554	−0.05158	−0.00214	−0.00902	0.05152	−0.00056
Hopping dist = 12.24745									
−0.00223	0.00022	0.00205	0.00010	0.00272	−0.00168	0.00000	−0.00038	0.00228	0.00001

Fitting rms error for 4 bands 75 mRy

13.2 Cerium Z = 58

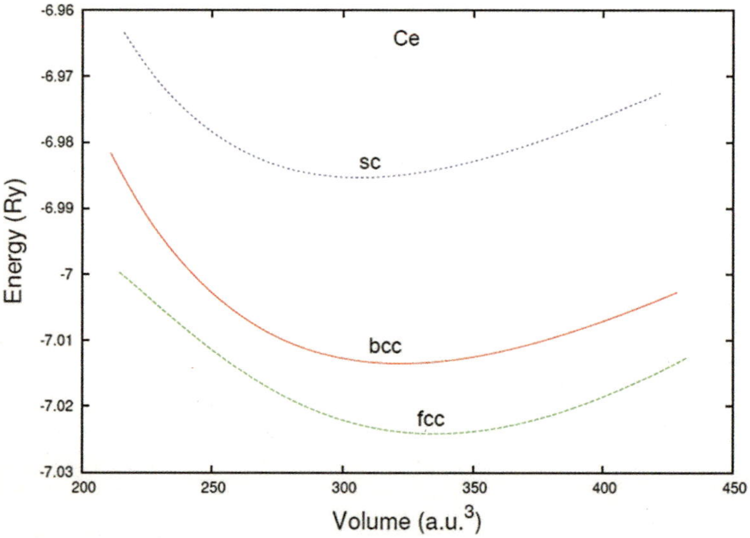

Fig. 13.3 Total energy of Ce

Table 13.4 Lattice contants and bulk modulus

	a (Bohr)	B (MBar)
sc	6.741	0.143
bcc	8.636	0.137
fcc	11.027	0.152
$\Delta E_{fcc-bcc} = 10.67$ mRy		

Table 13.5 Birch fit coefficients

	A_1	A_2	A_3	A_4
sc	−6.58872	−38.83308	1072.31343	−5753.56419
bcc	−6.55852	−48.25427	1521.44616	−12147.85661
fcc	−6.09195	−116.09939	4692.95763	−60860.67529

Cerium

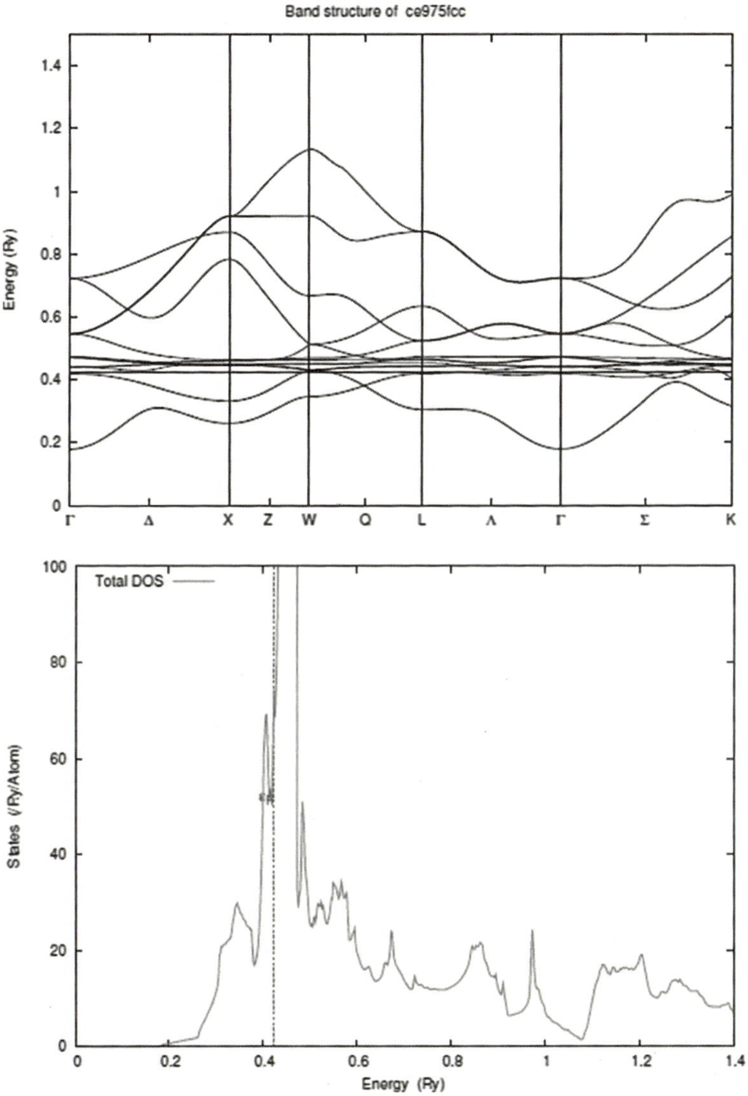

Fig. 13.4 Energy bands and density of states for Ce

Table 13.6 DOS at the Fermi level (APW results)

Energy	Total	s	p	e_g	t_{2g}	f	Velocity	Plasmon energy
Ry		States/Ry/atom					cm/s	eV
0.422	62.506	1.646	1.764	2.351	7.192	44.412	0.164 × 10E8	3.050

13.3 Praseodymium Z = 59

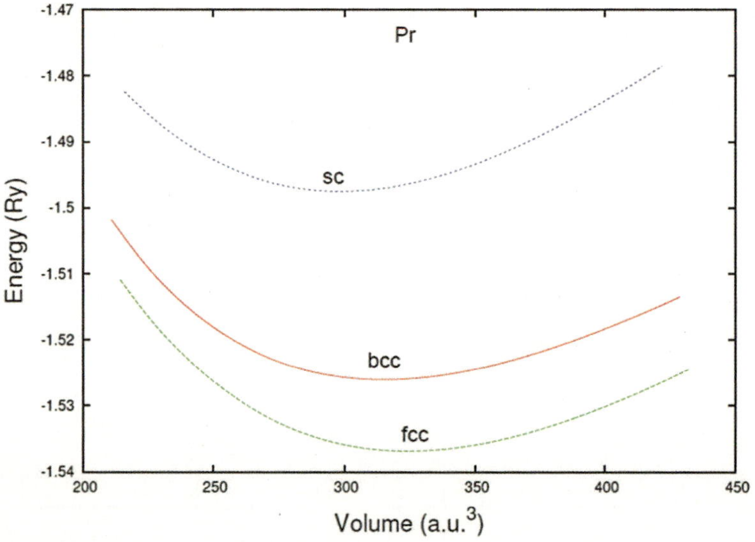

Fig. 13.5 Total energy for Pr

Table 13.7 Lattice constants and bulk modulus

	a (Bohr)	B (MBar)
sc	6.682	0.158
bcc	8.572	0.134
fcc	10.906	0.143
$\Delta E_{fcc-bcc} = 10.83$ mRy		

Table 13.8 Birch fit coefficients

	A_1	A_2	A_3	A_4
sc	−0.72045	−87.95795	3207.34421	−37020.92561
bcc	−0.99091	−59.34927	2054.81418	−21020.38835
fcc	−0.87447	−77.06659	2848.19692	−32392.74258

Praseodymium

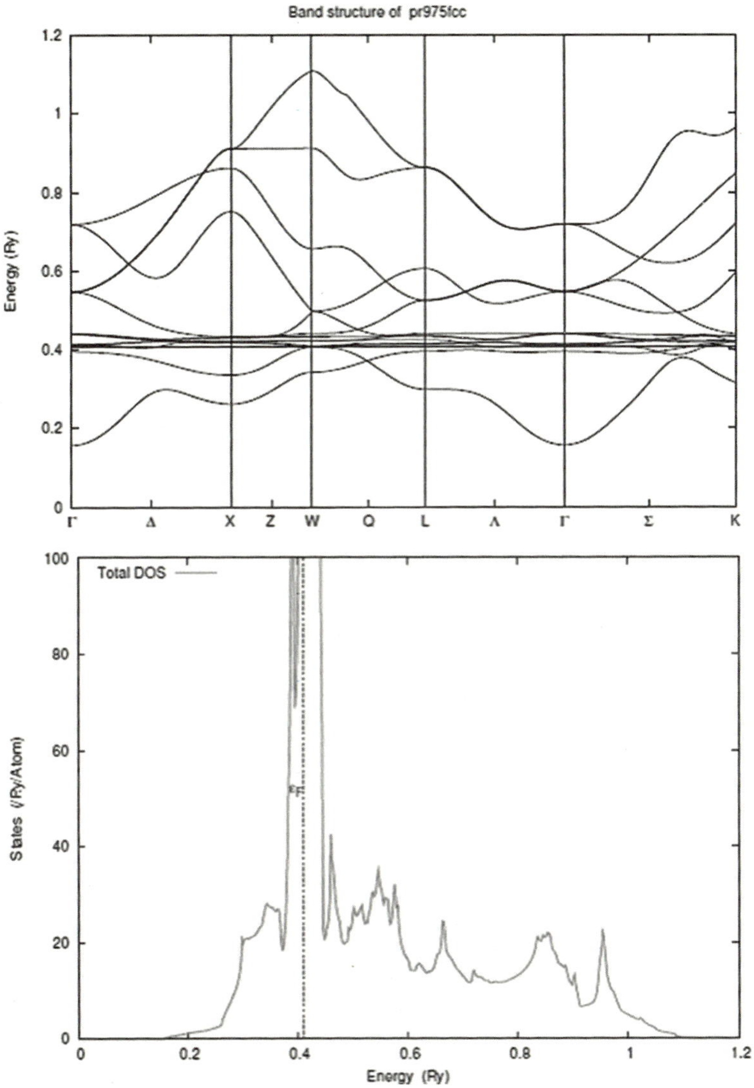

Fig. 13.6 Energy bands and density of states for Pr

Table 13.9 DOS at the Fermi level (APW results)

Energy	Total	s	p	e_g	t_{2g}	f	Velocity	Plasmon energy
Ry		States/Ry/atom					cm/s	eV
0.410	146.984	1.411	1.816	0.888	6.449	129.894	0.081 × 10E8	2.298

13.4 Neodymium Z = 60

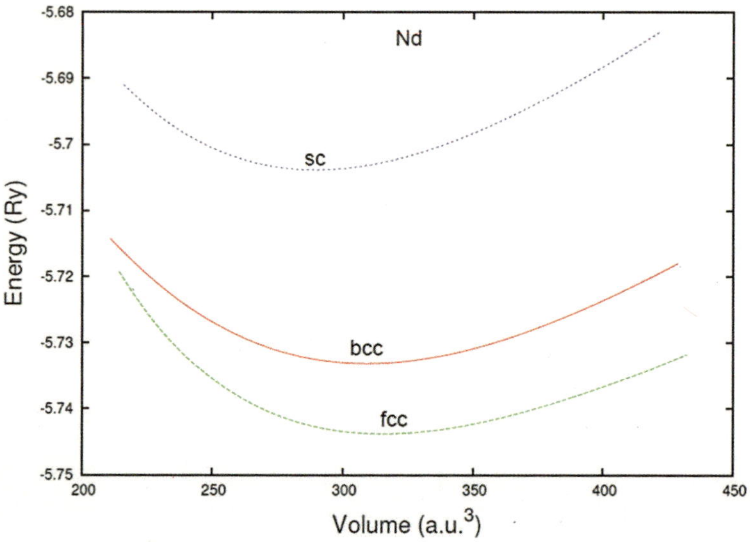

Fig. 13.7 Total energy of Nd

Table 13.10 Lattice constants and bulk modulus

	a (Bohr)	B (MBar)
sc	6.615	0.158
bcc	8.519	0.138
fcc	10.805	0.131
$\Delta E_{fcc-bcc} = 10.66$ mRy		

Table 13.11 Birch fit coefficients

	A_1	A_2	A_3	A_4
sc	−5.01696	−74.84425	2604.44877	−28209.59617
bcc	−5.05382	−78.27002	2897.11509	−33767.49214
fcc	−5.32284	−43.89216	1356.37349	−10489.50572

Neodymium

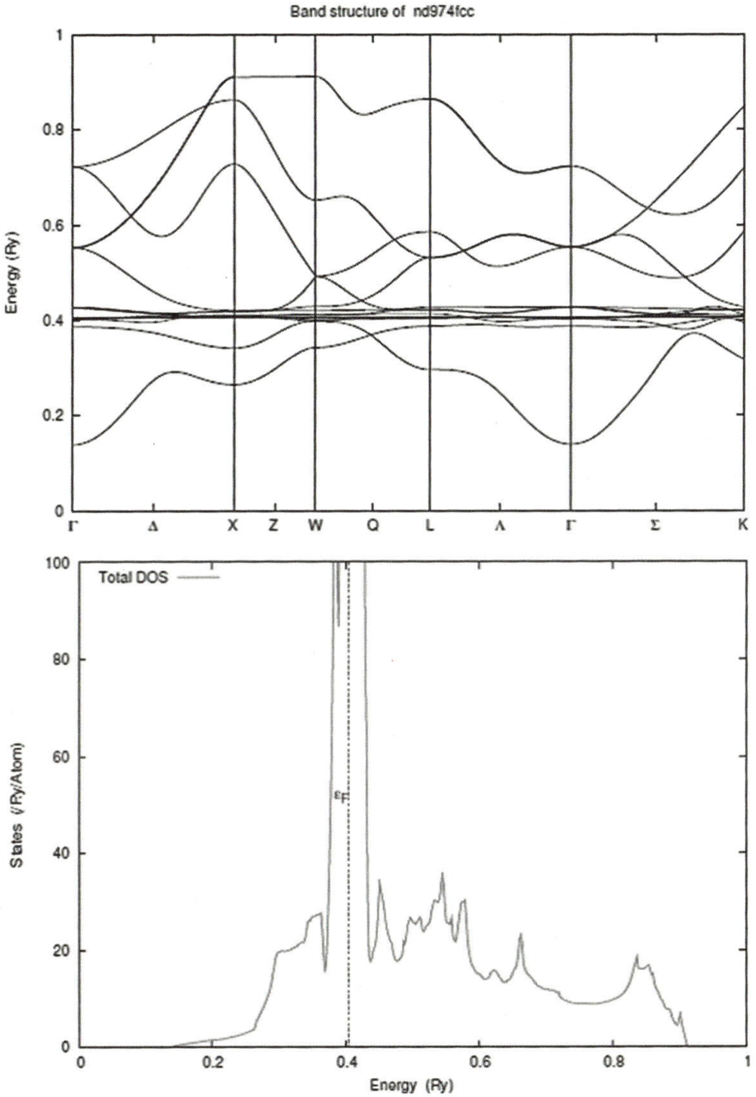

Fig. 13.8 Energy bands and density of states for Nd

Table 13.12 DOS at the Fermi level (APW results)

Energy	Total	s	p	e_g	t_{2g}	f	Velocity	Plasmon energy
Ry		States/Ry/atom					cm/s	eV
0.405	328.970	0.556	1.646	1.207	3.879	314.386	$0.038 \times 10E8$	1.539

13.5 Promethium Z = 61

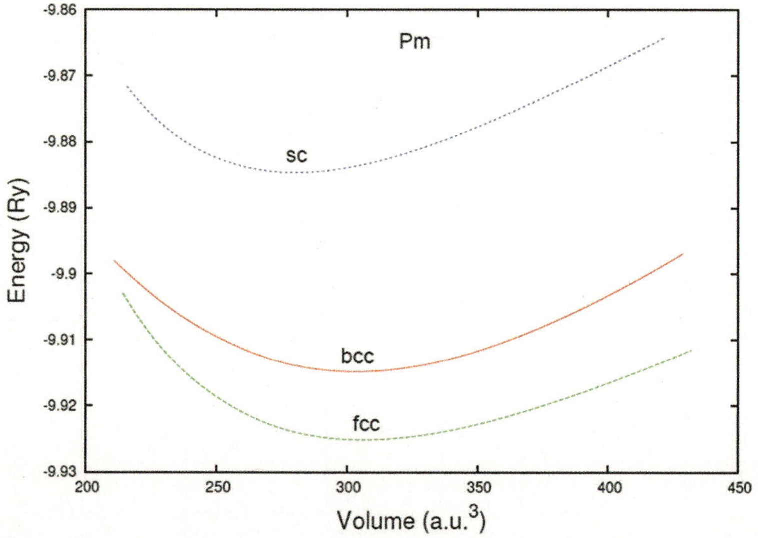

Fig. 13.9 Total energy of Pm

Table 13.13 Lattice constants and bulk modulus

	a (Bohr)	B (MBar)
sc	6.542	0.169
bcc	8.466	0.145
fcc	10.698	0.135
$\Delta E_{fcc-bcc} = 10.35$ mRy		

Table 13.14 Birch fit coefficients

	A_1	A_2	A_3	A_4
sc	−9.53262	−29.70987	608.51045	780.15461
bcc	−9.16361	−86.55913	3222.31064	−38172.58990
fcc	−9.57595	−33.19007	853.93789	−3034.58896

Promethium

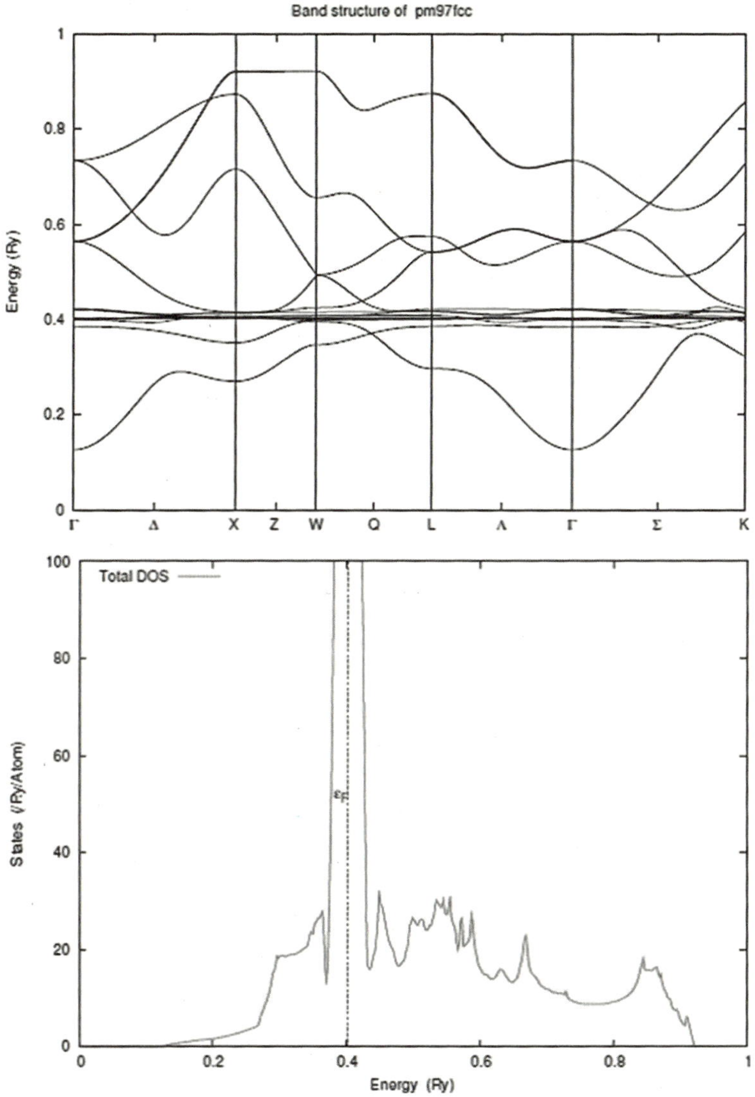

Fig. 13.10 Energy bands and density of states for Pm

Table 13.15 DOS at the Fermi level (APW results)

Energy Ry	Total	s States/Ry/atom	p	e_g	t_{2g}	f	Velocity cm/s	Plasmon energy eV
0.404	625.815	0.572	2.596	2.179	4.362	602.730	$0.025 \times 10E8$	1.462

13.6 Samarium Z = 62

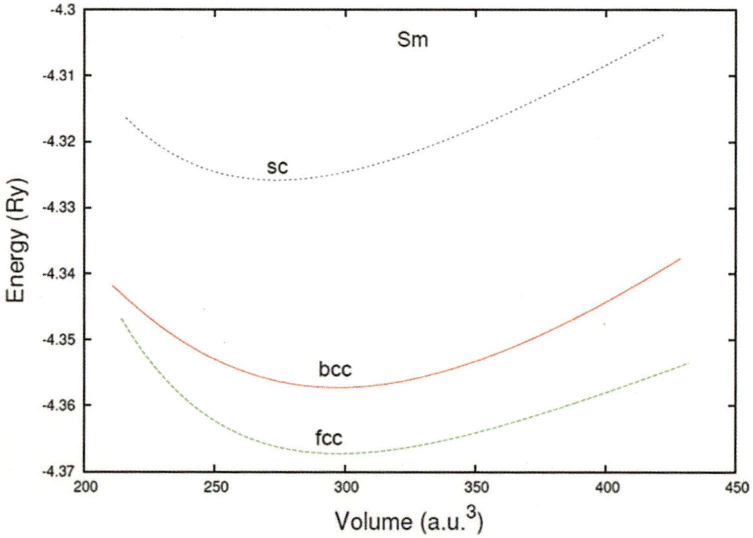

Fig. 13.11 Total energy of Sm

Table 13.16 Lattice constants and bulk modulus

	a (Bohr)	B (MBar)
sc	6.490	0.160
bcc	8.409	0.147
fcc	10.599	0.137
$\Delta E_{fcc-bcc} = 9.99$ mRy		

Table 13.17 Birch fit coefficients

	A_1	A_2	A_3	A_4
sc	−3.93361	−35.44047	898.10985	−4260.56713
bcc	−3.63343	−81.86705	2984.76283	−34488.43619
fcc	−4.17823	−11.40915	−109.66461	10817.42730

Samarium

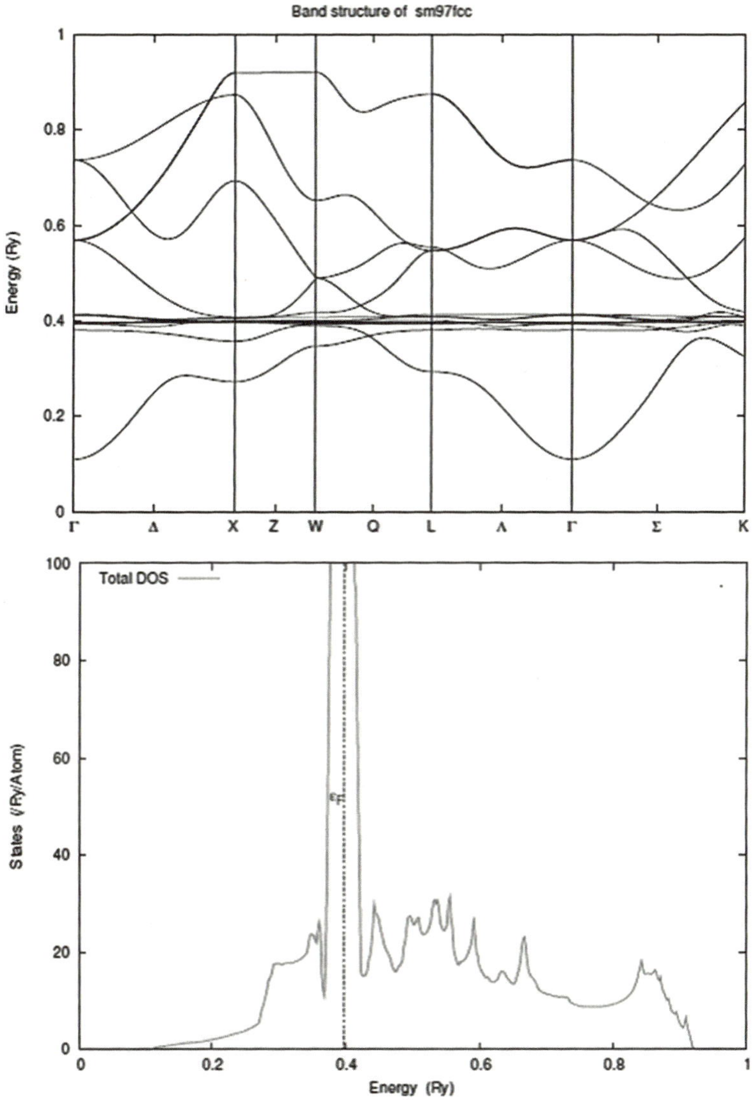

Fig. 13.12 Energy bands and density of states for Sm

Table 13.18 DOS at the Fermi level (APW results)

Energy Ry	Total	s States/Ry/atom	p	e_g	t_{2g}	f	Velocity cm/s	Plasmon energy eV
0.398	603.589	0.654	3.307	2.100	2.601	588.530	0.023 × 10E8	1.325

13.7 Europium Z = 63

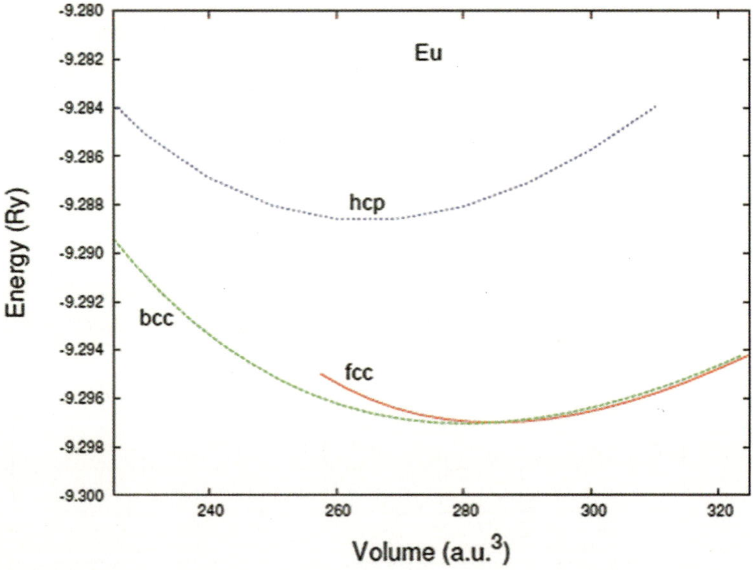

Fig. 13.13 Total energy of Eu

Table 13.19 Lattice constants and bulk modulus

	a (Bohr)	c (Bohr)	B (MBar)
hcp	5.721	9.343	0.207
fcc	10.443		0.187
bcc	8.244		0.151
$\Delta E_{bcc\text{–}fcc} = 0.045$ mRy			

Table 13.20 Birch fit coefficients

	A_1	A_2	A_3	A_4
hcp	−8.33808	−100.28241	3421.16614	−37206.13416
fcc	−9.05276	−14.09687	−152.25170	13195.61854
bcc	−8.87517	−40.34080	1134.52381	−7734.66400

Europium

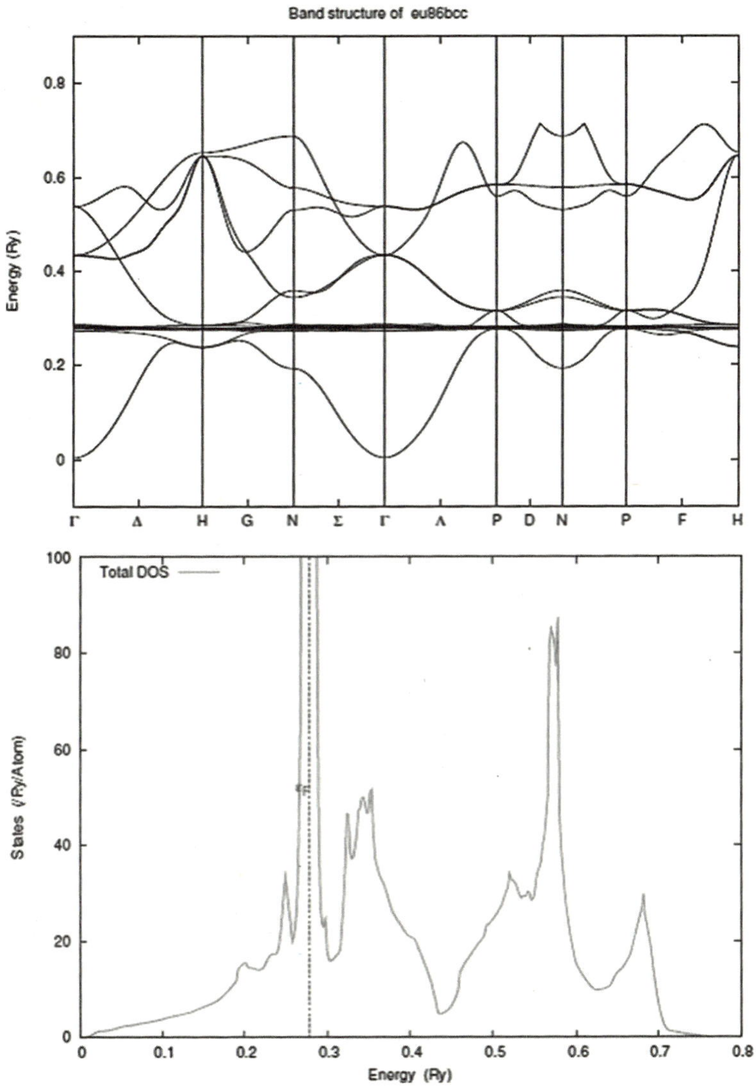

Fig. 13.14 Energy bands and density of states for Eu

Table 13.21 DOS at the Fermi level (APW results)

Energy	Total	s	p	e_g	t_{2g}	f	Velocity	Plasmon energy
Ry		States/Ry/atom					cm/s	eV
0.279	0.141	0.554	5.261	1.906	9.271	0.422	0.017 × 10E8	1.171

13.8 Gadolinium Z = 64

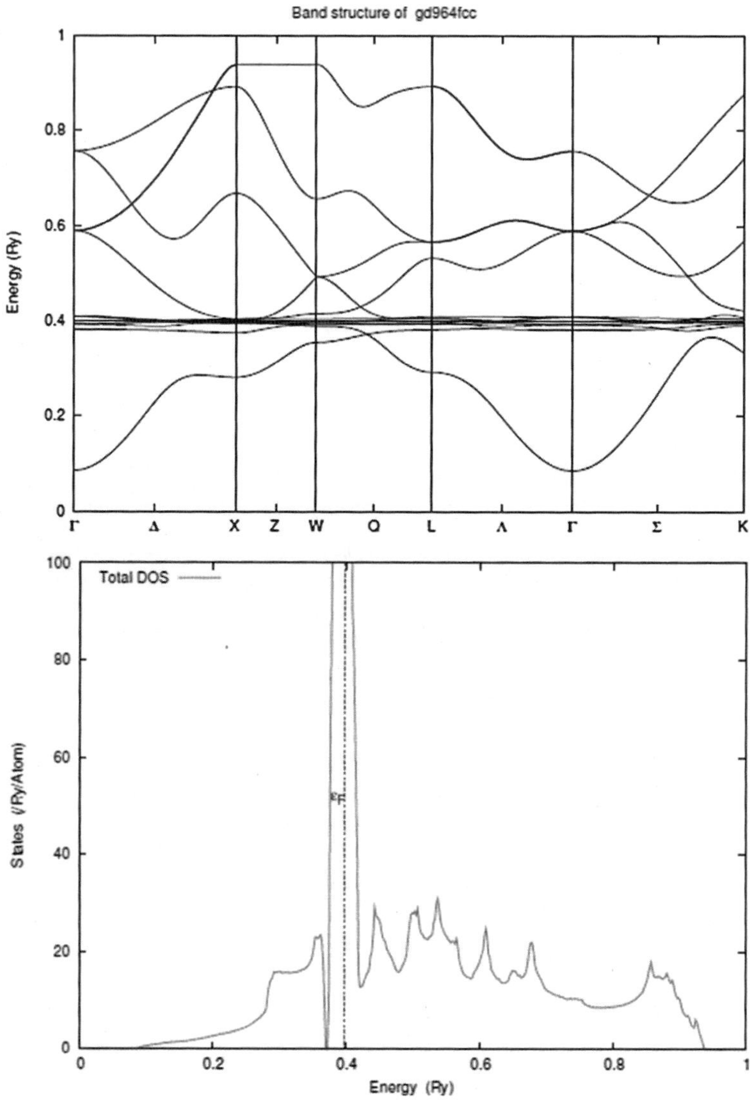

Fig. 13.15 Energy bands and density of states for Gd

Table 13.22 DOS at the Fermi level (APW results)

Energy	Total	s	p	e_g	t_{2g}	f	Velocity	Plasmon energy
Ry		States/Ry/atom					cm/s	eV
0.399	896.702	1.039	8.119	3.094	5.338	869.284	$0.021 \times 10E8$	1.406

13.9 Terbium Z = 65

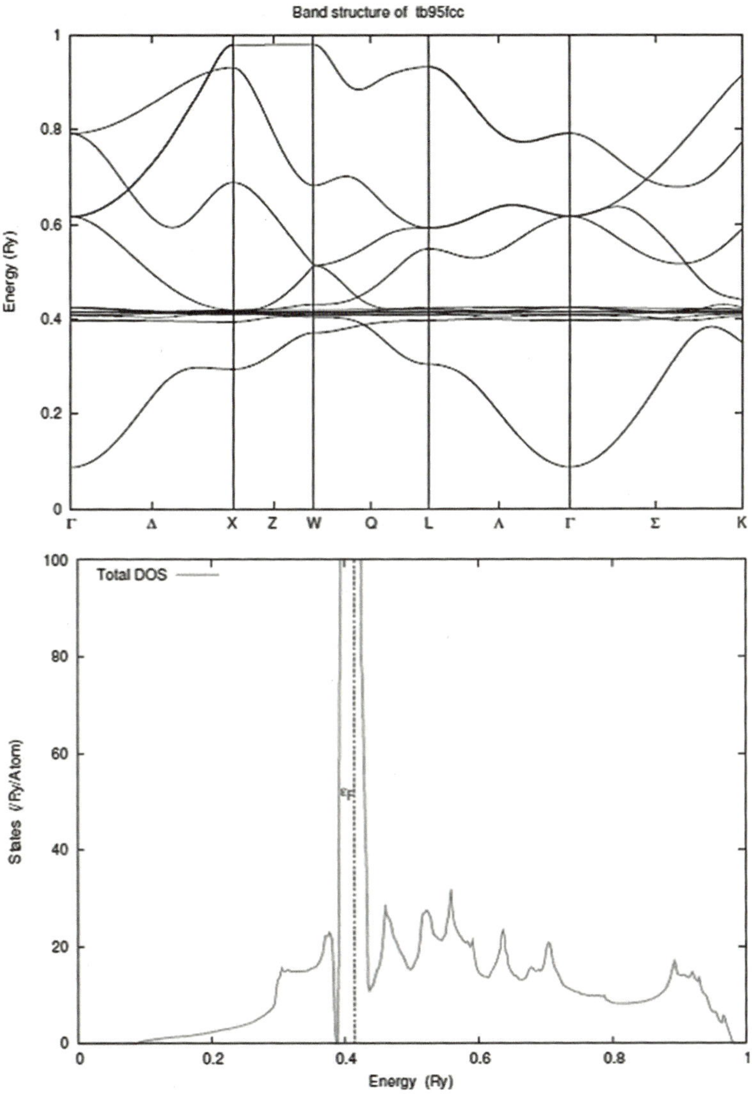

Fig. 13.16 Energy bands and density of states for Tb

Table 13.23 DOS at the Fermi level (APW results)

Energy	Total	s	p	e_g	t_{2g}	f	Velocity	Plasmon energy
Ry		States/Ry/atom					cm/s	eV
0.416	547.903	0.463	3.227	1.991	4.430	533.679	0.028 × 10E8	1.565

13.10 Dysprosium Z = 66

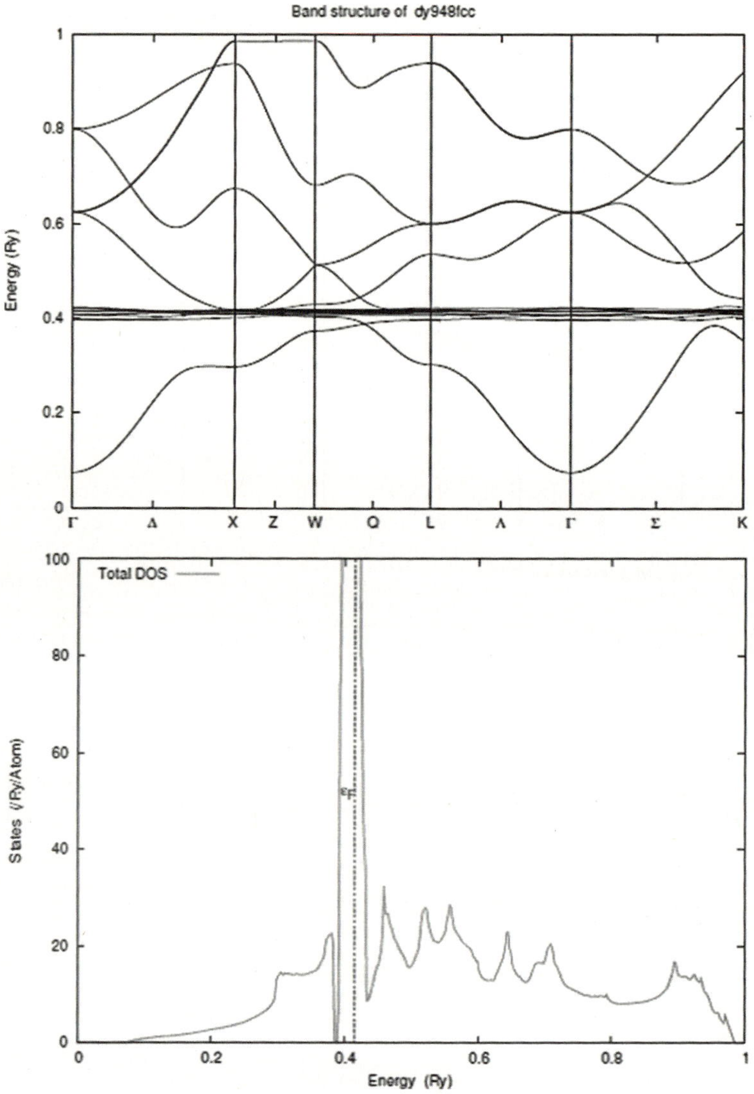

Fig. 13.17 Energy bands and density of states for Dy

Table 13.24 DOS at the Fermi level (APW results)

Energy Ry	Total	s States/Ry/atom	p	e_g	t_{2g}	f	Velocity cm/s	Plasmon energy eV
0.416	548.266	0.278	1.215	1.240	2.596	541.986	$0.029 \times 10E8$	1.695

13.11 Holmium Z = 67

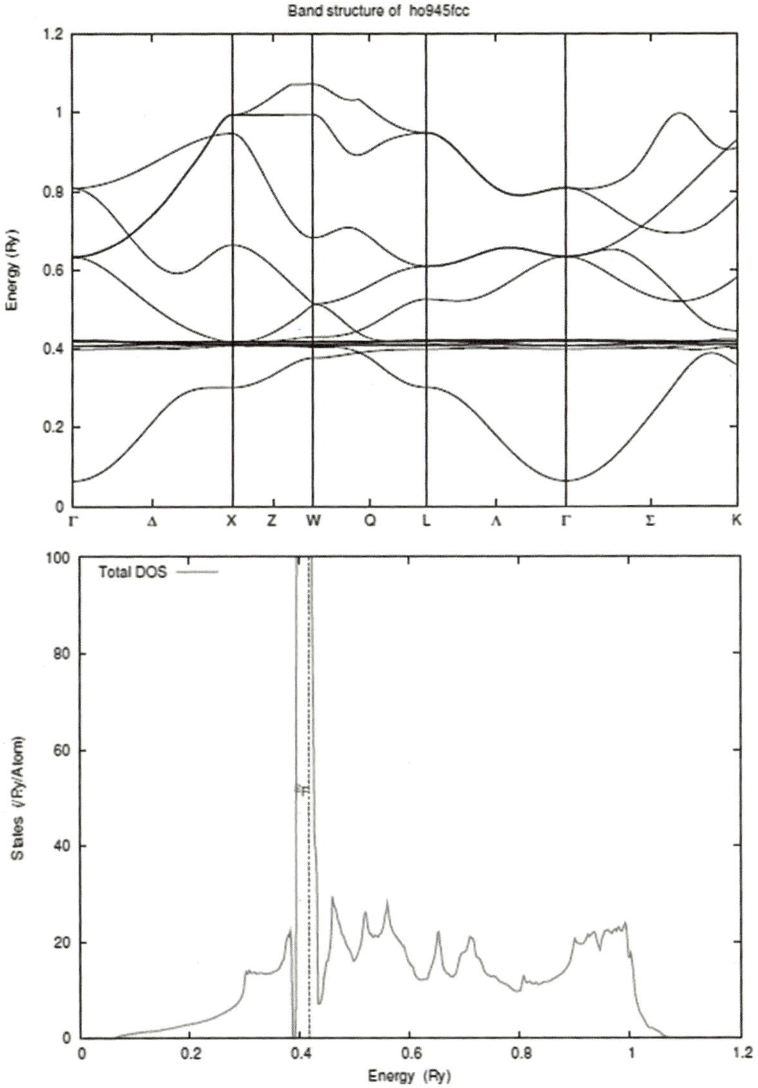

Fig. 13.18 Energy bands and density of states for Ho

Table 13.25 DOS at the Fermi level (APW results)

Energy Ry	Total	s States/Ry/atom	p	e_g	t_{2g}	f	Velocity cm/s	Plasmon energy eV
0.418	883.664	0.386	0.735	1.167	1.791	848.050	0.020 × 10E8	1.473

13.12 Erbium Z = 68

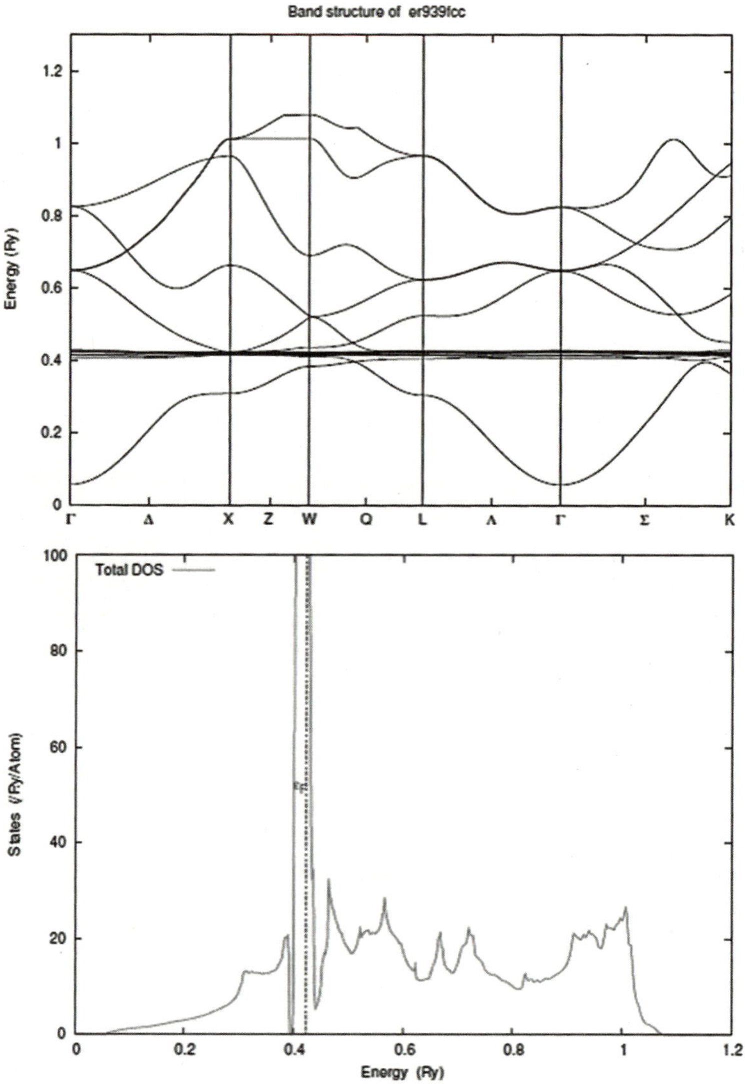

Fig. 13.19 Energy bands and density of states for Er

Table 13.26 DOS at the Fermi level (APW results)

Energy	Total	s	p	e_g	t_{2g}	f	Velocity	Plasmon energy
Ry		States/Ry/atom					cm/s	eV
0.424	482.826	0.545	0.641	1.161	1.203	475.896	0.025 × 10E8	1.345

13.13 Thulium Z = 69

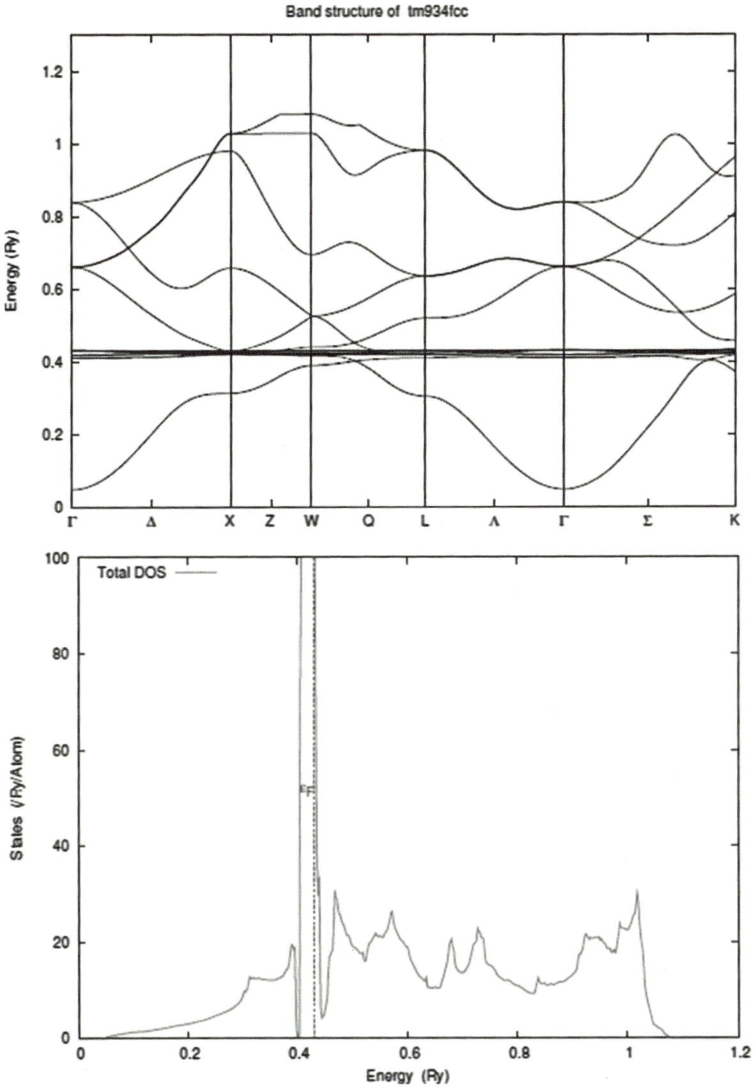

Fig. 13.20 Energy bands and density of states for Tm

Table 13.27 DOS at the Fermi level (APW results)

Energy	Total	s	p	e_g	t_{2g}	f	Velocity	Plasmon energy
Ry		States/Ry/atom					cm/s	eV
0.430	402.525	1.168	1.740	2.231	2.196	387.232	0.039 × 10E8	1.926

13.14 Ytterbium Z = 70

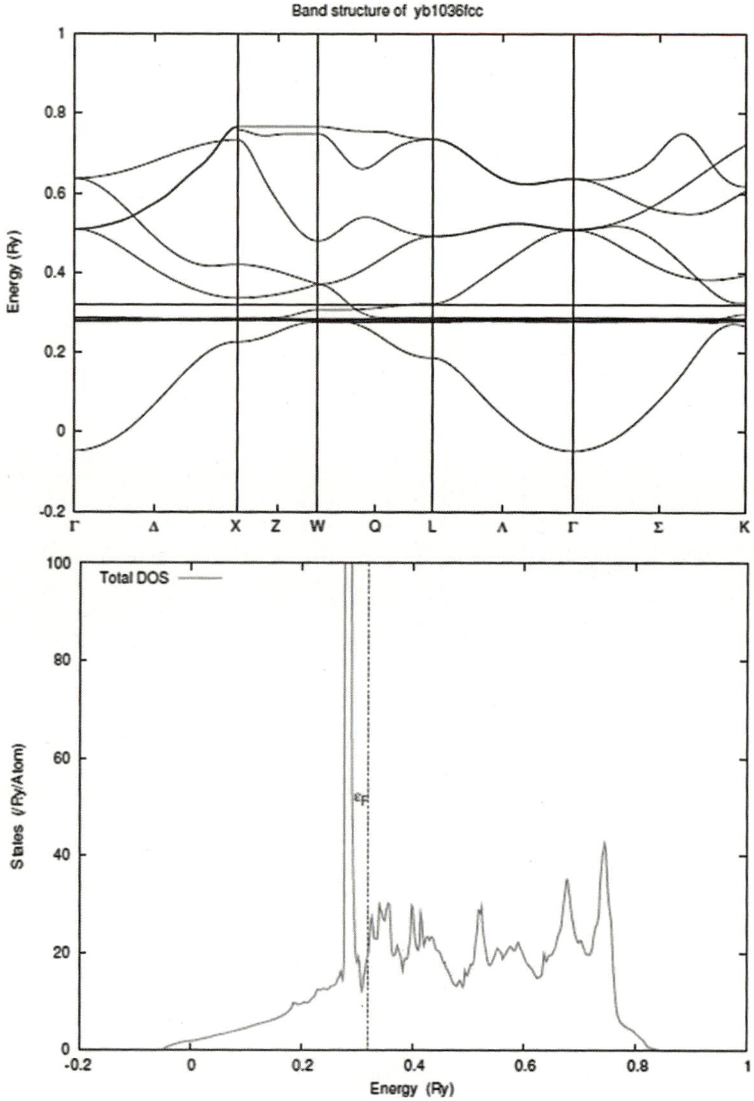

Fig. 13.21 Energy bands and density of states for Yb

Table 13.28 DOS at the Fermi level (APW results)

Energy	Total	s	p	e_g	t_{2g}	f	Velocity	Plasmon energy
Ry		States/Ry/atom					cm/s	eV
0.319	19.728	0.578	4.452	3.271	2.319	4.546	0.484 × 10E8	4.637

13.15 Lutetium Z = 71

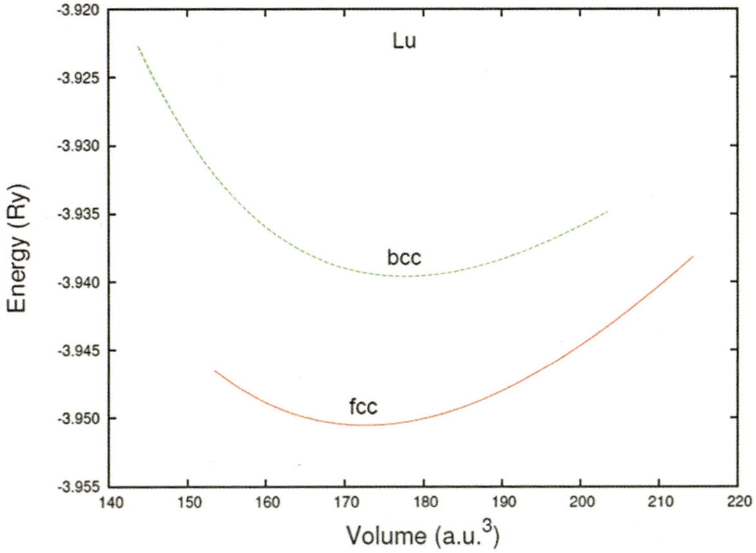

Fig. 13.22 Total energy of Lu

Table 13.29 Lattice constants and bulk modulus

	a (Bohr)	B (MBar)
bcc	7.082	0.495
fcc	8.839	0.484
$\Delta E_{fcc-bcc} = 10.91$ mRy		

Table 13.30 Birch fit coefficients

	A_1	A_2	A_3	A_4
bcc	−3.69662	−1.79756	−614.24555	13537.91948
fcc	−2.95591	−72.70133	1639.72165	−10595.72400

Lutetium

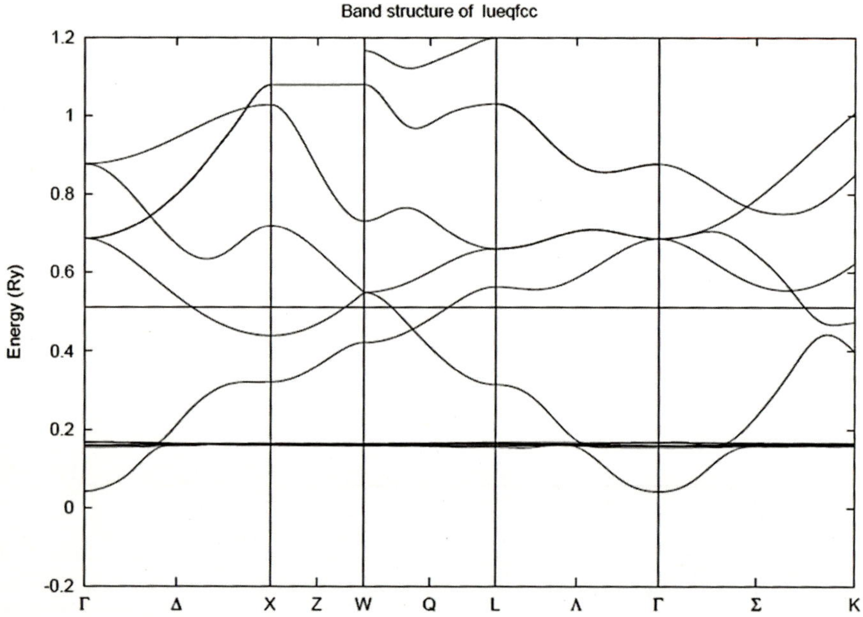

Fig. 13.23 Energy bands for Lu

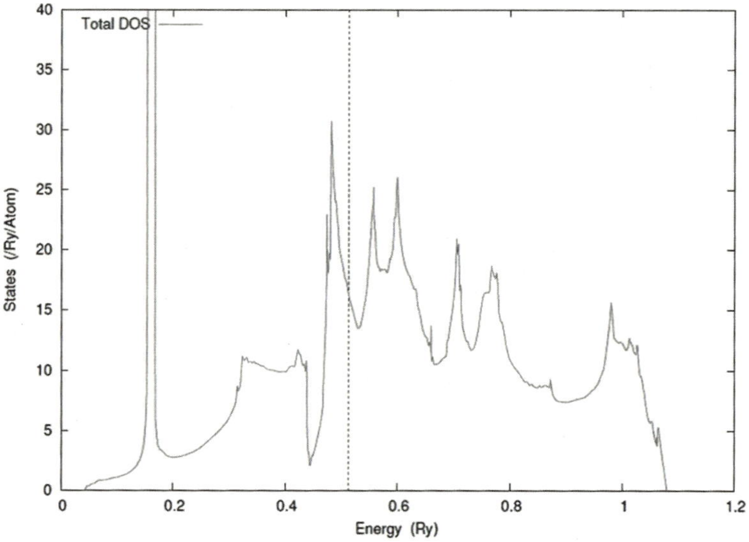

Fig. 13.24 Density of states for Lu

Table 13.31 DOS at the Fermi level (APW results)

Energy	Total	s	p	e_g	t_{2g}	f	Velocity	Plasmon energy
Ry		States/Ry/atom					cm/s	eV
0.512	16.525	0.291	2.462	2.242	7.297	0.393	0.448 × 10E8	4.988

Reference

1. L.W. Nixon, D.A. Papaconstantopoulos, Physica C. **470**, 709–715 (2010)

Chapter 14
Actinides

For Ac and Th the total energy graphs correctly predict the fcc structure as the ground state. The best agreement with experiment for the equilibrium lattice parameter corresponds to LDA calculations for Ac and GGA results for Th. From the energy bands and DOS in these two elements we show the f-bands to lie above the Fermi level (about 0.5 Ry for Ac and 0.2 Ry for Th). In addition, for Ac and Th we present TB parameters in an spd Hamiltonian by omitting the f-bands. For the rest of the elements in this row (Pa to Md) we found that the Fermi level starts entering the f-bands in Pa and gradually the f-DOS at E_F becomes very large reaching a maximum at the element Californium and the starts decreasing until the Fermi level moves above the f-bands in Nobelium. Comparing with the lanthanides we see that the f-bands are much wider in the actinides. However, as in the lanthanides standard density functional theory fails to give a total energy minimum in the elements Pa to Md and therefore we assumed the fcc structure and we present only the energy bands and DOS. For No and Lr for which the f-bands are away from E_F and do not hybridize with d-bands, it was possible to do an energy minimization for the fcc and bcc lattices.

D.A. Papaconstantopoulos, *Handbook of the Band Structure of Elemental Solids*,
DOI 10.1007/978-1-4419-8264-3_14

14.1 Actinium Z = 89

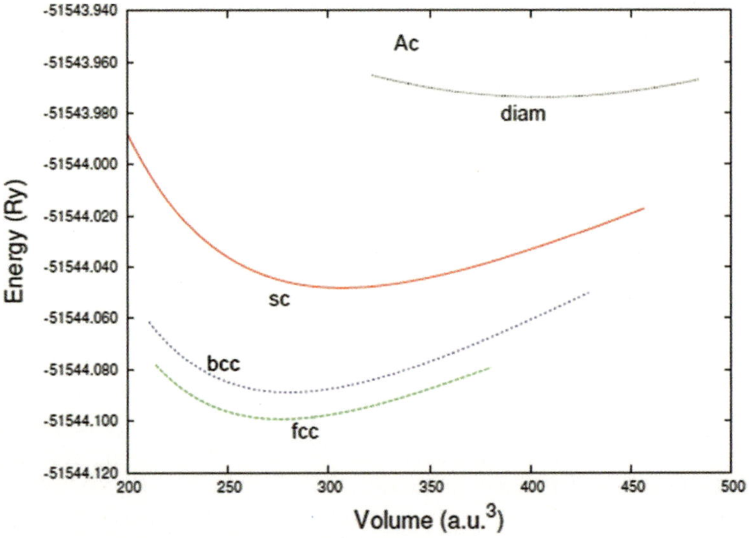

Fig. 14.1 Total energy of Ac

Table 14.1 Lattice constants and bulk modulus

	a (Bohr)	B (MBar)
diam	14.796	0.147
sc	6.746	0.241
bcc	8.247	0.296
fcc	10.345	0.276
fcc exp	10.038	
$\Delta E_{fcc-bcc} = 10.4$ mRy		

Table 14.2 Birch fit coefficients

	A_1	A_2	A_3	A_4
diam	−103085.32028	−605.93335	45791.97532	−1127715.08495
sc	−51543.44836	−56.13812	1382.99669	−3205.79857
bcc	−51543.43451	−56.93419	1274.86449	−1575.88635
fcc	−51543.67798	−28.91346	175.74508	12411.00380

Actinium

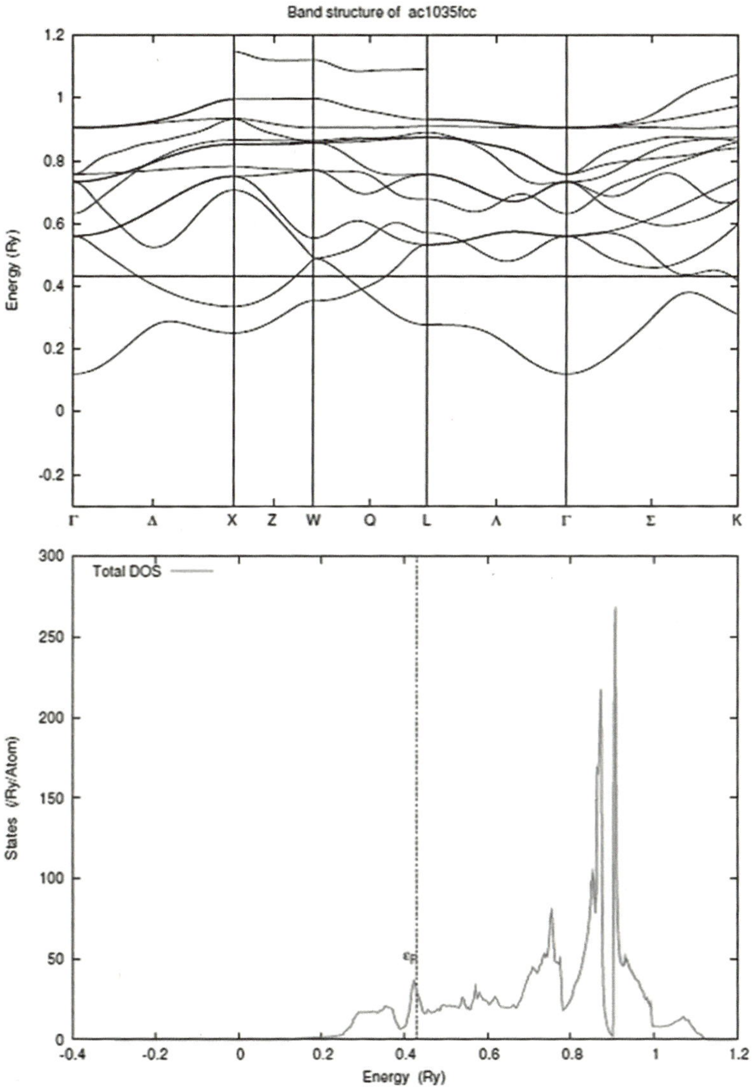

Fig. 14.2 Energy bands and density of states for Ac

Table 14.3 DOS at the Fermi level (APW results)

Energy Ry	Total	s States/Ry/atom	p	e_g	t_{2g}	f	Velocity cm/s	Plasmon energy eV
0.431	29.458	1.259	2.872	5.405	10.092	2.052	$0.407 \times 10E8$	4.773

Table 14.4 Actinium fcc z = 89 lattice constant = 10.03630 a.u.

Slater-Koster 3-center parameters	Orthogonal	Non-orthogonal	
	Energy integrals (Ry)	Energy integrals (Ry)	Overlap integrals
On site			
s,s(000)	0.68668	0.52913	
x,x(000)	0.77321	0.79496	
xy,xy(000)	0.64938	0.67143	
d2,d2(000)	0.66388	0.59553	
First neighbor			
s,s(110)	−0.04780	−0.04016	−0.01405
s,x(110)	−0.00448	−0.01469	−0.04160
s,xy(110)	−0.03237	0.01270	0.08680
s,d2(110)	0.00736	0.03945	0.02680
x,x(110)	0.00081	−0.15352	−0.19395
x,x(011)	−0.00544	0.02328	0.03962
x,y(110)	0.01976	−0.02763	−0.04338
x,xy(110)	0.00279	−0.02258	0.01157
x,xy(011)	−0.01650	−0.03144	−0.05755
z,d2(011)	0.00473	0.01676	0.01831
z,d1(011)	0.01870	−0.00694	−0.02260
xy,xy(110)	−0.04428	−0.02202	0.05563
xy,xy(011)	0.01643	−0.01191	−0.06431
xy,xz(011)	0.02274	0.04419	0.04656
xy,d2(110)	−0.02094	0.02648	0.01275
d2,d2(110)	−0.01562	−0.03483	−0.01105
dl,dl(110)	0.05012	−0.01625	−0.10770
Second neighbor			
s,s(200)	0.00704	0.00944	−0.03401
s,x(200)	0.01398	0.02851	0.05356
s,d2(002)	−0.04827	−0.01702	−0.03829
x,x(200)	0.03563	0.28654	0.34588
y,y(200)	−0.01157	0.02664	0.03222
x,xy(020)	−0.00956	0.02607	0.02615
z,d2(002)	−0.03882	−0.10459	−0.14355
xy,xy(200)	−0.00044	−0.00413	0.01234
xy,xy(002)	0.00907	0.02158	0.04229
d2,d2(002)	−0.00403	−0.03396	−0.02115
dl,dl(002)	−0.00731	0.03057	0.03535

Table 14.5 Actinium fcc

	Orthogonal			Non-orthogonal		
Band	RMS error mRy	Maximum k	Deviation mRy	RMS error mRy	Maximum k	Deviation mRy
1	13.8	(044)	38.3	4.0	(008)	8.0
2	16.9	(044)	45.3	7.0	(111)	15.2
3	20.4	(003)	45.4	4.1	(000)	9.3
4	12.0	(354)	46.1	7.8	(118)	23.6
5	28.1	(048)	89.0	5.0	(118)	13.4
6	29.0	(003)	75.2	27.1	(033)	128.6
1–6	21.1			12.3		

Table 14.6 Energy values in Ry at selected k-points

	Orthogonal	APW	Non-orthogonal
GAMMA 1	0.15533	0.17042	0.16555
GAMMA 12	0.83688	0.84322	0.84463
GAMMA 15	0.78291	0.79473	0.79416
GAMMA 25′	0.62002	0.62058	0.61132
X1 (008)	0.29237	0.28271	0.29072
X3 (008)	0.35717	0.37251	0.37697
X4′ (008)	0.77001	0.79370	0.79187
X5′ (008)	0.81995	0.82470	0.83114
L1 (444)	0.30354	0.32228	0.32218
L3 (444)	0.60755	0.58886	0.58807
L2′ (444)	0.59012	0.62895	0.63902
L3′ (444)	0.82726	0.82371	0.82353
W1 (048)	0.61261	0.61256	0.61581
W1 (048)	0.98777	1.00089	0.84571
W3 (048)	0.57908	0.54953	0.55162
W3 (048)	0.86396	0.84239	0.84606
W2′ (048)	0.38266	0.39828	0.39496
W2′ (048)	0.75338	0.93986	0.90654
EVEN(224)	0.36858	0.35387	0.35292
EVEN(224)	0.49412	0.52405	0.52671
EVEN(224)	0.60707	0.56481	0.57119
EVEN(224)	0.75798	0.76679	0.76337
ODD (224)	0.65375	0.78467	0.64450
ODD (224)	0.81855	0.93569	0.82536

Table 14.7 Fermi level quantities (non-orthogonal fit)

Energy		Densities of states				Velocity	Plasmon energy
	Total	s	p	t_{2g}	e_g		
Ry		States/Ry/atom				cm/s	eV
0.4870	21.43	1.74	2.68	10.08	6.93	0.63 × 10E8	6.82

Integrated densities of states				
Total	s	p	t_{2g}	e_g
Electrons				
3.00	0.88	0.16	0.94	1.03

Table 14.8 Actinium fcc

Slater-Koster 2-center parameters	Orthogonal	Non-orthogonal	
	Energy integrals (Ry)	Energy integrals (Ry)	Overlap integrals
On site			
s	0.65094	0.76266	
P	0.79283	0.81246	
dl	0.64460	0.64393	
d2	0.63416	0.64149	
First neighbor			
(sss)	−0.03562	−0.09063	−0.04542
(pps)	0.01142	−0.10030	−0.12858
(ppp)	−0.00707	−0.09834	−0.11548
(dds)	−0.06226	−0.06049	0.00673
(ddp)	0.04661	0.02712	−0.02645
(ddd)	−0.00935	−0.02046	−0.02318
(sps)	−0.02939	−0.01897	−0.07884
(sds)	0.03046	0.04629	0.17677
(pds)	−0.02498	0.02387	0.07473
(pdp)	0.01075	0.03887	0.03372
Second neighbor			
(sss)	−0.00719	0.07009	0.01736
(pps)	0.01876	0.20491	0.24242
(ppp)	−0.00481	0.06066	0.07541
(dds)	0.00065	−0.05691	−0.06990
(ddp)	0.00571	0.00643	0.00467
(ddd)	0.00073	0.00654	0.00691
(sps)	−0.01506	0.04591	0.16657
(sds)	0.02847	−0.10089	−0.21939
(pds)	−0.00747	−0.12850	−0.17779
(pdp)	0.00649	0.00377	−0.00731

	Orthogonal			Non-orthogonal		
Band	RMS error mRy	Maximum k	Deviation mRy	RMS error mRy	Maximum k	Deviation mRy
1	20.1	(055)	57.0	8.1	(003)	20.1
2	26.8	(008)	48.8	9.9	(224)	22.9
3	25.6	(042)	55.4	9.1	(042)	19.5
4	27.0	(084)	80.7	10.0	(033)	20.0
5	21.7	(084)	53.3	8.8	(022)	22.7
6	25.3	(264)	53.5	8.5	(033)	24.1
1–6	24.6			9.1		

14.2 Thorium Z = 90

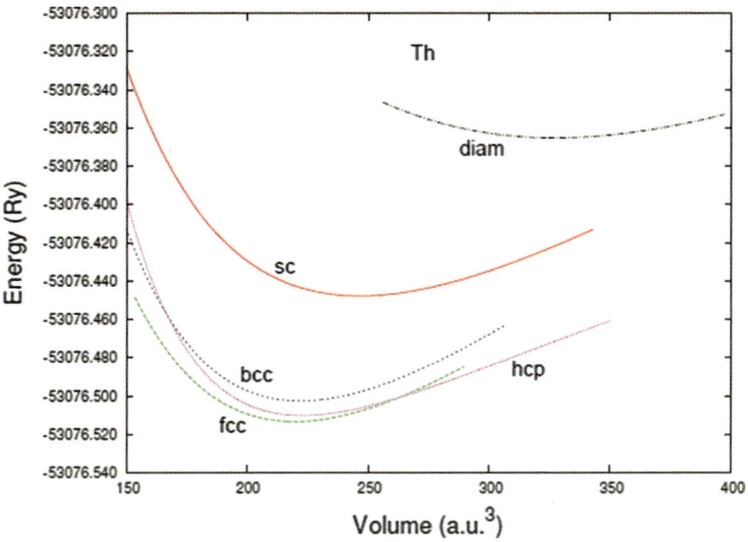

Fig. 14.3 Total energy of Th

Table 14.9 Lattice constants and bulk modulus

	a (Bohr)	c (Bohr)	B (MBar)
diam	13.778		0.293
sc	6.271		0.452
bcc	7.633		0.613
hcp	6.808	11.117	0.563
fcc	9.583		0.577
fcc exp	9.603		
$\Delta E_{fcc-hcp} = 3.2$ mRy			

Table 14.10 Birch fit coefficients

	A_1	A_2	A_3	A_4
diam	−106149.66868	−581.64565	35504.30662	−682779.98561
sc	−53075.39994	−90.10632	2225.39957	−11894.28503
bcc	−53075.32367	−91.57222	1957.73426	−6787.26687
hcp	−106152.84525	81.55758	−11314.58367	347768.54095
fcc	−53075.22037	−105.99381	2573.78909	−15606.32492

Thorium

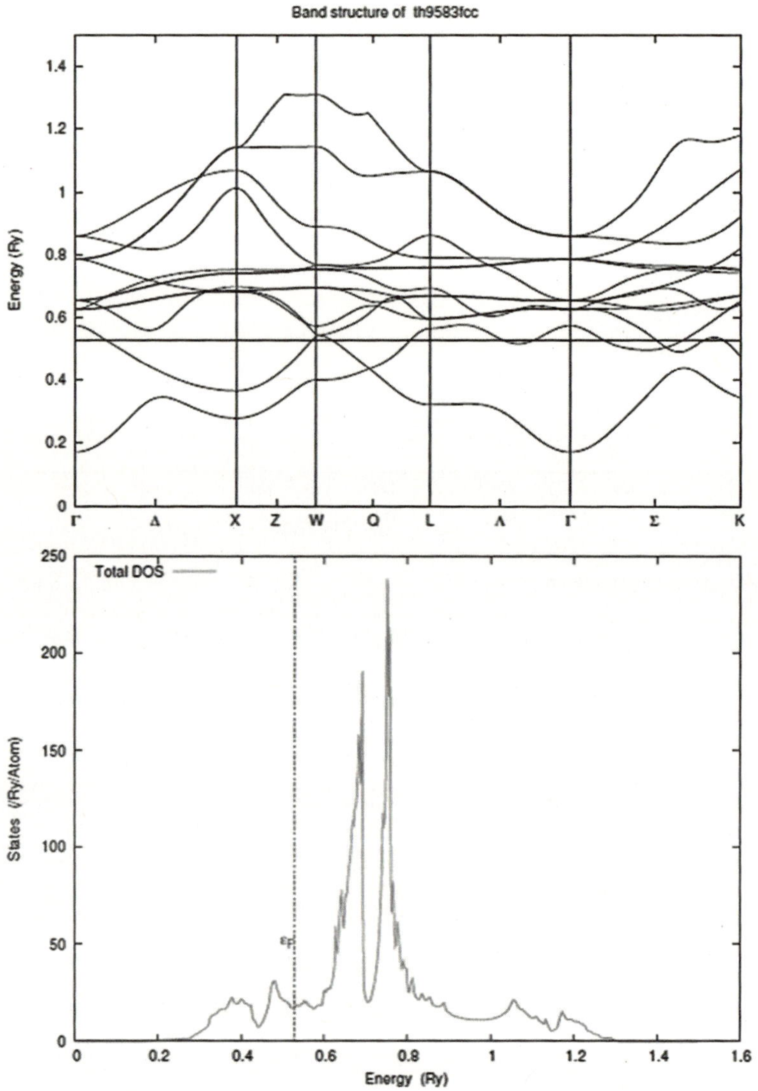

Fig. 14.4 Energy bands and density of states for Th

Table 14.11 DOS at the Fermi level (APW results)

Energy	Total	s	p	e_g	t_{2g}	f	Velocity	Plasmon energy
Ry		States/Ry/atom					cm/s	eV
0.528	16.704	0.727	1.053	2.019	4.673	4.894	0.553 × 10E8	5.490

Table 14.12 Thorium fcc Z = 90 lattice constant = 9.60790 a.u.

Slater-Koster 3-center parameters	Orthogonal	Non-orthogonal	
	Energy integrals (Ry)	Energy integrals (Ry)	Overlap integrals
On site			
s,s(000)	0.59036	0.51781	
x,x(000)	0.65967	0.78241	
xy,xy(000)	0.59626	0.59768	
d2,d2(000)	0.60042	0.59551	
First neighbor			
s,s(110)	−0.04012	−0.04269	−0.03865
s,x(110)	−0.00436	−0.00692	−0.01274
s,xy(110)	−0.02653	0.00282	0.04967
s,d2(110)	0.00456	0.04433	0.04627
x,x(110)	−0.00244	−0.17646	−0.20935
x,x(011)	−0.00130	0.05231	0.05700
x,y(110)	0.01130	−0.02692	−0.04344
x,xy(110)	−0.00328	−0.01422	−0.02063
x,xy (011)	−0.00088	−0.01538	−0.01883
z,d2(011)	0.01069	0.01793	0.01165
z,dl(011)	0.01395	0.00176	−0.01038
xy,xy(110)	−0.02912	−0.05474	−0.04246
xy,xy(011)	0.01581	−0.01329	−0.04958
xy,xz (011)	0.01580	0.04536	0.04953
xy,d2(110)	−0.00997	0.03106	0.03333
d2,d2(110)	−0.01347	−0.02678	−0.02261
dl,dl(110)	0.03705	−0.00489	−0.06111
Second neighbor			
s,s(200)	0.01016	0.01319	−0.00778
s,x(200)	0.00439	0.04360	0.06785
s,d2(002)	−0.03070	−0.01481	−0.02798
x,x(200)	0.01342	0.28294	0.35015
y,y(200)	−0.00061	0.02713	0.03135
x,xy(020)	−0.00861	0.00556	0.00524
z,d2(002)	−0.01775	−0.09460	−0.11080
xy,xy(200)	−0.00363	−0.01816	−0.02748
xy,xy(002)	0.00962	−0.00072	0.00122
d2,d2(002)	−0.01243	−0.02891	−0.01454
dl,dl(002)	−0.01106	0.00362	0.00039

Table 14.13 Thorium fcc

	Orthogonal			Non-orthogonal		
Band	RMS error mRy	Maximum k	Deviation mRy	RMS error mRy	Maximum k	Deviation mRy
1	10.3	(226)	28.2	4.6	(064)	12.6
2	20.8	(002)	42.3	11.8	(444)	35.4
3	15.8	(005)	33.5	8.9	(118)	26.0
4	12.3	(354)	31.6	9.6	(444)	26.7
1–4	15.3			9.1		

Table 14.14 Energy values in Ry at selected k-points

	Orthogonal	APW	Non-orthogonal
GAMMA 1	0.16991	0.16969	0.17303
GAMMA 15	0.65934	0.65779	0.67119
GAMMA 25′	0.61102	0.62383	0.60841
X1 (008)	0.27825	0.27761	0.27964
X3 (008)	0.35800	0.36447	0.36712
X4′ (008)	0.69836	0.69825	0.82547
X5′ (008)	0.68926	0.68942	0.78086
L1 (444)	0.31071	0.31981	0.32099
L3 (444)	0.61332	0.59292	0.60241
L2′ (444)	0.54493	0.56699	0.61961
L3′ (444)	0.68047	0.67198	0.66396
W1 (048)	0.57718	0.57254	0.57849
W3 (048)	0.55161	0.54007	0.54072
W3 (048)	0.69690	0.69684	0.80959
W2′ (048)	0.37770	0.39731	0.39062
W2′ (048)	0.68388	0.75738	0.67521
EVEN(224)	0.36733	0.35522	0.35245
EVEN(224)	0.48750	0.51322	0.51507
EVEN(224)	0.57083	0.55048	0.55481
EVEN(224)	0.66782	0.67145	0.68849
ODD (224)	0.64204	0.68131	0.62126
ODD (224)	0.67890	0.77519	0.68509

Table 14.15 Fermi level quantities (non-orthogonal fit)

Energy		Densities of states				Velocity	Plasmon energy
	Total	s	p	t_{2g}	e_g		
Ry		States/Ry/atom				cm/s	eV
0.5300	15.17	2.15	2.98	7.23	2.80	$0.65 \times 10E8$	6.33

Integrated densities of states Total Electrons	s	p	t_{2g}	e_g
4.00	0.90	0.61	1.53	0.96

Table 14.16 Thorium fcc

Slater-Koster 2-center parameters	Orthogonal	Non-orthogonal	
	Energy integrals (Ry)	Energy integrals (Ry)	Overlap integrals
On site			
s	0.53231	0.69563	
p	0.67762	0.68465	
dl	0.59737	0.61039	
d2	0.57596	0.60815	
First neighbor			
(sss)	−0.02977	−0.06809	−0.00752
(pps)	0.00109	−0.06702	−0.09772
(ppp)	−0.00451	−0.07980	−0.11259
(dds)	−0.03753	−0.12900	−0.14912
(ddp)	0.03638	0.02883	−0.00343
(ddd)	−0.00559	−0.04389	−0.06422
(sps)	−0.01138	−0.00563	−0.03903
(sds)	0.02617	0.05746	0.16846
(pds)	−0.00038	−0.04006	−0.04584
(pdp)	−0.00015	0.04321	0.05962
Second neighbor			
(sss)	−0.00012	0.03154	−0.08241
(pps)	0.00769	0.17214	0.24198
(ppp)	−0.00023	0.04401	0.06552
(dds)	−0.01861	−0.02393	−0.01433
(ddp)	−0.00368	0.01949	0.02813
(ddd)	0.00723	−0.00232	−0.00705
(sps)	−0.00130	0.01672	0.06140
(sds)	0.02207	−0.09565	−0.22269
(pds)	−0.00958	−0.09925	−0.14921
(pdp)	0.00805	0.02374	0.03235

	Orthogonal			Non-orthogonal		
Band	RMS error mRy	Maximum k	Deviation mRy	RMS error mRy	Maximum k	Deviation mRy
1	13.3	(055)	40.9	8.2	(224)	20.3
2	20.7	(111)	39.8	19.5	(444)	83.2
3	25.3	(042)	60.5	13.5	(022)	30.9
4	20.0	(084)	51.1	12.6	(354)	34.3
1–4	20.3			14.0		

14.3 Protactinium Z = 91

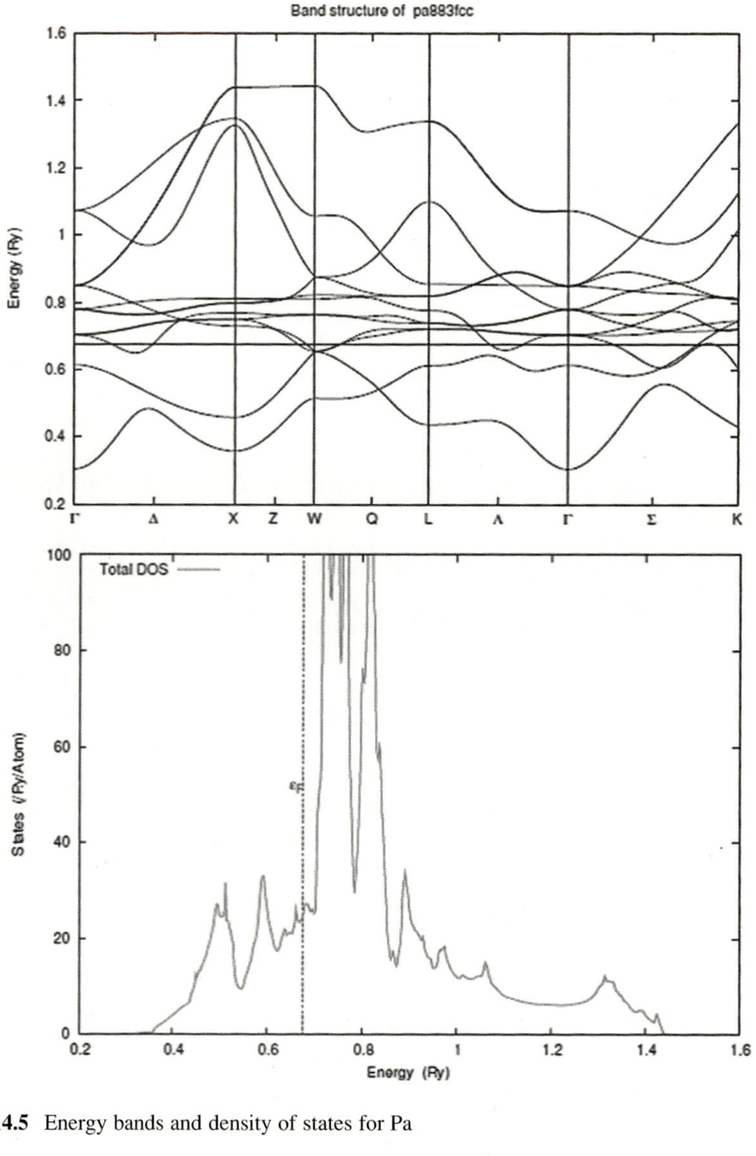

Fig. 14.5 Energy bands and density of states for Pa

Table 14.17 DOS at the Fermi level (APW results)

Energy Ry	Total	s States/Ry/atom	p	e_g	t_{2g}	f	Velocity cm/s	Plasmon energy eV
0.677	25.026	0.842	0.815	0.593	3.002	16.978	0.451 × 10E8	6.194

14.4 Uranium Z = 92

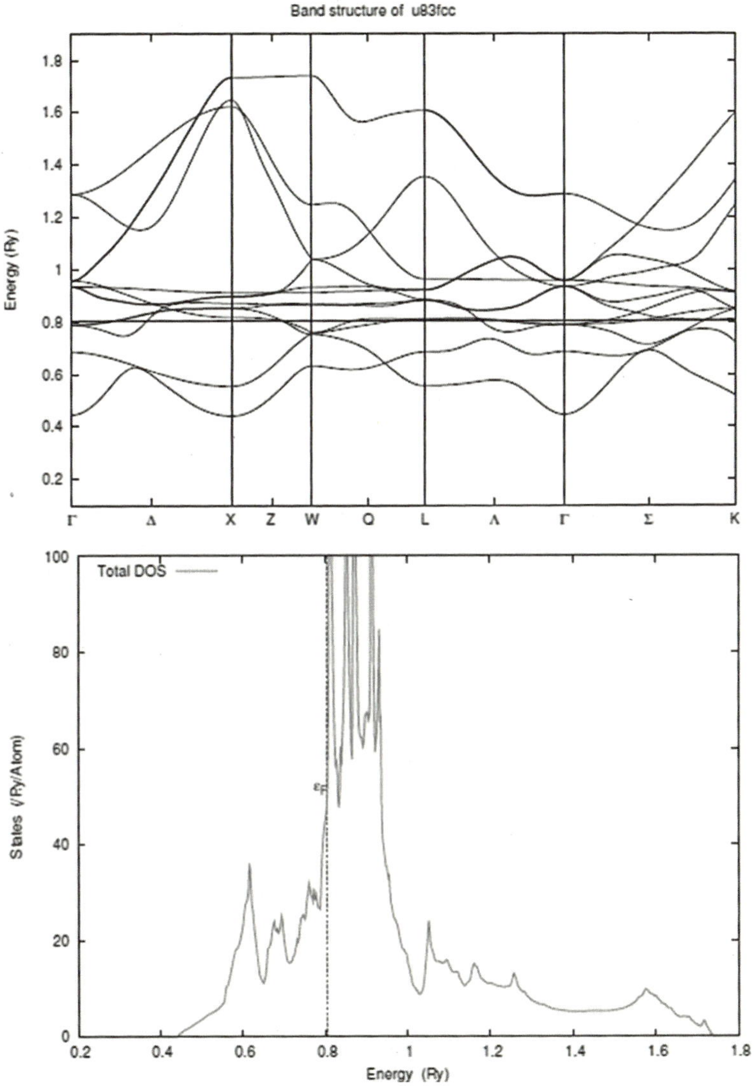

Fig. 14.6 Energy bands and density of states for U

Table 14.18 DOS at the Fermi level (APW results)

Energy Ry	Total	s States/Ry/atom	p	e_g	t_{2g}	f	Velocity cm/s	Plasmon energy eV
0.806	51.376	0.366	0.662	1.572	1.466	44.126	0.235 × 10E8	5.073

14.5 Neptunium Z = 93

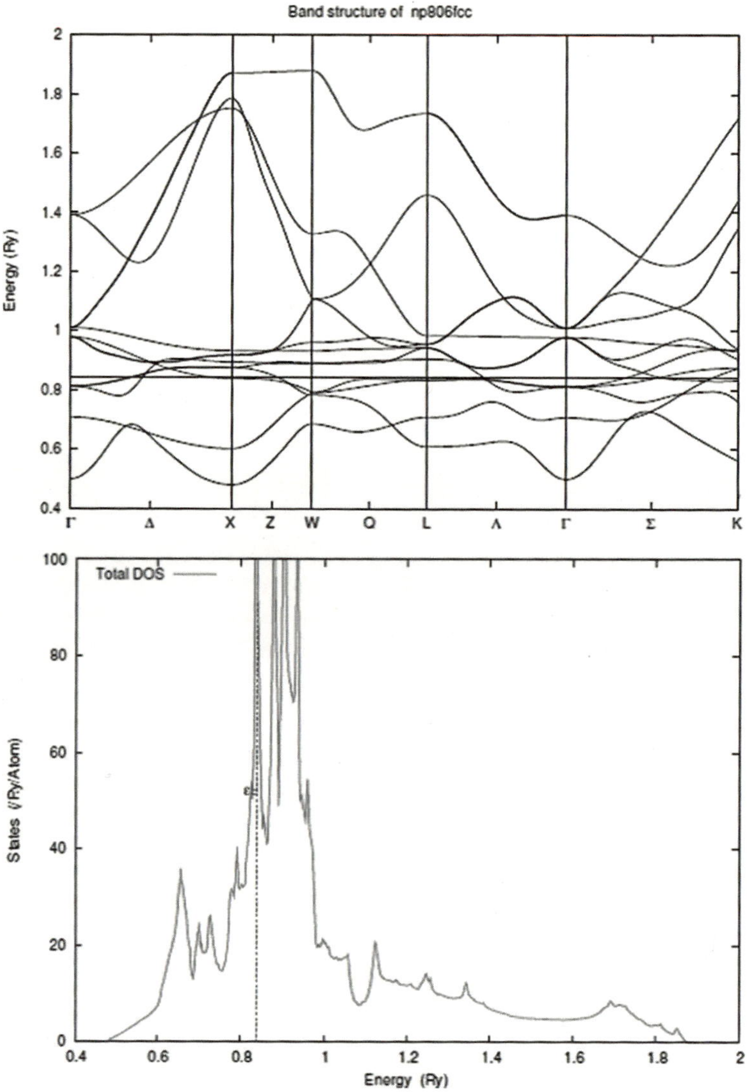

Fig. 14.7 Energy bands and density of states for Np

Table 14.19 DOS at the Fermi level (APW results)

Energy	Total	s	p	e_g	t_{2g}	f	Velocity	Plasmon energy
Ry		States/Ry/atom					cm/s	eV
0.841	91.971	0.303	0.762	2.071	1.923	82.779	0.182 × 10E8	5.490

14.6 Plutonium Z = 94

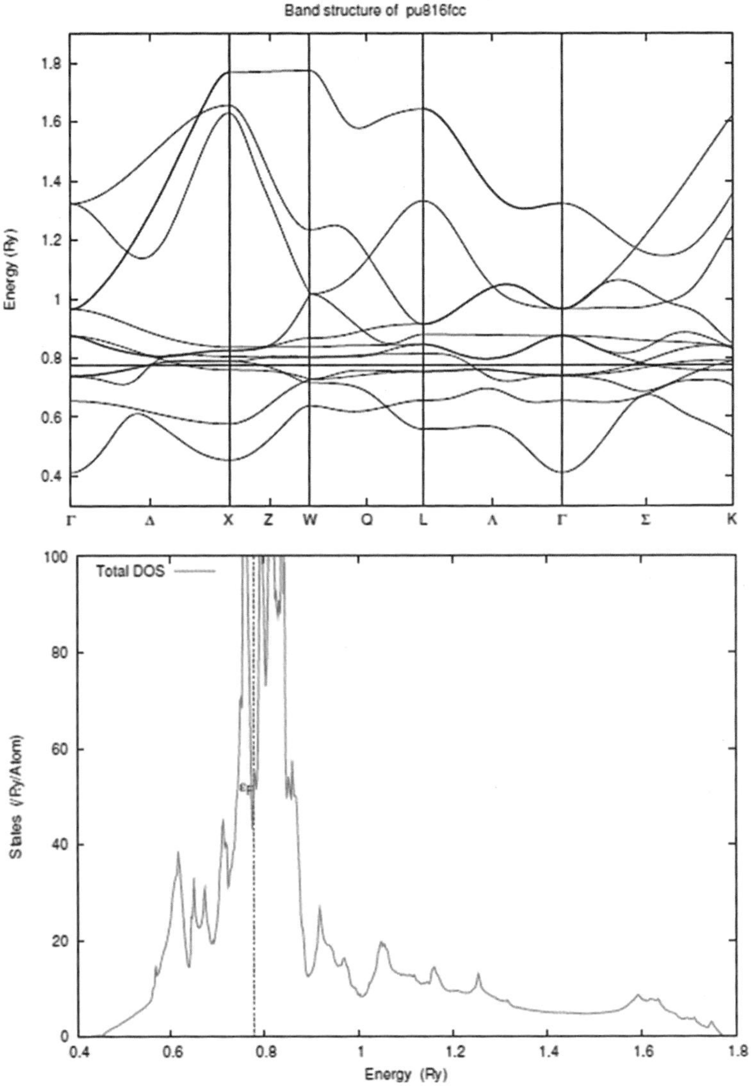

Fig. 14.8 Energy bands and density of states for Pu

Table 14.20 DOS at the Fermi level (APW results)

Energy	Total	s	p	e_g	t_{2g}	f	Velocity	Plasmon energy
Ry		States/Ry/atom					cm/s	eV
0.778	48.096	0.281	0.560	0.852	1.087	42.570	0.189 × 10E8	4.034

14.7 Americium Z = 95

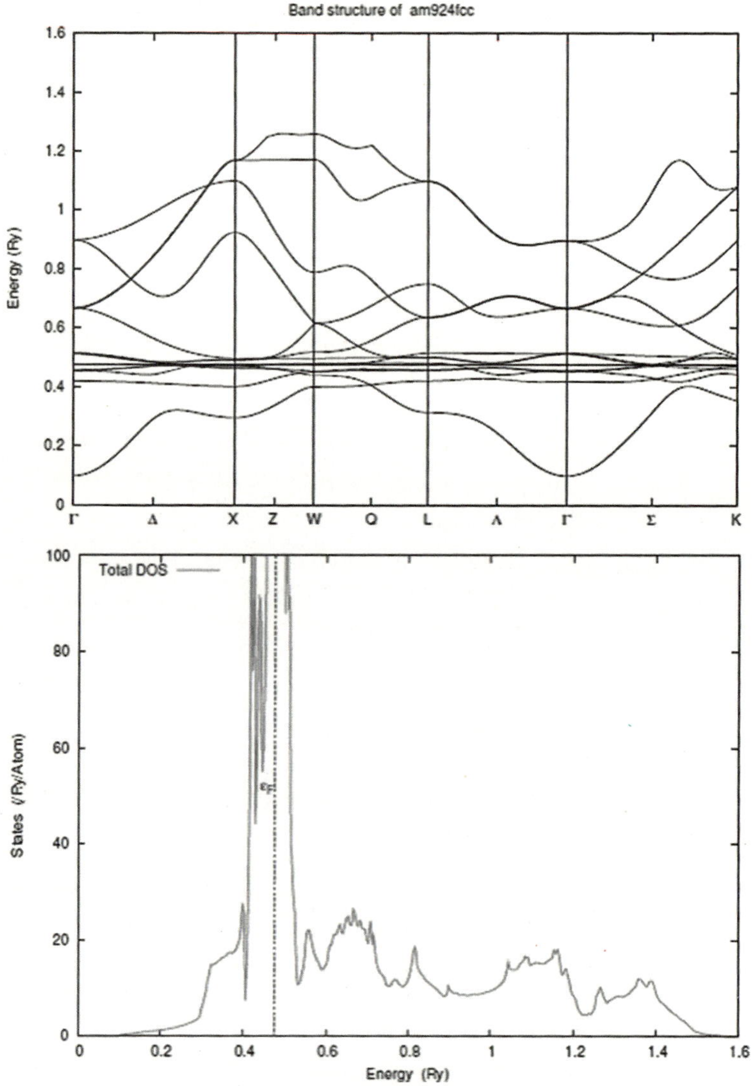

Fig. 14.9 Energy bands and density of states for Am

Table 14.21 DOS at the Fermi level (APW results)

Energy	Total	s	p	e_g	t_{2g}	f	Velocity	Plasmon energy
Ry		States/Ry/atom					cm/s	eV
0.475	216.703	0.634	3.101	1.703	2.778	203.384	0.064 × 10E8	2.399

14.8 Curium Z = 96

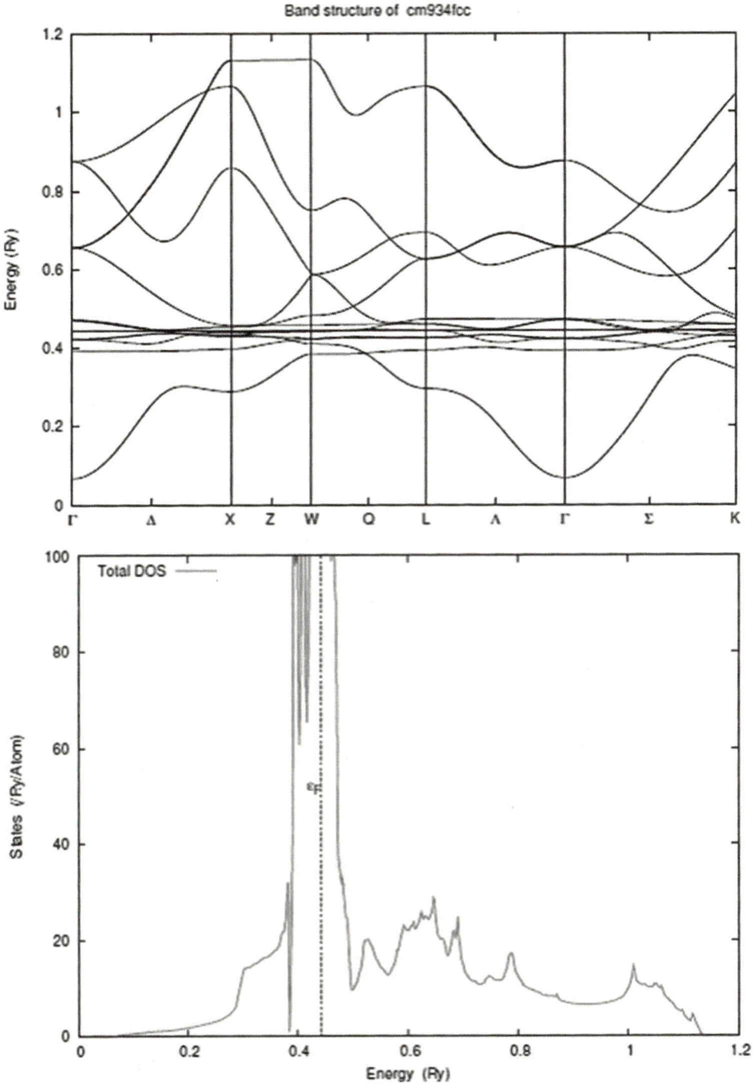

Fig. 14.10 Energy bands and density of states for Cm

Table 14.22 DOS at the Fermi level (APW results)

Energy	Total	s	p	e_g	t_{2g}	f	Velocity	Plasmon energy
Ry		States/Ry/atom					cm/s	eV
0.443	238.865	1.238	3.838	1.857	2.413	223.695	0.063 × 10E8	2.446

14.9 Berkelium Z = 97

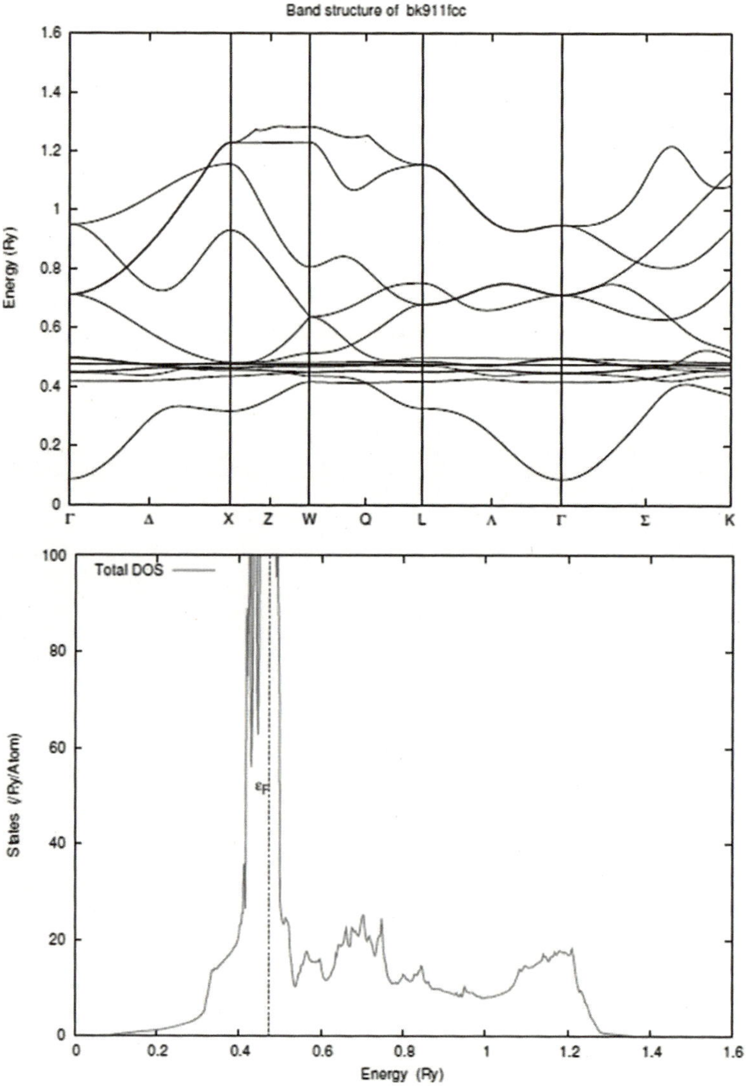

Fig. 14.11 Energy bands and density of states for Bk

Table 14.23 DOS at the Fermi level (APW results)

Energy	Total	s	p	e_g	t_{2g}	f	Velocity	Plasmon energy
Ry		States/Ry/atom					cm/s	eV
0.474	320.496	1.129	3.497	2.278	3.780	303.489	0.070 × 10E8	3.284

14.10 Californium Z = 98

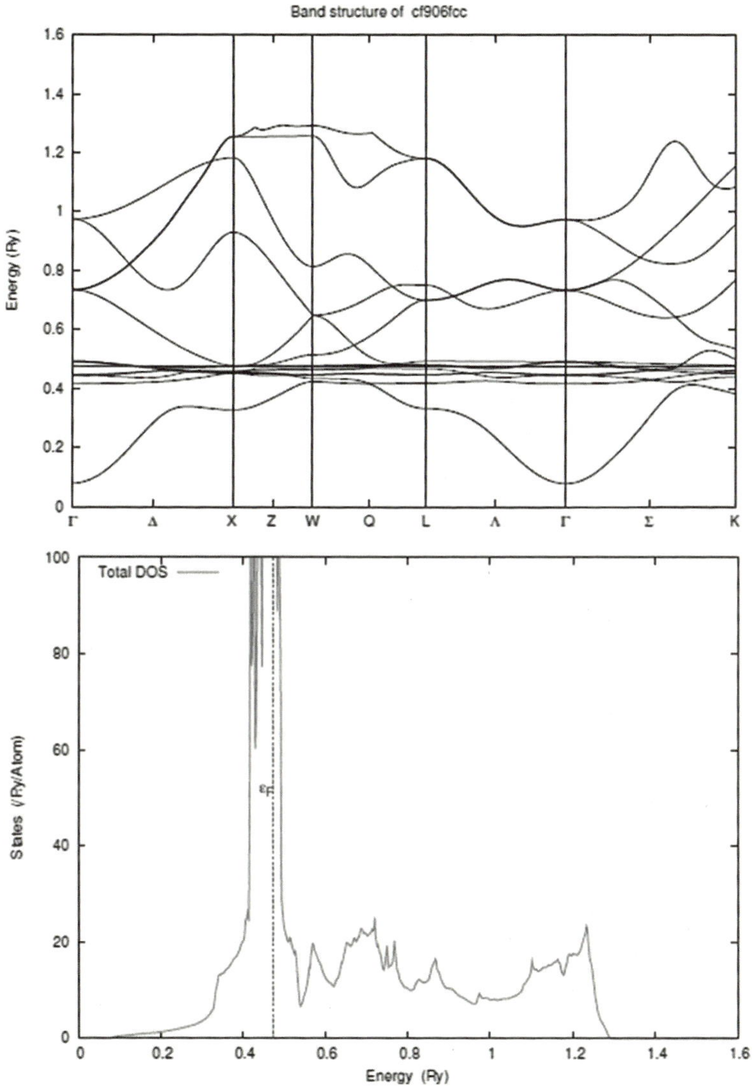

Fig. 14.12 Energy bands and density of states for Cf

Table 14.24 DOS at the Fermi level (APW results)

Energy Ry	Total	s States/Ry/atom	p	e_g	t_{2g}	f	Velocity cm/s	Plasmon energy eV
0.474	242.636	0.882	1.417	1.262	2.394	234.130	0.076 × 10E8	3.146

14.11 Einsteinium Z = 99

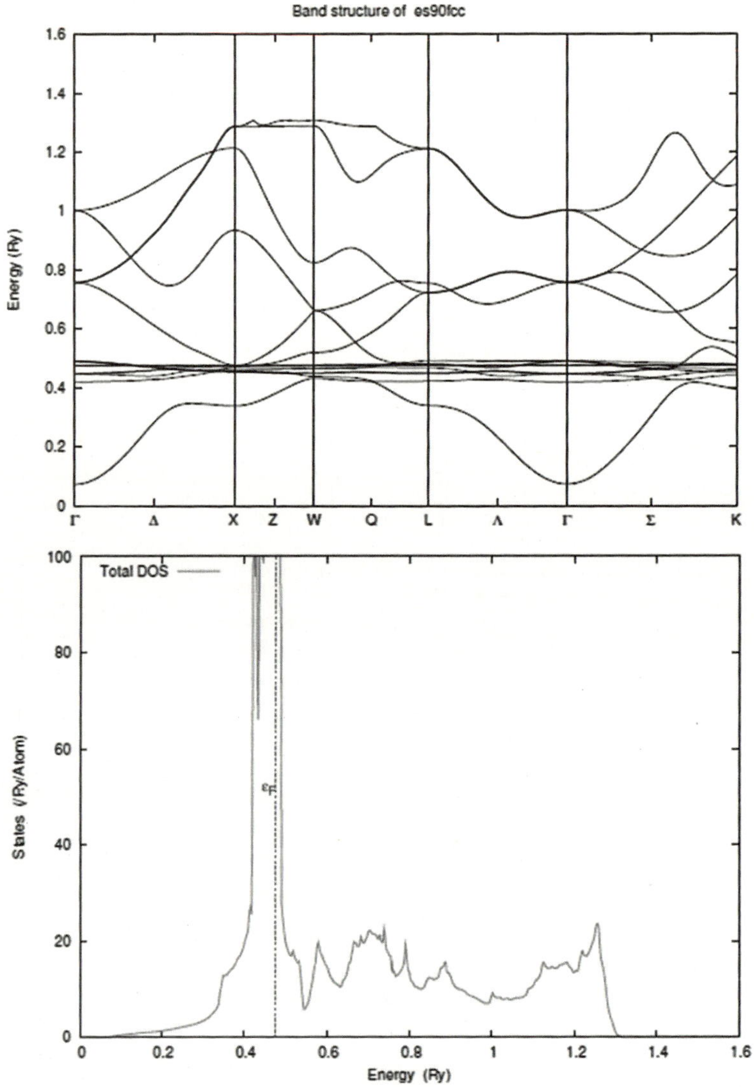

Fig. 14.13 Energy bands and density of states for Es

Table 14.25 DOS at the Fermi level (APW results)

Energy Ry	Total	s States/Ry/atom	p	e_g	t_{2g}	f	Velocity cm/s	Plasmon energy eV
0.477	350.699	1.130	1.049	0.989	1.465	325.858	0.057 × 10E8	2.717

14.12 Fermium Z = 100

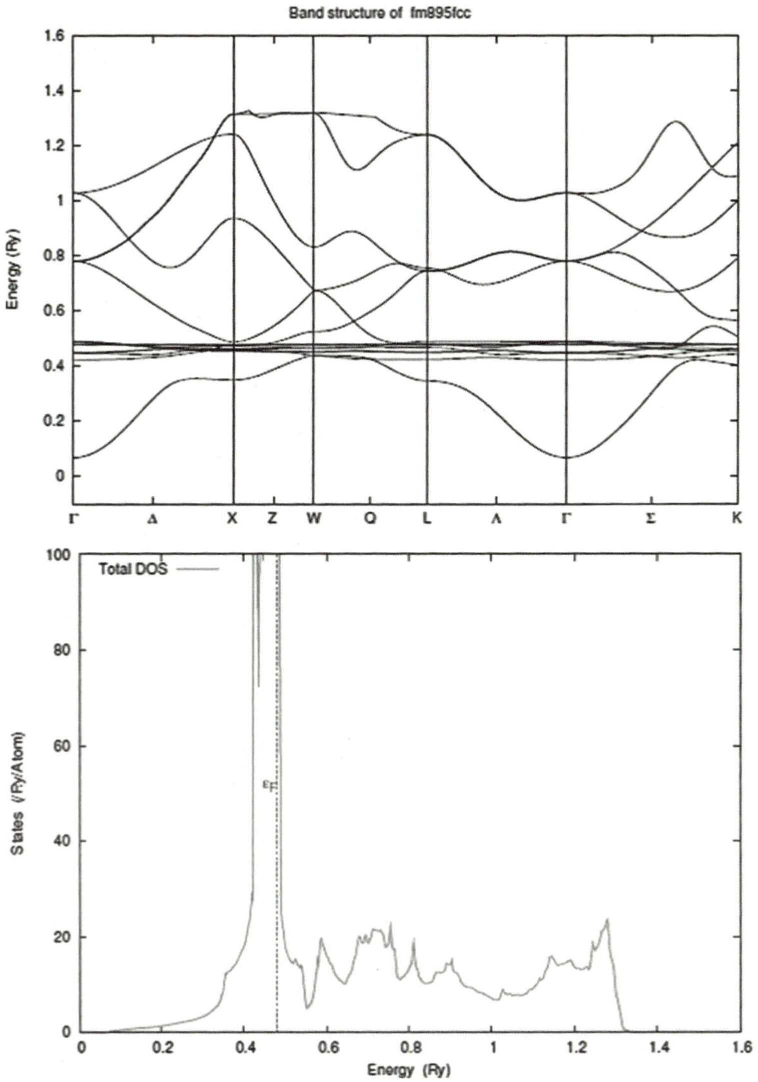

Fig. 14.14 Energy bands and density of states for Fm

Table 14.26 DOS at the Fermi level (APW results)

Energy	Total	s	p	e_g	t_{2g}	f	Velocity	Plasmon energy
Ry		States/Ry/atom					cm/s	eV
0.479	237.973	1.476	1.089	1.011	1.136	230.741	0.065 × 10E8	2.529

14.13 Mendelevium Z = 101

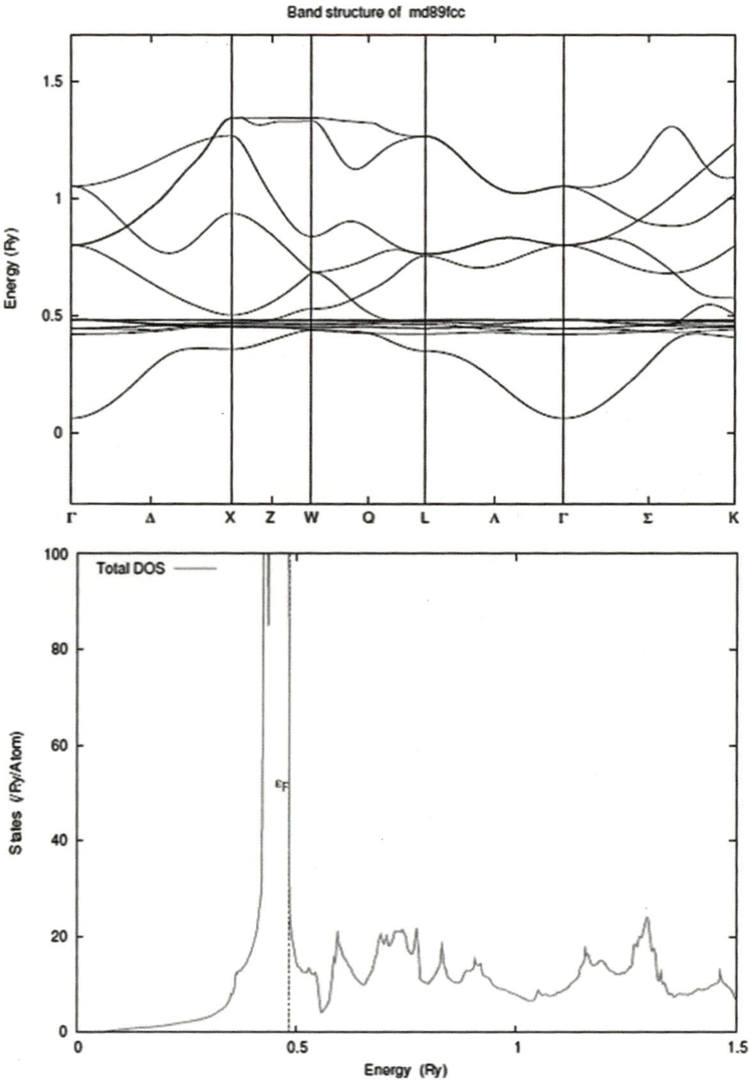

Fig. 14.15 Energy bands and density of states for Md

Table 14.27 DOS at the Fermi level (APW results)

Energy	Total	s	p	e_g	t_{2g}	f	Velocity	Plasmon energy
Ry		States/Ry/atom					cm/s	eV
0.483	86.562	1.187	0.609	1.256	1.128	80.238	0.194 × 10E8	4.851

14.14 Nobelium Z = 102

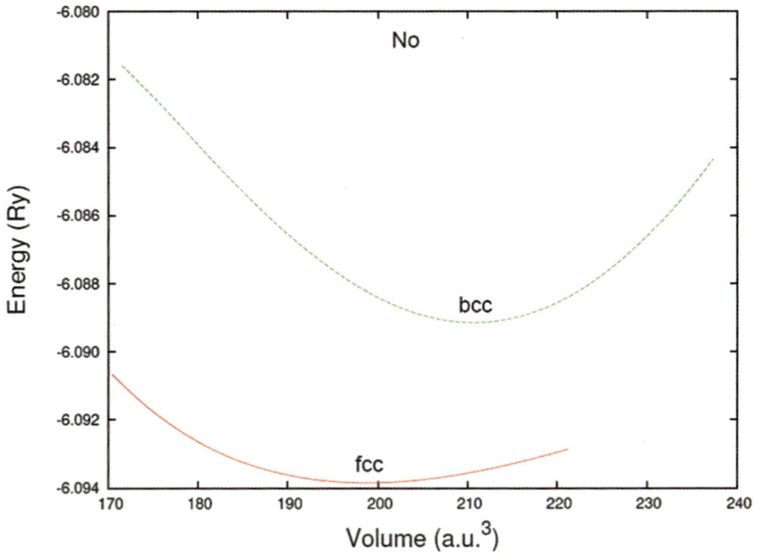

Fig. 14.16 Total energy of No

Table 14.28 Lattice constants and bulk modulus

	a (Bohr)	B (MBar)
bcc	7.497	0.412
fcc	9.264	0.153
$\Delta E_{fcc-bcc} = 4.7$ mRy		

Nobelium

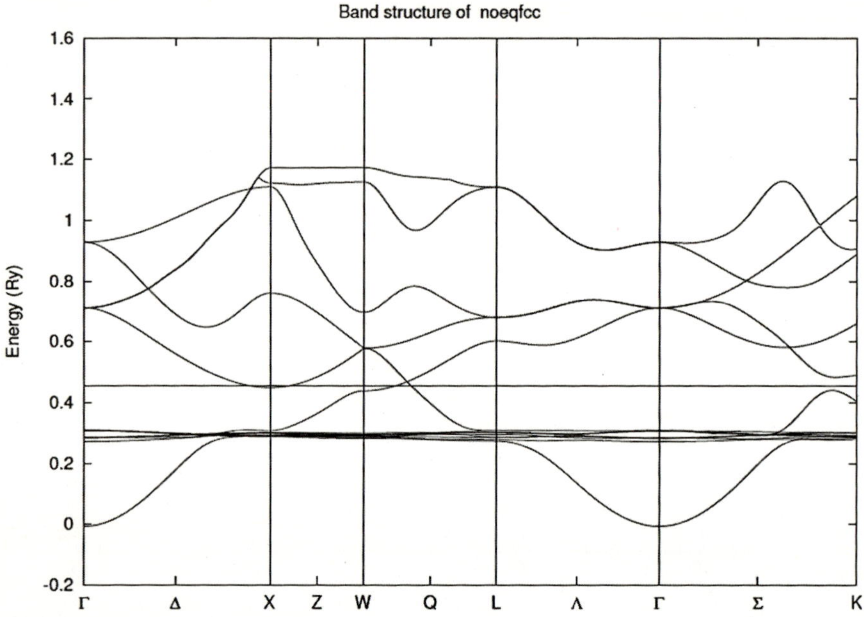

Fig. 14.17 Energy bands for No

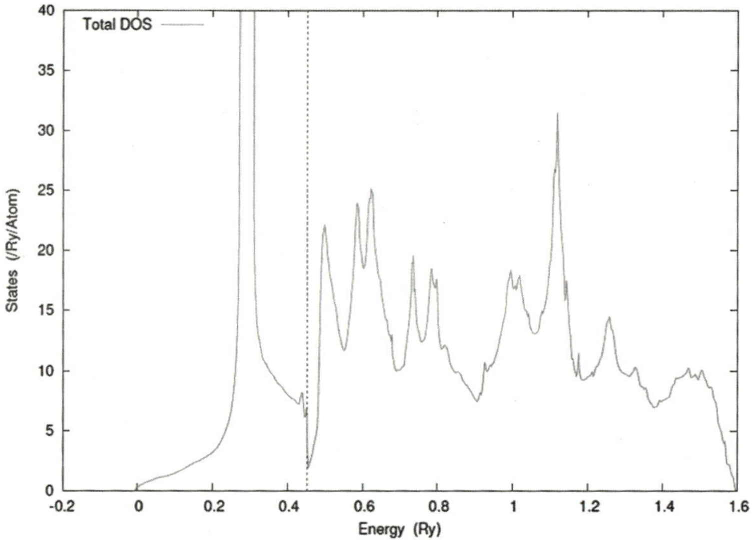

Fig. 14.18 Density of states for No

Table 14.29 DOS at the Fermi level (APW results)

Energy	Total	s	p	e_g	t_{2g}	f	Velocity	Plasmon energy
Ry		States/Ry/atom					cm/s	eV
0.454	4.204	0.168	0.411	1.099	1.215	0.161	$0.223 \times 10E8$	1.158

14.15 Lawrencium Z = 103

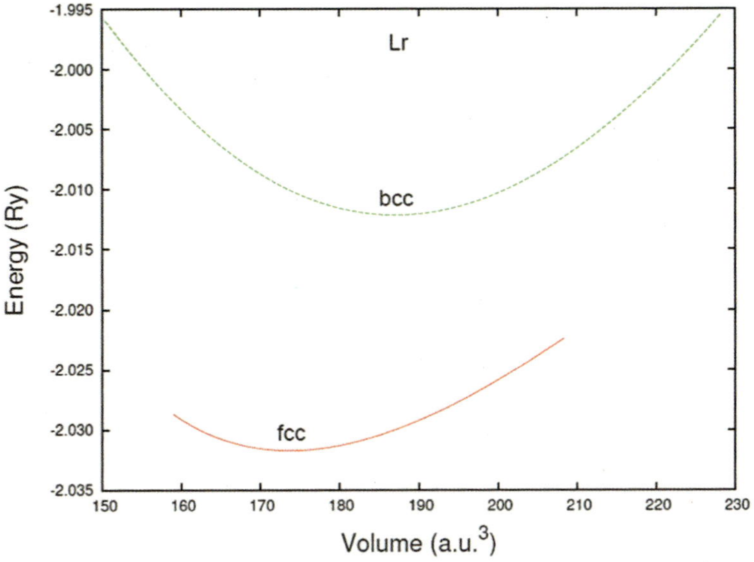

Fig. 14.19 Total energy of Lr

Table 14.30 Lattice constants and bulk modulus

	a (Bohr)	B (MBar)
bcc	7.206	0.627
fcc	8.859	0.588
$\Delta E_{fcc-bcc} = 19.61$ mRy		

Lawrencium

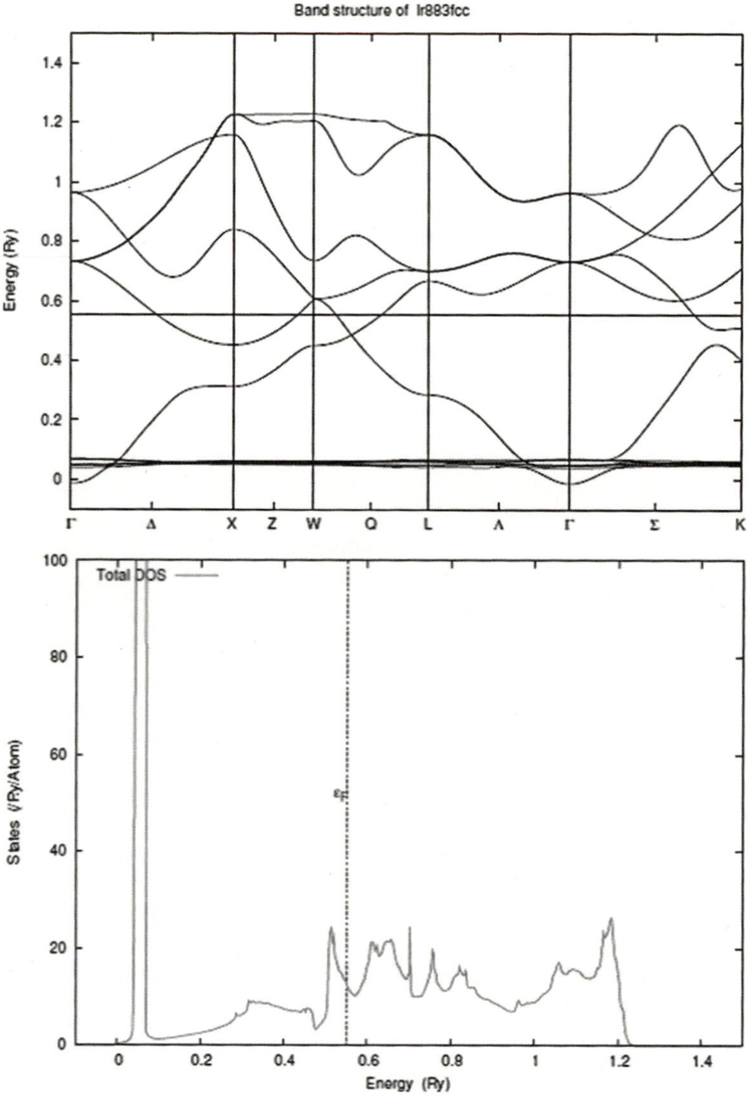

Fig. 14.20 Energy bands and density of states for Lr

Table 14.31 DOS at the Fermi level (APW results)

Energy Ry	Total	s States/Ry/atom	p	e_g	t_{2g}	f	Velocity cm/s	Plasmon energy eV
0.553	12.435	0.197	1.410	2.363	5.021	0.374	0.558 × 10E8	5.400

Chapter 15
Transactinides

These elements have electrons in the 6d subshell; they are radioactive and have only been synthesized in the laboratory. The total energy calculations done for the fcc and bcc lattices predict that the fcc is favored in Rutherfordium, Bohrium, Hassium, Meitnerium, Darmstadtium, and Copernicium, while the bcc structure is favored in Dubnium, Seaborgium and Roentgenium. An interesting characteristic of these materials is that the energy bands and DOS show a strong resemblance to those of the transition metals, while Copernicium is the only one which is not metallic. Three- and two-center orthogonal Slater-Koster parameters are also presented.

D.A. Papaconstantopoulos, *Handbook of the Band Structure of Elemental Solids*,
DOI 10.1007/978-1-4419-8264-3_15

15.1 Rutherfordium Z = 104

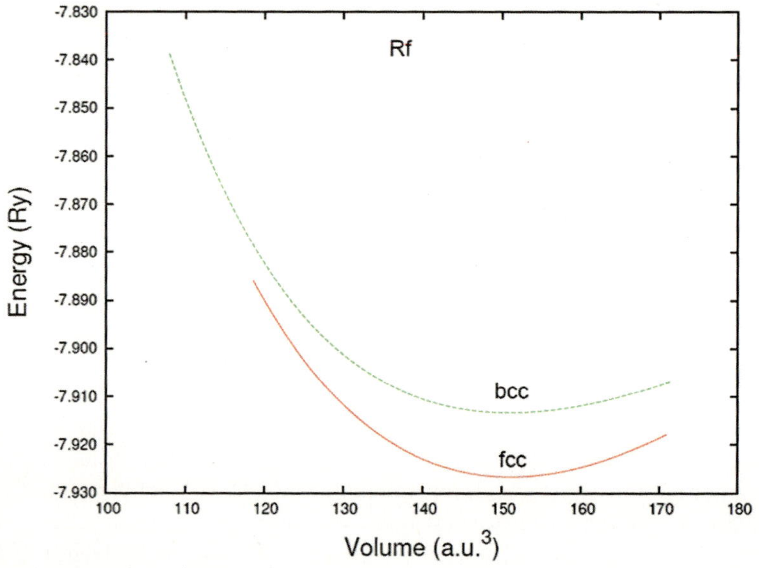

Fig. 15.1 Total energy of Rf

Table 15.1 Lattice constants and bulk modulus

	a (Bohr)	B (MBar)
bcc	6.709	0.901
fcc	8.455	1.207
$\Delta E_{fcc-bcc} = 13.35$ mRy		

Table 15.2 Rutherfordium fcc Z = 104 Orthogonal Slater-Koster parameters a = 8.46 a.u.

	Three-center (Ry)		Two-center (Ry)
On site			
s,s(000)	0.65578	s	0.82345
x,x(000)	1.42063	p	1.16809
xy,xy(000)	1.00368	d1	0.89890
d2,d2(000)	0.83318	d2	0.84551
First neighbor			
s,s(110)	−0.07528	(sss)	−0.06791
s,x(110)	0.06200	(pps)	0.05963
s,xy(110)	−0.05230	(ppp)	−0.05696
s,d2(110)	0.04455	(dds)	−0.11000
x,x(110)	0.05907	(ddp)	0.04845
x,x(011)	−0.00454	(ddd)	−0.00655
x,y(110)	0.09100	(sps)	0.07699
x,xy(110)	−0.09667	(sds)	−0.08605

(continued)

Table 15.2 (continued)

	Three-center (Ry)		Two-center (Ry)
x,xy(011)	0.00168	(pds)	−0.10786
z,d2(011)	0.00848	(pdp)	0.03438
z,d1(011)	0.05359		
xy,xy(110)	−0.08422		
xy,xy(011)	0.01871		
xy,xz(011)	−0.00948		
xy,d2(110)	0.04147		
d2,d2(110)	−0.03600		
d1,d1(110)	0.05570		
Second neighbor			
s,s(200)	0.03703	(sss)	−0.00612
s,x(200)	−0.05273	(pps)	0.04534
s,d2(002)	0.01964	(ppp)	−0.02393
x,x(200)	−0.03288	(dds)	0.00358
y,y(200)	0.01082	(ddp)	−0.00049
x,xy(020)	−0.01370	(ddd)	0.00144
z,d2(002)	0.01358	(sps)	−0.01073
xy,xy(200)	−0.02351	(sds)	0.01813
xy,xy(002)	0.00407	(pds)	−0.01049
d2,d2(002)	0.00269	(pdp)	0.01587
d1,d1(002)	0.00369		

Band	RMS error mRy	Maximum k	Deviation mRy	Band	RMS error mRy	Maximum k	Deviation mRy
1	4.5	(640)	13.4	1	24.4	(640)	47.0
2	6.7	(640)	16.8	2	17.8	(640)	44.0
3	5.2	(222)	9.6	3	11.6	(222)	27.5
4	4.7	(840)	9.7	4	14.8	(840)	27.1
5	6.1	(840)	13.9	5	29.5	(840)	81.3
6	6.8	(642)	20.1	6	28.1	(642)	80.6
1–6	5.7			1–6	22.1		

Rutherfordium

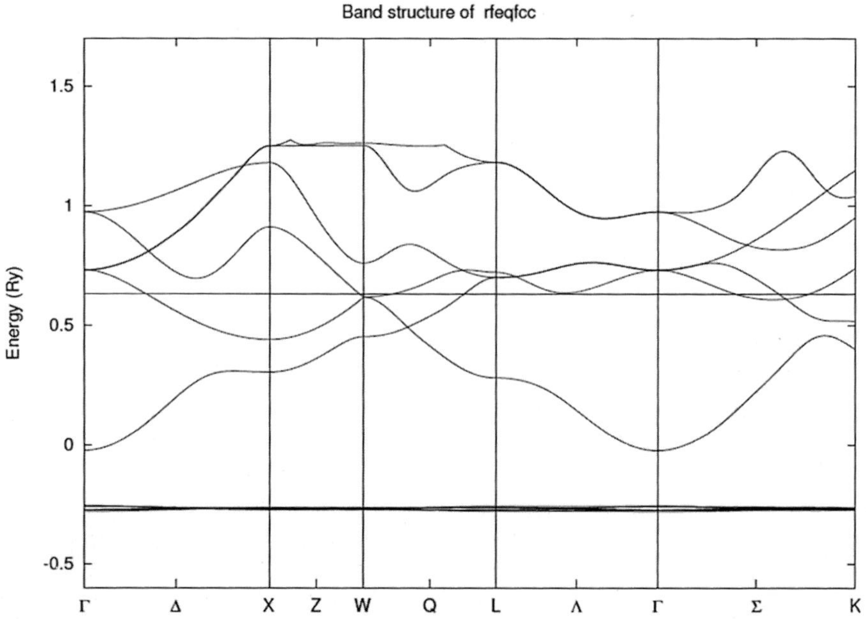

Fig. 15.2 Energy bands for Rf

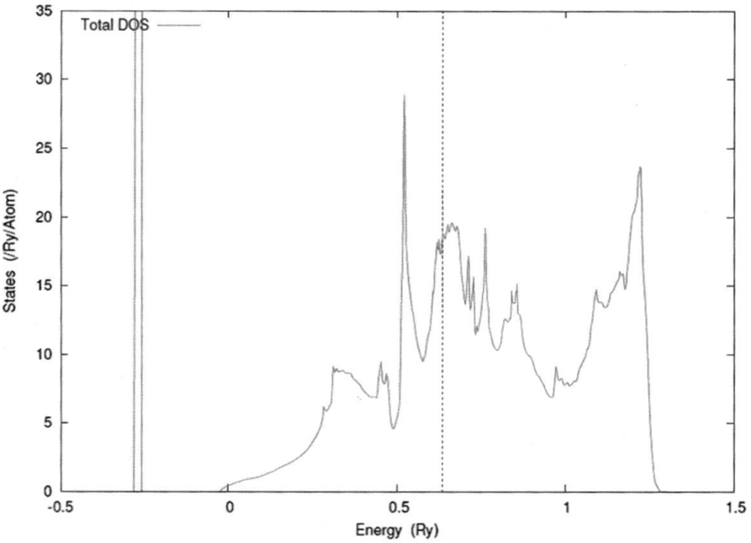

Fig. 15.3 Density of states for Rf

Table 15.3 DOS at the Fermi level (APW results)

Energy Ry	Total	s States/Ry/atom	p	e_g	t_{2g}	f	Velocity cm/s	Plasmon energy eV
0.633	18.36	0.41	2.49	2.38	8.61	0.56	0.588 × 10E8	7.380

15.2 Dubnium Z = 105

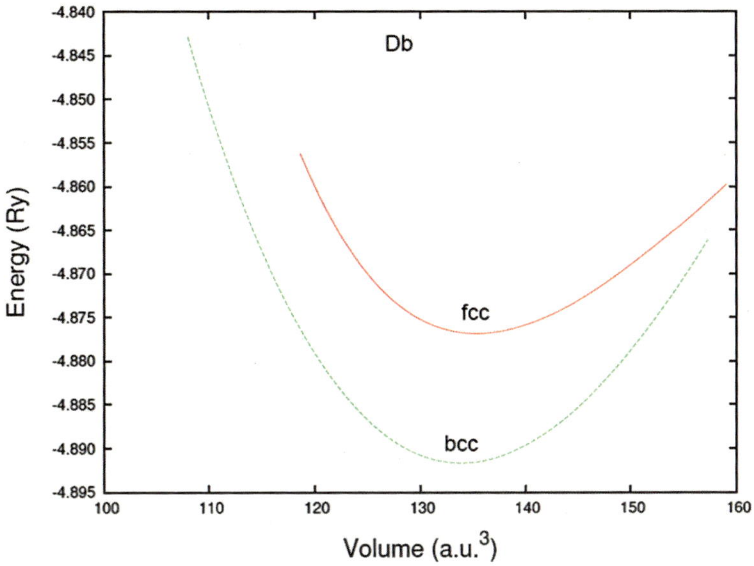

Fig. 15.4 Total energy of Db

Table 15.4 Lattice constants and bulk modulus

	a (Bohr)	B (MBar)
bcc	6.446	2.255
fcc	8.150	1.995
$\Delta E_{fcc-bcc} = 14.84$ mRy		

Table 15.5 Dubnium bcc Z = 105 Orthogonal Slater-Koster parameters a = 6.45 a.u

	Three-center (Ry)		Two-center (Ry)
On site			
s,s(000)	0.61009	s	0.68047
x,x(000)	1.46335	p	1.39783
xy,xy(000)	0.99758	d1	0.95733
d2,d2(000)	0.82694	d2	0.82438
First Neighbor			
s,s(111)	−0.08948	(sss)	−0.08984
s,x(111)	0.04977	(pps)	0.11691
s,xy(111)	0.03952	(ppp)	−0.00709
x,x(111)	0.06647	(dds)	−0.12216
x,y(111)	0.04129	(ddp)	0.05025
x,xy(111)	0.05406	(ddd)	0.00459
x,yz(111)	0.06144	(sps)	0.09340

(continued)

Table 15.5 (continued)

	Three-center (Ry)		Two-center (Ry)
x,d1(111)	0.01486	(sds)	−0.09187
xy,xy(111)	−0.02955	(pds)	−0.16315
xy,xz(111)	−0.04586	(pdp)	0.02473
xy,d2(111)	0.00211		
d2,d2(111)	0.03516		
Second neighbor			
s,s(200)	−0.01784	(sss)	−0.01910
s,x(200)	0.04732	(pps)	0.14803
s,d2(002)	−0.05555	(ppp)	−0.02103
x,x(200)	0.16492	(dds)	−0.07486
y,y(200)	0.02375	(ddp)	−0.01723
x,xy(020)	0.02557	(ddd)	0.00838
z,d2(002)	−0.11485	(sps)	0.04473
xy,xy(200)	−0.03495	(sds)	−0.04192
xy,xy(002)	0.00809	(pds)	−0.09432
d2,d2(002)	−0.07416	(pdp)	−0.00572
d1,d1(002)	0.00848		
Third neighbor			
s,s(220)	0.01516	(sss)	0.00907
s,x(220)	−0.01831	(pps)	−0.00930
s,xy(220)	−0.02039	(ppp)	−0.00285
s,d2(220)	−0.01104	(dds)	0.01410
x,x(220)	−0.01333	(ddp)	−0.00179
x,x(022)	0.00663	(ddd)	0.00118
x,y(220)	−0.00807	(sps)	−0.01575
x,xy(220)	−0.01241	(sds)	0.01178
x,xy(022)	0.00212	(pds)	0.00207
z,d2(022)	0.00373	(pdp)	0.00153
z,d1(022)	−0.00759		
xy,xy(220)	0.01811		
xy,xy(022)	0.00085		
xy,xz(022)	−0.00106		
yz,d2(220)	0.01400		
d2,d2(220)	0.00486		
d1,d1(220)	−0.00268		

Band	RMS error mRy	Maximum k	Deviation mRy	Band	RMS error mRy	Maximum k	Deviation mRy
1	4.0	(620)	11.4	1	5.1	(620)	13.0
2	3.5	(300)	11.3	2	9.9	(300)	24.7
3	2.8	(611)	8.2	3	7.6	(611)	25.4
4	4.2	(330)	13.3	4	11.6	(330)	28.1
5	4.3	(200)	15.8	5	8.2	(200)	25.4
6	6.1	(531)	14.7	6	12.0	(531)	26.0
1–6	4.3			1–6	9.4		

Dubnium

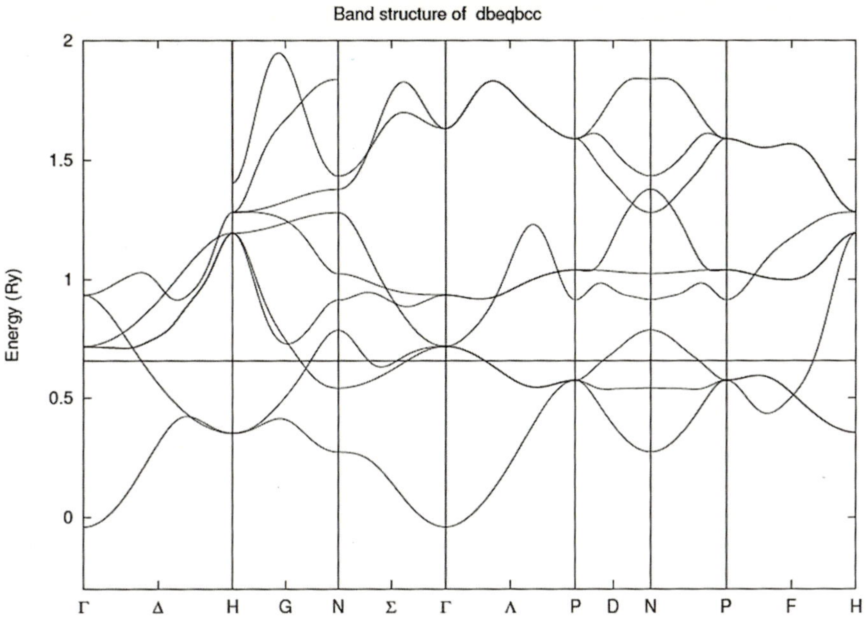

Fig. 15.5 Energy bands for Db

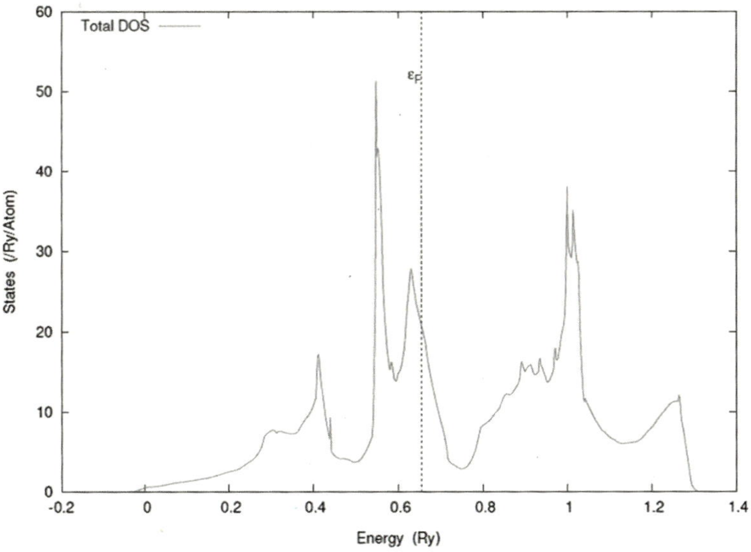

Fig. 15.6 Density of states for Db

Table 15.6 DOS at the Fermi level (APW results)

Energy	Total	s	p	e_g	t_{2g}	f	Velocity	Plasmon energy
Ry		States/Ry/atom					cm/s	eV
0.65533	20.569	0.343	2.320	2.360	10.081	0.459	0.620 × 10E8	8.755

15.3 Seaborgium Z = 106

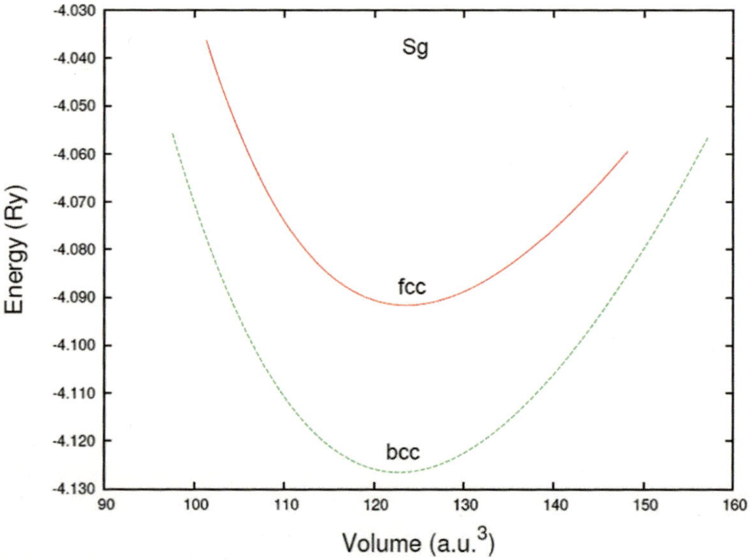

Fig. 15.7 Total energy of Sg

Table 15.7 Lattice constants and bulk modulus

	a (Bohr)	B (MBar)
bcc	6.263	2.999
fcc	7.907	2.757
$\Delta E_{fcc-bcc} = 34.9$ mRy		

Table 15.8 Seaborgium bcc Z = 106 Orthogonal Slater-Koster parameters a = 6.28 a.u

	Three-center (Ry)		Two-center (Ry)
On site			
s,s(000)	0.66153	s	0.77721
x,x(000)	1.65129	p	1.75731
xy,xy(000)	1.02459	d1	0.97812
d2,d2(000)	0.83678	d2	0.83387
First neighbor			
s,s(111)	−0.09758	(sss)	−0.10139
s,x(111)	0.05639	(pps)	0.25632
s,xy(111)	0.04308	(ppp)	−0.05155
x,x(111)	0.05721	(dds)	−0.13090
x,y(111)	0.06870	(ddp)	0.05306
x,xy(111)	0.05742	(ddd)	0.00379
x,yz(111)	0.07184	(sps)	0.14482
x,d1(111)	0.01778	(sds)	−0.10305

(continued)

Table 15.8 (continued)

	Three-center (Ry)		Two-center (Ry)
xy,xy(111)	−0.03160	(pds)	−0.18693
xy,xz(111)	−0.04909	(pdp)	0.04360
xy,d2(111)	0.00173		
d2,d2(111)	0.03715		
Second neighbor			
s,s(200)	−0.01955	(sss)	−0.01803
s,x(200)	0.04490	(pps)	0.04313
s,d2(002)	−0.06326	(ppp)	0.02616
x,x(200)	0.12386	(dds)	−0.07691
y,y(200)	0.03241	(ddp)	−0.01857
x,xy(020)	0.03264	(ddd)	0.00762
z,d2(002)	−0.12034	(sps)	0.02947
xy,xy(200)	−0.03941	(sds)	−0.04701
xy,xy(002)	0.00968	(pds)	−0.08957
d2,d2(002)	−0.07707	(pdp)	−0.02088
d1,d1(002)	0.00839		
Third neighbor			
s,s(220)	0.01693	(sss)	0.00848
s,x(220)	−0.02257	(pps)	−0.03089
s,xy(220)	−0.02328	(ppp)	−0.00334
s,d2(220)	−0.01058	(dds)	0.01643
x,x(220)	−0.02318	(ddp)	−0.00124
x,x(022)	0.00038	(ddd)	0.00041
x,y(220)	−0.01696	(sps)	−0.01808
x,xy(220)	−0.01695	(sds)	0.01298
x,xy(022)	−0.00024	(pds)	0.01391
z,d2(022)	0.00474	(pdp)	0.00224
z,d1(022)	−0.00865		
xy,xy(220)	0.02026		
xy,xy(022)	0.00057		
xy,xz(022)	−0.00112		
yz,d2(220)	0.01557		
d2,d2(220)	0.00485		
d1,d1(220)	−0.00165		

Band	RMS error mRy	Maximum k	Deviation mRy	Band	RMS error mRy	Maximum k	Deviation mRy
1	3.8	(000)	8.6	1	3.7	(000)	9.5
2	3.7	(300)	11.2	2	8.9	(300)	27.7
3	2.6	(611)	6.4	3	4.9	(611)	19.1
4	3.7	(330)	13.4	4	9.9	(330)	23.1
5	3.9	(200)	15.6	5	7.8	(200)	19.1
6	4.3	(611)	11.1	6	8.3	(611)	20.1
1–6	3.7			1–6	7.6		

Seaborgium

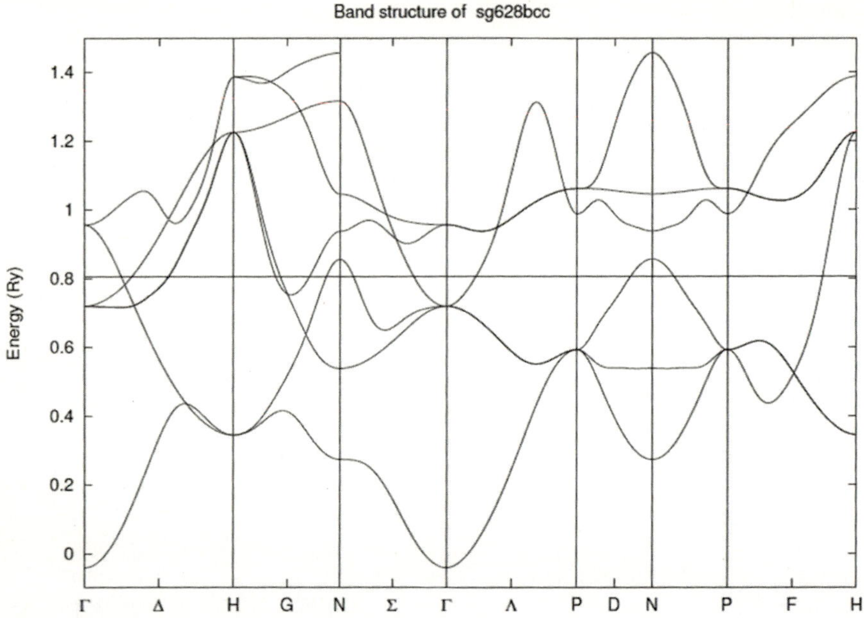

Fig. 15.8 Energy bands for Sg

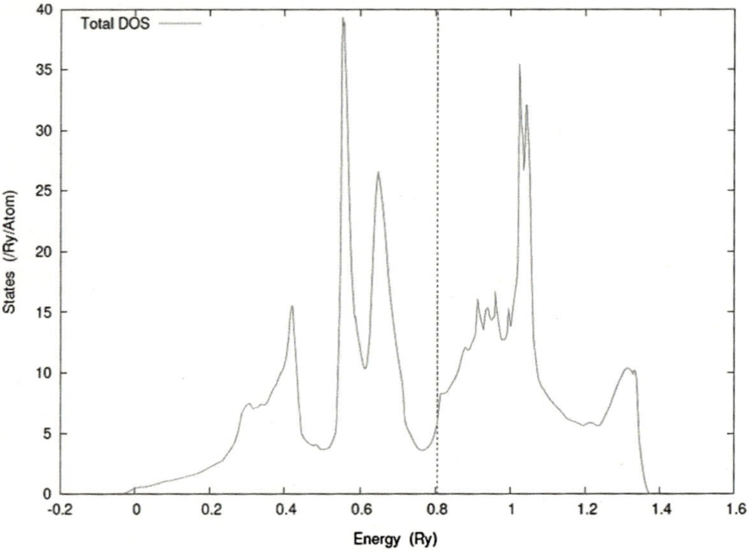

Fig. 15.9 Density of states for Sg

Table 15.9 DOS at the Fermi level (APW results)

Energy	Total	s	p	e_g	t_{2g}	f	Velocity	Plasmon energy
Ry		States/Ry/atom					cm/s	eV
0.806	6.031	0.226	0.407	0.891	3.185	0.233	0.992 × 10E8	7.878

15.4 Bohrium Z = 107

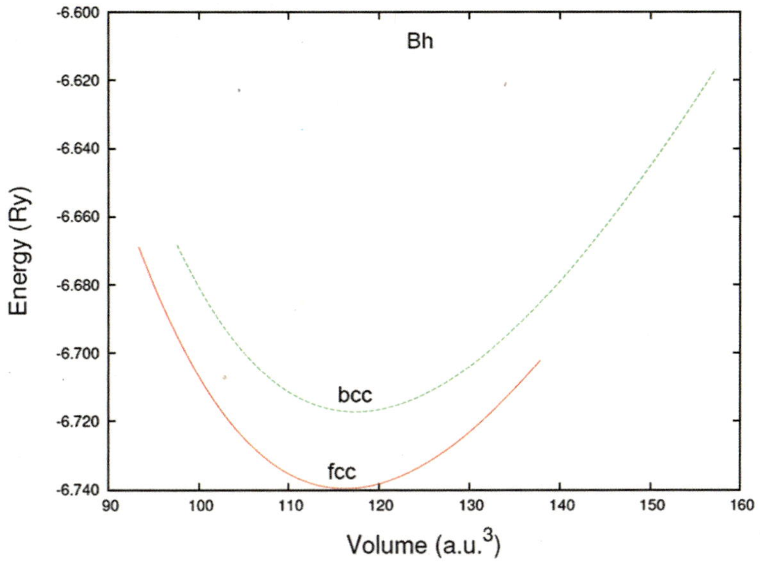

Fig. 15.10 Total energy of Bh

Table 15.10 Lattice constants and bulk modulus

	a (Bohr)	B (MBar)
bcc	6.170	3.330
fcc	7.748	3.472
$\Delta E_{fcc-bcc} = 22.37$ mRy		

Table 15.11 Bohrium fcc Z = 107 Orthogonal Slater-Koster parameters a = 7.75 a.u

	Three-center (Ry)		Two-center (Ry)
On site			
s,s(000)	0.78442	s	0.82404
x,x(000)	1.54634	p	1.16946
xy,xy(000)	0.79273	d1	0.81465
d2,d2(000)	0.88362	d2	0.77461
First neighbor			
s,s(110)	−0.06318	(sss)	−0.07102
s,x(110)	0.07496	(pps)	0.04048
s,xy(110)	−0.07654	(ppp)	−0.05798
s,d2(110)	0.07158	(dds)	−0.11326
x,x(110)	0.03588	(ddp)	0.04801
x,x(011)	−0.04732	(ddd)	−0.00660
x,y(110)	0.10518	(sps)	0.07876

(continued)

Table 15.11 (continued)

	Three-center (Ry)		Two-center (Ry)
x,xy(110)	−0.07492	(sds)	−0.09249
x,xy(011)	0.01535	(pds)	−0.11081
z,d2(011)	−0.03610	(pdp)	0.03748
z,d1(011)	0.08979		
xy,xy(110)	−0.08647		
xy,xy(011)	0.01942		
xy,xz(011)	0.02106		
xy,d2(110)	0.04984		
d2,d2(110)	−0.02572		
d1,d1(110)	0.02434		
Second neighbor			
s,s(200)	−0.02266	(sss)	−0.01303
s,x(200)	0.00583	(pps)	0.06091
s,d2(002)	−0.00272	(ppp)	−0.03228
x,x(200)	−0.00325	(dds)	0.00210
y,y(200)	−0.01809	(ddp)	0.00023
x,xy(020)	0.01518	(ddd)	0.00184
z,d2(002)	−0.00327	(sps)	0.00503
xy,xy(200)	0.00821	(sds)	0.01693
xy,xy(002)	0.00227	(pds)	−0.01262
d2,d2(002)	−0.00040	(pdp)	0.01403
d1,d1(002)	0.00532		

Band	RMS error mRy	Maximum k	Deviation mRy	Band	RMS error mRy	Maximum k	Deviation mRy
1	4.2	(550)	12.4	1	20.2	(550)	45.0
2	6.9	(300)	14.9	2	13.7	(300)	33.2
3	6.8	(330)	19.7	3	8.9	(330)	22.3
4	4.6	(400)	9.3	4	10.2	(400)	21.7
5	4.6	(400)	9.3	5	20.5	(400)	49.3
6	4.9	(550)	12.8	6	32.0	(550)	77.6
1–6	5.4			1–6	19.3		

Bohrium

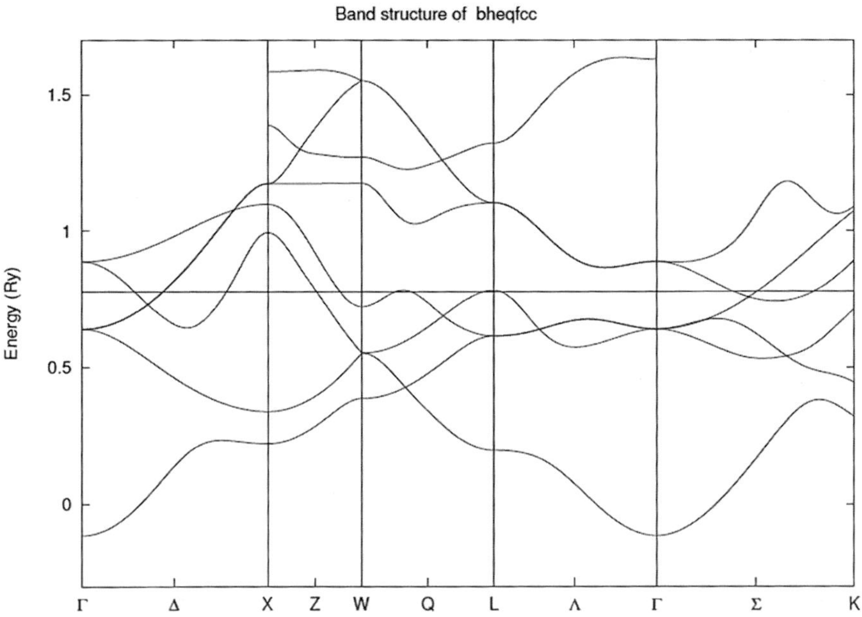

Fig. 15.11 Energy bands for Bh

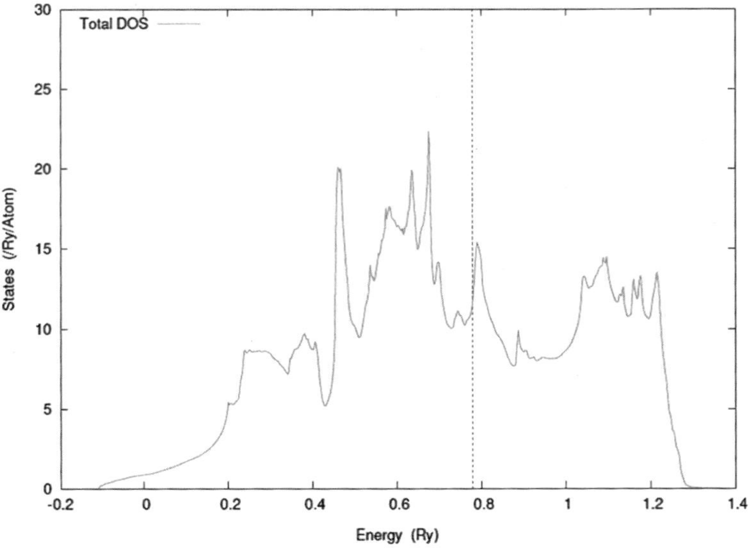

Fig. 15.12 Density of states for Bh

Table 15.12 DOS at the Fermi level (APW results)

Energy Ry	Total	s States/Ry/atom	p	e_g	t_{2g}	f	Velocity cm/s	Plasmon energy eV
0.778	11.530	0.316	0.656	6.142	2.673	0.367	$0.778 \times 10E8$	8.820

15.5 Hassium Z = 108

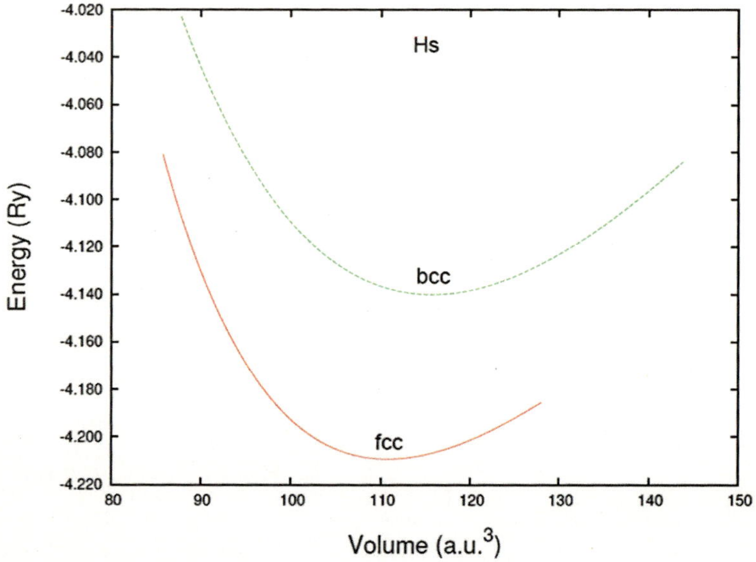

Fig. 15.13 Total energy of Hs

Table 15.13 Lattice constants and bulk modulus

	a (Bohr)	B (MBar)
bcc	6.139	3.392
fcc	7.624	3.682
$\Delta E_{fcc-bcc} = 69.18$ mRy		

Table 15.14 Hassium fcc Z = 108 Orthogonal Slater-Koster parameters a = 7.62 a.u

Three-center (Ry)		Two-center (Ry)	
On site			
s,s(000)	0.74718	s	0.80139
x,x(000)	1.57840	p	1.13531
xy,xy(000)	0.73076	d1	0.75526
d2,d2(000)	0.82352	d2	0.70720
First neighbor			
s,s(110)	−0.06397	(sss)	−0.07973
s,x(110)	0.07803	(pps)	0.03538
s,xy(110)	−0.07556	(ppp)	−0.06037
s,d2(110)	0.07123	(dds)	−0.10940
x,x(110)	0.04137	(ddp)	0.04795
x,x(011)	−0.04626	(ddd)	−0.00621
x,y(110)	0.11192	(sps)	0.08478
x,xy(110)	−0.07401	(sds)	−0.09037

(continued)

Table 15.14 (continued)

	Three-center (Ry)		Two-center (Ry)
x,xy(011)	0.01533	(pds)	−0.10476
z,d2(011)	−0.03772	(pdp)	0.03624
z,d1(011)	0.09041		
xy,xy(110)	−0.08379		
xy,xy(011)	0.01892		
xy,xz(011)	0.02137		
xy,d2(110)	0.04804		
d2,d2(110)	−0.02460		
d1,d1(110)	0.02272		
Second neighbor			
s,s(200)	−0.02337	(sss)	−0.00089
s,x(200)	0.00688	(pps)	0.04165
s,d2(002)	−0.00329	(ppp)	−0.02256
x,x(200)	−0.00460	(dds)	0.00184
y,y(200)	−0.01900	(ddp)	−0.00057
x,xy(020)	0.01536	(ddd)	0.00069
z,d2(002)	−0.00299	(sps)	−0.01319
xy,xy(200)	0.00837	(sds)	0.01077
xy,xy(002)	0.00199	(pds)	−0.01248
d2,d2(002)	−0.00041	(pdp)	0.01193
d1,d1(002)	0.00529		

Band	RMS error mRy	Maximum k	Deviation mRy	Band	RMS error mRy	Maximum k	Deviation mRy
1	4.1	(550)	12.9	1	22.4	(550)	42.9
2	6.5	(300)	13.9	2	15.6	(300)	39.9
3	6.4	(330)	17.2	3	9.6	(330)	23.7
4	4.3	(400)	8.5	4	14.4	(400)	36.7
5	4.1	(400)	8.5	5	21.0	(400)	51.2
6	4.4	(550)	11.6	6	26.9	(550)	73.4
1–6	5.1			1–6	19.2		

Hassium

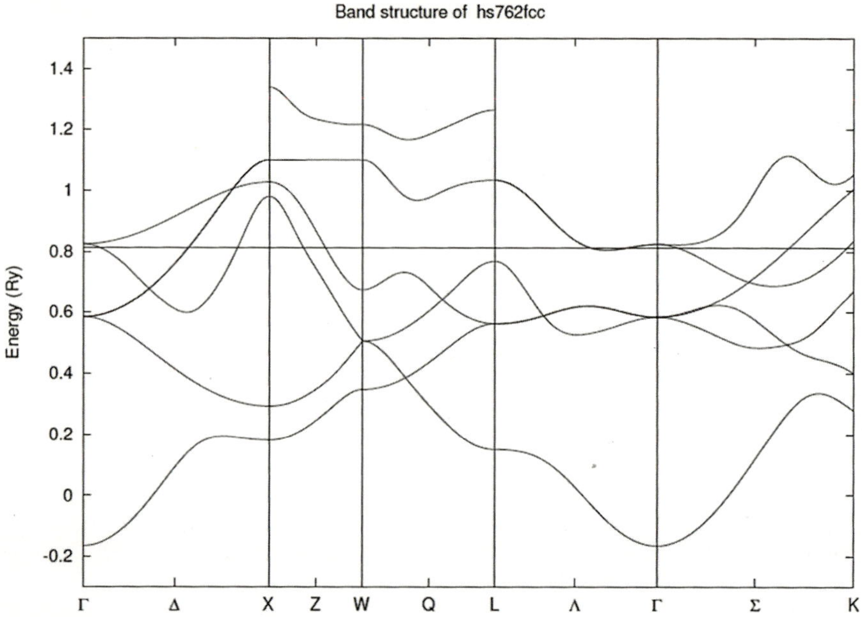

Fig. 15.14 Energy bands for Hs

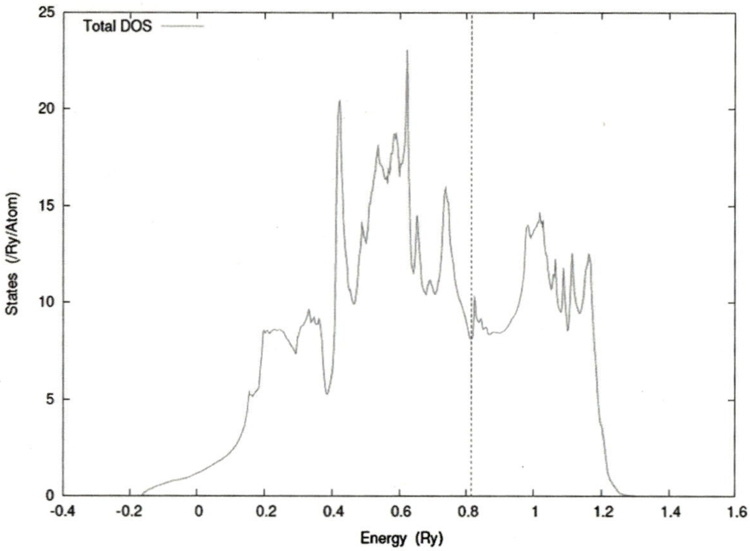

Fig. 15.15 Density of states for Hs

Table 15.15 DOS at the Fermi level (APW results)

Energy	Total	s	p	e_g	t_{2g}	f	Velocity	Plasmon energy
Ry		States/Ry/atom					cm/s	eV
0.814	8.113	0.130	0.425	3.469	2.978	0.249	1.109 × 10E8	10.817

15.6 Meitnerium Z = 109

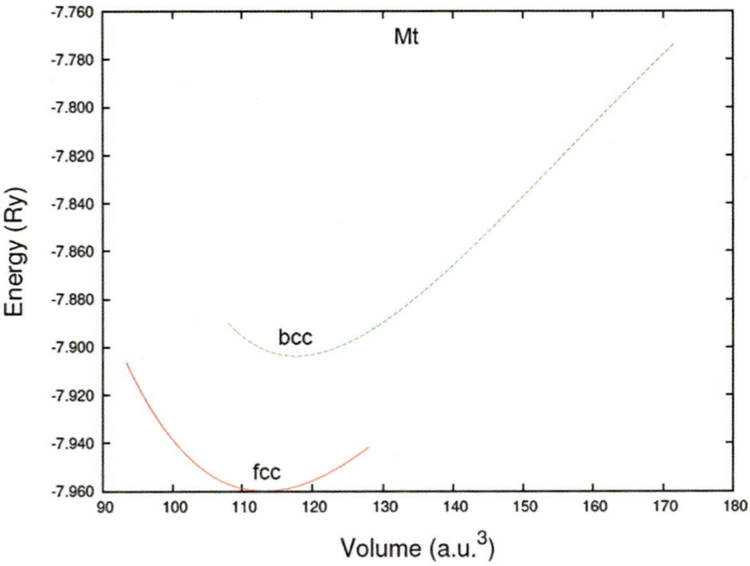

Fig. 15.16 Total energy of Mt

Table 15.16 Lattice constants and bulk modulus

	a (Bohr)	B (MBar)
bcc	6.174	4.082
fcc	7.679	3.370
$\Delta E_{fcc-bcc} = 56.31$ mRy		

Table 15.17 Meitnerium fcc Z = 109 Orthogonal Slater-Koster parameters a = 7.69 a.u

	Three-center (Ry)		Two-center (Ry)
On site			
s,s(000)	0.52016	s	0.64061
x,x(000)	1.49187	p	1.50575
xy,xy(000)	0.68920	d1	0.61702
d2,d2(000)	0.56749	d2	0.57189
First neighbor			
s,s(110)	−0.07688	(sss)	−0.07242
s,x(110)	0.07090	(pps)	0.17000
s,xy(110)	−0.06454	(ppp)	−0.02702
s,d2(110)	0.04263	(dds)	−0.09499
x,x(110)	0.06707	(ddp)	0.04091
x,x(011)	−0.01051	(ddd)	−0.00459
x,y(110)	0.10999	(sps)	0.11277

(continued)

Table 15.17 (continued)

	Three-center (Ry)		Two-center (Ry)
x,xy(110)	−0.10004	(sds)	−0.08201
x,xy(011)	0.00289	(pds)	−0.12011
z,d2(011)	0.00674	(pdp)	0.02567
z,d1(011)	0.04881		
xy,xy(110)	−0.07241		
xy,xy(011)	0.01652		
xy,xz(011)	−0.00356		
xy,d2(110)	0.03697		
d2,d2(110)	−0.02983		
d1,d1(110)	0.04608		
Second neighbor			
s,s(200)	0.02796	(sss)	−0.00019
s,x(200)	−0.03990	(pps)	0.00069
s,d2(002)	0.01285	(ppp)	−0.01054
x,x(200)	−0.02899	(dds)	0.00038
y,y(200)	0.00076	(ddp)	−0.00145
x,xy(020)	−0.00897	(ddd)	0.00216
z,d2(002)	0.00826	(sps)	−0.00171
xy,xy(200)	−0.01745	(sds)	0.00441
xy,xy(002)	0.00265	(pds)	−0.02485
d2,d2(002)	−0.00022	(pdp)	−0.00267
d1,d1(002)	0.00350		

Band	RMS error mRy	Maximum k	Deviation mRy	Band	RMS error mRy	Maximum k	Deviation mRy
1	2.9	(550)	10.1	1	17.9	(550)	36.2
2	4.5	(642)	8.7	2	15.4	(642)	28.7
3	4.0	(440)	11.9	3	20.3	(440)	63.0
4	2.9	(543)	7.9	4	24.0	(543)	92.0
5	3.1	(840)	7.9	5	17.5	(840)	36.9
6	4.0	(622)	8.5	6	24.4	(622)	69.8
1–6	3.6			1–6	20.2		

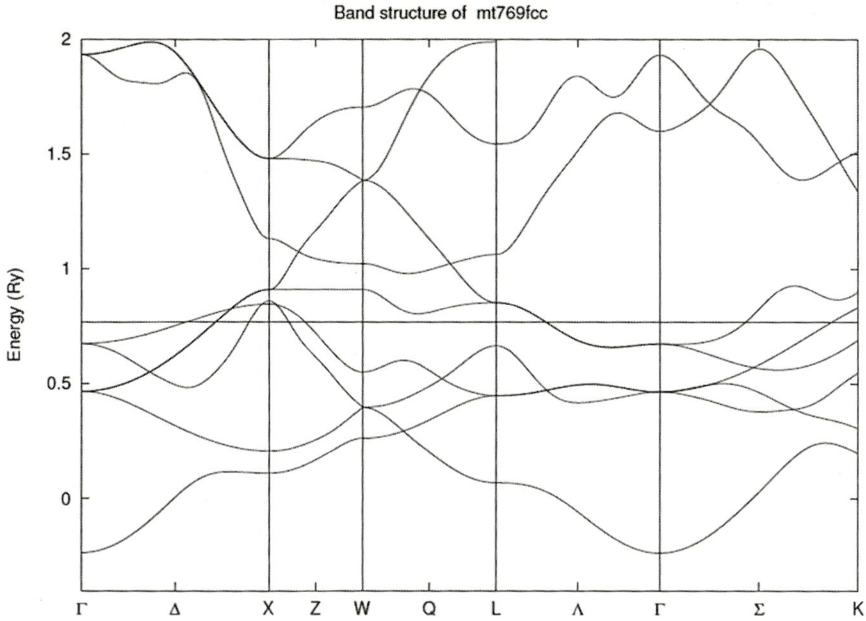

Fig. 15.17 Energy bands for Mt

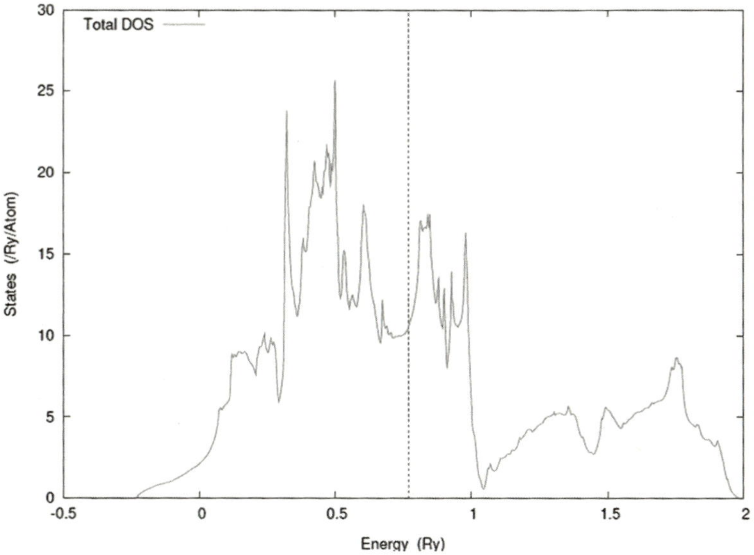

Fig. 15.18 Density of states for Mt

Table 15.18 DOS at the Fermi level (APW results)

Energy	Total	s	p	e_g	t_{2g}	f	Velocity	Plasmon energy
Ry		States/Ry/atom					cm/s	eV
0.769	10.468	0.111	0.274	3.828	5.429	0.230	$1.059 \times 10E8$	11.565

15.7 Darmstadtium Z = 110

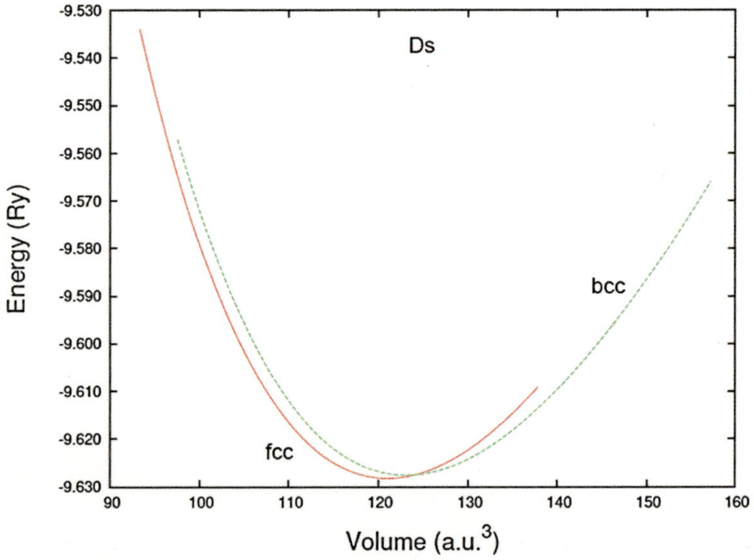

Fig. 15.19 Total energy of Ds

Table 15.19 Lattice constants and bulk modulus

	a (Bohr)	B (MBar)
bcc	6.268	2.781
fcc	7.853	2.965
$\Delta E_{fcc-bcc} = 0.67$ mRy		

Table 15.20 Darmstadtium fcc Z = 110 Orthogonal Slater-Koster parameters a = 7.85 a.u

Three-center (Ry)		Two-center (Ry)	
On site			
s,s(000)	0.36951	s	0.45721
x,x(000)	1.34860	p	1.35707
xy,xy(000)	0.50281	d1	0.45050
d2,d2(000)	0.41794	d2	0.42063
First neighbor			
s,s(110)	−0.06693	(sss)	−0.06389
s,x(110)	0.06590	(pps)	0.17084
s,xy(110)	−0.05694	(ppp)	−0.02295
s,d2(110)	0.03602	(dds)	−0.07762
x,x(110)	0.06913	(ddp)	0.03440
x,x(011)	−0.00715	(ddd)	−0.00428
x,y(110)	0.10690	(sps)	0.10115

(continued)

Table 15.20 (continued)

	Three-center (Ry)		Two-center (Ry)
x,xy(110)	−0.08559	(sds)	−0.06979
x,xy(011)	0.00276	(pds)	−0.10165
z,d2(011)	0.00515	(pdp)	0.02185
z,d1(011)	0.04106		
xy,xy(110)	−0.05912		
xy,xy(011)	0.01378		
xy,xz(011)	0.00065		
xy,d2(110)	0.03071		
d2,d2(110)	−0.02416		
d1,d1(110)	0.03783		
Second neighbor			
s,s(200)	0.01949	(sss)	−0.00052
s,x(200)	−0.02726	(pps)	0.00738
s,d2(002)	0.00843	(ppp)	−0.01303
x,x(200)	−0.01824	(dds)	−0.00093
y,y(200)	−0.00385	(ddp)	0.00000
x,xy(020)	−0.00596	(ddd)	0.00144
z,d2(002)	0.00518	(sps)	0.00312
xy,xy(200)	−0.01181	(sds)	0.00143
xy,xy(002)	0.00181	(pds)	−0.01871
d2,d2(002)	−0.00135	(pdp)	−0.00052
d1,d1(002)	0.00224		

Band	RMS error mRy	Maximum k	Deviation mRy	Band	RMS error mRy	Maximum k	Deviation mRy
1	2.0	(550)	5.5	1	12.8	(550)	30.2
2	3.4	(642)	6.7	2	11.2	(642)	22.6
3	3.2	(440)	7.9	3	13.5	(440)	38.9
4	1.9	(543)	4.9	4	20.4	(543)	72.2
5	2.0	(840)	5.0	5	11.9	(840)	25.0
6	3.1	(622)	6.9	6	19.6	(622)	50.8
1–6	2.7			1–6	15.4		

Darmstadtium

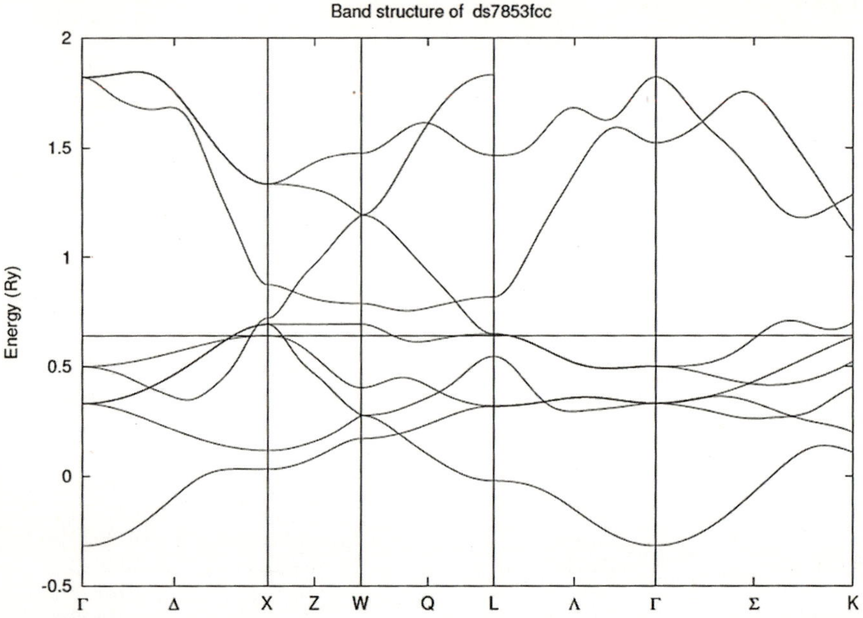

Fig. 15.20 Energy bands for Ds

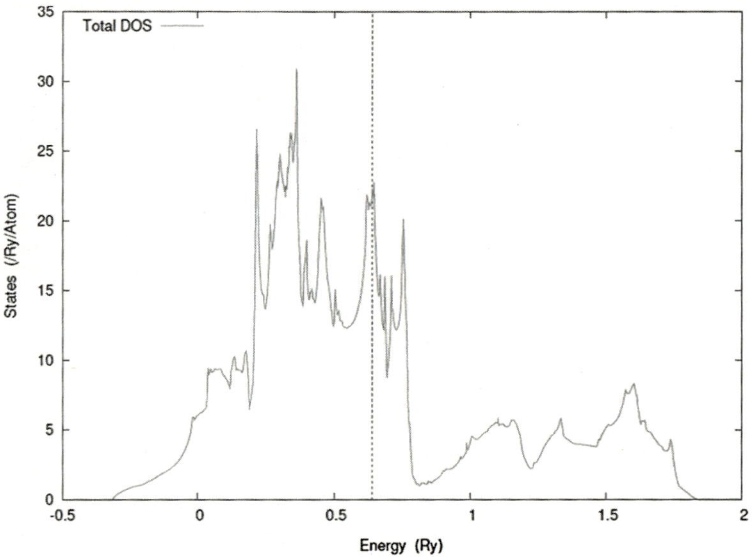

Fig. 15.21 Density of states for Ds

Table 15.21 DOS at the Fermi level (APW results)

Energy	Total	s	p	e_g	t_{2g}	f	Velocity	Plasmon energy
Ry		States/Ry/atom					cm/s	eV
0.639	22.082	0.352	0.246	6.154	14.604	0.186	$0.611 \times 10E8$	9.397

15.8 Roentgenium Z = 111

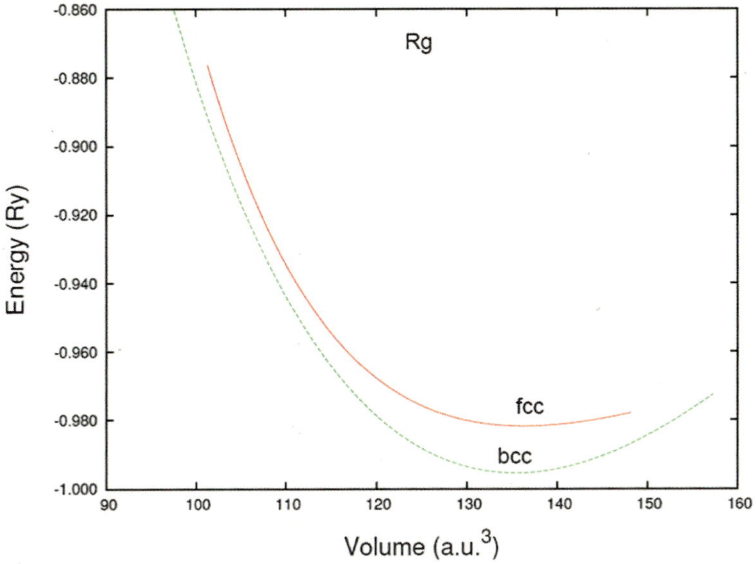

Fig. 15.22 Total energy of Rg

Table 15.22 Lattice constants and bulk modulus

	a (Bohr)	B (MBar)
bcc	6.469	2.375
fcc	8.170	1.440
$\Delta E_{fcc-bcc} = 13.66$ mRy		

Table 15.23 Roentgenium bcc Z = 111 Orthogonal Slater-Koster parameters a = 6.47 a.u

	Three-center (Ry)		Two-center (Ry)
On site			
s,s(000)	0.18737	s	0.21770
x,x(000)	1.08678	p	1.05826
xy,xy(000)	0.32434	d1	0.31142
d2,d2(000)	0.26604	d2	0.26227
First neighbor			
s,s(111)	−0.06612	(sss)	−0.06586
s,x(111)	0.05010	(pps)	0.11572
s,xy(111)	0.03356	(ppp)	−0.03975
x,x(111)	0.02684	(dds)	−0.06592
x,y(111)	0.05173	(ddp)	0.03040
x,xy(111)	0.03191	(ddd)	−0.00145
x,yz(111)	0.04968	(sps)	0.08837
x,d1(111)	0.01249	(sds)	−0.06097

(continued)

Table 15.23 (continued)

	Three-center (Ry)		Two-center (Ry)
xy,xy(111)	−0.01656	(pds)	−0.10693
xy,xz(111)	−0.02440	(pdp)	0.02677
xy,d2(111)	0.00765		
d2,d2(111)	0.01993		
Second neighbor			
s,s(200)	−0.01443	(sss)	−0.01486
s,x(200)	0.03518	(pps)	0.09558
s,d2(002)	−0.03180	(ppp)	−0.02595
x,x(200)	0.09325	(dds)	−0.03372
y,y(200)	−0.00453	(ddp)	−0.00352
x,xy(020)	0.00666	(ddd)	0.00242
z,d2(002)	−0.06117	(sps)	0.03459
xy,xy(200)	−0.00929	(sds)	−0.02688
xy,xy(002)	0.00302	(pds)	−0.04648
d2,d2(002)	−0.03339	(pdp)	−0.00182
d1,d1(002)	0.00214		
Third neighbor			
s,s(220)	0.00568	(sss)	0.00295
s,x(220)	−0.00866	(pps)	−0.02361
s,xy(220)	−0.00645	(ppp)	−0.00208
s,d2(220)	−0.00350	(dds)	0.00561
x,x(220)	−0.01632	(ddp)	0.00030
x,x(022)	0.00669	(ddd)	−0.00036
x,y(220)	−0.01401	(sps)	−0.00961
x,xy(220)	−0.00635	(sds)	0.00436
x,xy(022)	0.00007	(pds)	0.00305
z,d2(022)	0.00152	(pdp)	0.00228
z,d1(022)	−0.00317		
xy,xy(220)	0.00650		
xy,xy(022)	0.00012		
xy,xz(022)	0.00011		
yz,d2(220)	0.00367		
d2,d2(220)	0.00118		
d1,d1(220)	−0.00037		

Band	RMS error mRy	Maximum k	Deviation mRy	Band	RMS error mRy	Maximum k	Deviation mRy
1	0.9	(222)	2.3	1	2.3	(222)	6.5
2	1.1	(311)	2.8	2	3.3	(311)	8.0
3	0.9	(611)	2.4	3	2.5	(611)	6.4
4	1.3	(444)	6.5	4	2.9	(444)	6.4
5	1.1	(440)	4.1	5	2.2	(440)	6.4
6	1.8	(442)	5.0	6	3.0	(442)	9.0
1–6	1.2			1–6	2.7		

Roentgenium

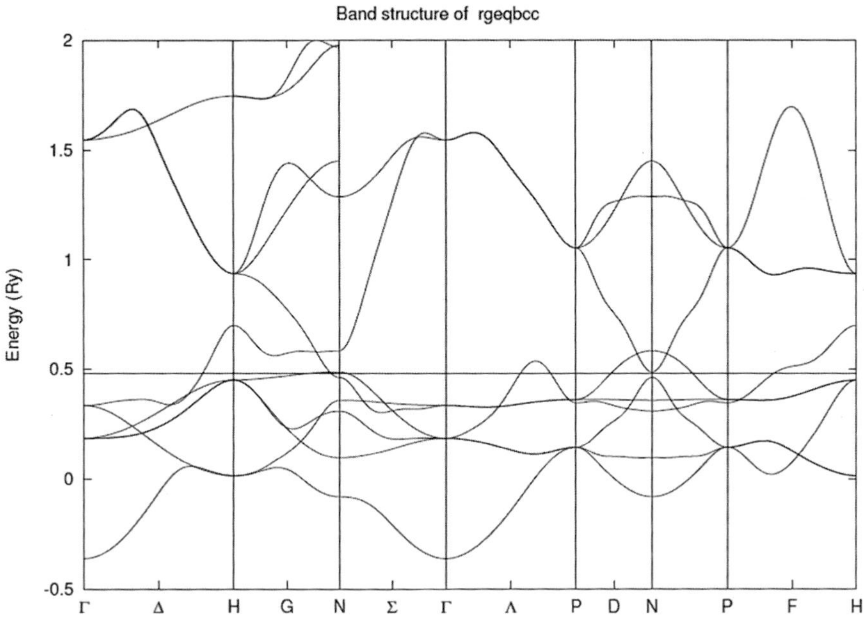

Fig. 15.23 Energy bands for Rg

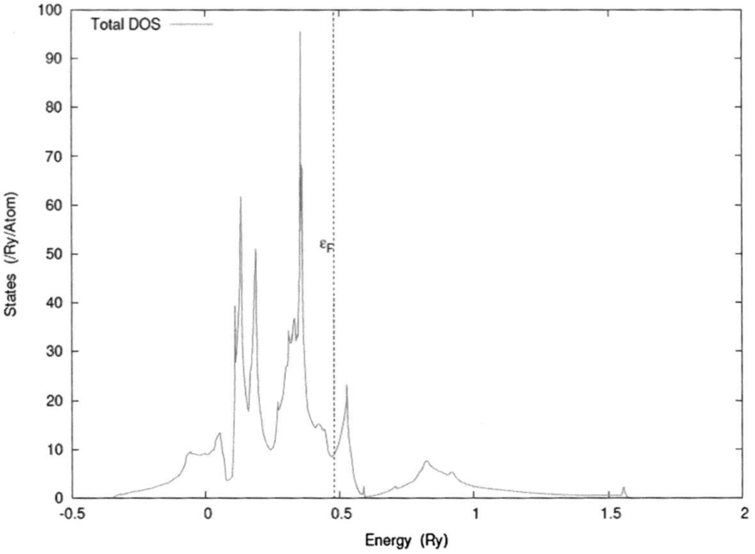

Fig. 15.24 Density of states for Rg

Table 15.24 DOS at the Fermi level (APW results)

Energy Ry	Total	s States/Ry/atom	p	e_g	t_{2g}	f	Velocity cm/s	Plasmon energy eV
0.480	8.979	1.960	0.238	0.215	6.061	0.097	$0.896 \times 10E8$	8.308

15.9 Copernicium Z = 112

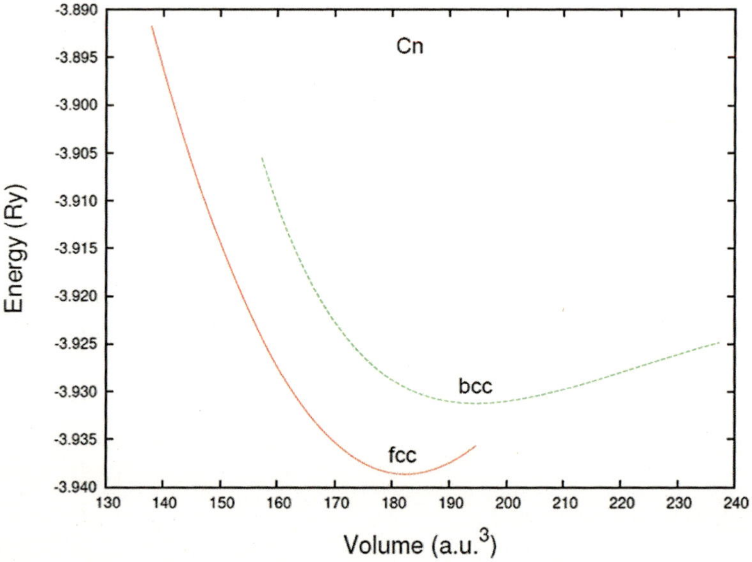

Fig. 15.25 Total energy of Cn

Table 15.25 Lattice constants and bulk modulus

	a (Bohr)	B (MBar)
bcc	7.301	0.485
fcc	9.002	1.110
$\Delta E_{fcc-bcc} = 7.39$ mRy		

Table 15.26 Copernicium fcc Z = 112 Orthogonal Slater-Koster parameters a = 9.00 a.u

	Three-center (Ry)		Two-center (Ry)
On site			
s,s(000)	−0.09513	s	−0.09537
x,x(000)	0.81450	p	0.81481
xy,xy(000)	−0.04260	d1	−0.04247
d2,d2(000)	−0.04250	d2	−0.04423
First neighbor			
s,s(110)	−0.03181	(sss)	−0.03189
s,x(110)	0.04130	(pps)	0.13949
s,xy(110)	−0.02607	(ppp)	−0.00650
s,d2(110)	0.01590	(dds)	−0.02870
x,x(110)	0.06187	(ddp)	0.01346
x,x(011)	0.00290	(ddd)	−0.00177
x,y(110)	0.07925	(sps)	0.05657
x,xy(110)	−0.02371	(sds)	−0.03055

(continued)

Table 15.26 (continued)

	Three-center (Ry)		Two-center (Ry)
x,xy(011)	0.00532	(pds)	−0.03829
z,d2(011)	−0.00154	(pdp)	0.00878
z,d1(011)	0.01698		
xy,xy(110)	−0.02176		
xy,xy(011)	0.00550		
xy,xz(011)	0.00751		
xy,d2(110)	0.01209		
d2,d2(110)	−0.00832		
d1,d1(110)	0.01340		
Second neighbor			
s,s(200)	−0.00111	(sss)	−0.00072
s,x(200)	0.00036	(pps)	0.01599
s,d2(002)	−0.00044	(ppp)	−0.01489
x,x(200)	0.00752	(dds)	−0.00184
y,y(200)	−0.01071	(ddp)	0.00105
x,xy(020)	0.00261	(ddd)	−0.00004
z,d2(002)	−0.00794	(sps)	0.00447
xy,xy(200)	0.00124	(sds)	−0.00043
xy,xy(002)	−0.00009	(pds)	−0.01212
d2,d2(002)	−0.00175	(pdp)	0.00086
d1,d1(002)	−0.00053		

Band	RMS error mRy	Maximum k	Deviation mRy	Band	RMS error mRy	Maximum k	Deviation mRy
1	2.9	(333)	6.3	1	3.5	(333)	8.8
2	2.4	(800)	4.7	2	3.7	(800)	8.7
3	2.6	(800)	4.3	3	2.7	(800)	6.1
4	2.2	(840)	8.1	4	3.6	(840)	11.1
5	1.2	(622)	2.6	5	2.0	(622)	4.6
6	3.2	(660)	7.9	6	4.0	(660)	10.1
1–6	2.5			1–6	3.3		

Copernicium

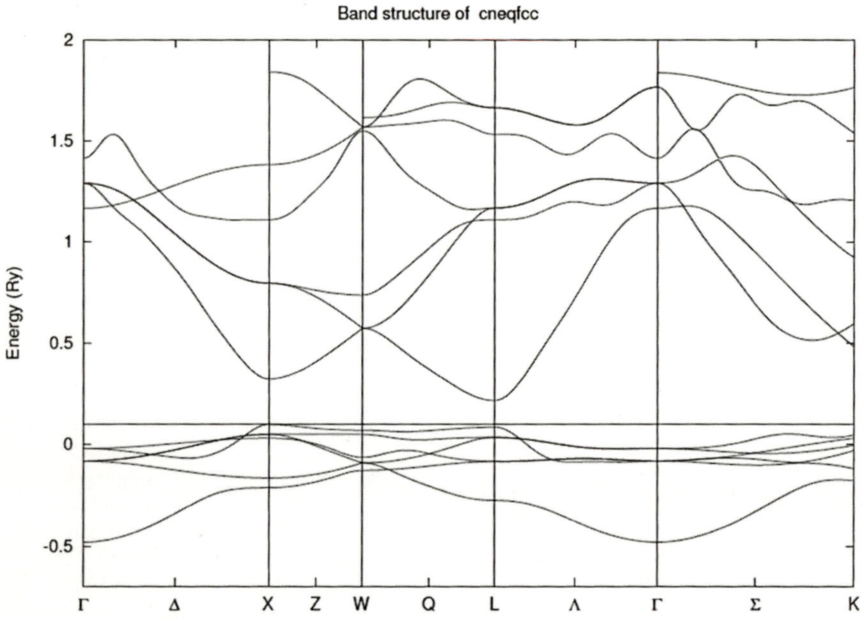

Fig. 15.26 Energy bands for Cn

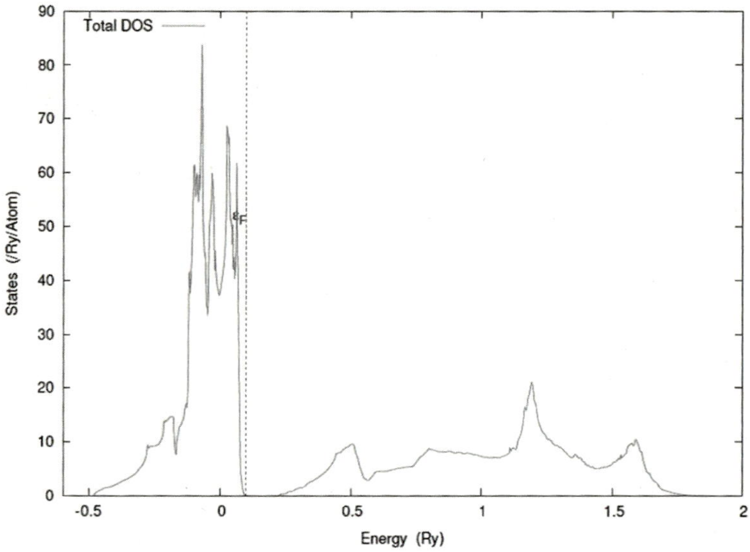

Fig. 15.27 Density of states for Cn

Table 15.27 Energy Gap

Ry
0.12

Appendix 1
Spin–orbit Coupling

The implementation of spin–orbit coupling in this TB approach requires doubling of the original 9×9 matrix and the introduction of two additional parameters ξ_p and ξ_d in the second 9×9 block corresponding to the p and d orbitals, respectively. We have followed here the formalism of Hass et al. [1] where the Hamiltonian has the form

$$\widetilde{H} + \widetilde{H}_{so} = \begin{pmatrix} \widetilde{H} + \widetilde{H}_{so}(\uparrow\uparrow) & \widetilde{H}_{so}(\uparrow\downarrow) \\ \widetilde{H}_{so}(\downarrow\uparrow) & \widetilde{H} + \widetilde{H}_{so}(\downarrow\downarrow) \end{pmatrix},$$

where H is the usual 9×9 Slater–Koster matrix and H_{so} is a 9×9 matrix that contains the spin–orbit parameters ξ. Although the determination of these two parameters could be made by fitting to first–principles calculations that include the spin–orbit interaction, we chose the simple approach of using the atomic values of Herman and Skillman [2] found by perturbation theory. In the present work [3] we have applied this methodology to obtain the spin–orbit effects in both the total energy and the band structure in the 5d series of transition metals. In the energy band figures we present below, which include the spin–orbit interaction, we follow the double group notation of Cornwell [4].

In Fig. A.1 we show the total energy of Au as a function of volume for both the fcc and bcc structures. The solid lines represent the results that include the spin–orbit interaction (WSO) and the dotted lines correspond to the results without the spin–orbit interaction (NSO). The results with the spin–orbit interaction maintain the result of the WSD case, predicting the fcc as the ground state in Au. It is also important to note that the predicted equilibrium lattice constant and bulk modulus are very close between NSO (a = 7.675 Bohr, B = 1.945 Mbar) and WSO (a = 7.682 Bohr, B = 1.936 Mbar) and in good agreement with experiment (a = 7.71 Bohr, B = 1.73 Mbar).

D.A. Papaconstantopoulos, *Handbook of the Band Structure of Elemental Solids*,
DOI 10.1007/978-1-4419-8264-3

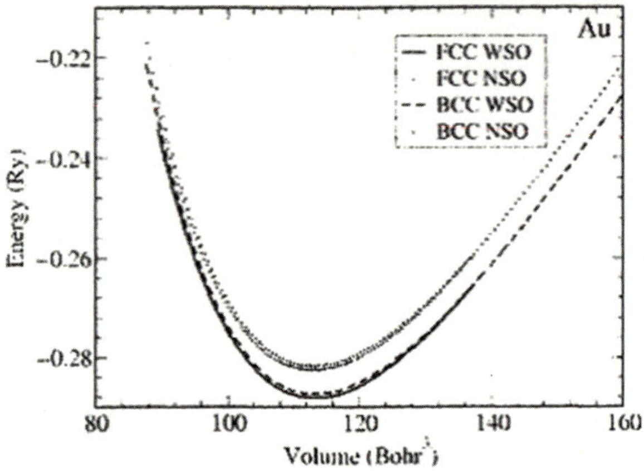

Fig. A.1 Au–total energy as a function of volume comparing spin–orbit and non–spin–orbit results

In Fig. A.2 we show the energy bands of the fcc metal Au. The panel on the left shows the results with the spin–orbit interaction included, and the panel on the right shows the results without the spin–orbit coupling. Note that the non–spin–orbit results are scalar–relativistic, i.e. they include the mass–velocity and Darwin corrections. We see that the lowest band is unchanged near Γ, since it has *s*–like symmetry, but shifts slightly when it approaches the boundaries of the Brillouin zone at X_1 which has mixed *s* and *d* symmetry. The next block of bands starting at about −0.4 Ry represents the *d*–bands. The $\Gamma_{25'}$, which is triply degenerate, splits in the double group notation, to a doubly degenerate state and a non–degenerate state. Note that the doubly degenerate Γ_{12} does not split but shifts to higher energy, and in the double group notation is called $\Gamma_8{}^-$. Similar splittings occur at the other high symmetry points X_5, W_3, and L_3 where each is represented by two non–degenerate states. Near the Fermi level, E_f, we observe no change of the band that crosses E_f. Although this band (sixth) has d–character at Γ_{12}, it develops to an sp band near E_f and is not effected by the spin–orbit coupling. Therefore, the spin–orbit coupling does not affect the Fermi surface or the density of states (DOS) at E_f.

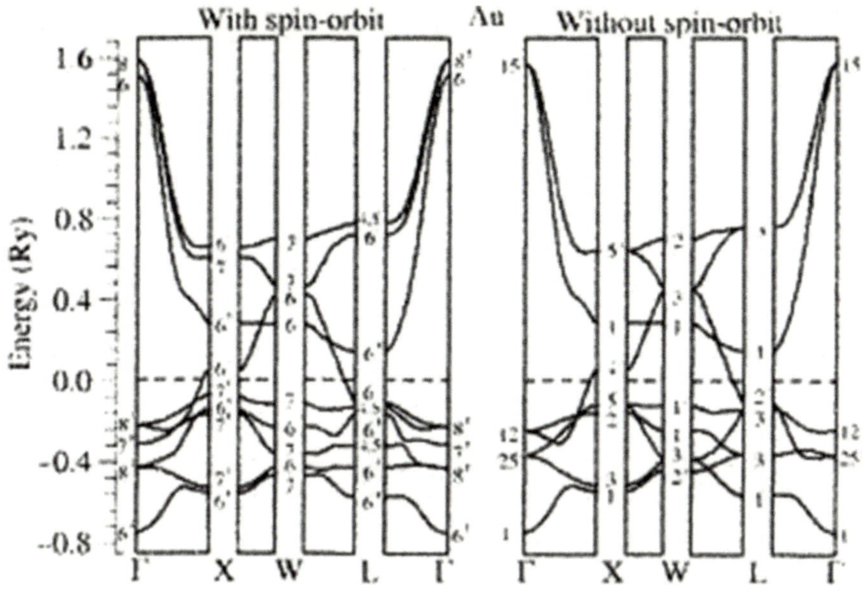

Fig. A.2 Comparison of the energy bands of Au with and without spin–orbit coupling

The DOS are shown in Fig. A.3. From this figure one can see that from E_f down to the first peak the DOS is not effected by the spin–orbit coupling. Further down in the region where the DOS is dominated by the d–bands there are pronounced spin–orbit effects. At the bottom where the *s*–like DOS is dominant, spin–orbit coupling has no effect as expected. We also note that because there are no significant differences at E_f the $N(E_f) = 4.28$ states/Ry/atom WSO is very close to the NSO result $N(E_f) = 3.98$ states/Ry/atom.

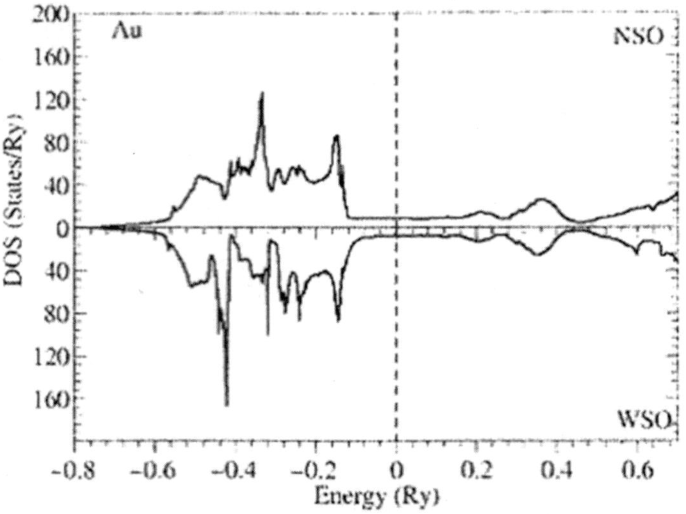

Fig. A.3 Comparison of the density of states of Au. *NSO* denotes the non–spin–orbit case and *WSO* refers to the DOS that includes spin–orbit coupling

We discuss here the bcc metal W. In Fig. A.4 the total energy results are shown. As in Au the WSO (a = 5.945 Bohr, B = 3.265 Mbar) and NSO (a = 5.935 Bohr, B = 3.340 Mbar) results are in close agreement and reproduce the value of the measure lattice parameter and bulk modulus very well (a = 5.97 Bohr, B = 3.230 Mbar). Regarding the energetic difference between fcc and bcc, we find a somewhat larger discrepancy between WSO and NSO, but the basic conclusion is that spin–orbit coupling does not produce significant differences.

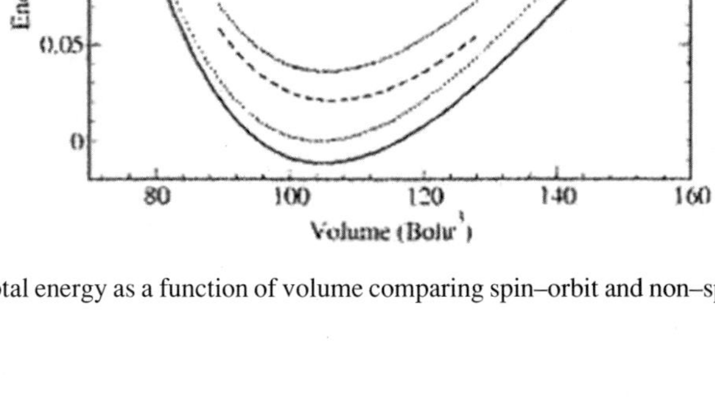

Fig. A.4 W–total energy as a function of volume comparing spin–orbit and non–spin–orbit results

In Fig. A.5 the energy bands of W are shown. At Γ the $\Gamma_{25'}$ state splits into a doubly degenerate Γ_{8+} and a non–degenerate Γ_{7+}.

Similarly, the triply degenerate P_4 state splits into a doubly degenerate P_8 and a non–degenerate P_7. There is no splitting due to spin–orbit coupling along the Γ–N direction. Looking near E_f we observe changes only in the Γ–H direction which will have an effect on the Fermi surface around Γ. In the other directions there are no changes in crossings with E_f that could cause changes to the Fermi surface.

In Fig. A.6 the DOS is shown for both the WSO and NSO cases. Here we see much smaller differences than those seen in Fig. A.3 for Au. The reason is that in the bcc structure away from the centre of the Brillouin zone there are not as many splittings due to spin–orbit coupling as is the case for the fcc structure. Also for W, with atomic number $Z=74$, we expect smaller splitting than in Au ($Z=79$). As a result at E_f we have $N(E_f)=4.10$ states/Ry/atom for WSO and $N(E_f)=4.09$ states/Ry/atom for NSO, a negligible difference.

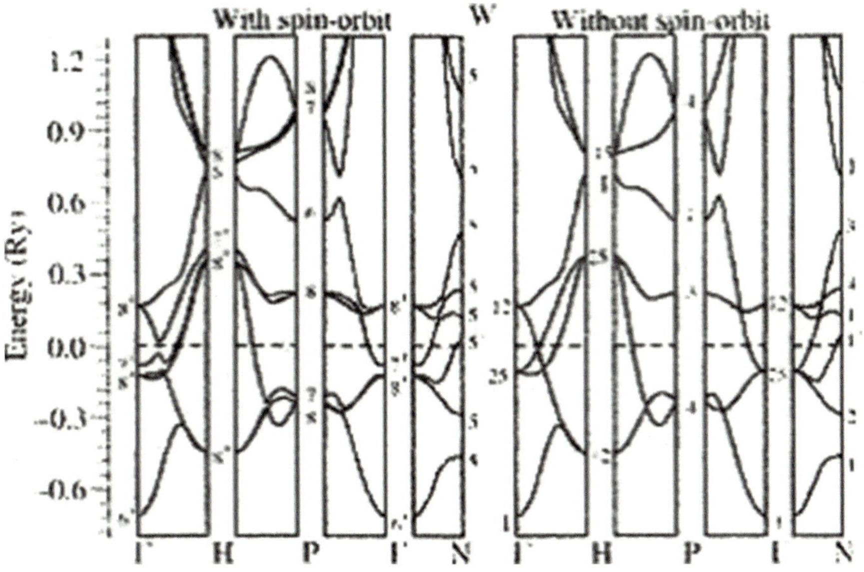

Fig. A.5 Comparison of the energy bands of W with and without spin–orbit coupling

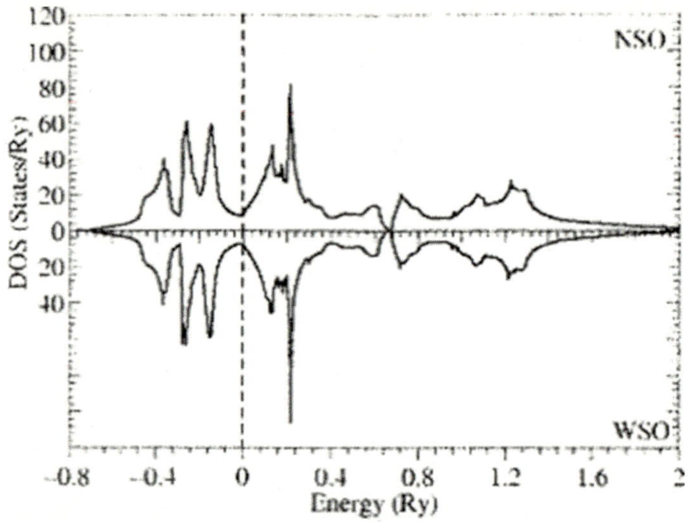

Fig. A.6 Comparison of the density of states of W. *NSO* denotes the non–spin–orbit case and *WSO* refers to the DOS that includes spin–orbit coupling

In Tables A.1 and A.2 we list the spin–orbit splittings for all fcc and bcc metals in the 5d row of the periodic table. We have chosen to show the splittings at the high symmetry points. As expected the bcc metals Ta and W show smaller energy splittings than the fcc metals Ir, Pt, and Au. It is worth noting from Tables A.1 and A.2 that the value of the spin–orbit parameter ξ increases with Z, a trend that is also followed in the values of the splittings.

Table A.1 FCC: spin–orbit parameter ξ_d and splittings at high symmetry k–points in Ry

	ξ_d	$\Gamma_{25'}$	X_5	$X_{5'}$	W_3	L_3	$L_{3'}$
Ir	0.021	0.073	0.059	0.041	0.046	0.075	0.041
Pt	0.023	0.086	0.067	0.057	0.056	0.088	0.050
Au	0.027	0.114	0.070	0.055	0.073	0.104	0.059

Table A.2 BCC: spin–orbit parameter ξ_d and splittings at high symmetry k–points in Ry

	ξ_d	$\Gamma_{25'}$	$H_{25'}$	H_{15}	P_4
Ta	0.012	0.035	0.044	0.035	0.026
W	0.014	0.045	0.052	0.052	0.032

Table A.3 Diamond: spin–orbit parameter ξ_p and splittings at high symmetry k–points in eV

	ξ_p	$\Gamma_{25'}$	Γ_{15}	$L_{3'}$	L_3
Si	0.012	0.036	0.036	0.024	0.024
Ge	0.070	0.210	0.210	0.141	0.138
Sn	0.159	0.422	0.476	0.250	0.309

Semiconductors

For the semiconductors we followed the same approach as for the transition metals. Since the Hamiltonian now involves only sp states the spin-orbit parameter of relevance is ξ_p. The values of ξ_p taken again from Ref. [2] are given in Table A.3 together with the splittings at the Γ and L symmetry points. As can be seen from the energy band figures below the effect of spin-orbit is small for Si, but it becomes significant for Ge and fairly large for Sn. We note that the semi-metallic behavior shown for Sn in the semiconductor chapter is removed by spin-orbit and a small gap is opening up.

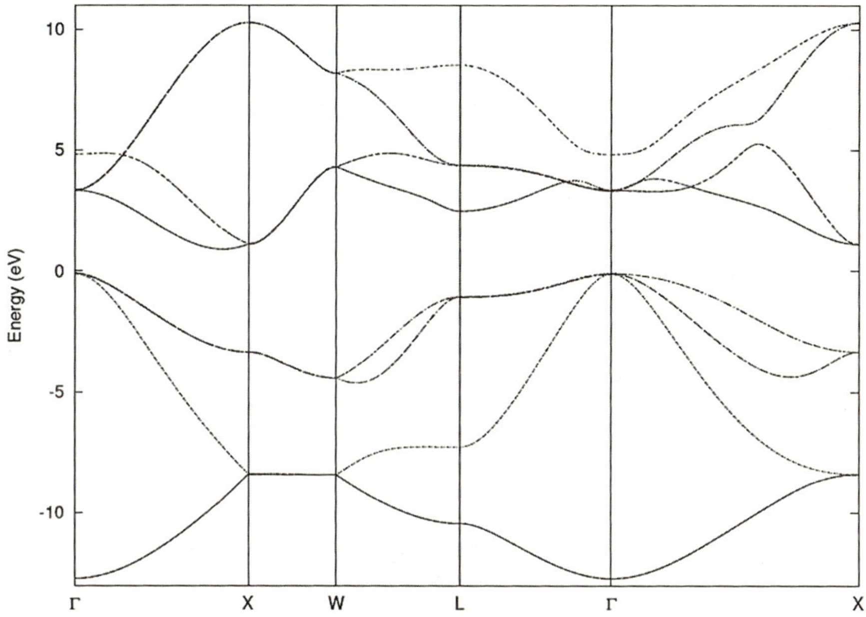

Fig. A.7 Energy bands of silicon

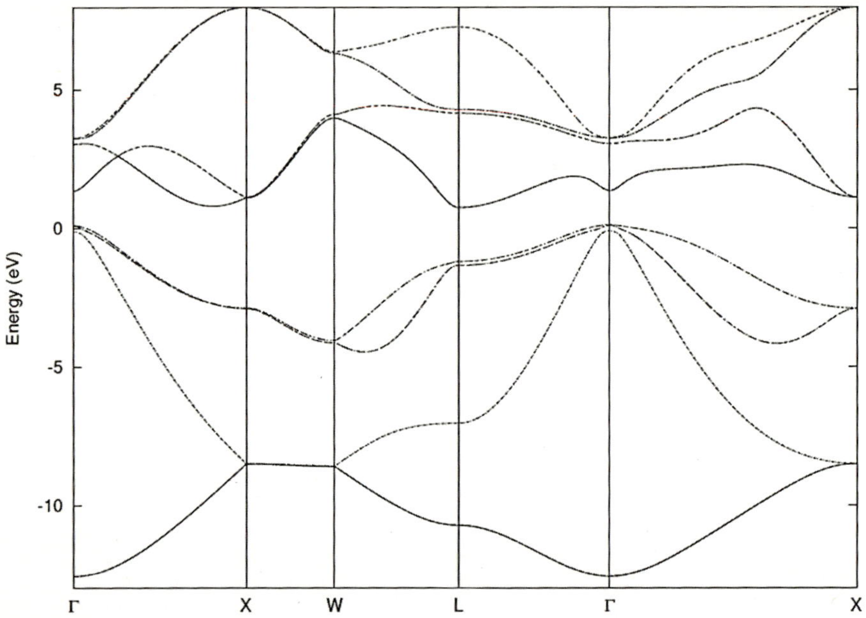

Fig. A.8 Energy bands of germanium

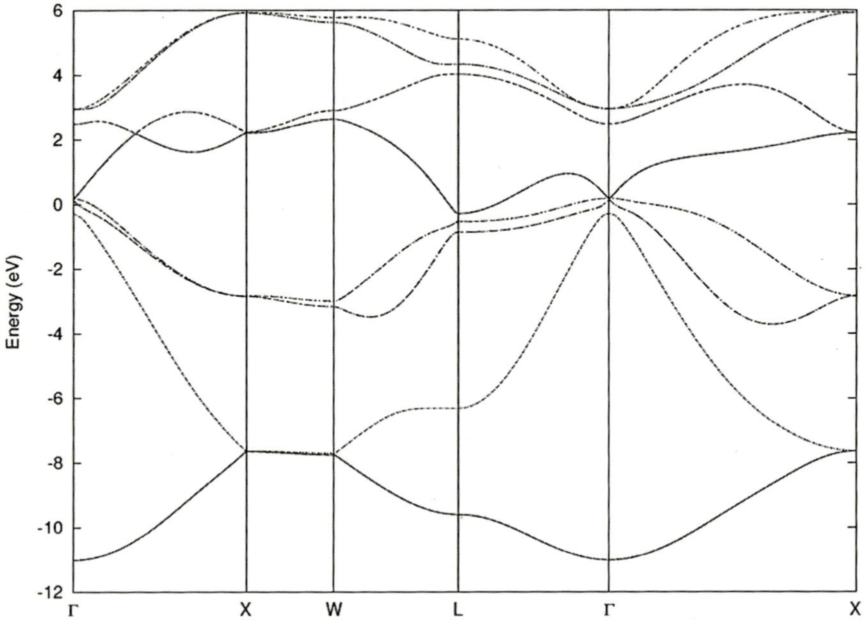

Fig. A.9 Energy bands of tin

References

1. K.C. Hass, H. Ehrenreich, B. Velicky, Phys. Rev. B **27**, 1088 (1983)
2. D. Herman, S. Skillman, *Atomic Structure Calculations* (Prentice Hall, Englewood Cliffs, 1963)
3. M.M. Lach–hab, D.A. Papaconstantopoulos, Phil. Mag. **88**, 2799 (2008)
4. J.F. Cornwell, *Group Theory in Electronic Energy Bands in Solids*, vol. 10 (Wiley Interscience, 1969), pp. 218–219

Appendix 2
Modifications and Extensions to Harrison's Tight-Binding Theory

Harrison's tight-binding theory provides an excellent qualitative description of the electronic structure of the elements across the periodic table. However, the resulting band structures are insignificant disagreement with those found by standard methods, particularly for the transition metals. For these systems we developed a new procedure to generate both the prefactors of Harrison's hopping parameters and the onsite energies. Our approach gives an impressive improvement and puts Harrison's theory on a quantitative basis. Our method retains the most attractive aspect of the theory, in using a revised set of universal prefactors for the hopping integrals.

To accomplish this we have made the following modifications to Harrison's theory: (1) We introduced a p onsite energy as an additional parameter to the s and d onsite energies used by Harrison, and fit them all to APW results. (2) We modified the sp hopping integrals of Harrison, by introducing a dimensionless parameter γ_s as follows:

$$V_{ll'm} = \eta_{ll'm} \frac{\hbar^2 \gamma_s}{m_e d^2} \tag{A.1}$$

The parameter γ_s provides more flexibility to fit the first and sixth bands. (3) For the sd, pd, and dd interactions we used the original form of Harrison's matrix elements as follows:

$$V_{ldm} = \eta_{ldm} \frac{\hbar^2 r_d^{3/2}}{m_e d^{7/2}} \tag{A.2}$$

$$V_{ddm} = \eta_{ddm} \frac{\hbar^2 r_d^3}{m_e d^5} \tag{A.3}$$

with r_d as a new element-dependent parameter. We obtained new hopping prefactors by simultaneously fitting the APW energy bands of the following

D.A. Papaconstantopoulos, *Handbook of the Band Structure of Elemental Solids*,
DOI 10.1007/978-1-4419-8264-3

12 transition metals: V, Cr, Ni, Cu, Nb, Mo, Pd, Ag, Ta, W, Pt and Au. In this fit, all 12 elements have the same common prefactors $\eta_{ll'm}$ but different (for each element) onsite energies s, p and d, and also different values for the parameters γ_s and r_d that appear in the hopping parameters. Our Hamiltonian corresponds to an orthogonal basis set as in Harrison. We did the above fitting at the equilibrium lattice constants of the structure, which is the ground state of each element, and included interactions of nearest, second-nearest, and third-nearest neighbors for the *bcc* structure and nearest and second-nearest for the *fcc* structure. Using the parameters determined with the above procedure, were produced APW energy bands and density of states (DOS) remarkably well, not only for the 12 elements originally fitted, but also for the rest of the transition metals, the alkaline earth and the noble metals, as seen in Fig. A.10 for four of the elements. Our new Hamiltonian prefactors, common for all metals, together with Harrison's original prefactors are shown in Table A.4. The onsite terms and the parameters γ_s and r_d for each element are shown in Table A.5.

We used the prefactors of Table A.4 with new onsite energies γ_s and r_d to fit the rest of the transition metals, including those with hcp ground states. For the hcp metals, we fitted energy bands of fcc structures at the equilibrium lattice, and found that our parameters produce good transferability, ie. reproduced the hcp energy bands very well without fitting them. The hcp energy bands of Ti and Ru are shown in Fig. A.11. We also fitted energy bands of the ferromagnetic elements Fe, Co and Ni, and calculated magnetic moments of the three elements at the experimental lattice constant. Table A.6 shows good agreement of magnetic moments of Fe, Co and Ni with experimental values.

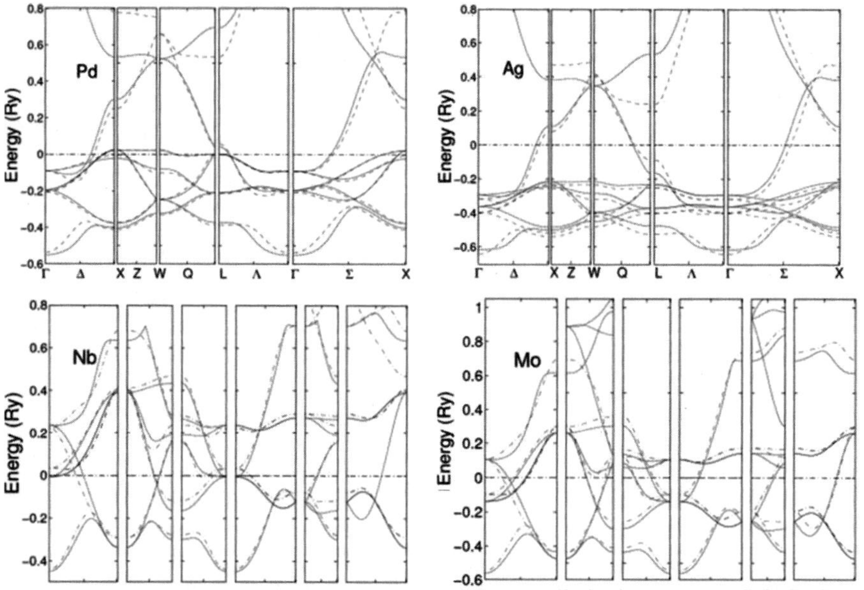

Fig. A.10 APW and Modified Harrison's energy bands of Ag, Pd, Nb and Mo. The *solid line* is the modified Harrison result and dash-dotted line is the APW result

Table A.4 Harrison's hopping prefactors and our modified values

	$\eta_{ss\sigma}$	$\eta_{sp\sigma}$	$\eta_{pp\sigma}$	$\eta_{pp\pi}$	$\eta_{sd\sigma}$
Harrison	−1.32	1.42	2.22	−0.63	−3.16
Modified Harrison	−0.90	1.44	2.19	−0.03	−3.12
	$\eta_{pd\sigma}$	$\eta_{pd\pi}$	$\eta_{dd\sigma}$	$\eta_{dd\pi}$	$\eta_{dd\delta}$
Harrison	−2.95	1.36	−16.2	8.75	−2.39
Modified Harrison	−4.26	2.08	−21.22	12.60	−2.29

To summarize, we have reevaluated the ten universal prefactors in Harrison's hopping parameters and redetermined the *s, p, d* onsite energies together with the parameters γ_s and r_d. This enables us to calculate very accurately the band structure of all the transition, alkaline earth and noble metals.

Table A.5 Onsite parameters, γ_s and r_d

Name	$s(Ry)$	$p(Ry)$	$d(Ry)$	γ_s	$r_d(a.u.)$
K	0.26067	0.22200	0.25426	1.13081	3.65981
Ca	0.11994	0.24522	0.03657	1.07535	2.66618
Sc	0.14809	0.38903	−0.06695	0.98860	2.12358
Ti	0.51352	0.79759	0.21879	0.92307	1.85267
V	0.64331	0.73136	0.04711	0.90164	1.65358
Cr	0.76372	0.86088	0.06389	0.87733	1.51087
Mn	0.47377	0.86874	−0.00300	0.82491	1.42366
Cu	0.54432	0.93013	−0.05425	0.92178	1.23548
Zn	0.44779	0.77968	−0.10598	0.78430	0.97054
Sr	0.32339	0.41296	0.22810	1.22463	3.29024
Y	0.29652	0.51778	−0.07450	1.19367	2.75073
Zr	0.54322	0.87432	0.17820	1.15726	2.40732
Nb	0.85097	0.99247	0.24572	1.08802	2.19244
Mo	0.83057	0.97345	0.10805	1.06314	2.01708
Tc	0.64629	1.08072	0.09302	1.00014	1.91300
Ru	0.65130	1.07465	0.04760	1.00001	1.80799
Rh	0.68579	1.06923	0.06445	0.99989	1.71702
Pd	0.57192	0.95218	0.04268	0.90172	1.63401
Ag	0.44541	0.79565	−0.04959	0.84306	1.52479
Ba	−0.04951	0.03502	−0.20497	1.07269	3.56198
Hf	0.29999	0.75423	0.32239	0.88204	2.50856
Ta	0.70455	0.92990	0.23577	1.12532	2.31790
W	0.64038	0.86882	0.09170	1.11008	2.17888
Re	0.60996	1.15988	0.15348	1.14822	2.08878
Os	0.53044	1.06428	0.05117	1.11453	2.01287
Ir	0.47125	1.01759	0.01404	1.06585	1.91597
Pt	0.43374	0.94903	0.00569	1.00933	1.83802
Au	0.37521	0.84519	−0.02211	0.94002	1.75060
Hg	0.36137	0.68747	−0.10952	0.90569	1.53337

(continued)

Table A.5 (continued)

Name	$s(Ry)$	$p(Ry)$	$d(Ry)$	γ_s	$r_d(a.u.)$
Fe[a]	0.87761	0.84369	0.02940	0.94826	1.33156
Fe[b]	0.84395	0.88024	0.19670	0.93012	1.43124
Ni[a]	0.45155	0.69040	−0.04560	0.72937	1.22004
Ni[b]	0.46394	0.70316	−0.00173	0.73568	1.24548
Co[a]	0.69846	0.68425	−0.06187	0.79695	1.26137
Co[b]	0.66026	0.70002	0.06184	0.77917	1.33184

[a]Ferromagnetic spin up
[b]Ferromagnetic spin down

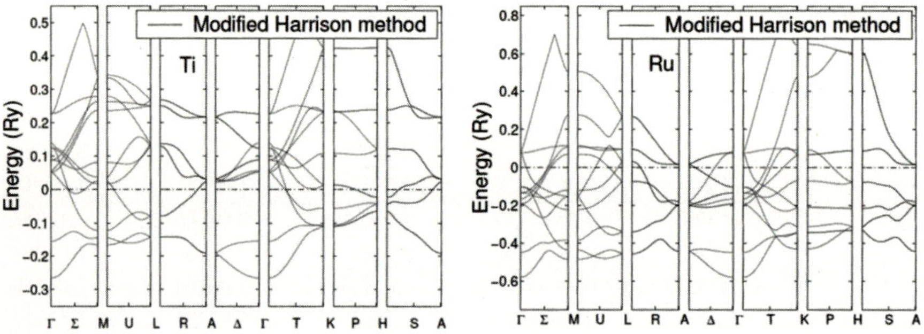

Fig. A.11 Energy bands of hcp Ti and Ru

Table A.6 Magnetic moments of Fe, Co and Ni

Element	Structure	TB(μ_B)	Exp.(μ_B)
Fe	bcc	2.21	2.22
Co	hcp	1.52	1.72
Ni	fcc	0.56	0.61

References

1. W.A. Harrison, *Electronic Structure and Properties of Solids* (Dover, 1980)
2. Lei Shi, D.A. Papaconstantopoulos, Phys. Rev. B **70**, 205101 (2004)

Appendix 3
Alternate Structures

In this appendix 15 additional tables are presented, including the hcp 3d and 4d transition metals calculated now in a cubic bcc or fcc structure. An fcc parametrization for Mn and tabulations for paramagnetic Fe, Co, and Ni are also given.

Table A.7 Scandium bcc z = 21 lattice constant = 6.70000 a.u.

Slater–Koster 3–center parameters	Orthogonal	Non–orthogonal	
	Energy integrals (Ry)	Energy integrals (Ry)	Overlap integrals
On site			
s, s (000)	0.79602	0.66148	
x, x (000)	1.25811	0.98077	
xy, xy (000)	0.73094	0.67174	
d2, d2 (000)	0.66480	0.63707	
First neighbor			
s, s (111)	−0.09004	−0.04596	0.09045
s, x (111)	0.06252	0.02620	−0.06448
s, xy (111)	0.03590	0.01993	−0.02382
x, x (111)	0.07076	−0.01978	−0.03489
x, y (111)	0.04736	0.02635	−0.04886
x, xy (111)	0.02823	0.00115	−0.01759
x, yz (111)	0.03822	0.01859	−0.03164
x, d1 (111)	0.00859	0.01174	−0.02249
xy, xy (111)	−0.01235	−0.00715	0.00573
xy, xz (111)	−0.02016	−0.00770	0.01762
xy, d2 (111)	0.00756	0.00941	−0.01484
d2, d2 (111)	0.02091	0.01105	−0.02378
Second neighbor			
s, s (200)	−0.01984	−0.00537	0.05996
s, x (200)	0.05074	−0.04503	−0.13038
s, d2 (002)	−0.04192	0.00608	0.06306
x, x (200)	0.12658	−0.16978	−0.27075
y, y (200)	0.03373	0.00366	0.01259
x, xy (020)	0.01536	0.00489	0.01715
z, d2 (002)	−0.06874	0.03394	0.11015

(continued)

D.A. Papaconstantopoulos, *Handbook of the Band Structure of Elemental Solids*,
DOI 10.1007/978-1-4419-8264-3

Table A.7 (continued)

Slater–Koster 3–center parameters	Orthogonal	Non–orthogonal	
	Energy integrals (Ry)	Energy integrals (Ry)	Overlap integrals
xy, xy (200)	−0.00922	−0.00183	−0.02044
xy, xy (002)	0.00280	0.00077	0.00054
d2, d2 (002)	−0.03633	−0.00068	0.05577
d1, d1 (002)	0.00323	−0.00099	−0.00014
Third neighbor			
s, s (220)	0.01405		
s, x (220)	−0.01468		
s, xy (220)	−0.01063		
s, d2 (220)	−0.00244		
x, x (220)	−0.01276		
x, x (022)	0.01343		
x, y (220)	−0.01360		
x, xy (220)	−0.00905		
x, xy (022)	0.00362		
z, d2 (022)	0.00075		
z, d1 (022)	−0.00313		
xy, xy (220)	0.00787		
xy, xy (022)	−0.00031		
xy, xz (022)	−0.00085		
xy, d2 (220)	0.00452		
d2, d2 (220)	0.00121		
d1, d1 (220)	−0.00053		

Table A.8 Scandium bcc

	Orthogonal			Non–orthogonal		
Band	RMS error mRy	Maximum k	Deviation mRy	RMS error mRy	Maximum k	Deviation mRy
1	0.9	(222)	2.9	0.4	(333)	1.2
2	0.8	(008)	2.2	0.2	(131)	1.1
3	1.2	(242)	3.5	0.3	(161)	1.6
4	1.3	(332)	3.2	0.3	(140)	1.2
5	1.2	(240)	3.6	0.4	(131)	2.2
6	2.8	(353)	8.9	0.6	(342)	2.4
1–6	1.5			0.4		

Table A.9 Energy values in Ry at selected k–points

	Orthogonal	APW	Non–orthogonal
GAMMA 1	0.12524	0.12507	0.12555
GAMMA12	0.73683	0.73735	0.73767
GAMMA 15	2.16397	2.16635	2.16633
GAMMA 25′	0.62988	0.63035	0.63074
H1 (008)	1.56582	1.56716	1.56704
Hl2 (008)	0.40231	0.40016	0.40064
H15 (008)	1.03175	1.03368	1.03310

(continued)

Table A.9 (continued)

	Orthogonal	APW	Non–orthogonal
H25′(008)	0.82743	0.82733	0.82789
NI (044)	0.37465	0.37420	0.37383
NI (044)	0.73831	0.74051	0.74081
NI (044)	1.09830	1.09554	1.09587
N2 (044)	0.53259	0.53387	0.53391
N3 (044)	0.85517	0.85251	0.85296
N4 (044)	0.76870	0.76613	0.76646
NI′ (044)	0.57233	0.57395	0.57420
N3′ (044)	1.53213	1.53504	1.53518
N4′ (044)	1.33012	1.33054	1.33065
P1 (444)	1.08361	1.08372	1.08357
P3 (444)	0.76818	0.77036	0.77067
P4 (444)	0.50030	0.50146	0.50109
P4 (444)	1.11263	1.11294	1.11233
(341)	0.38847	0.38883	0.38874
(341)	0.53249	0.53313	0.53330
(341)	0.54845	0.54706	0.54703
(341)	0.73484	0.73496	0.73508
(341)	0.77306	0.77256	0.77266
(341)	0.86792	0.86799	0.86827

Table A.10 Fermi level quantities (non–orthogonal fit)

Energy		Densities of states				Velocity	Plasmon energy
	Total	s	p	t_{2g}	e_g		
Ry		States/Ry/atom				cm/s	eV
0.5170	32.59	0.18	5.81	19.13	7.48	$0.27 \times 10E8$	4.57

Integrated densities of states				
Total	s	p	t_{2g}	e_g
Electrons				
3.00	0.76	0.41	0.83	1.00

Table A.11 Scandium bcc

Slater–Koster 2–center parameters	Orthogonal	Non–orthogonal	
	Energy integrals (Ry)	Energy integrals (Ry)	Overlap integrals
On site			
s	0.85459	0.66176	
p	1.25507	0.98610	
dl	0.70407	0.66288	
d2	0.66705	0.64236	
First neighbor			
(sss)	−0.08980	−0.03968	0.09245
(pps)	0.16525	0.03093	−0.13595
(ppp)	0.02308	−0.04112	0.01435
(dds)	−0.05340	−0.01109	0.05701

(continued)

Table A.11 (continued)

Slater–Koster 2–center parameters	Orthogonal Energy integrals (Ry)	Non–orthogonal Energy integrals (Ry)	Overlap integrals
(ddp)	0.03110	0.02181	−0.03133
(ddd)	−0.00131	−0.00975	−0.00249
(sps)	0.10682	−0.04136	0.11480
(sds)	−0.06417	−0.02882	0.04781
(pds)	−0.08823	0.00462	−0.08254
(pdp)	0.01673	−0.02678	0.03095
Second neighbor			
(sss)	−0.02016	−0.01452	0.05148
(pps)	0.14253	−0.15331	−0.25831
(ppp)	0.01465	0.01614	0.02072
(dds)	−0.03646	−0.01037	0.04099
(ddp)	0.00202	−0.00229	−0.02343
(ddd)	0.00239	−0.00017	0.00068
(sps)	0.05550	0.03046	0.11837
(sds)	−0.03485	0.00474	0.06001
(pds)	−0.06168	−0.02809	−0.10121
(pdp)	−0.00454	0.00990	0.02391
Third neighbor			
(sss)	0.00901		
(pps)	−0.00837		
(ppp)	0.00389		
(dds)	0.00450		
(ddp)	−0.00071		
(ddd)	−0.00011		
(sps)	−0.00940		
(sds)	0.00308		
(pds)	0.00278		
(pdp)	−0.00185		

	Orthogonal			Non–orthogonal		
Band	RMS error mRy	Maximum k	Deviation mRy	RMS error mRy	Maximum k	Deviation mRy
1	2.4	(330)	7.1	0.4	(333)	1.2
2	4.6	(044)	15.7	0.3	(252)	0.8
3	4.1	(350)	10.6	0.4	(161)	1.6
4	4.4	(260)	13.7	0.5	(140)	1.2
5	3.4	(008)	8.5	0.6	(131)	2.5
6	5.0	(353)	13.5	0.7	(342)	2.2
1–6	4.1			0.5		

Table A.12 Titanium fcc z = 22 lattice constant = 7.74000 a.u.

Slater–Koster 3–center parameters	Orthogonal Energy integrals (Ry)	Non–orthogonal Energy integrals (Ry)	Overlap integrals
On site			
s, s (000)	1.04668	0.78712	
x, x (000)	1.45429	1.11848	

(continued)

Table A.12 (continued)

Slater–Koster 3–center parameters	Orthogonal	Non–orthogonal	
	Energy integrals (Ry)	Energy integrals (Ry)	Overlap integrals
xy, xy (000)	0.69179	0.68615	
d2, d2 (000)	0.68321	0.67302	
First neighbor			
s, s (110)	−0.07452	−0.03797	0.06933
s, x (110)	0.08004	0.01493	−0.09013
s, xy (110)	−0.04132	−0.03996	0.03278
s, d2 (110)	0.02089	0.02797	−0.00251
x, x (110)	0.07775	−0.00406	−0.07492
x, x (011)	0.00677	0.05044	0.06993
x, y (110)	0.11001	−0.05900	−0.15219
x, xy (110)	−0.03828	−0.02128	0.03468
x, xy (011)	0.01582	−0.02123	−0.05196
z, d2 (011)	0.01384	−0.00995	−0.03227
z, d1 (011)	0.03016	0.00565	−0.03928
xy, xy (110)	−0.03882	−0.00456	0.04701
xy, xy (011)	0.01172	0.00232	−0.01925
xy, xz (011)	0.01945	−0.01323	−0.04098
xy, d2 (110)	0.02105	0.01160	−0.00921
d2, d2 (110)	−0.01775	−0.00992	0.01157
d1, d1 (110)	0.03549	0.01895	−0.02522
Second neighbor			
s, s (200)	−0.00324	−0.01510	−0.00898
s, x (200)	0.00315	0.04260	0.01274
s, d2 (002)	0.00105	−0.02311	−0.02495
x, x (200)	−0.00055	0.06339	−0.01400
y, y (200)	−0.01371	0.02762	0.02466
x, xy (020)	0.00355	−0.01781	−0.01777
z, d2 (002)	−0.00242	−0.03126	−0.01883
xy, xy (200)	0.00341	−0.01421	−0.01580
xy, xy (002)	0.00120	−0.00174	−0.00196
d2, d2 (002)	−0.00758	−0.00375	0.00808
d1, d1 (002)	−0.00214	0.00889	0.01103

Table A.13 Titanium fcc

	Orthogonal			Non–orthogonal		
Band	RMS error mRy	Maximum k	Deviation mRy	RMS error mRy	Maximum k	Deviation mRy
1	2.6	(333)	7.2	0.2	(022)	0.4
2	2.8	(055)	6.3	0.4	(226)	1.0
3	4.3	(008)	11.7	0.3	(055)	0.8
4	3.8	(048)	9.5	0.3	(226)	0.7
5	3.6	(044)	9.0	0.3	(224)	1.2
6	5.8	(066)	19.6	0.5	(055)	1.1
1–6	4.0			0.3		

Table A.14 Energy values in Ry at selected k–points

	Orthogonal	APW	Non–orthogonal
GAMMA 1	0.13295	0.13505	0.13548
GAMMA 12	0.76057	0.76102	0.76132
GAMMA 15	2.04740	2.03070	2.03074
GAMMA 25′	0.64629	0.64763	0.64757
X1 (008)	0.37618	0.37741	0.37761
X1 (008)	1.35474	1.35301	1.35325
X2 (008)	0.83154	0.83852	0.83824
X3 (008)	0.45884	0.46167	0.46200
X5 (008)	0.86310	0.86424	0.86468
X4′ (008)	0.80345	0.81518	0.81547
X5′ (008)	1.37123	1.46016	1.45901
Ll (444)	0.39759	0.39924	0.39915
Ll (444)	1.18873	1.18580	1.18582
L3 (444)	0.62776	0.63357	0.63372
L3 (444)	0.83812	0.84155	0.84135
L2′ (444)	0.63014	0.63088	0.63032
L3′ (444)	1.95031	2.01273	2.01265
Wl (048)	0.74244	0.73295	0.73323
Wl (048)	1.35121	1.35822	1.35841
W3 (048)	0.56606	0.56255	0.56254
W3 (048)	1.22347	1.21252	1.21333
Wl′ (048)	0.85829	0.86507	0.86495
W2′ (048)	0.45711	0.45597	0.45596
W2′ (048)	1.43699	1.34323	1.34365
EVEN (224)	0.36302	0.36539	0.36576
EVEN (224)	0.56179	0.55986	0.55939
EVEN (224)	0.62915	0.62996	0.63033
EVEN (224)	0.80745	0.80586	0.80709
ODD (224)	0.67218	0.67512	0.67445
ODD (224)	0.82597	0.82578	0.82565

Table A.15 Fermi level quantities (non–orthogonal fit)

Energy		Densities of states				Velocity	Plasmon energy
	Total	s	p	t_{2g}	e_g		
Ry		States/Ry/atom				cm/s	eV
0.5860	23.51	0.32	3.07	16.59	3.53	0.33 × 10E8	5.69

Integrated densities of states				
Total Electrons	s	p	t_{2g}	e_g
4.00	0.73	0.41	1.79	1.07

Table A.16 Titanium fcc

Slater–Koster 2–center parameters	Orthogonal	Non–orthogonal	
	Energy integrals (Ry)	Energy integrals (Ry)	Overlap integrals
On site			

(continued)

Table A.16 (continued)

Slater–Koster 2–center parameters	Orthogonal	Non–orthogonal	
	Energy integrals (Ry)	Energy integrals (Ry)	Overlap integrals
s	1.03558	0.62214	
P	1.41712	0.94286	
dl	0.69081	0.66501	
d2	0.67846	0.66340	
First neighbor			
(sss)	−0.07401	−0.01050	0.13248
(pps)	0.17076	−0.22041	−0.34928
(ppp)	−0.01111	−0.01049	0.05731
(dds)	−0.05135	−0.00641	0.06271
(ddp)	0.03367	0.01029	−0.04006
(ddd)	−0.00670	−0.00114	0.00708
(sps)	0.10645	0.05594	0.21449
(sds)	−0.04761	−0.01021	0.06991
(pds)	−0.06011	−0.03011	−0.11613
(pdp)	0.02290	−0.00361	0.05319
Second neighbor			
(sss)	−0.00327	−0.02352	0.00483
(pps)	0.03300	0.08840	−0.00337
(ppp)	−0.00407	−0.01445	−0.00180
(dds)	−0.00704	−0.02135	−0.01243
(ddp)	0.00328	−0.00044	−0.00348
(ddd)	0.00052	0.00080	0.00096
(sps)	0.00560	−0.04022	0.00443
(sds)	−0.00870	−0.01258	0.00359
(pds)	−0.01865	0.04523	0.01487
(pdp)	0.00559	−0.00184	0.00327

	Orthogonal			Non–orthogonal		
Band	RMS error mRy	Maximum k	Deviation mRy	RMS error mRy	Maximum k	Deviation mRy
1	7.9	(048)	16.6	0.7	(264)	1.8
2	11.5	(008)	25.3	0.7	(264)	1.6
3	9.5	(008)	31.4	1.1	(044)	3.3
4	7.1	(048)	19.2	0.8	(224)	2.5
5	7.0	(444)	14.6	0.7	(224)	1.9
6	10.7	(048)	26.1	0.9	(226)	2.5
1–6	9.1			0.8		

Table A.17 Titanium bcc z = 22 lattice constant = 6.24838 a.u.

Slater–Koster 3–center parameters	Orthogonal	Non–orthogonal	
	Energy integrals (Ry)	Energy integrals (Ry)	Overlap integrals
On site			
s, s (000)	0.92161	0.75885	
x, x (000)	1.42364	1.12159	
xy, xy (000)	0.71925	0.67788	

(continued)

Table A.17 (continued)

Slater–Koster 3–center parameters	Orthogonal	Non–orthogonal	
	Energy integrals (Ry)	Energy integrals (Ry)	Overlap integrals
d2, d2 (000)	0.67503	0.65170	
First neighbor			
s, s (111)	−0.10302	−0.04761	0.08894
s, x (111)	0.07599	0.02711	−0.06545
s, xy (111)	0.03924	0.02466	−0.01803
x, x (111)	0.08088	−0.02325	−0.03893
x, y (111)	0.05459	0.02409	−0.05239
x, xy (111)	0.02559	0.00513	−0.01324
x, yz (111)	0.04061	0.02066	−0.02771
x, d1 (111)	0.00733	0.01483	−0.01809
xy, xy (111)	−0.01251	−0.00824	0.00481
xy, xz (111)	−0.02029	−0.00919	0.01543
xy, d2 (111)	0.00932	0.01108	−0.01055
d2, d2 (111)	0.02105	0.01328	−0.01778
Second neighbor			
s, s (200)	−0.02270	−0.01918	0.04788
s, x (200)	0.05482	−0.02383	−0.10852
s, d2 (002)	−0.04319	−0.00172	0.04992
x, x (200)	0.13972	−0.14020	−0.23507
y, y (200)	0.03593	0.00378	0.01359
x, xy (020)	0.00961	0.00425	0.01798
z, d2 (002)	−0.07319	0.01585	0.08267
xy, xy (200)	−0.00295	0.00068	−0.01635
xy, xy (002)	−0.00029	0.00244	0.00308
d2, d2 (002)	−0.03523	−0.01108	0.03749
d1, d1 (002)	0.00242	−0.00138	−0.00081
Third neighbor			
s, s (220)	0.01404		
s, x (220)	−0.01459		
s, xy (220)	−0.00726		
s, d2 (220)	−0.00129		
x, x (220)	−0.01333		
x, x (022)	0.01694		
x, y (220)	−0.01756		
x, xy (220)	−0.00798		
x, xy (022)	0.00344		
z, d2 (022)	−0.00064		
z, d1 (022)	−0.00326		
xy, xy (220)	0.00562		
xy, xy (022)	0.00034		
xy, xz (022)	−0.00002		
xy, d2 (220)	0.00328		

(continued)

Table A.17 (continued)

Slater–Koster 3–center parameters	Orthogonal	Non–orthogonal	
	Energy integrals (Ry)	Energy integrals (Ry)	Overlap integrals
d2, d2 (220)	0.00124		
d1, d1 (220)	−0.00143		

Table A.18 Titanium bcc

	Orthogonal			Non–orthogonal		
Band	RMS error mRy	Maximum k	Deviation mRy	RMS error mRy	Maximum k	Deviation mRy
1	1.2	(222)	3.8	0.4	(333)	1.2
2	0.9	(350)	3.2	0.3	(131)	1.3
3	1.5	(242)	3.9	0.4	(161)	1.8
4	1.6	(350)	4.0	0.4	(140)	2.2
5	1.5	(240)	5.3	0.9	(131)	5.4
6	3.4	(353)	8.8	0.9	(140)	4.1
1–6	1.9			0.6		

Table A.19 Energy values in Ry at selected k–points

	Orthogonal	APW	Non–orthogonal
GAMMA 1	0.12975	0.13081	0.13150
GAMMA 12	0.74390	0.74409	0.74455
GAMMA 15	2.45501	2.45734	2.45735
GAMMA 25′	0.63195	0.63227	0.63273
H1 (008)	1.77805	1.78005	1.77992
H12 (008)	0.40703	0.40523	0.40570
H15 (008)	1.16089	1.16547	1.16374
H25′ (008)	0.83214	0.83228	0.83278
N1 (044)	0.38663	0.38596	0.38553
N1 (044)	0.74341	0.74715	0.74762
N1 (044)	1.17460	1.17108	1.17137
N2 (044)	0.53506	0.53647	0.53657
N3 (044)	0.85967	0.85725	0.85762
N4 (044)	0.77287	0.77009	0.77087
N1′ (044)	0.63968	0.63962	0.64006
N3′ (044)	1.73375	1.73747	1.73725
N4′ (044)	1.51317	1.51737	1.51734
P1 (444)	1.22625	1.22624	1.22613
P3 (444)	0.77230	0.77408	0.77429
P4 (444)	0.51858	0.51938	0.51902
P4 (444)	1.19986	1.19814	1.19844
(341)	0.40477	0.40509	0.40494
(341)	0.53907	0.53972	0.53990
(341)	0.58854	0.58746	0.58759
(341)	0.74187	0.74212	0.74226
(341)	0.78052	0.78027	0.78054
(341)	0.88565	0.88615	0.88641

Table A.20 Fermi level quantities (non–orthogonal fit)

Energy		Densities of states				Velocity	Plasmon energy
	Total	s	p	t_{2g}	e_g		
Ry		States/Ry/atom				cm/s	eV
0.5615	28.82	0.29	3.67	19.44	5.43	$0.31 \times 10E8$	5.34

Integrated densities of states				
Total	s	p	t_{2g}	e_g
Electrons				
4.00	0.67	0.43	1.67	1.23

Table A.21 Titanium bcc

Slater–Koster 2–center parameters	Orthogonal	Non–orthogonal	
	Energy integrals (Ry)	Energy integrals (Ry)	Overlap integrals
On site			
s	0.95493	0.73327	
p	1.42376	1.10489	
dl	0.70671	0.67252	
d2	0.67140	0.65546	
First neighbor			
(sss)	−0.10288	−0.05123	0.09351
(pps)	0.19121	0.02980	−0.14248
(ppp)	0.02539	−0.05759	0.00862
(dds)	−0.05393	−0.01962	0.04541
(ddp)	0.03164	0.02465	−0.02320
(ddd)	−0.00175	−0.00873	−0.00236
(sps)	0.12391	−0.05000	0.11737
(sds)	−0.06674	−0.03538	0.03946
(pds)	−0.09144	0.01408	−0.07055
(pdp)	0.01879	−0.03295	0.02296
Second neighbor			
(sss)	−0.02291	−0.00762	0.06078
(pps)	0.15698	−0.17948	−0.26041
(ppp)	0.02050	0.02348	0.02598
(dds)	−0.03543	−0.01823	0.02669
(ddp)	0.00282	−0.00130	−0.02032
(ddd)	0.00185	−0.00028	0.00043
(sps)	0.06144	0.04628	0.12739
(sds)	−0.03711	0.00359	0.05781
(pds)	−0.06164	−0.01736	−0.08539
(pdp)	−0.00538	0.01282	0.02842
Third neighbor			
(sss)	0.01114		
(pps)	−0.01617		
(ppp)	0.00642		
(dds)	0.00398		

(continued)

Table A.21 (continued)

Slater–Koster 2–center parameters	Orthogonal	Non–orthogonal	
	Energy integrals (Ry)	Energy integrals (Ry)	Overlap integrals
(ddp)	−0.00026		
(ddd)	−0.00031		
(sps)	−0.01318		
(sds)	0.00205		
(pds)	0.00282		
(pdp)	−0.00212		

	Orthogonal			Non–orthogonal		
Band	RMS error mRy	Maximum k	Deviation mRy	RMS error mRy	Maximum k	Deviation mRy
1	2.7	(330)	8.5	0.4	(333)	1.4
2	4.4	(044)	13.9	0.3	(131)	1.1
3	3.6	(226)	11.0	0.4	(161)	2.0
4	3.9	(260)	13.3	0.5	(140)	2.1
5	3.0	(070)	7.7	0.9	(131)	5.4
6	5.0	(353)	12.8	0.9	(140)	4.2
1–6	3.9			0.6		

Table A.22 Manganese fcc Z = 25 lattice constant = 6.80000 a.u.

Slater–Koster 3–center parameters	Orthogonal	Non–orthogonal	
	Energy integrals (Ry)	Energy integrals (Ry)	Overlap integrals
On site			
s, s (000)	1.19665	0.87679	
x, x (000)	1.59346	1.37446	
xy, xy (000)	0.66232	0.65407	
d2, d2 (000)	0.65126	0.64606	
First neighbor			
s, s (110)	−0.09038	−0.04393	0.09524
s, x (110)	0.08845	0.01160	−0.10104
s, xy (110)	−0.04020	−0.03547	0.02581
s, d2 (110)	−0.00133	0.03145	−0.00017
x, x (110)	0.10303	0.01673	−0.06945
x, x (011)	0.00199	0.03116	0.05152
x, y (110)	0.08726	−0.06922	−0.14781
x, xy (110)	−0.03621	−0.02116	0.02681
x, xy (011)	0.01672	−0.00658	−0.03958
z, d2 (011)	0.01269	−0.02270	−0.04908
z, d1 (011)	0.01233	0.00837	−0.03023
xy, xy (110)	−0.03668	−0.01513	0.03190
xy, xy (011)	0.01139	0.00556	−0.01255
xy, xz (011)	0.01861	0.00445	−0.01923
xy, d2 (110)	0.01962	0.01112	−0.01006
d2, d2 (110)	−0.01778	−0.00018	0.02311
d1, d1 (110)	0.03611	0.00920	−0.03744
Second neighbor			

(continued)

Table A.22 (continued)

Slater–Koster 3–center parameters	Orthogonal	Non–orthogonal	
	Energy integrals (Ry)	Energy integrals (Ry)	Overlap integrals
s, s (200)	−0.00118	−0.01837	0.00602
s, x (200)	0.02062	0.05582	0.00226
s, d2 (002)	−0.02772	−0.01968	−0.00617
x, x (200)	0.07008	0.18196	0.04759
y, y (200)	0.00924	0.02399	0.01505
x, xy (020)	0.00495	−0.01122	−0.01657
z, d2 (002)	−0.03478	−0.05824	−0.04428
xy, xy (200)	0.00378	−0.00193	−0.00382
xy, xy (002)	−0.00039	−0.00135	−0.00138
d2, d2 (002)	−0.01049	−0.01232	−0.00200
d1, d1 (002)	−0.00056	0.00721	0.01077

Table A.23 Manganese fcc

	Orthogonal			Non–orthogonal		
Band	RMS error mRy	Maximum k	Deviation mRy	RMS error mRy	Maximum k	Deviation mRy
1	4.4	(048)	11.8	0.4	(174)	1.1
2	3.2	(064)	6.6	0.6	(008)	1.7
3	2.6	(007)	6.0	0.8	(007)	2.8
4	3.1	(226)	7.1	1.3	(118)	6.0
5	3.2	(044)	8.0	0.6	(000)	1.8
6	4.3	(055)	13.2	0.7	(000)	1.8
1–6	3.5			0.8		

Table A.24 Manganese fcc

Energy values in Ry at selected k–points	Orthogonal	APW	Non–orthogonal
GAMMA 1	0.10508	0.10936	0.10987
GAMMA 12	0.72813	0.72644	0.72823
GAMMA 15	2.60277	2.59718	2.59717
GAMMA 25′	0.62261	0.62079	0.62195
X1 (008)	0.36578	0.37230	0.37320
X1 (008)	1.55119	1.50913	1.50935
X2 (008)	0.79699	0.79894	0.79998
X3 (008)	0.44040	0.44345	0.44515
X5 (008)	0.82490	0.82384	0.82420
X4′ (008)	0.95430	0.95202	0.95207
X5′ (008)	1.76260	1.80112	1.80104
L1 (444)	0.40095	0.39566	0.39555
L1 (444)	1.30033	1.29038	1.29058
L3 (444)	0.60478	0.60999	0.61086
L3 (444)	0.80049	0.80230	0.80302

(continued)

Table A.24 (continued)

Energy values in Ry at selected k–points	Orthogonal	APW	Non–orthogonal
L2′ (444)	0.71824	0.72441	0.72432
L3′ (444)	1.76540	2.51672	2.51672
W1 (048)	0.71513	0.71296	0.71476
W1 (048)	1.57186	1.62437	1.62411
W3 (048)	0.55877	0.55514	0.55602
W3 (048)	1.38027	1.37773	1.37842
W1′ (048)	0.82335	0.82414	0.82520
W2′ (048)	0.46630	0.45449	0.45527
W2′ (048)	1.49188	1.50811	1.50851
EVEN (224)	0.37557	0.37901	0.37939
EVEN (224)	0.56463	0.56165	0.56187
EVEN (224)	0.61081	0.61226	0.61311
EVEN (224)	0.79479	0.79251	0.79255
ODD (224)	0.65133	0.65219	0.65120
ODD (224)	0.78430	0.78989	0.79000

Table A.25 Fermi level quantities (non–orthogonal fit)

Energy	Densities of states					Velocity	Plasmon energy
	Total	s	p	t_{2g}	e_g		
Ry		States/Ry/atom				cm/s	eV
0.6945	23.44	0.17	0.63	7.84	14.80	0.39 × 10E8	7.65

Integrated densities of states				
Total	s	p	t_{2g}	e_g
Electrons				
7.00	0.62	0.36	3.80	2.22

Table A.26 Manganese fcc

Slater–Koster 2–center parameters	Orthogonal	Non–orthogonal	
	Energy integrals (Ry)	Energy integrals (Ry)	Overlap integrals
On site			
s	1.19916	0.72688	
p	1.61761	1.13468	
dl	0.66341	0.64999	
d2	0.65377	0.64893	
First neighbor			
(sss)	−0.08908	−0.02259	0.13404
(pps)	0.19941	−0.24732	−0.34629
(ppp)	0.00731	−0.02354	−0.05546
(dds)	−0.04750	−0.01997	0.04074
(ddp)	0.03191	0.01509	−0.02626
(ddd)	−0.00683	−0.00198	0.00442
(sps)	0.12874	−0.05419	−0.21777
(sds)	−0.04696	−0.02331	0.05122
(pds)	−0.05804	0.00380	0.08255

(continued)

Table A.26 (continued)

Slater–Koster 2–center parameters	Orthogonal	Non–orthogonal	
	Energy integrals (Ry)	Energy integrals (Ry)	Overlap integrals
(pdp)	0.02460	0.01201	−0.04420
Second neighbor			
(sss)	−0.00422	−0.02733	0.00927
(pps)	0.05911	0.07930	−0.02518
(ppp)	0.00316	−0.00964	0.00538
(dds)	−0.00687	−0.01964	−0.01482
(ddp)	0.00295	0.00010	−0.00241
(ddd)	0.00003	0.00102	0.00170
(sps)	0.01805	0.04173	−0.01905
(sds)	−0.01159	−0.14708	0.00332
(pds)	−0.01423	−0.03843	−0.00556
(pdp)	0.00456	0.00079	−0.00642

	Orthogonal			Non–orthogonal		
Band	RMS error mRy	Maximum k	Deviation mRy	RMS error mRy	Maximum k	Deviation mRy
1	4.3	(066)	8.7	0.5	(280)	0.9
2	7.9	(007)	18.8	1.0	(008)	2.1
3	3.4	(044)	9.9	1.1	(007)	2.8
4	5.1	(084)	13.0	1.4	(118)	6.4
5	6.0	(044)	12.9	1.0	(000)	2.9
6	6.1	(174)	12.3	0.9	(000)	2.9
1–6	5.7			1.0		

Table A.27 Iron bcc Z = 26 lattice constant = 5.30000 a.u.

Slater–Koster 3–center parameters	Orthogonal	Non–orthogonal	
	Energy integrals (Ry)	Energy integrals (Ry)	Overlap integrals
On site			
s, s (000)	1.24680	0.86866	
x, x (000)	1.87370	1.32862	
xy, xy (000)	0.66810	0.66316	
d2, d2 (000)	0.67377	0.65539	
First neighbor			
s, s (111)	−0.13952	−0.06349	0.10714
s, x (111)	0.11529	0.03368	−0.08095
s, xy (111)	0.03196	0.02254	−0.01673
x, x (111)	0.10922	−0.02558	−0.02960
x, y (111)	0.07938	0.02738	−0.07544
x, xy (111)	0.02131	0.00284	−0.00386
x, yz (111)	0.02821	0.01842	−0.03479
x, d1 (111)	0.00619	0.03201	0.00217
xy, xy (111)	−0.01196	−0.00343	0.01122

(continued)

Table A.27 (continued)

Slater–Koster 3–center parameters	Orthogonal	Non–orthogonal	
	Energy integrals (Ry)	Energy integrals (Ry)	Overlap integrals
xy, xz (111)	−0.01904	−0.00810	0.01557
xy d2 (111)	0.01364	0.01657	0.00007
d2, d2 (111)	0.02005	0.01835	−0.00578
Second neighbor			
s, s (200)	−0.03164	−0.01425	0.06771
s, x (200)	0.06228	−0.06607	−0.14970
s, d2 (002)	−0.04087	0.00950	0.06800
x, x (200)	0.16405	−0.34092	−0.33982
y, y (200)	0.04308	0.03079	0.02278
x, xy (020)	−0.01389	0.01133	0.03093
z, d2 (002)	−0.07101	0.06309	0.15071
xy, xy (200)	0.01569	0.00464	−0.01118
xy, xy (002)	−0.00247	0.00123	0.00437
d2, d2 (002)	−0.03085	−0.01212	0.02864
d1, d1 (002)	−0.00017	−0.00012	0.00019
Third neighbor			
s, s (220)	0.01464		
s, x (220)	−0.01519		
s, xy (220)	0.00326		
s, d2 (220)	0.00331		
x, x (220)	−0.01647		
x, x (022)	0.02954		
x, y (220)	−0.03130		
x, xy (220)	0.00210		
x, xy (022)	0.00288		
z, d2 (022)	−0.00332		
z, d1 (022)	0.00037		
xy, xy (220)	−0.00327		
xy, xy (022)	0.00065		
xy, xz (022)	0.00097		
xy, d2 (220)	−0.00019		
d2, d2 (220)	0.00089		
d1, d1 (220)	−0.00201		

Table A.28 Iron bcc

	Orthogonal			Non–orthogonal		
Band	RMS error mRy	Maximum k	Deviation mRy	RMS error mRy	Maximum k	Deviation mRy
1	2.3	(222)	7.1	0.7	(008)	2.1
2	1.4	(353)	3.5	0.7	(008)	2.1
3	2.0	(442)	6.3	0.6	(044)	2.6
4	1.3	(330)	3.1	0.8	(044)	2.1
5	1.6	(044)	8.3	0.7	(131)	1.6
6	6.0	(008)	20.1	1.2	(170)	5.1
1–6	2.9			0.8		

Table A.29 Energy values in Ry at selected k–points

	Orthogonal	APW	Non–orthogonal
GAMMA 1	0.11643	0.12049	0.12161
GAMMA12	0.73437	0.73494	0.73580
GAMMA 15	3.23432	3.23662	3.23661
GAMMA 25′	0.62238	0.62269	0.62318
H1 (008)	2.34876	2.35130	2.35128
Hl2 (008)	0.41352	0.41446	0.41658
H15 (008)	1.48680	1.50692	1.50325
H25′ (008)	0.81368	0.81355	0.81402
N1 (044)	0.40588	0.40423	0.40389
N1 (044)	0.73357	0.73661	0.73924
N1 (044)	1.35146	1.34684	1.34709
N2 (044)	0.53382	0.53387	0.53404
N3 (044)	0.83845	0.83616	0.83708
N4 (044)	0.75681	0.75435	0.75649
N1′ (044)	0.79238	0.78411	0.78462
N3′ (044)	2.27939	2.28429	2.28420
N4′ (044)	2.06247	2.07212	2.07180
P1 (444)	1.61231	1.60717	1.60699
P3 (444)	0.76006	0.75723	0.75766
P4 (444)	0.54028	0.53780	0.53748
P4 (444)	1.42186	1.41112	1.41202
(341)	0.42792	0.42814	0.42775
(341)	0.53990	0.54002	0.54016
(341)	0.64920	0.65050	0.65153
(341)	0.73208	0.73298	0.73346
(341)	0.77296	0.77346	0.77375
(341)	0.92387	0.92515	0.92494

Table A.30 Fermi level quantities (non–orthogonal fit)

Energy	Densities of states					Velocity	Plasmon energy
	Total	s	p	t_{2g}	e_g		
Ry		States/Ry/atom				cm/s	eV
0.7415	48.31	0.13	1.24	16.05	30.90	0.23 × 10E8	6.54

Integrated densities of states				
Total	s	p	t_{2g}	e_g
Electrons				
8.00	0.58	0.37	4.40	2.65

Table A.31 Iron bcc

Slater–Koster 2–center parameters	Orthogonal	Non–orthogonal	
	Energy integrals (Ry)	Energy integrals (Ry)	Overlap integrals
On site			
s	1.20177	0.89436	
p	1.87251	1.39705	
dl	0.68817	0.66684	

(continued)

Table A.31 (continued)

Slater–Koster 2–center parameters	Orthogonal	Non–orthogonal	
	Energy integrals (Ry)	Energy integrals (Ry)	Overlap integrals
d2	0.66437	0.66367	
First neighbor			
(sss)	−0.13944	−0.06429	0.10279
(pps)	0.26810	0.03367	−0.16522
(ppp)	0.02971	−0.07088	0.02495
(dds)	−0.05086	−0.02595	0.03467
(ddp)	0.03096	0.02919	−0.00653
(ddd)	−0.00303	−0.01186	−0.00868
(sps)	0.17780	−0.06645	0.12635
(sds)	−0.06781	−0.04290	0.02468
(pds)	−0.09308	0.03056	−0.04333
(pdp)	0.02089	−0.04031	0.01787
Second neighbor			
(sss)	−0.03141	−0.01874	0.06227
(pps)	0.18848	−0.25287	−0.28026
(ppp)	0.03907	0.01006	0.01670
(dds)	−0.03125	−0.03283	−0.00254
(ddp)	0.00618	0.01204	−0.00009
(ddd)	0.00071	−0.00201	−0.00207
(sps)	0.07354	0.05706	0.13709
(sds)	−0.03884	0.00378	0.05643
(pds)	−0.06028	−0.00292	−0.06786
(pdp)	−0.00383	0.00398	0.02643
Third neighbor			
(sss)	0.01818		
(pps)	−0.04447		
(ppp)	0.01641		
(dds)	0.00168		
(ddp)	0.00037		
(ddd)	−0.00056		
(sps)	−0.02567		
(sds)	−0.00039		
(pds)	0.00069		
(pdp)	−0.00282		

	Orthogonal			Non–orthogonal		
Band	RMS error	Maximum	Deviation	RMS error	Maximum	Deviation
1	3.3	(330)	9.3	0.8	(008)	2.3
2	3.2	(044)	9.3	0.8	(008)	2.3
3	3.5	(330)	9.6	0.7	(161)	2.5
4	2.6	(260)	8.9	0.5	(140)	1.9
5	2.3	(044)	10.1	0.7	(131)	1.6
6	6.8	(008)	21.2	2.2	(171)	9.2
1–6	3.9			1.1		

Table A.32 Cobalt hcp z = 27 lattice constants = 4.7377 a.u. 7.6885 a.u.

(continued)

Table A.32 (continued)

Slater–Koster 2–center parameters			
	Orthogonal	Non–orthogonal	
	Energy integrals (Ry)	Energy integrals (Ry)	Overlap integrals
Slater–Koster 2–center parameters			
	Orthogonal	Non–orthogonal	
	Energy integrals (Ry)	Energy integrals (Ry)	Overlap integrals
On site			
s	1.00263	0.67965	
p	1.38462	1.29572	
d0	0.58496	0.57253	
d1	0.58345	0.57267	
d2	0.58651	0.57473	
pd	−0.01555	0.00297	0.01500
First neighbor			
(sss)	−0.07879	−0.04717	0.08951
(sps)	0.11154	0.05414	−0.09421
(sds)	0.04839	0.01904	−0.05429
(pps)	0.15771	0.16401	−0.00625
(ppp)	−0.00092	0.13934	0.11818
(pds)	0.05816	0.04509	0.01077
(pdp)	−0.01975	0.02955	0.08862
(dds)	−0.04111	−0.01747	0.03883
(ddp)	0.02543	0.01730	−0.01836
(ddd)	−0.00407	−0.00180	0.00442
Second neighbor			
(sss)	−0.00681	−0.01585	0.01277
(sps)	0.02067	0.02247	−0.03897
(sds)	0.00916	0.02691	0.02227
(pps)	0.05683	0.07051	−0.02435
(ppp)	0.00039	0.02735	0.02349
(pds)	0.01689	0.04295	0.03078
(pdp)	−0.00285	0.00166	0.00801
(dds)	−0.00717	−0.01726	−0.01410
(ddp)	0.00181	0.00472	0.00531
(ddd)	0.00022	−0.00048	−0.00086
Third neighbor			
(sss)	−0.00178	0.00548	0.01540
(sps)	0.00454	−0.03021	−0.04279
(sds)	−0.00055	−0.00111	0.00115
(pps)	0.00505	−0.02068	−0.02605
(ppp)	0.00313	−0.01904	−0.01732
(pds)	−0.00318	−0.00344	0.00019
(pdp)	0.00097	−0.00771	−0.01329
(dds)	−0.00068	0.00001	0.00126
(ddp)	0.00035	−0.00057	−0.00151
(ddd)	0.00010	0.00241	0.00356

Table A.33 Cobalt hcp

(continued)

Table A.33 (continued)

	Orthogonal			Non–orthogonal		
Band	RMS error mRy	Maximum k	Deviation mRy	RMS error mRy	Maximum k	Deviation mRy
	Orthogonal			Non–orthogonal		
Band	RMS error mRy	Maximum k	Deviation mRy	RMS error mRy	Maximum k	Deviation mRy
1	2.7	(0 24 12)	5.5	0.9	(0 18 0)	2.5
2	3.4	(24 0 3)	7.3	0.8	(0 18 0)	1.8
3	2.9	(24 0 9)	7.8	0.7	(0 24 12)	1.1
4	2.8	(0 24 12)	7.7	0.8	(8 0 0)	1.5
5	1.2	(8 0 12)	3.1	0.8	(0 12 12)	1.7
6	1.6	(8 0 12)	3.1	0.8	(8 0 0)	1.7
7	2.2	(8 0 12)	4.9	0.7	(8 0 12)	1.8
8	2.7	(16 0 9)	5.2	0.8	(24 0 0)	2.2
9	2.7	(0 12 0)	5.1	0.6	(0 12 12)	1.3
10	2.2	(0 0 12)	4.9	0.7	(8 0 0)	1.6
11	3.8	(16 0 12)	9.9	1.1	(16 0 6)	3.6
12	4.9	(12 0 0)	15.8	2.2	(8 0 0)	6.8
1–12	2.91			1.00		

Table A.34 Energy values in Ry at selected k–points

	Orthogonal	APW	Non–orthogonal
GAMMA 1+	0.01269	0.01281	0.01353
GAMMA 1+	0.55089	0.55119	0.55232
GAMMA 3+	0.64501	0.64688	0.64644
GAMMA 5+	0.57957	0.57776	0.57771
GAMMA 6+	0.60724	0.60632	0.60679
GAMMA 4–	0.33838	0.34077	0.34156
GAMMA 4–	1.16100	1.15924	1.16156
GAMMA 5–	0.68922	0.68939	0.68896
GAMMA 6–	0.55176	0.55226	0.55110
M1+	0.36506	0.37457	0.37349
M1+	0.51558	0.51407	0.51402
M1+	0.81767	0.81180	0.81241
M2+	0.70250	0.70729	0.70738
M3+	0.39377	0.39378	039285
M4+	0.71015	0.71224	0.71150
M1–	0.54107	0.54090	0.54232
M2–	0.37042	0.36819	0.36804
M2–	0.57746	0.57920	0.57866
M2–	0.95210	0.94882	0.94811
M3–	0.70032	0.70097	0.70320
M4–	0.54890	0.55159	0.55121
A1	0.17857	0.17908	0.17909
A1	0.51957	0.51861	0.51865
A3	0.54872	0.54800	0.54833
A3	0.66144	0.66634	0.66633
L1	0.35334	0.34785	0.34836
L1	0.45263	0.46032	0.45925

(continued)

Table A.34 (continued)

	Orthogonal	APW	Non–orthogonal
L1	0.65428	0.65435	0.65420
L1	0.80299	0.80264	0.80278
L2	0.61613	0.61395	0.61276
L2	0.70674	0.70767	0.70837

Table A.35 Fermi level quantities (non–orthogonal fit)

Energy		Densities of states			
	Total	s	p	t_{2g}	e_g
Ry		States/Ry/atom			
0.6882	45.00	0.25	0.22	23.02	21.51

Integrated densities of states				
Total	s	p	t_{2g}	e_g
Electrons				
9.00	0.73	0.40	5.03	2.84

Table A.36 Cobalt fcc z = 27 lattice constant = 6.54000 a.u.

Slater–Koster 3–center parameters	Orthogonal	Non–orthogonal	
	Energy integrals (Ry)	Energy integrals (Ry)	Overlap integrals
On site			
s, s (000)	1.13427	0.84581	
x, x (000)	1.80880	1.30733	
xy, xy (000)	0.60873	0.61173	
d2, d2 (000)	0.60301	0.60183	
First neighbor			
s, s (110)	−0.09088	−0.03960	0.09743
s, x (110)	0.10157	0.02929	−0.09890
s, xy (110)	−0.03341	−0.04638	0.01276
s, d2 (110)	−0.00461	0.02327	−0.00473
x, x (110)	0.11407	0.03232	−0.06282
x, x (011)	0.02431	0.06785	0.06789
x, y (110)	0.14797	−0.10472	−0.19416
x, xy (110)	−0.02460	−0.01310	0.04186
x, xy (011)	0.02330	−0.00932	−0.04354
z, d2 (011)	0.01438	−0.03684	−0.07762
z, d1 (011)	0.00338	0.00322	−0.04083
xy, xy (110)	−0.03202	−0.01847	0.02194
xy, xy (011)	0.01011	0.00591	−0.00935
xy, xz (011)	0.01903	0.00390	−0.01633
xy, d2 (110)	0.01683	0.01422	−0.00329
d2, d2 (110)	−0.01632	0.00164	0.02414
d1, d1 (110)	0.03226	0.01218	−0.02794
Second neighbor			
s, s (200)	−0.00260	−0.04643	−0.01488
s, x (200)	−0.00299	0.07769	0.01447

(continued)

Table A.36 (continued)

Slater–Koster 3–center parameters	Orthogonal	Non–orthogonal	
	Energy integrals (Ry)	Energy integrals (Ry)	Overlap integrals
s, d2 (002)	−0.02975	−0.03342	−0.02803
x, x (200)	−0.02208	0.10012	−0.01811
y, y (200)	−0.00286	0.04258	0.01567
x, xy (020)	0.00235	−0.01727	−0.02096
z d2 (002)	−0.04576	−0.05522	−0.03562
xy, xy (200)	0.00375	−0.00266	−0.00475
xy, xy (002)	0.00085	−0.00053	−0.00024
d2, d2 (002)	−0.01011	−0.01387	−0.00733
d1, d1 (002)	0.00060	0.00392	0.00722

	Orthogonal			Non–orthogonal		
Band	RMS error mRy	Maximum k	Deviation mRy	RMS error mRy	Maximum k	Deviation mRy
1	7.7	(000)	15.0	0.7	(044)	1.5
2	3.2	(022)	7.4	0.7	(008)	2.1
3	3.8	(005)	9.1	0.7	(007)	2.3
4	6.0	(444)	22.7	1.2	(118)	5.5
5	4.2	(044)	11.6	0.6	(000)	1.5
6	5.4	(226)	15.2	0.7	(000)	1.5
1–6	5.3			0.8		

Table A.37 Cobalt fcc

Energy values in Ry at selected k–points	Orthogonal	APW	Non–orthogonal
GAMMA 1	0.02808	0.04308	0.04427
GAMMA 12	0.67012	0.66890	0.67043
GAMMA 15	2.76304	2.77572	2.77568
GAMMA 25′	0.57820	0.57594	0.57714
X1 (008)	0.34705	0.35163	0.35267
X1 (008)	1.48337	1.48007	1.47967
X2 (008)	0.73693	0.73268	0.73405
X3 (008)	0.41649	0.41686	0.41891
X5 (008)	0.75353	0.75472	0.75485
X4′ (008)	0.93790	0.93926	0.93901
X5′ (008)	1.65595	1.86437	1.86313
L1 (444)	0.37100	0.36524	0.36525
L1 (444)	1.21869	1.23630	1.23723
L3 (444)	0.56537	0.56710	0.56771
L3 (444)	0.73431	0.73594	0.73627
L2′ (444)	0.68060	0.70335	0.70226
L3′ (444)	2.45629	2.63026	2.63021
W1 (048)	0.65758	0.65972	0.66154

(continued)

Table A.37 (continued)

Energy values in Ry at selected k–points	Orthogonal	APW	Non–orthogonal
W1 (048)	1.51521	1.62824	1.62722
W3 (048)	0.52323	0.51775	0.51849
W3 (048)	1.35513	1.36483	1.36462
W1′ (048)	0.75012	0.75493	0.75640
W2′ (048)	0.44246	0.42927	0.42968
W2′ (048)	1.74484	1.50257	1.50430
EVEN(224)	0.33229	0.33756	0.33830
EVEN(224)	0.52904	0.52552	0.52492
EVEN(224)	0.57014	0.56968	0.57042
EVEN(224)	0.74362	0.73374	0.73424
ODD (224)	0.60323	0.60477	0.60414
ODD (224)	0.71949	0.72523	0.72484

Table A.38 Fermi level quantities (non–orthogonal fit)

Energy		Densities of states				Velocity	Plasmon energy
	Total	s	p	t_{2g}	e_g		
Ry		States/Ry/atom				cm/s	eV
0.7140	30.21	0.17	0.15	16.52	13.37	0.30 × 10E8	7.47

Integrated densities of states Total Electrons	s	p	t_{2g}	e_g
9.00	0.64	0.34	4.63	3.39

Table A.39 Cobalt fcc

Slater–Koster 2–center parameters	Orthogonal	Non–orthogonal	
	Energy integrals (Ry)	Energy integrals (Ry)	Overlap integrals
On site			
s	1.12946	0.69940	
p	1.75262	1.12429	
dl	0.60547	0.60507	
d2	0.60445	0.60453	
First neighbor			
(sss)	−0.09043	−0.03628	0.13218
(pps)	0.23748	−0.24739	−0.35422
(ppp)	−0.00142	−0.02469	0.05975
(dds)	−0.04213	−0.02133	0.03321
(ddp)	0.02976	0.01553	−0.01997
(ddd)	−0.00684	−0.00219	0.00307
(sps)	0.13649	−0.03494	−0.21526
(sds)	−0.03806	−0.02654	0.04473
(pds)	−0.04069	−0.00172	0.07731
(pdp)	0.02797	0.01402	−0.04303
Second neighbor			

(continued)

Table A.39 (continued)

Slater–Koster 2–center parameters	Orthogonal	Non–orthogonal	
	Energy integrals (Ry)	Energy integrals (Ry)	Overlap integrals
(sss)	−0.00337	−0.02456	0.01319
(pps)	0.02849	0.04619	−0.05358
(ppp)	0.01099	0.00588	0.01630
(dds)	−0.00759	−0.01997	−0.01909
(ddp)	0.00495	0.00050	−0.00158
(ddd)	−0.00016	0.00132	0.00240
(sps)	0.00135	0.03471	−0.02563
(sds)	−0.01119	−0.01161	0.00549
(pds)	−0.01061	−0.03864	−0.00841
(pdp)	0.01134	−0.00322	−0.01278

	Orthogonal			Non–orthogonal		
Band	RMS error mRy	Maximum k	Deviation mRy	RMS error mRy	Maximum k	Deviation mRy
1	8.8	(000)	19.1	0.5	(048)	1.2
2	8.1	(007)	17.9	1.0	(008)	2.5
3	3.8	(000)	7.5	1.0	(007)	2.3
4	8.4	(333)	38.1	1.3	(118)	5.8
5	6.5	(444)	16.0	0.9	(000)	2.4
6	15.0	(174)	41.1	0.9	(174)	2.5
1–6	9.1			1.0		

Table A.40 Nickel fcc z = 28 lattice constant = 6.55000 a.u.

Slater–Koster 3–center parameters	Orthogonal	Non–orthogonal	
	Energy integrals (Ry)	Energy integrals (Ry)	Overlap integrals
On site			
s, s (000)	1.12904	0.83661	
x, x (000)	1.85069	1.35892	
xy, xy (000)	0.56113	0.55869	
d2, d2 (000)	0.55808	0.55480	
First neighbor			
s, s (110)	−0.09529	−0.05505	0.09263
s, x (110)	0.11361	0.02980	−0.09350
s, xy (110)	−0.03364	−0.03811	0.00815
s, d2 (110)	−0.00471	0.03258	0.01014
x, x (110)	0.11628	0.02760	−0.07579
x, x (011)	0.01241	0.04074	0.06087
x, y (110)	0.15930	−0.05574	−0.14325
x, xy (110)	−0.02760	−0.02116	0.01991
x, xy (011)	0.01732	−0.00372	−0.04316
z, d2 (011)	0.00794	−0.01755	−0.05011

(continued)

Table A.40 (continued)

Slater–Koster 3–center parameters	Orthogonal	Non–orthogonal	
	Energy integrals (Ry)	Energy integrals (Ry)	Overlap integrals
z, d1 (011)	−0.00400	0.01210	−0.02134
xy, xy (110)	−0.02887	−0.01236	0.02876
xy, xy (011)	0.00926	0.00607	−0.00816
xy, xz (011)	0.01590	0.00890	−0.00913
xy, d2 (110)	0.01505	0.01020	−0.00700
d2, d2 (110)	−0.01483	−0.00133	0.01928
d1, d1 (110)	0.02911	0.00803	−0.03212
Second neighbor			
s, s (200)	0.00378	−0.02416	0.00525
s, x (200)	−0.02097	0.06231	0.00727
s, d2 (002)	−0.03470	−0.01131	0.00223
x, x (200)	−0.06560	0.20754	0.06446
y, y (200)	0.01472	0.02320	0.01341
x, xy (020)	0.00540	−0.00991	−0.01933
z d2 (002)	−0.05429	−0.05403	−0.05684
xy, xy (200)	0.00359	0.00003	−0.00112
xy, xy (002)	0.00039	−0.00113	−0.00137
d2, d2 (002)	−0.01004	−0.00651	−0.00275
d1, d1 (002)	0.00093	0.00382	−0.00722

Table A.41 Nickel fcc

	Orthogonal			Non–orthogonal		
Band	RMS error mRy	Maximum k	Deviation mRy	RMS error mRy	Maximum k	Deviation mRy
1	5.5	(333)	11.3	0.4	(174)	1.2
2	3.1	(055)	7.5	0.6	(008)	1.5
3	3.2	(004)	10.1	0.7	(007)	2.6
4	4.3	(444)	12.2	1.0	(118)	5.0
5	3.5	(224)	9.2	0.6	(000)	1.6
6	9.4	(174)	33.6	0.6	(000)	1.6
1–6	5.3			0.7		

Table A.42 Energy values in Ry at selected k–points

	Orthogonal	APW	Non–orthogonal
GAMMA 1	0.00825	0.01391	0.01450
GAMMA 12	0.61642	0.61429	0.61588
GAMMA 15	2.75828	2.74549	2.74548
GAMMA 25′	0.53488	0.53191	0.53294
X1 (008)	0.32523	0.33007	0.33068
X1 (008)	1.53408	1.40494	1.40512
X2 (008)	0.67798	0.67054	0.67150
X3 (008)	0.38676	0.38945	0.39092
X5 (008)	0.69176	0.69005	0.69043
X4′ (008)	0.89773	0.89061	0.89006
X5′ (008)	1.72872	1.81441	1.81433
Ll (444)	0.34719	0.33942	0.33954

(continued)

Table A.42 (continued)

	Orthogonal	APW	Non–orthogonal
LI (444)	1.17789	1.16004	1.16026
L3 (444)	0.52278	0.52447	0.52521
L3 (444)	0.67224	0.67409	0.67458
L2′ (444)	0.64860	0.66078	0.66055
L3′ (444)	2.56022	2.56978	2.56977
W1 (048)	0.60308	0.60714	0.60865
W1 (048)	1.54493	1.57600	1.57595
W3 (048)	0.48303	0.47944	0.48045
W3 (048)	1.29620	1.29801	1.29856
W1′ (048)	0.69020	0.69031	0.69124
W2′ (048)	0.40994	0.40111	0.40172
W2′ (048)	1.99178	1.43501	1.43540
EVEN (224)	0.29925	0.30644	0.30662
EVEN (224)	0.49153	0.48700	0.48729
EVEN (224)	0.52673	0.52671	0.52730
EVEN(224)	0.67896	0.67446	0.67452
ODD (224)	0.55497	0.55772	0.55719
ODD (224)	0.65512	0.66433	0.66421

Table A.43 Fermi level quantities (non–orthogonal fit)

Energy		Densities of states				Velocity	Plasmon energy
	Total	s	p	t_{2g}	e_g		
Ry		States/Ry/atom				cm/s	eV
0.6750	55.14	0.30	0.18	41.30	13.37	0.20 × 10E8	6.29

Integrated densities of states				
Total Electrons	s	p	t_{2g}	e_g
10.00	0.59	0.33	5.34	3.74

Table A.44 Nickel fcc

Slater–Koster 2–center parameters	Orthogonal	Non–orthogonal	
	Energy integrals (Ry)	Energy integrals (Ry)	Overlap integrals
On site			
s	1.15526	0.66135	
p	1.60587	1.09927	
dl	0.56068	0.55851	
d2	0.55753	0.55828	
First neighbor			
(sss)	−0.09525	−0.03817	0.13354
(pps)	0.21708	−0.21703	−0.34437
(ppp)	0.01660	−0.02535	0.05872
(dds)	−0.03712	−0.01997	0.03000

(continued)

Table A.44 (continued)

Slater–Koster 2–center parameters	Orthogonal	Non–orthogonal	
	Energy integrals (Ry)	Energy integrals (Ry)	Overlap integrals
(ddp)	0.02629	0.01339	−0.01966
(ddd)	−0.00600	−0.00144	0.00385
(sps)	0.14003	−0.02765	−0.21562
(sds)	−0.03880	−0.02854	0.03944
(pds)	−0.04400	−0.01120	0.06411
(pdp)	0.02377	0.01218	−0.04374
Second neighbor			
(sss)	−0.00065	−0.02760	0.01118
(pps)	0.06220	0.06663	−0.03786
(ppp)	0.00682	−0.00498	0.00876
(dds)	−0.00651	−0.01468	−0.01345
(ddp)	0.00344	0.00008	−0.00214
(ddd)	−0.00027	0.00033	0.00098
(sps)	0.01441	0.03980	−0.02446
(sds)	−0.01015	−0.01070	0.00618
(pds)	−0.01012	−0.02965	−0.00020
(pdp)	0.00510	−0.00090	−0.00998

	Orthogonal			Non–orthogonal		
Band	RMS error mRy	Maximum k	Deviation mRy	RMS error mRy	Maximum k	Deviation mRy
1	4.1	(444)	11.8	0.5	(264)	0.9
2	6.2	(007)	15.1	0.8	(008)	2.0
3	2.1	(000)	5.8	0.8	(007)	2.5
4	4.3	(048)	9.7	1.1	(118)	5.4
5	4.9	(444)	11.0	0.7	(000)	2.2
6	6.1	(174)	18.2	0.7	(000)	2.2
1–6	4.8			0.8		

Table A.45 Zinc fcc z = 30 lattice constant = 7.25000 a.u.

Slater–Koster 3–center parameters	Orthogonal	Non–orthogonal	
	Energy integrals (Ry)	Energy integrals (Ry)	Overlap integrals
On site			
s,s(000)	0.56945	0.37456	
x,x (000)	1.16229	0.84845	
xy,xy(000)	0.02443	0.02251	
d2,d2(000)	0.01955	0.01980	
First neighbor			
s,s (110)	−0.06308	−0.04954	0.09585

(continued)

Table A.45 (continued)

Slater–Koster 3–center parameters	Orthogonal	Non–orthogonal	
	Energy integrals (Ry)	Energy integrals (Ry)	Overlap integrals
s,x (110)	−0.07928	0.03828	−0.08909
s,xy (110)	−0.02621	−0.01927	−0.02144
s,d2(110)	0.00722	0.01358	0.00864
x, x (110)	0.12599	0.00062	−0.11515
x,x (011)	0.04345	0.05593	0.10949
x,y (110)	0.07512	−0.01381	−0.13703
x, xy (110)	0.02491	−0.01794	0.00720
x, xy (011)	0.00192	0.01009	−0.06688
z, d2 (011)	0.01301	−0.00015	0.00103
z, d1 (011)	−0.00443	0.00870	−0.00242
xy, xy (110)	−0.01087	−0.01125	−0.01757
xy, xy (011)	0.00331	0.00298	−0.00069
xy, xz (011)	0.00303	0.00418	0.006 46
xy, d2 (110)	−0.00537	0.00517	−0.00629
d2, d2 (110)	−0.00492	−0.00346	0.02402
d1, d1 (110)	0.00899	0.00749	−0.02401
Second neighbor			
s, s (200)	−0.00192	−0.03047	−0.01530
s, x (200)	−0.01098	0.04875	0.02111
s, d2 (002)	0.01010	−0.00699	0.00795
x, x (200)	0.05233	0.10701	0.05055
y, y (200)	0.02573	0.00831	0.01112
x, xy (020)	−0.00236	0.00438	−0.02155
z, d2 (002)	−0.00999	−0.01059	−0.02869
xy, xy (200)	0.00000	0.00116	0.00734
xy, xy (002)	−0.00032	−0.00016	−0.00018
d2,d2 (002)	−0.00264	−0.00228	−0.00103
d1, d1 (002)	0.00011	0.00045	0.01987

Table A.46 Zinc fcc

	Orthogonal			Non–orthogonal		
Band	RMS error mRy	Maximum k	Deviation mRy	RMS error mRy	Maximum k	Deviation mRy
1	1.6	(004)	3.7	0.4	(354)	0.7
2	1.3	(007)	3.0	0.4	(008)	0.9
3	1.7	(118)	4.1	0.6	(118)	2.7
4	2.1	(222)	7.9	0.3	(280)	0.6
5	1.5	(226)	3.4	0.3	(000)	0.9

(continued)

Table A.46 (continued)

Band	Orthogonal RMS error mRy	Maximum k	Deviation mRy	Non–orthogonal RMS error mRy	Maximum k	Deviation mRy
6	7.4	(174)	26.6	1.0	(048)	2.2
1–6	3.4			0.6		

Table A.47 Energy values in Ry at selected k–points

	Orthogonal	APW	Non–orthogonal
GAMMA 1	−0.19896	−0.19624	−0.19567
GAMMA 12	0.03637	0.03553	0.03644
GAMMA 15	2.55160	1.99971	1.99975
GAMMA 25′	0.00678	0.00536	0.00591
X1 (008)	−0.05570	−0.05486	−0.05426
X1 (008)	0.81410	0.74848	0.74932
X2 (008)	0.05945	0.05742	0.05789
X3 (008)	−0.04618	−0.04440	−0.04355
X5 (008)	0.06729	0.06524	0.06534
X4′ (008)	0.53574	0.52672	0.52480
X5′ (008)	1.19606	1.24662	1.23233
Ll (444)	−0.05133	−0.05204	−0.05189
L1 (444)	0.63312	0.51643	0.51653
L3 (444)	0.00538	0.00392	0.00432
L3 (444)	0.05896	0.05886	0.05923
L2′ (444)	0.35378	0.36546	0.36483
L3′ (444)	1.25519	1.78234	1.78220
W1 (048)	0.03606	0.03707	0.03742
W1 (048)	0.82406	0.87511	0.87550
W3 (048)	−0.00053	−0.00227	−0.00145
W3 (048)	0.77406	0.77853	0.77637
W1′ (048)	0.06857	0.06522	0.06587
W2′ (048)	−0.02569	−0.02622	−0.02600
W2′ (048)	0.98798	0.89123	0.89370
EVEN (224)	−0.04201	−0.03892	−0.03831
EVEN (224)	0.00268	0.00497	0.00543
EVEN (224)	0.02226	0.02009	0.01971
EVEN (224)	0.14930	0.14928	0.14912
ODD (224)	0.01651	0.01696	0.01663
ODD (224)	0.05402	0.05587	0.05564

Table A.48 Fermi level quantities (non–orthogonal fit)

Energy		Densities of states				Velocity	Plasmon energy
	Total	s	p	t_{2g}	e_g		
Ry		States/Ry/atom				cm/s	eV
0.6130	3.94	1.46	2.14	0.33	0.01	1.61 × 10E8	11.76

Integrated densities of states				
Total	s	p	t_{2g}	e_g
Electrons				

(continued)

Table A.48 (continued)

Integrated densities of states Total Electrons	s	p	t_{2g}	e_g
12.00	1.05	0.74	6.22	3.99

Table A.49 Zinc fcc

Slater–Koster 2–center parameters	Orthogonal	Non–orthogonal	
	Energy integrals (Ry)	Energy integrals (Ry)	Overlap integrals
On site			
s	0.54206	0.31482	
p	1.09049	0.60111	
d1	0.01696	0.02040	
d2	0.01910	0.02088	
First neighbor			
(sss)	−0.06257	−0.05519	0.10980
(pps)	0.17246	−0.15087	−0.39507
(ppp)	0.01605	0.00182	0.08925
(dds)	−0.01349	−0.01346	0.00926
(ddp)	0.00899	0.00748	−0.00825
(ddd)	−0.00201	−0.00105	0.00185
(sps)	0.09868	0.03050	−0.20540
(sds)	−0.01546	−0.02365	0.01706
(pds)	−0.00454	−0.02033	0.07720
(pdp)	0.01084	0.01549	−0.05784
Second neighbor			
(sss)	0.00113	−0.01900	0.00615
(pps)	0.04112	0.05128	−0.04791
(ppp)	0.01023	0.00254	0.00612
(dds)	−0.00230	−0.00166	−0.00254
(ddp)	0.00156	0.00043	−0.00010
(ddd)	−0.00026	−0.00021	−0.00219
(sps)	0.00321	0.03299	−0.02140
(sds)	−0.00229	−0.00487	0.00042
(pds)	0.00543	−0.01036	−0.00058
(pdp)	−0.00095	0.00049	−0.01745

	Orthogonal			Non–orthogonal		
Band	RMS error mRy	Maximum k	Deviation mRy	RMS error mRy	Maximum k	Deviation mRy
1	2.4	(000)	5.7	0.4	(174)	1.2
2	2.3	(055)	5.6	0.4	(008)	1.1
3	2.3	(044)	4.2	0.7	(118)	3.0

(continued)

Table A.49 (continued)

	Orthogonal			Non–orthogonal		
Band	RMS error mRy	Maximum k	Deviation mRy	RMS error mRy	Maximum k	Deviation mRy
4	2.1	(222)	6.1	0.3	(000)	0.6
5	1.9	(224)	4.6	0.4	(000)	1.2
6	8.4	(174)	25.4	0.8	(048)	2.9
1–6	4.0			0.5		

Table A.50 Yttrium fcc z = 39 lattice constant = 9.34000 a.u.

Slater–Koster 3–center parameters	Orthogonal	Non–orthogonal	
	Energy integrals (Ry)	Energy integrals (Ry)	Overlap integrals
On site			
s, s (000)	0.66791	0.74824	
x, x (000)	1.04760	0.87828	
xy, xy (000)	0.71316	0.62169	
d2, d2 (000)	0.61168	0.59090	
First neighbor			
s, s (110)	−0.06097	−0.06812	−0.05146
s, x (110)	0.05541	0.04682	−0.02904
s, xy (110)	−0.04847	−0.02621	0.05779
s, d2 (110)	0.02919	0.04056	0.01412
x, x (110)	0.04649	0.00656	−0.05060
x, x(011)	−0.00436	0.01834	0.04302
x, y (110)	0.07544	−0.01532	−0.11035
x, xy (110)	−0.06417	−0.00856	0.06814
x, xy (011)	0.00244	−0.00861	−0.03375
z, d2 (011)	0.00838	−0.01259	−0.02777
z, d1 (011)	0.03421	0.00900	−0.04456
xy, xy (110)	−0.05002	−0.01357	0.06272
xy, xy (011)	0.01258	−0.00188	−0.03838
xy, xz (011)	−0.00257	−0.00178	−0.02889
xy, d2 (110)	0.02409	0.01941	−0.00793
d2, d2 (110)	−0.02196	−0.00794	0.02126
d1, d1 (110)	0.03796	0.00522	−0.06158
Second neighbor			
s, s (200)	0.02478	0.01392	−0.03612
s, x (200)	−0.03173	0.01193	0.03370
s, d2 (002)	0.01410	−0.02065	−0.04100
x, x (200)	−0.02793	0.02964	0.00316
y, y (200)	0.00423	0.01313	0.01168
x, xy (020)	−0.00840	−0.00350	0.00106

(continued)

Table A.50 (continued)

Slater–Koster 3–center parameters	Orthogonal	Non–orthogonal	
	Energy integrals (Ry)	Energy integrals (Ry)	Overlap integrals
z, d2 (002)	0.00984	−0.03369	−0.02133
xy, xy (200)	−0.01434	0.00149	0.01118
xy, xy (002)	0.00278	−0.00199	−0.00184
d2, d2 (002)	−0.00185	−0.01941	−0.00952
d1, d1 (002)	0.00342	0.00189	0.00191

Table A.51 Yttrium fcc

	Orthogonal			Non–orthogonal		
Band	RMS error mRy	Maximum k	Deviation mRy	RMS error mRy	Maximum k	Deviation mRy
1	1.9	(055)	5.7	0.4	(006)	1.2
2	3.8	(055)	11.3	0.4	(008)	1.5
3	3.6	(044)	9.7	0.5	(226)	1.1
4	3.3	(118)	10.1	1.4	(118)	6.7
5	3.0	(048)	8.2	0.7	(048)	1.5
6	3.3	(004)	7.4	0.7	(048)	1.4
1–6	3.2			0.8		

Table A.52 Energy values in Ry at selected k–points

	Orthogonal	APW	Non–orthogonal
GAMMA 1	0.08496	0.08567	0.08604
GAMMA 12	0.71240	0.70871	0.70997
GAMMA 15	1.36313	1.36020	1.36024
GAMMA 25′	0.56192	0.56229	0.56290
X1 (008)	0.27529	0.27846	0.27901
X1 (008)	1.12995	1.15099	1.15091
X2 (008)	0.82406	0.81948	0.82097
X3 (008)	0.36066	0.36400	0.36550
X5 (008)	0.86141	0.85755	0.85794
X4′ (008)	0.61928	0.62031	0.62026
X5′ (008)	1.02612	1.04218	1.04061
L1 (444)	0.29110	0.29091	0.29152
L1 (444)	1.01364	1.03528	1.03476
L3 (444)	0.54161	0.54319	0.54352
L3 (444)	0.82004	0.82247	0.82287
L2′ (444)	0.48302	0.48583	0.48585
L3′ (444)	1.38831	1.39507	1.39499
W1 (048)	0.62300	0.62642	0.62739
W1 (048)	1.04997	1.02892	1.02834
W3 (048)	0.46198	0.45876	0.45874
W3 (048)	1.01222	1.01364	1.01501
W1′ (048)	0.85029	0.85849	0.86000
W2′ (048)	0.35520	0.35368	0.35368
W2′ (048)	1.23352	1.22284	1.22341
EVEN (224)	0.26522	0.26583	0.26547
EVEN (224)	0.45409	0.45283	0.45235

(continued)

Table A.52 (continued)

	Orthogonal	APW	Non–orthogonal
EVEN (224)	0.53565	0.53249	0.53286
EVEN (224)	0.74102	0.73939	0.73946
ODD (224)	0.58379	0.59083	0.59038
ODD (224)	0.79924	0.79680	0.79623

Table A.53 Fermi level quantities (non–orthogonal fit)

Energy		Densities of states				Velocity	Plasmon energy
	Total	s	p	t_{2g}	e_g		
Ry		States/Ry/atom				cm/s	eV
0.4295	21.10	0.36	3.89	12.04	4.82	0.40 × 10E8	4.64

Integrated densities of states				
Total	s	p	t_{2g}	e_g
Electrons				
3.00	0.73	0.48	0.96	0.82

Table A.54 Yttrium fcc

Slater–Koster 2–center parameters	Orthogonal	Non–orthogonal	
	Energy integrals (Ry)	Energy integrals (Ry)	Overlap integrals
On site			
s	0.73462	0.53669	
p	1.04983	0.72670	
d1	0.65132	0.57102	
d2	0.62251	0.56616	
First neighbor			
(sss)	−0.04813	−0.03416	0.06553
(pps)	0.11361	−0.15454	−0.32265
(ppp)	−0.02258	−0.02228	0.03624
(dds)	−0.06535	0.00697	0.11137
(ddp)	0.03210	0.00959	−0.06534
(ddd)	−0.00422	−0.00463	0.00509
(sps)	0.07556	−0.00352	−0.15583
(sds)	−0.05876	0.00222	0.10444
(pds)	−0.08429	0.07011	0.19819
(pdp)	0.01942	0.00774	−0.05195
Second neighbor			
(sss)	−0.01240	0.00524	0.00634
(pps)	0.02847	0.01606	−0.03188
(ppp)	−0.01006	0.01222	0.01818
(dds)	−0.00114	−0.03264	−0.02455

(continued)

Table A.54 (continued)

Slater–Koster 2–center parameters	Orthogonal	Non–orthogonal	
	Energy integrals (Ry)	Energy integrals (Ry)	Overlap integrals
(ddp)	−0.00072	0.00154	−0.00187
(ddd)	0.00169	0.00167	0.00235
(sps)	0.00702	−0.00635	−0.01814
(sds)	0.00782	−0.00220	0.00791
(pds)	−0.01399	−0.04638	−0.02297
(pdp)	0.00384	−0.00362	−0.00936

	Orthogonal			Non–orthogonal		
Band	RMS error mRy	Maximum k	Deviation mRy	RMS error mRy	Maximum k	Deviation mRy
1	12.7	(084)	31.5	1.0	(264)	2.0
2	11.7	(008)	24.0	1.4	(055)	4.0
3	10.9	(044)	29.6	2.1	(044)	7.0
4	10.9	(005)	19.5	2.1	(118)	7.0
5	12.1	(044)	24.8	1.2	(022)	3.3
6	16.7	(066)	39.1	1.1	(000)	2.9
1–6	12.7			1.5		

Table A.55 Zirconium fcc z = 40 lattice constant = 8.31550 a.u.

Slater–Koster 3–center parameters	Orthogonal	Non–orthogonal	
	Energy integrals (Ry)	Energy integrals (Ry)	Overlap integrals
On site			
s, s (000)	0.93238	0.84929	
x, x (000)	1.37107	1.15020	
xy, xy (000)	0.88720	0.75226	
d2, d2 (000)	0.74947	0.71494	
First neighbor			
s, s (110)	−0.07824	−0.05568	0.02288
s, x (110)	0.06718	0.02461	−0.05958
s, xy (110)	−0.06834	−0.01880	0.07407
s, d2 (110)	0.03758	0.04167	−0.00072
x, x (110)	0.05214	−0.00664	−0.05103
x,x (011)	−0.00900	0.02282	0.04068
x, y (110)	0.08953	−0.04444	−0.12721
x,xy (110)	−0.08440	0.00157	0.07515
x, xy (011)	0.00194	−0.01023	−0.03152
z, d2 (011)	0.01154	−0.02168	−0.03741
z, d1 (011)	0.04331	0.00639	−0.05197

(continued)

Table A.55 (continued)

Slater–Koster 3–center parameters	Orthogonal	Non–orthogonal	
	Energy integrals (Ry)	Energy integrals (Ry)	Overlap integrals
xy, xy (110)	−0.06484	−0.01237	0.07120
xy, xy (011)	0.01625	0.00071	−0.03560
xy, xz (011)	−0.00628	0.00258	−0.02670
xy, d2 (110)	0.03013	0.02195	−0.01185
d2, d2(110)	−0.02780	−0.00944	0.02443
d1, d1 (110)	0.04714	0.00119	−0.07442
Second neighbor			
s, s (200)	0.03206	0.00442	−0.02719
s, x (200)	−0.04131	0.01734	0.02427
s, d2 (002)	0.01593	−0.02241	−0.02565
x, x (200)	−0.03014	0.00856	−0.02495
y, y (200)	0.00565	0.01727	0.01107
x, xy (020)	−0.01009	−0.00691	−0.00116
z, d2 (002)	0.01026	−0.02713	−0.00016
xy, xy (200)	−0.02030	−0.00067	0.00809
xy, xy (002)	0.00300	−0.00014	0.00039
d2, d2 (002)	−0.00093	−0.01794	0.00140
d1, d1 (002)	0.00451	0.00439	0.00384

Table A.56 Zirconium fcc

	Orthogonal			Non–orthogonal		
Band	RMS error mRy	Maximum k	Deviation mRy	RMS error mRy	Maximum k	Deviation mRy
1	3.1	(055)	12.7	0.6	(006)	1.1
2	5.8	(044)	11.6	0.8	(008)	2.7
3	4.1	(048)	7.9	0.6	(033)	1.3
4	4.4	(118)	14.1	2.0	(118)	9.7
5	4.0	(380)	9.1	0.9	(048)	2.2
6	4.9	(055)	12.3	0.9	(000)	1.9
1–6	4.5			1.1		

Table A.57 Energy values in Ry at selected k–points

	Orthogonal	APW	Non–orthogonal
GAMMA 1	0.18584	0.18717	0.18683
GAMMA 12	0.87622	0.87098	0.87287
GAMMA 15	1.71452	1.70175	1.70185
GAMMA 25′	0.68262	0.68210	0.68287
X1 (008)	0.33947	0.34017	0.34090
X1 (008)	1.52000	1.56599	1.56546
X2 (008)	1.02128	1.01644	1.01794
X3 (008)	0.42269	0.43043	0.43308
X5 (008)	1.07138	1.06694	1.06741
X4′ (008)	0.88023	0.88250	0.88209
X5′ (008)	1.36938	1.37747	1.37689
L1 (444)	0.38367	0.38213	0.38278

(continued)

Table A.57 (continued)

	Orthogonal	APW	Non–orthogonal
L1 (444)	1.36902	1.38301	1.38282
L3 (444)	0.66010	0.65828	0.65902
L3 (444)	1.01588	1.01960	1.02023
L2′ (444)	0.69253	0.70181	0.70159
L3′ (444)	1.76688	1.78439	1.78412
W1 (048)	0.78737	0.78842	0.78971
W1 (048)	1.39722	1.38658	1.38672
W3 (048)	0.59368	0.58582	0.58584
W3 (048)	1.33675	1.33842	1.33888
W1′ (04B)	1.05939	1.06814	1.07036
W2′ (048)	0.43933	0.44116	0.44161
W2′ (048)	1.60421	1.57597	1.57700
EVEN (224)	0.39149	0.39382	0.39383
EVEN (224)	0.58225	0.58074	0.58011
EVEN (224)	0.65818	0.65143	0.65167
EVEN (224)	0.92807	0.92679	0.92779
ODD (224)	0.71717	0.72282	0.72236
ODD (224)	0.98703	0.98751	0.98679

Table A.58 Fermi level quantities (non–orthogonal fit)

Energy		Densities of states				Velocity	Plasmon energy
	Total	s	p	t_{2g}	e_g		
Ry		States/Ry/atom				cm/s	eV
0.6090	18.53	0.56	2.96	11.91	3.24	0.50 × 10E8	6.89

Integrated densities of states				
Total	s	p	t_{2g}	e_g
Electrons				
4.00	0.75	0.38	1.72	1.15

Table A.59 Zirconium fcc

Slater–Koster 2–center parameters	Orthogonal	Non–orthogonal	
	Energy integrals (Ry)	Energy integrals (Ry)	Overlap integrals
On site			
s	1.06596	0.62743	
P	1.32528	1.14191	
d1	0.80094	0.70690	
d2	0.75723	0.70319	
First neighbor			
(sss)	−0.06800	−0.00200	0.12522

(continued)

Table A.59 (continued)

Slater–Koster 2–center parameters	Orthogonal	Non–orthogonal	
	Energy integrals (Ry)	Energy integrals (Ry)	Overlap integrals
(pps)	0.12087	−0.09772	−0.20325
(ppp)	−0.02246	−0.04819	0.00367
(dds)	−0.08459	−0.00301	0.10194
(ddp)	0.04152	0.01588	−0.05855
(ddd)	−0.00602	−0.00533	0.00459
(sps)	−0.09756	−0.05171	−0.15038
(sds)	−0.07321	0.00625	0.10711
(pds)	0.09627	0.04893	0.14450
(pdp)	−0.02581	0.00865	−0.03903
Second neighbor			
(sss)	−0.01173	−0.02113	−0.00186
(pps)	0.02678	0.18003	0.09854
(ppp)	−0.00611	−0.06394	−0.03434
(dds)	−0.00054	−0.03259	−0.01805
(ddp)	−0.00111	0.00084	−0.00276
(ddd)	0.00212	0.00249	0.00276
(sps)	0.00138	0.00872	−0.06151
(sds)	0.00497	−0.01143	0.00425
(pds)	0.02104	−0.05737	−0.02755
(pdp)	−0.00376	0.01095	0.00314

	Orthogonal			Non–orthogonal		
Band	RMS error mRy	Maximum k	Deviation mRy	RMS error mRy	Maximum k	Deviation mRy
1	17.7	(048)	35.4	3.1	(444)	6.5
2	16.3	(008)	31.5	2.7	(442)	6.8
3	15.7	(033)	40.1	4.2	(033)	11.5
4	15.9	(005)	29.8	3.6	(118)	12.0
5	18.0	(044)	37.8	2.0	(000)	4.4
6	20.2	(066)	65.7	8.8	(174)	43.5
1–6	17.4			4.6		

Table A.60 Zirconium bcc Z = 40 lattice constant = 6.60000 a.u.

Slater–Koster 3–center parameters	Orthogonal	Non–orthogonal	
	Energy integrals (Ry)	Energy integrals (Ry)	Overlap integrals
On site			
s, s(000)	0.91715	0.69622	

(continued)

Table A.60 (continued)

Slater–Koster 3–center parameters	Orthogonal	Non–orthogonal	
	Energy integrals (Ry)	Energy integrals (Ry)	Overlap integrals
x, x (000)	1.39921	1.00099	
xy, xy(000)	0.87142	0.73150	
d2, d2(000)	0.76876	0.67222	
First neighbor			
s, s(111)	−0.09760	−0.01702	0.11642
s, x(111)	0.06175	−0.02612	−0.10951
s, xy(111)	0.04964	0.00756	−0.04974
x, x(111)	0.07570	−0.05232	−0.05233
x, y(111)	0.03925	−0.08186	−0.13331
x, xy (111)	0.04140	−0.01575	−0.03993
x, yz(111)	0.05178	−0.02314	−0.08621
x, d1(111)	0.00771	−0.00243	−0.04896
xy, xy (111)	−0.02190	−0.00947	0.01367
xy, xz(111)	−0.03417	−0.00425	0.03800
xy, d2(111)	0.00296	0.00754	−0.03208
d2, d2(111)	0.02933	0.00820	−0.04996
Second neighbor			
s, s(200)	−0.01858	−0.01685	0.05754
s, x(200)	0.05516	−0.01737	−0.11877
s, d2(002)	−0.05836	0.00999	0.08474
x, x(200)	0.16571	−0.10787	−0.24098
y, y(200)	0.02786	−0.00862	0.00959
x, xy(020)	0.01892	−0.01581	0.00546
z, d2(002)	−0.10851	0.06664	0.16431
xy, xy(200)	−0.02235	0.02188	−0.00296
xy, xy(002)	0.00256	−0.00797	−0.01256
d2, d2(002)	−0.05842	0.01378	0.10301
d1, d1(002)	0.00739	−0.00675	−0.00549
Third neighbor			
s, s(220)	0.01499		
s, x(220)	−0.01377		
s, xy(220)	−0.01373		
s, d2(220)	−0.00836		
x, x(220)	−0.00776		
x, x(022)	0.00637		
x, y(220)	−0.00446		
x, xy(220)	−0.01055		
x, xy(022)	0.00188		
z, d2(022)	0.00092		
z, d1(022)	−0.00651		
xy, xy(220)	0.01437		
xy, xy(022)	0.00099		
xy, xz(022)	−0.00063		
xy, d2(220)	0.00992		
d2, d2(220)	0.00448		
d1, d1(220)	−0.00539		

Table A.61 Zirconium bcc

	Orthogonal			Non–orthogonal		
	RMS error	Maximum	Deviation	RMS error	Maximum	Deviation
Band	mRy	k	mRy	mRy	k	mRy
1	2.9	(260)	7.9	0.8	(000)	2.5
2	2.6	(030)	7.7	0.6	(252)	1.2
3	2.3	(044)	6.5	0.7	(333)	2.8
4	2.3	(332)	5.2	0.7	(230)	2.2
5	3.2	(002)	9.8	1.0	(131)	3.7
6	4.6	(226)	15.8	1.4	(443)	3.3
1–6	3.1			0.9		

Table A.62 Energy values in Ry at selected k–points

	Orthogonal	APW	Non–orthogonal
GAMMA 1	0.20479	0.19906	0.20157
GAMMA 12	0.84490	0.85087	0.84995
GAMMA 15	2.41111	2.41140	2.41140
GAMMA 25′	0.67735	0.67892	0.67825
H1 (008)	1.76634	1.76704	1.76705
H12 (008)	0.37555	0.37009	0.37095
H15 (008)	1.19993	1.19893	1.19933
H25′(008)	1.02783	1.02772	1.02940
N1 (044)	0.37163	0.37362	0.37328
N1 (044)	0.85924	0.86310	0.86372
N1 (044)	1.40839	1.41278	1.41260
N2 (044)	0.53544	0.53719	0.53678
N3 (044)	1.08217	1.08242	1.08363
N4 (044)	0.91377	0.91029	0.91049
N1′ (044)	0.72832	0.73481	0.73439
N3′ (044)	1.70672	1.70879	1.70882
N4′ (044)	1.35626	1.35238	1.35246
P1 (444)	1.20857	1.21608	1.21775
P3 (444)	0.91637	0.91989	0.92040
P4 (444)	0.55311	0.55463	0.55373
P4 (444)	1.38784	1.40047	1.39950
(341)	0.40563	0.40514	0.40459
(341)	0.54332	0.54524	0.54525
(341)	0.66152	0.66178	0.66187
(341)	0.85048	0.85063	0.85045
(341)	0.92330	0.92418	0.92441
(341)	1.09991	1.09899	1.09920

Table A.63 Fermi level quantities (non–orthogonal fit)

Energy		Densities of states				Velocity	Plasmon energy
	Total	s	p	t_{2g}	e_g		
Ry		States/Ry/atom				cm/s	eV
0.5920	19.17	0.62	2.37	12.06	4.13	$0.44 \times 10E8$	6.12

Integrated densities of states				
Total Electrons	s	p	t_{2g}	e_g
4.00	0.68	0.22	1.74	1.37

Table A.64 Zirconium bcc

Slater–Koster 2–center parameters	Orthogonal	Non–orthogonal	
	Energy integrals (Ry)	Energy integrals (Ry)	Overlap integrals
On site			
s	0.99646	0.69689	
p	1.39706	0.98300	
d1	0.83845	0.71671	
d2	0.75923	0.68582	
First neighbor			
(sss)	−0.09760	−0.02468	0.11413
(pps)	0.15445	−0.19747	−0.30782
(ppp)	0.03615	0.02451	0.07790
(dds)	−0.09095	−0.00183	0.10555
(ddp)	0.04137	0.01943	−0.06820
(ddd)	0.00219	−0.00938	0.00463
(sps)	0.09238	0.03380	0.19121
(sds)	−0.09637	−0.00533	0.09626
(pds)	−0.13356	−0.08741	−0.19171
(pdp)	0.01913	−0.00216	0.07906
Second neighbor			
(sss)	−0.01945	−0.00518	0.06693
(pps)	0.18503	−0.16501	−0.28079
(ppp)	−0.00001	0.02537	0.02893
(dds)	−0.05918	−0.00141	0.08181
(ddp)	−0.00814	0.00599	−0.02486
(ddd)	0.00646	−0.00612	−0.00586
(sps)	0.06463	0.04584	0.14575
(sds)	−0.04477	0.00650	0.07910
(pds)	−0.08865	−0.04926	−0.14511
(pdp)	−0.00661	0.00359	0.02249
Third neighbor			
(sss)	0.00858		
(pps)	0.01972		
(ppp)	−0.00520		
(dds)	0.01134		
(ddp)	−0.00255		
(ddd)	0.00072		
(sps)	−0.00023		
(sds)	0.00654		
(pds)	0.00121		
(pdp)	−0.00001		

	Orthogonal			Non–orthogonal		
Band	RMS error mRy	Maximum k	Deviation mRy	RMS error mRy	Maximum k	Deviation mRy
1	4.4	(044)	11.2	0.9	(000)	3.3
2	10.1	(044)	29.6	1.0	(121)	2.2
3	6.6	(008)	20.3	1.4	(226)	3.7
4	10.0	(260)	21.4	2.1	(231)	5.0
5	8.5	(170)	21.3	1.9	(330)	5.2
6	10.1	(442)	24.8	2.1	(050)	5.3
1–6	8.6			1.6		

Table A.65 Technetium fcc Z = 43 lattice constant = 7.25000 a.u.

Slater–Koster 3–center parameters	Orthogonal	Non–orthogonal	
	Energy integrals (Ry)	Energy integrals (Ry)	Overlap integrals
On site			
s, s(000)	1.22903	1.07591	
x, x(000)	1.70551	1.47686	
xy, xy(000)	0.75393	0.68692	
d2, d2(000)	0.68755	0.66286	
First neighbor			
s, s(110)	−0.09236	−0.06451	0.02801
s, x(110)	0.08870	0.04007	−0.05449
s, xy(110)	−0.07939	−0.05340	0.04587
s, d2(110)	0.04136	0.03993	0.00113
x, x(110)	0.07129	0.02216	−0.04404
x, x(011)	−0.00602	0.06576	0.06277
x, y(110)	0.11851	−0.01531	−0.09630
x, xy (110)	−0.08250	−0.03449	0.04037
x, xy(011)	0.00641	−0.01632	−0.03834
z, d2(011)	0.01032	0.00060	−0.01758
z, d1 (011)	0.04618	0.01357	−0.03886
xy, xy (110)	−0.06317	−0.01540	0.06604
xy, xy (011)	0.01603	0.00914	−0.02164
xy, xz (011)	0.00906	0.00090	−0.02880
xy, d2(110)	0.03316	0.02610	−0.00579
d2, d2(110)	−0.02639	−0.01849	0.01308
dl, d1(110)	0.04426	0.02925	−0.03267
Second neighbor			
s, s(200)	0.01334	−0.01026	−0.02408
s, x(200)	−0.01635	0.05047	0.03334
s, d2 (002)	0.01100	−0.01781	−0.02509
x, x (200)	−0.00975	0.07793	0.00815
y, y(200)	−0.01240	0.01137	0.01479
x, xy(020)	−0.00190	−0.01560	−0.00904
z, d2(002)	0.00364	−0.01788	−0.00699
xy, xy(200)	−0.00797	−0.01385	−0.00671
xy, xy(002)	0.00223	0.00034	0.00075
d2, d2(002)	−0.00319	−0.00606	0.00629
dl, d1(002)	0.00105	0.00853	0.01025

Table A.66 Technetium fcc

	Orthogonal			Non–orthogonal		
Band	RMS error mRy	Maximum k	Deviation mRy	RMS error mRy	Maximum k	Deviation mRy
1	2.7	(055)	11.0	0.6	(174)	1.7
2	4.3	(048)	9.0	0.9	(008)	2.8
3	3.6	(048)	9.0	1.2	(007)	5.3
4	3.3	(118)	13.8	2.4	(118)	12.4
5	3.4	(380)	7.6	1.0	(000)	2.8
6	4.3	(044)	9.8	1.2	(062)	3.7
1–6	3.7			1.3		

Table A.67 Energy values in Ry at selected k–points

	Orthogonal	APW	Non–orthogonal
GAMMA 1	0.20069	0.20130	0.20155
GAMMA 12	0.78832	0.78557	0.78834
GAMMA 15	2.18268	2.17456	2.17467
GAMMA 25′	0.60207	0.60260	0.60404
X1 (008)	0.28429	0.28450	0.28550
X1 (008)	1.75703	1.78043	1.78001
X2 (008)	0.92802	0.92570	0.92705
X3 (008)	0.34567	0.34950	0.35231
X5 (008)	0.97920	0.97560	0.97644
X4′ (008)	1.04198	1.05024	1.05071
X5′ (008)	1.66052	1.71870	1.71664
L1 (444)	0.33653	0.34006	0.34100
L1 (444)	1.52137	1.53334	1.53313
L3 (444)	0.58194	0.58452	0.58600
L3 (444)	0.92956	0.93117	0.93232
L2′ (444)	0.82651	0.82544	0.82479
L3′ (444)	2.24865	2.25131	2.25129
W1 (048)	0.74939	0.74956	0.75164
W1 (048)	1.67524	1.65033	1.65068
W3 (048)	0.54629	0.53734	0.53852
W3 (048)	1.54885	1.54879	1.54952
W1′ (048)	0.97027	0.97639	0.97869
W2′ (048)	0.39528	0.39518	0.39603
W2′ (048)	1.80414	1.75800	1.75901
EVEN(224)	0.38619	0.38508	0.38545
EVEN(224)	0.56022	0.55760	0.55861
EVEN(224)	0.60396	0.60190	0.60114
EVEN(224)	0.89863	0.89891	0.89983
ODD (224)	0.64782	0.65342	0.65282
ODD (224)	0.90653	0.90437	0.90382

Table A.68 Fermi level quantites (non–orthogonal fit)

Energy		Densities of states				Velocity	Plasmon energy
	Total	s	p	t_{2g}	e_g		
Ry		States/Ry/atom				cm/s	eV
0.7320	16.71	0.27	1.04	5.97	9.43	0.64 × 10E8	10.43

Integrated densities of states				
Total	s	p	t_{2g}	e_g
Electrons				
7.00	0.63	0.44	3.63	2.34

Table A.69 Technetium fcc

Slater–Koster 2–center parameters	Orthogonal	Non–orthogonal	
	Energy integrals (Ry)	Energy integrals (Ry)	Overlap integrals
On site			
s	1.27403	0.82159	
p	1.75676	1.26591	
d1	0.70073	0.65006	
d2	0.67934	0.64638	
First neighbor			
(sss)	−0.08985	−0.01956	0.10656
(pps)	0.18823	−0.22638	−0.29155
(ppp)	−0.03714	−0.04767	0.02420
(dds)	−0.08323	−0.02817	0.07549
(ddp)	0.04493	0.02548	−0.04232
(ddd)	−0.00780	−0.00476	0.00443
(sps)	−0.12253	−0.05840	−0.18150
(sds)	−0.06191	−0.01379	0.08618
(pds)	0.08719	0.04025	0.13788
(pdp)	−0.02805	0.01359	−0.04287
Second neighbor			
(sss)	−0.00020	−0.02089	−0.00027
(pps)	0.01681	0.08591	0.00181
(ppp)	−0.00749	−0.01046	0.00476
(dds)	−0.00453	−0.02912	−0.01945
(ddp)	0.00437	0.00073	−0.00224
(ddd)	−0.00028	0.00288	0.00367
(sps)	0.00458	0.04336	0.00582
(sds)	−0.01462	−0.02282	−0.00746
(pds)	0.02186	−0.05815	−0.02508
(pdp)	−0.01141	0.00069	−0.00467

	Orthogonal			Non–orthogonal		
Band	RMS error mRy	Maximum k	Deviation mRy	RMS error mRy	Maximum k	Deviation mRy
1	13.7	(084)	30.6	1.8	(055)	6.6
2	15.8	(008)	37.8	2.0	(044)	4.8
3	11.0	(006)	30.9	2.4	(044)	6.4
4	14.2	(333)	36.2	3.0	(118)	12.4
5	12.6	(044)	26.4	1.9	(000)	4.0
6	18.8	(008)	42.8	1.4	(000)	4.0
1–6	14.6			2.2		

Table A.70 Ruthenium fcc Z = 44 lattice constant = 7.20000 a.u.

Slater–Koster 3–center parameters	Orthogonal	Non–orthogonal	
	Energy integrals (Ry)	Energy integrals (Ry)	Overlap integrals
On site			
s, s(000)	1.16552	1.02792	
x, x (000)	1.66204	1.45628	
xy, xy(000)	0.62603	0.58613	
d2, d2(000)	0.58465	0.57317	
First neighbor			
s, s (110)	−0.08938	−0.06391	0.03043
s, x (110)	0.08988	0.05088	−0.05178
s, xy (110)	−0.06914	−0.05990	0.03382
s, d2 (110)	0.03620	0.03889	0.00906
x, x (110)	0.07712	0.05240	−0.03223
x, x(011)	−0.00297	0.08462	0.07120
x, y (110)	0.12007	−0.00329	−0.09369
x, xy (110)	−0.06996	−0.04470	0.02574
x, xy (011)	0.00877	−0.01980	−0.04691
z, d2 (011)	0.00917	−0.00123	−0.02211
z, d1 (011)	0.04161	0.01875	−0.02966
xy, xy (110)	−0.05513	−0.01781	0.06005
xy, xy (011)	0.01419	0.00950	−0.01873
xy, xz (011)	0.01238	0.00073	−0.02904
xy, d2 (110)	0.02960	0.02677	−0.00123
d2, d2(110)	−0.02316	−0.01657	0.00996
d1, d1(110)	0.03945	0.02985	−0.01890
Second neighbor			
s, s (200)	0.00753	−0.01481	−0.02349
s, x(200)	−0.01082	0.05664	0.03668
s, d2 (002)	0.00652	−0.01852	−0.02912
x, x (200)	−0.00693	0.09164	0.01833
y, y (200)	−0.01152	0.00725	0.01002
x, xy (020)	−0.00131	−0.01366	−0.00668
z, d2 (002)	0.00203	−0.02886	−0.01922
xy, xy (200)	−0.00370	−0.01325	−0.00908
xy, xy (002)	0.00139	−0.00055	−0.00023
d2, d2 (002)	−0.00397	−0.01099	−0.00760
d1, d1 (002)	−0.00005	0.00545	0.00823

Table A.71 Ruthenium fcc

	Orthogonal			Non–orthogonal		
Band	RMS error mRy	Maximum k	Deviation mRy	RMS error mRy	Maximum k	Deviation mRy
1	2.6	(055)	8.5	0.6	(174)	1.8
2	3.6	(055)	7.5	0.9	(008)	2.4
3	3.3	(048)	6.9	1.2	(007)	5.1
4	2.9	(118)	11.3	2.2	(118)	11.6
5	3.1	(380)	7.2	1.1	(000)	3.0
6	4.1	(066)	15.7	1.2	(000)	3.0
1–6	3.3			1.3		

Table A.72 Energy value in Ry at seleected k-points

	Orthogonal	APW	Non–orthogonal
GAMMA 1	0.13812	0.13976	0.14059
GAMMA 12	0.67034	0.66800	0.67095
GAMMA 15	2.20718	2.19819	2.19823
GAMMA 25′	0.50700	0.50787	0.50946
X1 (008)	0.22709	0.22769	0.22884
X1 (008)	1.63076	1.65311	1.65270
X2 (008)	0.79045	0.78997	0.79132
X3 (008)	0.27998	0.28199	0.28436
X5 (008)	0.83452	0.83367	0.83434
X4′ (008)	0.97322	0.97930	0.98050
X5′ (008)	1.61398	1.68176	1.67956
L1 (444)	0.26938	0.27356	0.27457
L1 (444)	1.38999	1.40236	1.40200
L3 (444)	0.49019	0.49344	0.49482
L3 (444)	0.79407	0.79543	0.79649
L2′ (444)	0.76147	0.76079	0.75939
L3′ (444)	2.20227	2.24455	2.24405
W1 (048)	0.64829	0.64484	0.64683
W1 (048)	1.57093	1.56689	1.56653
W3 (048)	0.45939	0.45247	0.45380
W3 (048)	1.45235	1.45587	1.45723
W1′ (048)	0.82896	0.83426	0.83631
W2′ (048)	0.33049	0.32901	0.32910
W2′ (048)	1.72624	1.65091	1.65295
EVEN(224)	0.31708	0.31639	0.31674
EVEN(224)	0.47931	0.47617	0.47692
EVEN(224)	0.51539	0.51405	0.51365
EVEN(224)	0.78540	0.78558	0.78784
ODD (224)	0.55051	0.55509	0.55439
ODD (224)	0.77477	0.77305	0.77190

Table A.73 Fermi level quantities (non–orthogonal. fit)

Energy		Densities of states				Velocity	Plasmon energy
	Total	s	p	t_{2g}	e_g		
Ry		States/Ry/atom				cm/s	eV
0.6830	15.71	0.19	0.65	6.05	8.82	0.63 × 10E8	10.08

Integrated densities of states				
Total	s	p	t_{2g}	e_g
Electrons				
8.00	0.61	0.44	4.03	2.93

Table A.74 Ruthenium fcc

Slater–Koster 2–center parameters	Orthogonal	Non–orthogonal	
	Energy integrals (Ry)	Energy integrals (Ry)	Overlap integrals
On site			
s	1.21270	0.74682	
p	1.63368	1.20952	
d1	0.59665	0.55479	
d2	0.57910	0.55274	
First neighbor			
(sss)	−0.08754	−0.02496	0.11195
(pps)	0.18006	−0.20513	−0.29345
(ppp)	−0.02338	−0.04782	0.02723
(dds)	−0.07279	−0.03167	0.06662
(ddp)	0.03899	0.02499	−0.03764
(ddd)	−0.00665	−0.00440	0.00411
(sps)	−0.12287	−0.05051	−0.18817
(sds)	−0.06121	−0.02019	0.07938
(pds)	0.08256	0.01988	0.12250
(pdp)	−0.02538	0.01766	−0.03918
Second neighbor			
(sss)	−0.00504	−0.01926	0.00329
(pps)	0.03401	0.08637	0.00002
(ppp)	−0.00485	−0.01630	0.00194
(dds)	−0.00448	−0.02226	−0.01562
(ddp)	0.00291	0.00101	−0.00184
(ddd)	0.00006	0.00152	0.00244
(sps)	−0.00147	0.04022	−0.00013
(sds)	−0.01189	−0.01920	−0.00429
(pds)	0.01891	−0.04523	−0.01528
(pdp)	−0.00830	0.00236	−0.00322

	Orthogonal			Non–orthogonal		
	RMS error	Maximum	Deviation	RMS error	Maximum	Deviation
Band	mRy	k	mRy	mRy	k	mRy
1	10.9	(048)	22.4	1.2	(055)	4.4
2	12.2	(007)	28.8	1.5	(224)	3.3
3	7.2	(033)	15.0	1.7	(007)	4.4
4	10.4	(333)	31.6	2.6	(118)	11.9
5	9.4	(044)	20.9	1.6	(000)	3.9
6	11.6	(006)	25.4	1.3	(000)	3.9
1–6	10.4			1.7		

Table A.75 Cadmium fcc Z = 48 lattice constant = 8.20000 a.u.

Slater–Koster 3–center parameters	Orthogonal	Non–orthogonal	
	Energy integrals (Ry)	Energy integrals (Ry)	Overlap integrals
On site			
s, s (000)	0.36367	0.19685	
x, x (000)	0.93593	0.62172	
xy, xy (000)	−0.27418	−0.24317	
d2, d2(000)	−0.26983	−0.27476	
First neighbor			
s, s (110)	−0.04888	−0.04015	0.11700
s, x (110)	0.05787	0.01761	−0.13620
s, xy (110)	−0.00833	−0.01575	−0.04532
s, d2 (110)	−0.00657	−0.00720	0.05453
x, x (110)	0.07557	0.01144	−0.07093
x, x (011)	0.02036	−0.03503	0.00380
x, y (110)	0.08749	−0.03999	−0.19310
x, xy (110)	0.00162	−0.04883	0.01695
x, xy (011)	0.00566	0.00890	0.00480
z, d2 (011)	−0.00361	−0.00652	0.02512
z, d1 (011)	0.00179	0.00609	0.00106
xy, xy (110)	−0.01133	0.02782	−0.15310
xy, xy (011)	0.00311	−0.00094	0.00841
xy, xz (011)	0.00482	−0.02617	0.08583
xy, d2 (110)	0.00608	0.00811	−0.00657
d2, d2 (110)	−0.00483	−0.01331	0.03433
d1, d1 (110)	0.00785	0.02282	−0.05551
Second neighbor			
s, s (200)	0.00137	−0.03371	−0.02077
s, x (200)	0.00033	0.06775	0.06068
s, d2 (002)	−0.01593	0.00030	−0.02954
x, x (200)	−0.00352	0.10591	0.05820
y, y (200)	−0.00585	0.01701	−0.00469
x, xy (020)	−0.00145	0.01032	−0.01496
z, d2 (002)	−0.02524	0.02935	−0.14112
xy, xy (200)	0.00121	−0.01063	0.04257
xy, xy (002)	−0.00002	−0.00965	0.03047
d2, d2 (002)	−0.00180	0.00036	−0.01174
d1, d1 (002)	−0.00037	−0.00791	0.03065

Table A.76 Cadmium fcc

	Orthogonal			Non–orthogonal		
Band	RMS error mRy	Maximum k	Deviation mRy	RMS error mRy	Maximum k	Deviation mRy
1	1.7	(048)	4.0	2.0	(222)	8.2
2	1.8	(005)	4.3	1.2	(055)	2.9
3	2.1	(118)	4.6	1.5	(118)	3.5
4	2.0	(048)	4.8	1.4	(044)	3.8
5	0.8	(000)	1.5	1.6	(048)	4.6
6	8.7	(048)	40.6	1.2	(222)	2.6
1–6	3.9			1.5		

Table A.77 Energy values in Ry at selected k–points

	Orthogonal	APW	Non–orthogonal
GAMMA 1	−0.21463	−0.21216	−0.21373
GAMMA 12	−0.25826	−0.25980	−0.25855
GAMMA 15	1.59152	1.56366	1.56381
GAMMA 25′	−0.28981	−0.29239	−0.29329
X1 (008)	−0.33615	−0.33532	−0.33441
X1 (008)	0.57044	0.55561	0.55426
X2 (008)	−0.23166	−0.23387	−0.23247
X3 (008)	−0.33957	−0.33889	−0.33762
X5 (008)	−0.22402	−0.22392	−0.22577
X4′ (008)	0.38238	0.39781	0.39872
X5′ (008)	0.82405	0.96444	0.96238
L1 (444)	−0.32247	−0.32466	−0.32509
L1 (444)	0.36038	0.35419	0.35508
L3 (444)	−0.29137	−0.29307	−0.29226
L3 (444)	−0.23167	−0.23199	−0.23029
L2′ (444)	0.26645	0.27148	0.27042
L3′ (444)	1.31631	1.40016	1.40011
W1 (048)	−0.25967	−0.25490	−0.25419
W1 (048)	0.57177	0.61240	0.61048
W3 (048)	−0.28694	−0.29040	−0.29015
W3 (048)	0.62689	0.61511	0.61714
W1′ (048)	−0.22396	−0.22395	−0.21934
W2′ (048)	−0.30644	−0.31044	−0.31141
W2′ (048)	0.83830	0.72655	0.72922
EVEN (224)	−0.31368	−0.31184	−0.31488
EVEN (224)	−0.28934	−0.28963	−0.28769
EVEN (224)	−0.25826	−0.25967	−0.25653
EVEN (224)	0.04773	0.05186	0.04965
ODD (224)	−0.28079	−0.27856	−0.28101
ODD (224)	−0.23664	−0.23567	−0.23724

Table A.78 Fermi level quantities (non–orthogonal fit)

Energy		Densities of states				Velocity	Plasmon energy
	Total	s	p	t_{2g}	e_g		
Ry		States/Ry/atom				cm/s	V
0.454	4.71	2.74	1.79	0.12	0.07	1.61 × 10E8	0.92

Integrated densities of states				
Total	s	p	t_{2g}	e_g
Electrons				
2.00	1.38	0.54	0.04	0.05

Table A.79 Cadmium fcc

Slater–Koster 2–center parameters	Orthogonal	Non–orthogonal	
	Energy integrals (Ry)	Energy integrals (Ry)	Overlap integrals
On site			
s	0.35903	0.17101	
p	0.94214	0.45590	
d1	−0.27438	−0.27140	
d2	−0.27190	−0.27075	
First neighbor			
(sss)	−0.04957	−0.04895	0.11050
(pps)	0.16102	−0.11481	−0.39746
(ppp)	−0.00137	−0.00285	0.08790
(dds)	−0.01471	−0.01671	0.00665
(ddp)	0.00781	0.00754	−0.00241
(ddd)	−0.00145	−0.00064	−0.00078
(sps)	0.07376	0.02843	−0.20712
(sds)	−0.01120	−0.02901	0.03370
(pds)	0.00544	−0.04725	0.09750
(pdp)	0.00144	0.02284	−0.03268
Second neighbor			
(sss)	0.00322	−0.01377	0.00462
(pps)	0.03957	0.03562	−0.05746
(ppp)	−0.01650	0.00349	0.00984
(dds)	−0.00158	0.00029	−0.00431
(ddp)	0.00130	0.00037	−0.00006
(ddd)	−0.00023	−0.00031	0.00065
(sps)	0.00570	0.02617	−0.01287
(sds)	−0.00272	−0.00336	0.00075
(pds)	−0.00982	−0.01046	0.00746
(pdp)	−0.00551	0.00260	−0.00456

	Orthogonal			Non–orthogonal		
Band	RMS error mRy	Maximum k	Deviation mRy	RMS error mRy	Maximum k	Deviation mRy
1	1.5	(007)	3.4	0.5	(008)	1.1
2	1.6	(006)	3.6	0.5	(008)	1.3
3	1.9	(064)	4.2	0.9	(118)	4.0
4	1.7	(044)	4.9	0.4	(000)	1.4
5	1.5	(354)	3.4	0.4	(000)	1.4
6	26.6	(008)	86.6	0.2	(042)	0.7
1–6	11.0			0.5		

Appendix 4
Computer Programs

In this appendix we list computer programs that generate eigenvalues and eigenvectors by diagonalizing the TB–Hamiltonian for the four different structures we used. A diagonalization routine is required in order to execute the programs listed in this section.

The Fortran programs listed below can be requested from the author at dpapacon@gmu.edu to avoid the need to retype them.

A. **Body-centered cubic structure**

This program generates eigenvalues and eigenvectors for the 9×9 spd Hamiltonian; it diagonalizes a real matrix. The user is expected to add a diagonalization subroutine named HDIAG. The Slater–Koster parameters should be read in the order given for example in the manganese Tables.

```
      PROGRAM BCCSK
C AUTHOR: D.A.PAPACONSTANTOPOULOS
C  CREATES INPUT FOR CALCULATING THE DOS
C  BCC MONATOMIC-REAL, BUT EQUIVALENT TO COMPLEX-OVERLAP VERSION
C   CAUTION: THE DIMENSIONING OF Q10,Q11,Q12,Q13, SHOULD BE
C   CONSISTENT WITH THE NUMER OF K-POINTS CREATED BY THE
C   ASSIGNED VALUE TO MESH
C A diagonalization subroutine needs to be added. CALL HDIAG
      IMPLICIT REAL*8 (A-H,O-Z)
      CHARACTER*80 TITLE
      CHARACTER*10 INA,INB
      COMMON/BLOC1/Q10(9,969),Q11(9,969),Q12(9,969),Q13(9,969)
      DIMENSION H(9,9),U(9,9),S(9,9),HNEW(9,9),UU(9,9),T(9,9)
      DIMENSION TITLE(20),THIRD(17),PARAM(44)
      DIMENSION A(3)
      INA = 'par.in'
      OPEN(10,FILE=INA,BLANK='ZERO')
```


D.A. Papaconstantopoulos, *Handbook of the Band Structure of Elemental Solids*,
DOI 10.1007/978-1-4419-8264-3

```
c   OPEN(11,FILE='dosdat',BLANK='ZERO')
    INB = 'bcc.out'
    OPEN(15,FILE=INB,BLANK='ZERO')
    READ (10,35) TITLE
  35 FORMAT(20A4)
    WRITE (6,335) TITLE
 335 FORMAT (1H1,20A4)
C   ICENTER=3 THREE-CENTER APPROXIMATION
C   ICENTER=2 AND NPAR=44 TWO-CENTER NON-ORTHOGONAL
C   ICENTER=2 AND NPAR=34 TWO-CENTER ORTHOGONAL
C   MESH  K-POINT MESH GENERATOR
    READ (10,9999) MESH,ICENTER,NPAR
 9999 FORMAT (3I5)
    WRITE(6,9998) MESH,ICENTER,NPAR
 9998 FORMAT (5X,'MESH=',I5,10X,'ICENTER=',I5,10X,'NPAR=',I5)
    N=9
    NS=9
    SR2=SQRT(2.)
    SR3=SQRT(3.)
    IF (ICENTER.EQ.3)
    1READ(10,55)A1,A5,A11,A15,A2,A3,A4,A6,A7,A8,A9,A10,A12,
    A13,A14,A16
    1,B1,B2,B3,B4,B5,B6,B7,B8,B9,B10,B11,C1,C2,C3,C4,C5,C6,
    C7,C8,
    1C9,C10,C11,C12,C13,C14,C15,C16,C17,S2,S3,S4,S6,S7,S8,S9,
    S10,
    1S12,S13,S14,S16,T1,T2,T3,T4,T5,T6,T7,T8,T9,T10,T11
  55 FORMAT(F12.10)
C   THE ORDER OF THE SK-PARAMETERS HERE IS THE SAME AS IN THE TABLES
C   CAUTION, THE 3RD NEIGHBOR PARAMETERS C1,C2,...,C17 MUST BE SET
C   EQUAL TO ZERO FOR THE NON-ORTHOGONAL CASES.
    IF (ICENTER.EQ.3)
    1WRITE(6,999)A1,A5,A11,A15,A2,A3,A4,A6,A7,A8,A9,A10,A12,
    A13,A14,
    1A16,B1,B2,B3,B4,B5,B6,B7,B8,B9,B10,B11,C1,C2,C3,C4,C5,
    C6,C7,
    1C8,C9,C10,C11,C12,
    1C13,C14,C15,C16,C17,S2,S3,S4,S6,S7,S8,S9,S10,S12,S13,
    S14,S16,
    1T1,T2,T3,T4,T5,T6,T7,T8,T9,T10,T11
  999 FORMAT(//10X,'VALUES OF THE SK PARAMETERS',//67(25X,F15.8/))
    IF  (ICENTER.EQ.2.AND.NPAR.EQ.44)  READ(10,55)  (PARAM(I),
    I=1,44)
    IF (ICENTER.EQ.2.AND.NPAR.EQ.44) WRITE(15,999) (PARAM(I),
    I=1,44)
```

```
      IF (ICENTER.EQ.2.AND.NPAR.EQ.34) READ(10,55) (PARAM(I),
      I=1,34)
      IF (ICENTER.EQ.2.AND.NPAR.EQ.34) WRITE(6,999) (PARAM(I),
      I=1,34)
      IF (ICENTER.EQ.3) GO TO 500
      IF (NPAR.EQ.34) GO TO 499
C     THIS SECTION IS FOR THE TWO-CENTER CALCULATIONS
      Q1=PARAM(1)
      Q2=PARAM(2)
      Q3=PARAM(3)
      Q4=PARAM(4)
      P1=PARAM(5)
      P2=PARAM(6)
      P3=PARAM(7)
      P4=PARAM(8)
      P5=PARAM(9)
      P6=PARAM(10)
      P12=PARAM(11)
      P14=PARAM(12)
      P24=PARAM(13)
      P35=PARAM(14)
      R1=PARAM(15)
      R2=PARAM(16)
      R3=PARAM(17)
      R4=PARAM(18)
      R5=PARAM(19)
      R6=PARAM(20)
      R12=PARAM(21)
      R14=PARAM(22)
      R24=PARAM(23)
      R35=PARAM(24)
      PP1=PARAM(25)
      PP2=PARAM(26)
      PP3=PARAM(27)
      PP4=PARAM(28)
      PP5=PARAM(29)
      PP6=PARAM(30)
      PP12=PARAM(31)
      PP14=PARAM(32)
      PP24=PARAM(33)
      PP35=PARAM(34)
      RR1=PARAM(35)
      RR2=PARAM(36)
      RR3=PARAM(37)
      RR4=PARAM(38)
      RR5=PARAM(39)
```

```
RR6=PARAM(40)
RR12=PARAM(41)
RR14=PARAM(42)
RR24=PARAM(43)
RR35=PARAM(44)
A1=Q1
A5=Q2
A11=Q3
A15=Q4
B1=R1
B2=R12
B3=R14
B4=R2
B5=R3
B6=R35
B7=R24
B8=R5
B9=R6
B10=R4
B11=R6
A2=P1
A3=P12/SR3
A4=P14/SR3
A6=(P2+2.*P3)/3.0
A7=(P2-P3)/3.0
A8=(P24+P35/SR3)/3.0
A9=(P24-2.*P35/SR3)/3.
A10=P35/SR3
A12=P4/3.+2.*P5/9.+4.*P6/9.
A13=P4/3.-P5/9.-2.*P6/9.
A14=-2./(3.*SR3)*(P5-P6)
A16=(2.*P5+P6)/3.0
T1=RR1
T2=RR12
T3=RR14
T4=RR2
T5=RR3
T6=RR35
T7=RR24
T8=RR5
T9=RR6
T10=RR4
T11=RR6
S2=PP1
S3=PP12/SR3
S4=PP14/SR3
```

```
      S6=(PP2+2.*PP3)/3.0
      S7=(PP2-PP3)/3.0
      S8=(PP24+PP35/SR3)/3.0
      S9=(PP24-2.*PP35/SR3)/3.
      S10=PP35/SR3
      S12=PP4/3.+2.*PP5/9.+4.*PP6/9.
      S13=PP4/3.-PP5/9.-2.*PP6/9.
      S14=-2./(3.*SR3)*(PP5-PP6)
      S16=(2.*PP5+PP6)/3.0
      C1=0.
      C2=0.
      C3=0.
      C4=0.
      C5=0.
      C6=0.
      C7=0.
      C8=0.
      C9=0.
      C10=0.
      C11=0.
      C12=0.
      C13=0.
      C14=0.
      C15=0.
      C16=0.
      C17=0.
  499 CONTINUE
      IF (NPAR.EQ.44) GO TO 500
      Q1=PARAM(1)
      Q2=PARAM(2)
      Q3=PARAM(3)
      Q4=PARAM(4)
      P1=PARAM(5)
      P2=PARAM(6)
      P3=PARAM(7)
      P4=PARAM(8)
      P5=PARAM(9)
      P6=PARAM(10)
      P12=PARAM(11)
      P14=PARAM(12)
      P24=PARAM(13)
      P35=PARAM(14)
      R1=PARAM(15)
      R2=PARAM(16)
      R3=PARAM(17)
```

```
R4=PARAM(18)
R5=PARAM(19)
R6=PARAM(20)
R12=PARAM(21)
R14=PARAM(22)
R24=PARAM(23)
R35=PARAM(24)
TC1=PARAM(25)
TC2=PARAM(26)
TC3=PARAM(27)
TC4=PARAM(28)
TC5=PARAM(29)
TC6=PARAM(30)
TC12=PARAM(31)
TC14=PARAM(32)
TC24=PARAM(33)
TC35=PARAM(34)
A1=Q1
A5=Q2
A11=Q3
A15=Q4
B1=R1
B2=R12
B3=R14
B4=R2
B5=R3
B6=R35
B7=R24
B8=R5
B9=R6
B10=R4
B11=R6
A2=P1
A3=P12/SR3
A4=P14/SR3
A6=(P2+2.*P3)/3.0
A7=(P2-P3)/3.0
A8=(P24+P35/SR3)/3.0
A9=(P24-2.*P35/SR3)/3.
A10=P35/SR3
A12=P4/3.+2.*P5/9.+4.*P6/9.
A13=P4/3.-P5/9.-2.*P6/9.
A14=-2./(3.*SR3)*(P5-P6)
A16=(2.*P5+P6)/3.0
C1=TC1
```

```
      C2=TC12/SR2
      C3=SR3*TC14/2.
      C4=-TC14/2.
      C5=(TC2+TC3)/2.
      C6=TC3
      C7=(TC2-TC3)/2.
      C8=SR3*SR2*TC24/4.
      C9=TC35/SR2
      C10=(SR2*SR3*TC35+TC24/SR2)/4.
      C11=(-SR3*TC24+2.*TC35)/(4.*SR2)
      C12=(3.*TC4+TC6)/4.
      C13=(TC5+TC6)/2.
      C14=(TC5-TC6)/2.
      C15=SR3*(TC6-TC4)/4.
      C16=(TC4+3.*TC6)/4.
      C17=TC5
      S2=0.0
      S3=0.0
      S4=0.0
      S6=0.0
      S7=0.0
      S8=0.0
      S9=0.0
      S10=0.0
      S12=0.0
      S13=0.0
      S14=0.0
      S16=0.0
      T1=0.0
      T2=0.0
      T3=0.0
      T4=0.0
      T5=0.0
      T6=0.0
      T7=0.0
      T8=0.0
      T9=0.0
      T10=0.0
      T11=0.0
  500 CONTINUE
      JK=0
      IMESH=2*MESH+1
C     THE K-VECTOR UNITS ARE CHOSEN SO THAT THE POINT H HAS
C     CORDINATES (8,0,0)
      SCALE= 8./FLOAT(2*MESH)
```

```
      DO 451  IYBK=1,IMESH
      DO 451  IXBK=1,IYBK
      DO 451  IZBK=1,IXBK
      IT= IXBK+IYBK-1
      IF (IMESH.LT.IT) GO TO 451
      JK=JK+1
      A(1)= FLOAT(IXBK-1)
      A(2)= FLOAT(IYBK-1)
      A(3)= FLOAT(IZBK-1)
      XX =A(1)*SCALE
      YY =A(2)*SCALE
      ZZ =A(3)*SCALE
      PI = 3.14159265
      X1 = PI*XX/8.
      Y1 = PI*YY/8.
      Z1 = PI*ZZ/8.
       COSX=COS(X1)
       COSY=COS(Y1)
       COSZ=COS(Z1)
       SINX=SIN(X1)
       SINY=SIN(Y1)
       SINZ=SIN(Z1)
      COS2X=COS(2.*X1)
      COS2Y=COS(2.*Y1)
      COS2Z=COS(2.*Z1)
      SIN2X=SIN(2.*X1)
      SIN2Y=SIN(2.*Y1)
      SIN2Z=SIN(2.*Z1)
      H(1,1)=A1+8.*A2*COSX*COSY*COSZ+2.*B1*(COS2X+COS2Y+COS2Z)
     1+4.*C1*(COS2X*COS2Y+COS2Y*COS2Z+COS2Z*COS2X)
C
C+++++H(1,2),H(1,3),H(1,4) ARE REVERSED IN SIGN SO THAT REAL
MATRIX IS
C   EQUIVALENT TO COMPLEX MATRIX
C
      H(1,2)=8.*A3*SINX*COSY*COSZ+2.*B2*SIN2X
     1+4.*C2*SIN2X*(COS2Y+COS2Z)
      H(1,3)=8.*A3*COSX*SINY*COSZ+2.*B2*SIN2Y
     1+4.*C2*SIN2Y*(COS2Z+COS2X)
      H(1,4)=8.*A3*COSX*COSY*SINZ+2.*B2*SIN2Z
     1+4.*C2*SIN2Z*(COS2X+COS2Y)
      H(1,2)=-H(1,2)
      H(1,3)=-H(1,3)
      H(1,4)=-H(1,4)
      H(1,5)=-8.*A4*SINX*SINY*COSZ-4.*C3*SIN2X*SIN2Y
```

```
 H(1,6)=-8.*A4*COSX*SINY*SINZ-4.*C3*SIN2Y*SIN2Z
 H(1,7)=-8.*A4*SINX*COSY*SINZ-4.*C3*SIN2Z*SIN2X
 H(1,8)=SR3*B3*(COS2X-COS2Y)+2.*SR3*C4*COS2Z*(COS2Y-COS2X)
 H(1,9)=B3*(-COS2X-COS2Y+2.*COS2Z)-2.*C4*(-2.*COS2X*COS2Y
 +COS2X*
1COS2Z+COS2Y*COS2Z)
 H(2,2)=A5+8.*A6*COSX*COSY*COSZ+2.*B4*COS2X+2.*B5*(COS2Y
 +COS2Z)
1+4.*C5*COS2X*(COS2Y+COS2Z)+4.*C6*COS2Y*COS2Z
 H(2,3)=-8.*A7*SINX*SINY*COSZ-4.*C7*SIN2X*SIN2Y
 H(2,4)=-8.*A7*SINX*COSY*SINZ-4.*C7*SIN2X*SIN2Z
 H(2,5)=8.*A8*COSX*SINY*COSZ+2.*B6*SIN2Y+4.*C8*COS2X*SIN2Y
 +4.*C9
1*SIN2Y*COS2Z
 H(2,6)=-8.*A9*SINX*SINY*SINZ
 H(2,7)=8.*A8*COSX*COSY*SINZ+2.*B6*SIN2Z+4.*C8*COS2X*SIN2Z
 +4.*C9
1*SIN2Z*COS2Y
 H(2,8)=8.*A10*SINX*COSY*COSZ+SR3*B7*SIN2X+2.
 *SR3*C10*SIN2X*(COS2Y
1+COS2Z)+2.*C11*SIN2X*(COS2Y-COS2Z)
 H(2,9)=-(8./SR3)*A10*SINX*COSY*COSZ-B7*SIN2X-2.
 *C10*SIN2X*(COS2Y+
1COS2Z)+2.*SR3*C11*SIN2X*(COS2Y-COS2Z)
 H(3,3)=A5+8.*A6*COSX*COSY*COSZ+2.*B4*COS2Y+2.*B5*(COS2Z
 +COS2X)+
14.*C5*COS2Y*(COS2Z+COS2X)+4.*C6*COS2Z*COS2X
 H(3,4)=-8.*A7*COSX*SINY*SINZ-4.*C7*SIN2Y*SIN2Z
 H(3,5)=8.*A8*SINX*COSY*COSZ+2.*B6*SIN2X+4.*C8*COS2Y*SIN2X
 +4.*C9
1*SIN2X*COS2Z
 H(3,6)=8.*A8*COSX*COSY*SINZ+2.*B6*SIN2Z+4.*C8*COS2Y*SIN2Z
 +4.*C9
1*SIN2Z*COS2X
 H(3,7)=-8.*A9*SINX*SINY*SINZ
 H(3,8)=-8.*A10*COSX*SINY*COSZ-SR3*B7*SIN2Y-2.
 *SR3*C10*SIN2Y*
1(COS2Z+COS2X)-2.*C11*SIN2Y*(COS2X-COS2Z)
 H(3,9)=-(8./SR3)*A10*COSX*SINY*COSZ-B7*SIN2Y-2.
 *C10*SIN2Y*
1(COS2Z+COS2X)-2.*SR3*C11*SIN2Y*(COS2Z-COS2X)
 H(4,4)=A5+8.*A6*COSX*COSY*COSZ+2.*B5*(COS2X+COS2Y)+2.
 *B4*COS2Z
1+4.*C5*COS2Z*(COS2X+COS2Y)+4.*C6*COS2X*COS2Y
 H(4,5)=-8.*A9*SINX*SINY*SINZ
```

```
H(4,6)=8.*A8*COSX*SINY*COSZ+2.*B6*SIN2Y+4*C8*COS2Z*SIN2Y
+4.*C9
1*SIN2Y*COS2X
H(4,7)=8.*A8*SINX*COSY*COSZ+2.*B6*SIN2X+4.*C8*COS2Z*SIN2X
+4.*C9
1*SIN2X*COS2Y
H(4,8)=4.*C11*SIN2Z*(COS2Y-COS2X)
H(4,9)=   (16./SR3)*A10*COSX*COSY*SINZ+2.*B7*SIN2Z+4*C10*
SIN2Z*
1(COS2X+COS2Y)
H(5,5)=A11+8.*A12*COSX*COSY*COSZ+2.*B8*(COS2X+COS2Y)+2.
*B9*COS2Z
1+4.*C12*COS2X*COS2Y+4.*C13*COS2Z*(COS2X+COS2Y)
H(5,6)=-8.*A13*SINX*COSY*SINZ-4.*C14*SIN2X*SIN2Z
H(5,7)=-8.*A13*COSX*SINY*SINZ-4.*C14*SIN2Y*SIN2Z
H(5,8)=0.0
H(5,9)=-8.*A14*SINX*SINY*COSZ-4.*C15*SIN2X*SIN2Y
H(6,6)=A11+8.*A12*COSX*COSY*COSZ+2.*B8*(COS2Z+COS2Y)+2.
*B9*COS2X
1+4.*C12*COS2Z*COS2Y+4.*C13*COS2X*(COS2Z+COS2Y)
H(6,7)=-8.*A13*SINX*SINY*COSZ-4.*C14*SIN2Y*SIN2X
H(6,8)=-4.*SR3*A14*COSX*SINY*SINZ-2.*SR3*C15*SIN2Y*SIN2Z
H(6,9)=4.*A14*COSX*SINY*SINZ+2.*C15*SIN2Y*SIN2Z
H(7,7)=A11+8.*A12*COSX*COSY*COSZ+2.*B8*(COS2Z+COS2X)+2.
*B9*COS2Y
1+4.*C12*COS2Z*COS2X+4.*C13*COS2Y*(COS2X+COS2Z)
H(7,8)=4.*SR3*A14*SINX*COSY*SINZ+2.*SR3*C15*SIN2X*SIN2Z
H(7,9)=4.*A14*SINX*COSY*SINZ+2.*C15*SIN2X*SIN2Z
H(8,8)=A15+8.*A16*COSX*COSY*COSZ+(3./2.)*B10*(COS2X
+COS2Y)+2.*B11
1*(.25*COS2X+.25*COS2Y+COS2Z)+3.*C16*COS2Z*(COS2X+COS2Y)
+C17*(4.*
1COS2X*COS2Y+COS2X*COS2Z+COS2Y*COS2Z)
H(8,9)=(1./2.)*SR3*B10*(-COS2X+COS2Y)-.5*SR3*B11*(-COS2X+
+COS2Y)+SR3*(C16-C17)*COS2Z*(COS2X-COS2Y)
H(9,9)=A15+8.*A16*COSX*COSY*COSZ+2.*B10*(.25*COS2X
+.25*COS2Y+
1COS2Z)+(3./2.)*B11*(COS2X+COS2Y)+C16*(4.*COS2X*COS2Y
+COS2X*COS2Z
2+COS2Y*COS2Z)+3.*C17*COS2Z*(COS2X+COS2Y)
S(1,1)=1.0+8.*S2*COSX*COSY*COSZ+2.*T1*(COS2X+COS2Y+COS2Z)
S(2,2)=1.0+8.*S6*COSX*COSY*COSZ+2.*T4*COS2X+2.*T5*(COS2Y
+COS2Z)
S(3,3)=1.0+8.*S6*COSX*COSY*COSZ+2.*T4*COS2Y+2.*T5*(COS2Z
+COS2X)
```

```
      S(4,4)=1.0+8.*S6*COSX*COSY*COSZ+2.*T5*(COS2X+COS2Y)+2.
      *T4*COS2Z
      S(5,5)=1.0+8.*S12*COSX*COSY*COSZ+2.*T8*(COS2X+COS2Y)+2.
      *T9*COS2Z
      S(6,6)=1.0+8.*S12*COSX*COSY*COSZ+2.*T8*(COS2Z+COS2Y)+2.
      *T9*COS2X
      S(7,7)=1.0+8.*S12*COSX*COSY*COSZ+2.*T8*(COS2Z+COS2X)+2.
      *T9*COS2Y
      S(8,8)=1.0+8.*S16*COSX*COSY*COSZ+(3./2.)*T10*(COS2X
      +COS2Y)+2.*T11
     1*(.25*COS2X+.25*COS2Y+COS2Z)
      S(9,9)=1.0+8.*S16*COSX*COSY*COSZ+2.*T10*(.25*COS2X
      +.25*COS2Y+
     1COS2Z)+(3./2.)*T11 *(COS2X+COS2Y)
C
C+++++S(1,2),S(1,3),S(1,4) ARE REVERSED IN SIGN SO THAT REAL
MATRIX IS
C   EQUIVALENT TO COMPLEX MATRIX
C
      S(1,2)=8.*S3*SINX*COSY*COSZ+2.*T2*SIN2X
      S(1,2)=-S(1,2)
      S(1,3)=8.*S3*COSX*SINY*COSZ+2.*T2*SIN2Y
      S(1,3)=-S(1,3)
      S(1,4)=8.*S3*COSX*COSY*SINZ+2.*T2*SIN2Z
      S(1,4)=-S(1,4)
      S(1,5)=-8.*S4*SINX*SINY*COSZ
      S(1,6)=-8.*S4*COSX*SINY*SINZ
      S(1,7)=-8.*S4*SINX*COSY*SINZ
      S(1,8)=SR3 *T3*(COS2X-COS2Y)
      S(1,9)=T3*(-COS2X-COS2Y+2.*COS2Z)
      S(2,3)=-8.*S7*SINX*SINY*COSZ
      S(2,4)=-8.*S7*SINX*COSY*SINZ
      S(2,5)=8.*S8*COSX*SINY*COSZ+2.*T6*SIN2Y
      S(2,6)=-8.*S9*SINX*SINY*SINZ
      S(2,7)=8.*S8*COSX*COSY*SINZ+2.*T6*SIN2Z
      S(2,8)=8.*S10*SINX*COSY*COSZ+SR3*T7*SIN2X
      S(2,9)=-(8./SR3)*S10*SINX*COSY*COSZ-T7*SIN2X
      S(3,4)=-8.*S7*COSX*SINY*SINZ
      S(3,5)=8.*S8*SINX*COSY*COSZ+2.*T6*SIN2X
      S(3,6)=8.*S8*COSX*COSY*SINZ+2.*T6*SIN2Z
      S(3,7)=-8.*S9*SINX*SINY*SINZ
      S(3,8)=-8.*S10*COSX*SINY*COSZ-SR3*T7*SIN2Y
      S(3,9)=-(8./SR3)*S10*COSX*SINY*COSZ-T7*SIN2Y
      S(4,5)=-8.*S9*SINX*SINY*SINZ
      S(4,6)=8.*S8*COSX*SINY*COSZ+2.*T6*SIN2Y
```

```
      S(4,7)=8.*S8*SINX*COSY*COSZ+2.*T6*SIN2X
      S(4,8)=0.0
      S(4,9)= (16./SR3)*S10*COSX*COSY*SINZ+2.*T7*SIN2Z
      S(5,6)=-8.*S13*SINX*COSY*SINZ
      S(5,7)=-8.*S13*COSX*SINY*SINZ
      S(5,8)=0.0
      S(5,9)=-8.*S14*SINX*SINY*COSZ
      S(6,7)=-8.*S13*SINX*SINY*COSZ
      S(6,8)=-4.*SR3*S14*COSX*SINY*SINZ
      S(6,9)=4.*S14*COSX*SINY*SINZ
      S(7,8)=4.*SR3*S14*SINX*COSY*SINZ
      S(7,9)=4.*S14*SINX*COSY*SINZ
      S(8,9)=(1./2.)*SR3*T10*(-COS2X+COS2Y)-.5*SR3*T11*(-COS2X+
     +COS2Y)
      DO 310 L=1,N
      DO 310 K=1,N
      H(K,L)=H(L,K)
      S(K,L)=S(L,K)
  310 CONTINUE
c     WRITE(15,325) XX,YY,ZZ
  325 FORMAT(3(F6.2,2X))
      NR=0
      IEGEN=0
C     DIAGONALIZES THE OVERLAP MATRIX S
      CALL HDIAG(S,NS,IEGEN, U,NR,NS)
      DO 8111 M=1,NS
      DO 8111 J=1,M
      HNEW(M,J)=0.0
      DO 8200 K=1,NS
      DO 8200 L=1,NS
      HNEW(M,J)=HNEW(M,J)+U(K,M)*H(K,L)*U(L,J)
 8200 CONTINUE
      HNEW(M,J)=HNEW(M,J)/(SQRT(S(M,M))*SQRT(S(J,J)))
       HNEW(J,M)=HNEW(M,J)
 8111 CONTINUE
C     DIAGONALIZES THE EFFECTIVE HAMILTONIAN HNEW
      CALL HDIAG(HNEW,NS,IEGEN,UU,NR,NS)
      DO 850 I1=1,NS
      DO 850 I2=1,I1
      T(I1,I2)=0.0
      DO 860 J1=1,NS
      T(I1,I2)=T(I1,I2)+U(I1,J1)*UU(J1,I2)/SQRT(S(J1,J1))
  860 CONTINUE
  850 CONTINUE
      DO 851 I1=1,NS
```

```
      DO 851 I2=I1,NS
      T(I1,I2)=0.0
      DO 861 J1=1,NS
      T(I1,I2)=T(I1,I2)+U(I1,J1)*UU(J1,I2)/SQRT(S(J1,J1))
  861 CONTINUE
  851 CONTINUE
      DO 390 J=1,NS
      ANORM=0.0
      DO 865 LNORM=1,NS
      ANORM=ANORM+(UU(LNORM,J)**2)/S(LNORM,LNORM)
  865 CONTINUE
      ANORM=SQRT(ANORM)
      DO 866 LNORM=1,NS
      T(LNORM,J)=T(LNORM,J)/ANORM
      U(LNORM,J)=T(LNORM,J)
  866 CONTINUE
  390 CONTINUE
      CALL SORT(HNEW,U,NS)
C     THE Q10,Q11,Q12,Q13 ARE THE SUMS OF THE SQUARES OF THE
C     EIGENVECTORS OVER 9 BANDS. THESE ARE USED TO FIND THE
C     ANGULAR MOMENTUM CHARACTER OF EACH STATE AND THUS
C     DECOMPOSE THE DOS
      DO 729  J=1,NS
      Q10(J,JK )=U(1,J)**2
      Q11(J,JK )= U(2,J)**2 + U(3,J)**2 + U(4,J)**2
      Q12(J,JK )= U(5,J)**2 + U(6,J)**2 + U(7,J)**2
      Q13(J,JK )= U(8,J)**2 + U(9,J)**2
  729 CONTINUE
  332 FORMAT(1X,5F10.5)
      DO 5554 J=1,NS
        WRITE(15,330)HNEW(J,J),Q10(J,JK),Q11(J,JK),Q12(J,JK),Q13
     (J,JK)
 5554 CONTINUE
  330 FORMAT(F9.5,3X,4F8.5)
  451 CONTINUE
      WRITE(15,452) JK
  452 FORMAT(1X,'NUMBER OF K- POINTS',2X,I5)
      STOP
      END
      SUBROUTINE SORT(H,U,N)
      implicit real*8 (A-H,O-Z)
C     THIS ROUTINE ORDERS EIGENVALUES AND EIGENVECTORS
      DIMENSION H(9,9),U(9,9),T(9)
      NN=N-1
      DO 50 L=1,NN
```

```
    LL=L+1
    DO 60 M=LL,N
    IF (H(L,L)-H(M,M)) 60,60,70
 70 TEMP=H(L,L)
    H(L,L)=H(M,M)
    H(M,M)=TEMP
    DO 100 MM=1,N
    T(MM)=U(MM,L)
    U(MM,L)=U(MM,M)
100 U(MM,M)=T(MM)
60  CONTINUE
50  CONTINUE
    RETURN
    END
```

```
 INPUT
MANGANESE BCC
  2  3  67
 0.8952054381  0 1        MANGANESE, BCC H-L POTENTIAL (R
 1.3517218828  0 2        MANGANESE, BCC H-L POTENTIAL (R
 0.7006170154  0 3        MANGANESE, BCC H-L POTENTIAL (R
 0.6870977283  0 4        MANGANESE, BCC H-L POTENTIAL (R
-0.0502326749  0 5        MANGANESE, BCC H-L POTENTIAL (R
 0.0301787984  0 6        MANGANESE, BCC H-L POTENTIAL (R
 0.0255550239  0 7        MANGANESE, BCC H-L POTENTIAL (R
-0.0005011358  0 8        MANGANESE, BCC H-L POTENTIAL (R
 0.0073730783  0 9        MANGANESE, BCC H-L POTENTIAL (R
 0.0066117458  0 10       MANGANESE, BCC H-L POTENTIAL (R
 0.0116945766  0 11       MANGANESE, BCC H-L POTENTIAL (R
 0.0236831941  0 12       MANGANESE, BCC H-L POTENTIAL (R
-0.0065287831  0 13       MANGANESE, BCC H-L POTENTIAL (R
-0.0082228919  0 14       MANGANESE, BCC H-L POTENTIAL (R
 0.0159323364  0 15       MANGANESE, BCC H-L POTENTIAL (R
 0.0192361511  0 16       MANGANESE, BCC H-L POTENTIAL (R
-0.0306956321  0 17       MANGANESE, BCC H-L POTENTIAL (R
-0.0341562591  0 18       MANGANESE, BCC H-L POTENTIAL (R
 0.0131229106  0 19       MANGANESE, BCC H-L POTENTIAL (R
-0.2447142154  0 20       MANGANESE, BCC H-L POTENTIAL (R
 0.0096795037  0 21       MANGANESE, BCC H-L POTENTIAL (R
-0.0012423185  0 22       MANGANESE, BCC H-L POTENTIAL (R
 0.0557515845  0 23       MANGANESE, BCC H-L POTENTIAL (R
 0.0111683132  0 24       MANGANESE, BCC H-L POTENTIAL (R
 0.0013539762  0 25       MANGANESE, BCC H-L POTENTIAL (R
-0.0073515503  0 26       MANGANESE, BCC H-L POTENTIAL (R
 0.0002250212  0 27       MANGANESE, BCC H-L POTENTIAL (R
```

```
 0.0000000000  1 28          MANGANESE, BCC H-L POTENTIAL (R
 0.0000000000  1 29          MANGANESE, BCC H-L POTENTIAL (R
 0.0000000000  1 30          MANGANESE, BCC H-L POTENTIAL (R
 0.0000000000  1 31          MANGANESE, BCC H-L POTENTIAL (R
 0.0000000000  1 32          MANGANESE, BCC H-L POTENTIAL (R
 0.0000000000  1 33          MANGANESE, BCC H-L POTENTIAL (R
 0.0000000000  1 34          MANGANESE, BCC H-L POTENTIAL (R
 0.0000000000  1 35          MANGANESE, BCC H-L POTENTIAL (R
 0.0000000000  1 36          MANGANESE, BCC H-L POTENTIAL (R
 0.0000000000  1 37          MANGANESE, BCC H-L POTENTIAL (R
 0.0000000000  1 38          MANGANESE, BCC H-L POTENTIAL (R
 0.0000000000  1 39          MANGANESE, BCC H-L POTENTIAL (R
 0.0000000000  1 40          MANGANESE, BCC H-L POTENTIAL (R
 0.0000000000  1 41          MANGANESE, BCC H-L POTENTIAL (R
 0.0000000000  1 42          MANGANESE, BCC H-L POTENTIAL (R
 0.0000000000  1 43          MANGANESE, BCC H-L POTENTIAL (R
 0.0000000000  1 44          MANGANESE, BCC H-L POTENTIAL (R
 0.1051913500  0 45          MANGANESE, BCC H-L POTENTIAL (R
-0.0757101029  0 46          MANGANESE, BCC H-L POTENTIAL (R
-0.0146760857  0 47          MANGANESE, BCC H-L POTENTIAL (R
-0.0203320775  0 48          MANGANESE, BCC H-L POTENTIAL (R
-0.0811426714  0 49          MANGANESE, BCC H-L POTENTIAL (R
-0.0027773958  0 50          MANGANESE, BCC H-L POTENTIAL (R
-0.0420891047  0 51          MANGANESE, BCC H-L POTENTIAL (R
-0.0096733952  0 52          MANGANESE, BCC H-L POTENTIAL (R
 0.0082451766  0 53          MANGANESE, BCC H-L POTENTIAL (R
 0.0167889241  0 54          MANGANESE, BCC H-L POTENTIAL (R
-0.0032028167  0 55          MANGANESE, BCC H-L POTENTIAL (R
-0.0079976860  0 56          MANGANESE, BCC H-L POTENTIAL (R
 0.0535156950  0 57          MANGANESE, BCC H-L POTENTIAL (R
-0.1322071999  0 58          MANGANESE, BCC H-L POTENTIAL (R
 0.0704541206  0 59          MANGANESE, BCC H-L POTENTIAL (R
-0.2975141406  0 60          MANGANESE, BCC H-L POTENTIAL (R
 0.0097368984  0 61          MANGANESE, BCC H-L POTENTIAL (R
 0.0158128161  0 62          MANGANESE, BCC H-L POTENTIAL (R
 0.1355598867  0 63          MANGANESE, BCC H-L POTENTIAL (R
-0.0036504590  0 64          MANGANESE, BCC H-L POTENTIAL (R
 0.0033772474  0 65          MANGANESE, BCC H-L POTENTIAL (R
 0.0386965871  0 66          MANGANESE, BCC H-L POTENTIAL (R
 0.0006326744  0 67          MANGANESE, BCC H-L POTENTIAL (R
```

B. Face-centered cubic structure

This program generates eigenvalues and eigenvectors for the 9×9 spd Hamiltonian; it diagonalizes a real matrix. The user is expected to add a diagonalization subroutine named HDIAG. The Slater–Koster parameters should be read in the order given for example in the platinum Tables.

```
      PROGRAM FCCSK
C  AUTHOR: D.A.PAPACONSTANTOPOULOS
C  FCC MONATOMIC-REAL, BUT EQUIVALENT TO COMPLEX-OVERLAP VERSION
C  CREATES INPUT FOR CALCULATING THE DOS
C   CAUTION: THE DIMENSIONING OF Q10,Q11,Q12,Q13, SHOULD BE
C   CONSISTENT WITH THE NUMER OF K-POINTS CREATED BY THE
C   ASSIGNED VALUE TO MESH
C A diagonalization subroutine needs to be added. CALL HDIAG
      IMPLICIT REAL*8 (A-H,O-Z)
      CHARACTER*10 INA,INB
      CHARACTER*80 TITLE
      DIMENSION  UU(9,9),T(9,9),H(9,9),U(9,9),S(9,9),HNEW(9,9)
      DIMENSION TITLE(20),PARAM(44)
      DIMENSION A(3)
      COMMON/BLOC1/Q10(9,505 ),Q11(9,505 ),Q12(9,505),Q13(9,505)
      INA = 'par.in'
      OPEN(10,FILE=INA,BLANK='ZERO')
c     OPEN(11,FILE='dosdat',BLANK='ZERO')
      INB = 'fcc.out'
      OPEN(15,FILE=INB,BLANK='ZERO')
      READ (10,35) TITLE
   35 FORMAT(20A4)
c     WRITE (15,335) TITLE
  335 FORMAT (1H1,20A4)
C     ICENTER=3 THREE-CENTER APPROXIMATION
C     ICENTER=2 TWO-CENTER APPROXIMATION
C     MESH  K-POINT MESH GENERATOR
      READ (10,9999) MESH,ICENTER
c     WRITE(15,9998) MESH,ICENTER
 9999 FORMAT (5I5)
 9998 FORMAT (5X,'MESH=',I5,10X,'ICENTER=',I5)
      NS=9
      SR2=SQRT(2.)
      SR3=SQRT(3.)
   55 FORMAT(F15.0)
      IF (ICENTER.EQ.3)
     1READ(10,55)A1,A6,A14,A19,A2,A3,A4,A5,A7,A8,A9,A10,A11,
      A12,A13,
```

```
     C  A15,A16,A17,A18,A20,A21,B1,B2,B3,B4,B5,B6,B7,B8,B9,B10,
     B11
     C              ,S2,S3,S4,S5,S7,S8,S9,S10,S11,S12,S13,
     C  S15,S16,S17,S18,S20,S21,T1,T2,T3,T4,T5,T6,T7,T8,T9,T10,
     T11
C    THE ORDER OF THE SK-PARAMETERS HERE IS THE SAME AS IN THE TABLES
c     IF (ICENTER.EQ.3)
c    1WRITE(15,999)A1,A6,A14,A19,A2,A3,A4,A5,A7,A8,A9,A10,A11,
      A12,A13,
c     C A15,A16,A17,A18,A20,A21,B1,B2,B3,B4,B5,B6,B7,B8,B9,B10,
      B11
c    C              ,S2,S3,S4,S5,S7,S8,S9,S10,S11,S12,S13,
c     C S15,S16,S17,S18,S20,S21,T1,T2,T3,T4,T5,T6,T7,T8,T9,T10,
      T11
     IF (ICENTER.EQ.2) READ(10,55)  (PARAM(I),I=1,44)
     IF (ICENTER.EQ.2) WRITE(15,999)  (PARAM(I),I=1,44)
     IF (ICENTER.EQ.3) GO TO 500
C     THIS SECTION IS FOR THE TWO-CENTER CALCULATIONS
     Q1=PARAM(1)
     Q2=PARAM(2)
     Q3=PARAM(3)
     Q4=PARAM(4)
     P1=PARAM(5)
     P2=PARAM(6)
     P3=PARAM(7)
     P4=PARAM(8)
     P5=PARAM(9)
     P6=PARAM(10)
     P12=PARAM(11)
     P14=PARAM(12)
     P24=PARAM(13)
     P35=PARAM(14)
     R1=PARAM(15)
     R2=PARAM(16)
     R3=PARAM(17)
     R4=PARAM(18)
     R5=PARAM(19)
     R6=PARAM(20)
     R12=PARAM(21)
     R14=PARAM(22)
     R24=PARAM(23)
     R35=PARAM(24)
     PP1=PARAM(25)
     PP2=PARAM(26)
     PP3=PARAM(27)
```

```
PP4=PARAM(28)
PP5=PARAM(29)
PP6=PARAM(30)
PP12=PARAM(31)
PP14=PARAM(32)
PP24=PARAM(33)
PP35=PARAM(34)
RR1=PARAM(35)
RR2=PARAM(36)
RR3=PARAM(37)
RR4=PARAM(38)
RR5=PARAM(39)
RR6=PARAM(40)
RR12=PARAM(41)
RR14=PARAM(42)
RR24=PARAM(43)
RR35=PARAM(44)
A1=Q1
A6=Q2
A14=Q3
A19=Q4
B1=R1
B2=R12
B3=R14
B4=R2
B5=R3
B6=R35
B7=R24
B8=R5
B9=R6
B10=R4
B11=R6
A2=P1
A3=P12/SR2
A4=SR3*P14/2.
A5=-P14/2.
A7=(P2+P3)/2.
A8=P3
A9=(P2-P3)/2.
A10=SR3*SR2*P24/4.
A11=P35/SR2
A12=(SR2*SR3*P35+P24/SR2)/4.
A13=(-SR3*P24+2.*P35)/(4.*SR2)
A15=(3.*P4+P6)/4.
A16=(P5+P6)/2.
```

```
      A17=(P5-P6)/2.
      A18=SR3*(P6-P4)/4.
      A20=(P4+3.*P6)/4.
      A21=P5
      T1=RR1
      T2=RR12
      T3=RR14
      T4=RR2
      T5=RR3
      T6=RR35
      T7=RR24
      T8=RR5
      T9=RR6
      T10=RR4
      T11=RR6
      S2=PP1
      S3=PP12/SR2
      S4=SR3*PP14/2.
      S5=-PP14/2.
      S7=(PP2+PP3)/2.
      S8=PP3
      S9=(PP2-PP3)/2.
      S10=SR3*SR2*PP24/4.
      S11=PP35/SR2
      S12=(SR2*SR3*PP35+PP24/SR2)/4.
      S13=(-SR3*PP24+2.*PP35)/(4.*SR2)
      S15=(3.*PP4+PP6)/4.
      S16=(PP5+PP6)/2.
      S17=(PP5-PP6)/2.
      S18=SR3*(PP6-PP4)/4.
      S20=(PP4+3.*PP6)/4.
      S21=PP5
  500 CONTINUE
  999 FORMAT(//10X,'VALUES OF THE SK PARAMETERS',//60(25X,F15.8/))
      N=9
      JK=0
      IMESH=2*MESH+1
C     THE K-VECTOR UNITS ARE CHOSEN SO THAT THE POINT X HAS
C     CORDINATES (8,0,0)
      SCALE= 8./FLOAT(2*MESH)
      DO 451  IYBK=1,IMESH
      DO 451  IXBK=1,IYBK
      DO 451  IZBK=1,IXBK
      IT= IXBK+IYBK+IZBK-3
      IF(MESH*3.LT.IT) GO TO 451
```

```
      JK=JK+1
      A(1)= FLOAT(IXBK-1)
      A(2)= FLOAT(IYBK-1)
      A(3)= FLOAT(IZBK-1)
      XX =A(1)*SCALE
      YY =A(2)*SCALE
      ZZ =A(3)*SCALE
      PI = 3.14159265
      X1 = PI*XX/8.
      Y1 = PI*YY/8.
      Z1 = PI*ZZ/8.
      COSX=COS(X1)
      COSY=COS(Y1)
      COSZ=COS(Z1)
      SINX=SIN(X1)
      SINY=SIN(Y1)
      SINZ=SIN(Z1)
      COS2X=COS(2.*X1)
      COS2Y=COS(2.*Y1)
      COS2Z=COS(2.*Z1)
      SIN2X=SIN(2.*X1)
      SIN2Y=SIN(2.*Y1)
      SIN2Z=SIN(2.*Z1)
      H(1,1)=A1+4.*A2*(COSX*COSY+COSY*COSZ+COSZ*COSX)+2.*B1*
       (COS2X+COS2Y
     1+COS2Z)
      H(1,2)=4.*A3*SINX*(COSY+COSZ)+2.*B2*SIN2X
      H(1,3)=4.*A3*SINY*(COSZ+COSX)+2.*B2*SIN2Y
      H(1,4)=4.*A3*SINZ*(COSX+COSY)+2.*B2*SIN2Z
C
C+++++H(1,2),H(1,3),H(1,4) ARE REVERSED IN SIGN SO THAT THE REAL
 MATRIX
C   EQUIVALENT TO THE COMPLEX MATRIX
C
      H(1,2)=-H(1,2)
      H(1,3)=-H(1,3)
      H(1,4)=-H(1,4)
      H(1,5)=-4.*A4*SINX*SINY
      H(1,6)=-4.*A4*SINY*SINZ
      H(1,7)=-4.*A4*SINZ*SINX
      H(1,8)=2.*SR3*A5*COSZ*(COSY-COSX)+SR3*B3*(COS2X-COS2Y)
      H(1,9)=-2.*A5*(-2.*COSX*COSY+COSY*COSZ+COSZ*COSX)+B3*
       (-COS2X-COS2Y
     1+2.*COS2Z)
```

```
H(2,2)=A6+4.*A7*COSX*(COSY+COSZ)+4.*A8*COSY*COSZ+2.
 *B4*COS2X+2.*B5
1*(COS2Y+COS2Z)
 H(2,3)=-4.*A9*SINX*SINY
 H(2,4)=-4.*A9*SINX*SINZ
 H(2,5)=4.*SINY*(A10*COSX+A11*COSZ)+2.*B6*SIN2Y
 H(2,6)=0.
 H(2,7)=4.*SINZ*(A10*COSX+A11*COSY)+2.*B6*SIN2Z
 H(2,8)=2.*SR3*A12*SINX*(COSY+COSZ)+2.*A13*SINX*(COSY-
 COSZ)+SR3*B7*
1SIN2X
 H(2,9)=-2.*A12*SINX*(COSY+COSZ)+2.*SR3*A13*SINX*(COSY-
 COSZ)-B7*
1SIN2X
 H(3,3)=A6+4.*A7*COSY*(COSZ+COSX)+4.*A8*COSZ*COSX+2.
 *B4*COS2Y+
12.*B5*(COS2Z+COS2X)
 H(3,4)=-4.*A9*SINY*SINZ
 H(3,5)=4.*SINX*(A10*COSY+A11*COSZ)+2.*B6*SIN2X
 H(3,6)=4.*SINZ*(A10*COSY+A11*COSX)+2.*B6*SIN2Z
 H(3,7)=0.
 H(3,8)=-2.*SR3*A12*SINY*(COSZ+COSX)-2.*A13*SINY*(COSX-
 COSZ)-SR3*B7
1*SIN2Y
 H(3,9)=-2.*A12*SINY*(COSZ+COSX)-2.*SR3*A13*SINY*(COSZ-
 COSX)-B7*
1SIN2Y
 H(4,4)=A6+4.*A7*COSZ*(COSX+COSY)+4.*A8*COSX*COSY+2.
 *B4*COS2Z+2.*B5
1*(COS2X+COS2Y)
 H(4,5)=0.
 H(4,6)=4.*SINY*(A10*COSZ+A11*COSX)+2.*B6*SIN2Y
 H(4,7)=4.*SINX*(A10*COSZ+A11*COSY)+2.*B6*SIN2X
 H(4,8)=4.*A13*SINZ*(COSY-COSX)
 H(4,9)=4.*A12*SINZ*(COSX+COSY)+2.*B7*SIN2Z
 H(5,5)=A14+4.*A15*COSX*COSY+4.*A16*COSZ*(COSX+COSY)+2.
 *B8*(COS2X+
1COS2Y)+2.*B9*COS2Z
 H(5,6)=-4.*A17*SINX*SINZ
 H(5,7)=-4.*A17*SINY*SINZ
 H(5,8)=0.
 H(5,9)=-4.*A18*SINX*SINY
 H(6,6)=A14+4.*A15*COSY*COSZ+4.*A16*COSX*(COSY+COSZ)+2.
 *B8*(COS2Z
1+COS2Y)+2.*B9*COS2X
```

```
      H(6,7)=-4.*A17*SINX*SINY
      H(6,8)=-2.*SR3*A18*SINY*SINZ
      H(6,9)=2.*A18*SINY*SINZ
      H(7,7)=A14+4.*A15*COSX*COSZ+4.*A16*COSY*(COSX+COSZ)+2.
       *B8*(COS2X+
     1COS2Z)+2.*B9*COS2Y
      H(7,8)=2.*SR3*A18*SINX*SINZ
      H(7,9)=2.*A18*SINX*SINZ
      H(8,8)=A19+3.*A20*COSZ*(COSX+COSY)+4.*A21*(COSX*COSY
       +COSX*COSZ/4.+
     1COSY*COSZ/4.)+(3./2.)*B10*(COS2X+COS2Y)+2.*B11*(COS2X/4.
       +COS2Y/4.+
     2COS2Z)
      H(8,9)=SR3*COSZ*(COSX-COSY)*(A20-A21)+(SR3/2.)*(COS2Y-
       COS2X)*(B10-
     1B11)
      H(9,9)=A19+4.*A20*(COSX*COSY+COSX*COSZ/4.+COSY*COSZ/4.)
       +3.*A21*
     1COSZ*(COSX+COSY)+2.*B10*(COS2X/4.+COS2Y/4.+COS2Z)+(3./
       2.)*B11*
     2(COS2X+COS2Y)
      S(1,1)=1.+4.*S2*(COSX*COSY+COSY*COSZ+COSZ*COSX)+2.*T1*
       (COS2X+COS2Y
     1+COS2Z)
C
C+++++S(1,2),S(1,3),S(1,4) ARE REVERSED IN SIGN SO THAT THE REAL
 MATRIX
C   EQUIVALENT TO COMPLEX MATRIX
C
      S(1,2)=4.*S3*SINX*(COSY+COSZ)+2.*T2*SIN2X
      S(1,3)=4.*S3*SINY*(COSZ+COSX)+2.*T2*SIN2Y
      S(1,4)=4.*S3*SINZ*(COSX+COSY)+2.*T2*SIN2Z
      S(1,2)=-S(1,2)
      S(1,3)=-S(1,3)
      S(1,4)=-S(1,4)
      S(1,5)=-4.*S4*SINX*SINY
      S(1,6)=-4.*S4*SINY*SINZ
      S(1,7)=-4.*S4*SINZ*SINX
      S(1,8)=2.*SR3*S5*COSZ*(COSY-COSX)+SR3*T3*(COS2X-COS2Y)
      S(1,9)=-2.*S5*(-2.*COSX*COSY+COSY*COSZ+COSZ*COSX)+T3*
       (-COS2X-COS2Y
     1+2.*COS2Z)
      S(2,2)=1.+4.*S7*COSX*(COSY+COSZ)+4.*S8*COSY*COSZ+2.
       *T4*COS2X+2.*T5
     1*(COS2Y+COS2Z)
```

```
 S(2,3)=-4.*S9*SINX*SINY
 S(2,4)=-4.*S9*SINX*SINZ
 S(2,5)=4.*SINY*(S10*COSX+S11*COSZ)+2.*T6*SIN2Y
 S(2,6)=0.
 S(2,7)=4.*SINZ*(S10*COSX+S11*COSY)+2.*T6*SIN2Z
 S(2,8)=2.*SR3*S12*SINX*(COSY+COSZ)+2.*S13*SINX*(COSY-
 COSZ)+SR3*T7*
1SIN2X
 S(2,9)=-2.*S12*SINX*(COSY+COSZ)+2.*SR3*S13*SINX*(COSY-
 COSZ)-T7*
1SIN2X
 S(3,3)=1.+4.*S7*COSY*(COSZ+COSX)+4.*S8*COSZ*COSX+2.
 *T4*COS2Y+
12.*T5*(COS2Z+COS2X)
 S(3,4)=-4.*S9*SINY*SINZ
 S(3,5)=4.*SINX*(S10*COSY+S11*COSZ)+2.*T6*SIN2X
 S(3,6)=4.*SINZ*(S10*COSY+S11*COSX)+2.*T6*SIN2Z
 S(3,7)=0.
 S(3,8)=-2.*SR3*S12*SINY*(COSZ+COSX)-2.*S13*SINY*(COSX-
 COSZ)-SR3*T7
1*SIN2Y
 S(3,9)=-2.*S12*SINY*(COSZ+COSX)-2.*SR3*S13*SINY*(COSZ-
 COSX)-T7*
1SIN2Y
 S(4,4)=1.+4.*S7*COSZ*(COSX+COSY)+4.*S8*COSX*COSY+2.
 *T4*COS2Z+2.*T5
1*(COS2X+COS2Y)
 S(4,5)=0.
 S(4,6)=4.*SINY*(S10*COSZ+S11*COSX)+2.*T6*SIN2Y
 S(4,7)=4.*SINX*(S10*COSZ+S11*COSY)+2.*T6*SIN2X
 S(4,8)=4.*S13*SINZ*(COSY-COSX)
 S(4,9)=4.*S12*SINZ*(COSX+COSY)+2.*T7*SIN2Z
 S(5,5)=1.+4.*S15*COSX*COSY+4.*S16*COSZ*(COSX+COSY)+2.*T8*
 (COS2X+
1COS2Y)+2.*T9*COS2Z
 S(5,6)=-4.*S17*SINX*SINZ
 S(5,7)=-4.*S17*SINY*SINZ
 S(5,8)=0.
 S(5,9)=-4.*S18*SINX*SINY
 S(6,6)=1.+4.*S15*COSY*COSZ+4.*S16*COSX*(COSY+COSZ)+2.*T8*
 (COS2Z
1+COS2Y)+2.*T9*COS2X
 S(6,7)=-4.*S17*SINX*SINY
 S(6,8)=-2.*SR3*S18*SINY*SINZ
 S(6,9)=2.*S18*SINY*SINZ
```

```
      S(7,7)=1.+4.*S15*COSX*COSZ+4.*S16*COSY*(COSX+COSZ)+2.*T8*
      (COS2X+
     1COS2Z)+2.*T9*COS2Y
      S(7,8)=2.*SR3*S18*SINX*SINZ
      S(7,9)=2.*S18*SINX*SINZ
      S(8,8)=1.+3.*S20*COSZ*(COSX+COSY)+4.*S21*(COSX*COSY
      +COSX*COSZ/4.+
      1COSY*COSZ/4.)+(3./2.)*T10*(COS2X+COS2Y)+2.*T11*(COS2X/4.
      +COS2Y/4.+
     2COS2Z)
      S(8,9)=SR3*COSZ*(COSX-COSY)*(S20-S21)+(SR3/2.)*(COS2Y-
      COS2X)*(T10-
     1T11)
      S(9,9)=1.+4.*S20*(COSX*COSY+COSX*COSZ/4.+COSY*COSZ/4.)+3.
      *S21*
      1COSZ*(COSX+COSY)+2.*T10*(COS2X/4.+COS2Y/4.+COS2Z)+(3./
      2.)*T11*
     2(COS2X+COS2Y)
      DO 310 L=1,N
      DO 310 K=1,N
      H(K,L)=H(L,K)
      S(K,L)=S(L,K)
  310 CONTINUE
c     WRITE(15,325) XX,YY,ZZ
  325 FORMAT(3(F6.2,2X))
      NR=0
      IEGEN=0
C     DIAGONALIZES THE OVERLAP MATRIX S
      CALL HDIAG(S,NS,IEGEN, U,NR,NS)
      DO 8111 M=1,NS
      DO 8111 J=1,M
      HNEW(M,J)=0.0
      DO 8200 K=1,NS
      DO 8200 L=1,NS
      HNEW(M,J)=HNEW(M,J)+U(K,M)*H(K,L)*U(L,J)
 8200 CONTINUE
c     S(M,M)=ABS(S(M,M))
c     S(J,J)=ABS(S(J,J))
      HNEW(M,J)=HNEW(M,J)/(SQRT(S(M,M))*SQRT(S(J,J)))
       HNEW(J,M)=HNEW(M,J)
 8111 CONTINUE
C     DIAGONALIZES THE EFFECTIVE HAMILTONIAN HNEW
      CALL HDIAG(HNEW,NS,IEGEN,UU,NR,NS)
      DO 850 I1=1,NS
      DO 850 I2=1,I1
```

```
      T(I1,I2)=0.0
      DO 860 J1=1,NS
c      S(J1,J1)=ABS(S(J1,J1))
      T(I1,I2)=T(I1,I2)+U(I1,J1)*UU(J1,I2)/SQRT(S(J1,J1))
 860 CONTINUE
 850 CONTINUE
      DO 851 I1=1,NS
      DO 851 I2=I1,NS
      T(I1,I2)=0.0
      DO 861 J1=1,NS
c      S(J1,J1)=ABS(S(J1,J1))
      T(I1,I2)=T(I1,I2)+U(I1,J1)*UU(J1,I2)/SQRT(S(J1,J1))
 861 CONTINUE
 851 CONTINUE
      DO 390 J=1,NS
      ANORM=0.0
      DO 865 LNORM=1,NS
      ANORM=ANORM+(UU(LNORM,J)**2)/S(LNORM,LNORM)
 865 CONTINUE
      ANORM=SQRT(ANORM)
      DO 866 LNORM=1,NS
      T(LNORM,J)=T(LNORM,J)/ANORM
      U(LNORM,J)=T(LNORM,J)
 866 CONTINUE
 390 CONTINUE
      CALL SORT(HNEW,U,NS)
C      THE Q10,Q11,Q12,Q13 ARE THE SUMS OF THE SQUARES OF THE
C      EIGENVECTORS OVER 9 BANDS. THESE ARE USED TO FIND THE
C      ANGULAR MOMENTUM CHARACTER OF EACH STATE AND THUS
C      DECOMPOSE THE DOS
      DO 729  J=1,NS
      Q10(J,JK )=U(1 ,J)**2
      Q11(J,JK )= U(2,J)**2 + U(3,J)**2 + U(4,J)**2
      Q12(J,JK )= U(5,J)**2 + U(6,J)**2 + U(7,J)**2
      Q13(J,JK )= U(8,J)**2 + U(9,J)**2
 729 CONTINUE
 332 FORMAT(1X,5F10.5)
      DO 5554 J=1,NS
      WRITE(15,330)HNEW(J,J),Q10(J,JK),Q11(J,JK),Q12(J,JK),Q13
       (J,JK)
 5554 CONTINUE
 330 FORMAT(F9.5,3X,4F8.5)
 451 CONTINUE
      WRITE(15,452) JK
 452 FORMAT(1X,'NUMBER OF K-POINTS',2X,I5)
```

```
      STOP
      END
      SUBROUTINE SORT(H,U,N)
      implicit real*8 (A-H,O-Z)
C     THIS ROUTINE ORDERS EIGENVALUES AND EIGENVECTORS
      DIMENSION H(9,9),U(9,9),T(9)
      NN=N-1
      DO 50 L=1,NN
      LL=L+1
      DO 60 M=LL,N
      IF (H(L,L)-H(M,M)) 60,60,70
   70 TEMP=H(L,L)
      H(L,L)=H(M,M)
      H(M,M)=TEMP
      DO 100 MM=1,N
      T(MM)=U(MM,L)
      U(MM,L)=U(MM,M)
  100 U(MM,M)=T(MM)
  60  CONTINUE
  50  CONTINUE
      RETURN
      END
```

```
INPUT
 Platinum
 4  3
 0.6416314244   1          PLATINUM
 1.2589205503   2          PLATINUM
 0.4430816174   3          PLATINUM
 0.4320656657   4          PLATINUM
-0.0639230013   5          PLATINUM
 0.0492522493   6          PLATINUM
-0.0555137061   7          PLATINUM
 0.0289367940   8          PLATINUM
 0.0154130906   9          PLATINUM
 0.0117542716  10          PLATINUM
 0.0131195020  11          PLATINUM
-0.0474181995  12          PLATINUM
-0.0015973587  13          PLATINUM
 0.0140918558  14          PLATINUM
 0.0167330708  15          PLATINUM
-0.0289147515  16          PLATINUM
 0.0106134648  17          PLATINUM
 0.0062306612  18          PLATINUM
 0.0233031027  19          PLATINUM
```

```
-0.0139374454  20          PLATINUM
 0.0359028243  21          PLATINUM
-0.0037766495  22          PLATINUM
 0.0537076704  23          PLATINUM
-0.0149604334  24          PLATINUM
 0.1269278228  25          PLATINUM
 0.0003684824  26          PLATINUM
-0.0050850390  27          PLATINUM
-0.0390631817  28          PLATINUM
-0.0067338832  29          PLATINUM
-0.0005522325  30          PLATINUM
-0.0131061170  31          PLATINUM
 0.0062670521  32          PLATINUM
 0.0151817501  33          PLATINUM
-0.0557834059  34          PLATINUM
 0.0425040275  35          PLATINUM
 0.0033541066  36          PLATINUM
-0.0625335872  37          PLATINUM
 0.0410989411  38          PLATINUM
-0.0823340639  39          PLATINUM
 0.0273388494  40          PLATINUM
-0.0314611457  41          PLATINUM
 0.0007443229  42          PLATINUM
-0.0339633077  43          PLATINUM
 0.0507927202  44          PLATINUM
-0.0176315289  45          PLATINUM
-0.0234415922  46          PLATINUM
-0.0067630056  47          PLATINUM
 0.0141386883  48          PLATINUM
-0.0034284843  49          PLATINUM
-0.0343889184  50          PLATINUM
 0.0504596978  51          PLATINUM
-0.0400902368  52          PLATINUM
 0.0462294593  53          PLATINUM
 0.0155499419  54          PLATINUM
-0.0065606399  55          PLATINUM
-0.0506199598  56          PLATINUM
-0.0031998244  57          PLATINUM
-0.0001028079  58          PLATINUM
-0.0186690204  59          PLATINUM
 0.0134765841  60          PLATINUM
```

C. Diamond structure

This program generates eigenvalues and eigenvectors for the 8 × 8 sp Hamiltonian; it diagonalizes a complex matrix. The user is expected to add a diagonalization subroutine for a complex matrix named CEIGEN which in turn calls the subroutines HTRIDI and HTRIBK not given here. The Slater–Koster parameters should be read in the order given, for example, in the silicon Tables.

```
      PROGRAM SLAKOS
C  AUTHOR: D.A.PAPACONSTANTOPOULOS
C  DIAMOND STRUCTURE WITH OVERLAP 3-CENTER AND 2-CENTER
C  COMPLEX MATRICES
C   CREATES INPUT FOR CALCULATING THE DOS TETRAHEDRON
C   CAUTION: THE DIMENSIONING OF Q10,Q11,Q12,Q13, SHOULD BE
C   CONSISTENT WITH THE NUMBER OF K-POINTS CREATED BY THE
C   ASSIGNED VALUE TO "MESH".
C A diagonalization package needs to be added in order to
diagonalze a
C complex matrix. CALL CEIGEN
      IMPLICIT REAL*8 (A-H,O-Z)
      DIMENSION H(8,8),G(8,8)
      DIMENSION TITLE(20),EVR(8,8),EVI(8,8)
      DIMENSION A(3),PARAM(38)
      DIMENSION  Z(8,8),ZI(8,8),U(8,8),UI(8,8)
      DIMENSION Q10(8,505 ),Q11(8,505 ),Q20(8,505 ),Q21(8,505 )
      DIMENSION HPR(8,8),HPI(8,8)
      DIMENSION EVAL(8),EVAL1(8),EVEC(8,8),EVEC1(8,8)
      DIMENSION T(8,8),BAND(17),R(8,8)
      DIMENSION TT(8,8),RR(8,8)
      DIMENSION S(8,8),P(8,8),CMHG(8,8),HNEW(8,8)
      COMPLEX COM,EVEC,CMHG,HNEW,EVEC1,T,U
      OPEN(10,FILE='par.in',BLANK='ZERO')
      OPEN(15,FILE='dmd.out',BLANK='ZERO')
      READ(10,35) TITLE
   35 FORMAT(20A4)
      WRITE (6,335) TITLE
  335  FORMAT (1H ,20A4)
C   ICENTER=3 THREE-CENTER APPROXIMATION
C   ICENTER=2 TWO-CENTER APPROXIMATION
C   MESH  K-POINT MESH GENERATOR
      READ(10,9999)MESH,ICENTER,NPAR
      WRITE (6,9998)MESH,ICENTER,NPAR
 9999 FORMAT (3I5)
 9998 FORMAT (5X,'MESH=',I5,10X,'ICENTER=',I5,10X,'NPAR=',I5)
      SR3=SQRT(3.0)
      PI =3.14159265
```

```
 110 FORMAT(F13.10)
    IF (ICENTER.EQ.3)
    1READ(10,110)A1,A7,A5,A6,A12,A13,A2,A3,A4,A8,A9,A10,A11,
    A14,A15,
    1A16,A17,A18,A19,A20,B5,B6,B12,B13,B2,B3,B4,B8,B9,B10,
    B11,B14,B15,
   2B16,B17,B18,B19,B20
C   THE ORDER OF THE SK-PARAMETERS HERE IS THE SAME AS IN THE TABLES
c    IF (ICENTER.EQ.3)
    WRITE(6,999)A1,A7,A5,A6,A12,A13,A2,A3,A4,A8,A9,A10,A11,
    A14,A15,
    1A16,A17,A18,A19,A20,B5,B6,B12,B13,B2,B3,B4,B8,B9,B10,
    B11,B14,B15,
   2B16,B17,B18,B19,B20
 999 FORMAT(//10X,'VALUES OF THE SK PARAMETERS',//40(25X,F15.8/))
    IF (ICENTER.EQ.2) READ(10,110) (PARAM(I),I=1,26)
    IF (ICENTER.EQ.2) WRITE(15,999) (PARAM(I),I=1,26)
    IF (ICENTER.EQ.3) GO TO 500
C    THIS SECTION IS FOR THE TWO-CENTER CALCULATIONS
     A1 =PARAM(1)
     A7 =PARAM(2)
     SSS1 =PARAM(3)
     SPS1=PARAM(4)
     PPS1=PARAM(5)
     PPP1=PARAM(6)
     SSS2=PARAM(7)
     SPS2=PARAM(8)
     PPS2=PARAM(9)
     PPP2=PARAM(10)
     SSS3=PARAM(11)
     SPS3=PARAM(12)
     PPS3=PARAM(13)
     PPP3=PARAM(14)
     SSS4=PARAM(15)
     SPS4=PARAM(16)
     PPS4=PARAM(17)
     PPP4=PARAM(18)
     SSS5=PARAM(19)
     SPS5=PARAM(20)
     PPS5=PARAM(21)
     PPP5=PARAM(22)
    SSS6=PARAM(23)
     SPS6=PARAM(24)
     PPS6=PARAM(25)
     PPP6=PARAM(26)
```

```
      A2=SSS2
      A3=0.0
      A4=SPS2/SQRT(2.)
      A5=SSS1
      A6=SPS1/SQRT(3.)
      A8=(PPS2+PPP2)/2.
      A9=PPP2
      A10=(PPS2-PPP2)/2.
      A11=0.0
      A12=(PPS1+2.*PPP1)/3.
      A13=(PPS1-PPP1)/3.
      A14=SSS3
      A15=3.*SPS3/SQRT(11.)
      A16=SPS3/SQRT(11.)
      A17=(9.*PPS3+2.*PPP3)/11.
      A18=(PPS3+10.*PPP3)/11.
      A19=-3.*(PPS3-PPP3)/11.
      A20=-(PPS3-PPP3)/11.
      B2=SSS5
      B3=0.0
      B4=SPS5/SQRT(2.)
      B5=SSS4
      B6=SPS4/SQRT(3.)
      B8=(PPS5+PPP5)/2.
      B9=PPP5
      B10=(PPS5-PPP5)/2.
      B11=0.0
      B12=(PPS4+2.*PPP4)/3.
      B13=(PPS4-PPP4)/3.
      B14=SSS6
      B15=3.*SPS6/SQRT(11.)
      B16=SPS6/SQRT(11.)
      B17=(9.*PPS6+2.*PPP6)/11.
      B18=(PPS6+10.*PPP6)/11.
      B19=-3.*(PPS6-PPP6)/11.
      B20=-(PPS6-PPP6)/11.
  500 CONTINUE
      DO 10000 I=1,8
      DO 10000 J=1,8
10000 HNEW(I,J)=CMPLX(0.0,0.0)
      N=8
      NS=N
      JK=0
      IMESH=2*MESH+1
C     THE K-VECTOR UNITS ARE CHOSEN SO THAT THE POINT X HAS
```

```
C   COORDINATES (8,0,0)
    SCALE=8./FLOAT(2*MESH)
    DO 451 IYBK=1,IMESH
    DO 451 IXBK=1,IYBK
    DO 451 IZBK=1,IXBK
    IT= IXBK+IYBK+IZBK-3
    IF(MESH*3.LT.IT) GO TO 451
    JK=JK+1
    A(1)= FLOAT(IXBK-1)
    A(2)= FLOAT(IYBK-1)
    A(3)= FLOAT(IZBK-1)
    XX =A(1)*SCALE
    YY =A(2)*SCALE
    ZZ =A(3)*SCALE
    PI = 3.14159265
C   CHECK UNITS HERE
    X1 = PI*XX/8.
    Y1 = PI*YY/8.
    Z1 = PI*ZZ/8.
    COSX=COS(X1/2.)
    COSY=COS(Y1/2.)
    COSZ=COS(Z1/2.)
    SINX=SIN(X1/2.)
    SINY=SIN(Y1/2.)
    SINZ=SIN(Z1/2.)
    COS2X=COS(X1)
    COS2Y=COS(Y1)
    COS2Z=COS(Z1)
    SIN2X=SIN(X1)
    SIN2Y=SIN(Y1)
    SIN2Z=SIN(Z1)
    COS3X=COS(3.*X1/2.)
    COS3Y=COS(3.*Y1/2.)
    COS3Z=COS(3.*Z1/2.)
    SIN3X=SIN(3.*X1/2.)
    SIN3Y=SIN(3.*Y1/2.)
    SIN3Z=SIN(3.*Z1/2.)
    H(1,1)=A1+4.0*A2*(COS2X*COS2Y+COS2Y*COS2Z+COS2Z*COS2X)
    G(1,1)=0.0
    H(1,2)=-4.0*A3*SIN2Z*SIN2Y
    G(1,2)=      4.0*A4*SIN2X*(COS2Y+COS2Z)
    H(1,3)=-4.*A3*SIN2X*SIN2Z
    G(1,3)=  4.0*A4*SIN2Y*(COS2Z+COS2X)
    H(1,4)=-4.0*A3*SIN2Y*SIN2X
    G(1,4)=  4.0*A4*SIN2Z*(COS2X+COS2Y)
```

```
 H(1,5)=4.0*A5*(COSX*COSY*COSZ)
1+4.*A14*(COS3X*COSY*COSZ+COSX*COS3Y*COSZ+COSX*COSY*COS3Z)
 G(1,5)=-4.*A5*(SINX*SINY*SINZ)
1+4.*A14*(SIN3X*SINY*SINZ+SINX*SIN3Y*SINZ+SINX*SINY*SIN3Z)
 H(1,6)=-4.*A6*COSX*SINY*SINZ
 1+4.0*A15*COS3X*SINY*SINZ+4.0*A16*COSX*(SINY*SIN3Z
 +SIN3Y*SINZ)
 G(1,6)=4.*A6*(SINX*COSY*COSZ)
 1+4.0*A15*SIN3X*COSY*COSZ+4.0*A16*SINX*(COSY*COS3Z
 +COS3Y*COSZ)
 H(1,7)=-4.0*A6*COSY*SINZ*SINX
 1+4.0*A15*COS3Y*SINZ*SINX+4.0*A16*COSY*(SINZ*SIN3X
 +SIN3Z*SINX)
 G(1,7)=4.0*A6*(SINY*COSZ*COSX)
 1+4.0*A15*SIN3Y*COSZ*COSX+4.0*A16*SINY*(COSZ*COS3X
 +COS3Z*COSX)
 H(1,8)=-4.0*A6*COSZ*SINX*SINY
 1+4.0*A15*COS3Z*SINX*SINY+4.0*A16*COSZ*(SINX*SIN3Y
 +SIN3X*SINY)
 G(1,8)=4.0*A6*(SINZ*COSX*COSY)
 1+4.0*A15*SIN3Z*COSX*COSY+4.0*A16*SINZ*(COSX*COS3Y
 +COS3X*COSY)
 H(2,2)=A7+4.0*A8*COS2X*(COS2Y+COS2Z)+4.0*A9*COS2Y*COS2Z
 G(2,2)=0.0
 H(2,3)=-4.0*A10*SIN2X*SIN2Y
 G(2,3)=  -4.0*A11*SIN2Z*(COS2X-COS2Y)
 H(2,4)=-4.0*A10*SIN2X*SIN2Z
 G(2,4)=   4.0*A11*SIN2Y*(COS2Z-COS2X)
 H(2,5)=-H(1,6)
 G(2,5)=-G(1,6)
 H(2,6)=4.0*A12*(COSX*COSY*COSZ)
 1+4.0*A17*COS3X*COSY*COSZ+4.0*A18*COSX*(COSY*COS3Z
 +COS3Y*COSZ)
 G(2,6)=-4.0*A12* SINX*SINY*SINZ
 1+4.0*A17*SIN3X*SINY*SINZ+4.0*A18*SINX*(SINY*SIN3Z
 +SIN3Y*SINZ)
 H(2,7)=-4.*A13* SINX*SINY*COSZ
 1+4.0*A19*COSZ*(SIN3X*SINY+SINX*SIN3Y)
 +4.0*A20*SINX*SINY*COS3Z
 G(2,7)=4.*A13*(COSX*COSY*SINZ)
 1+4.0*A19*SINZ*(COS3X*COSY+COSX*COS3Y)
 +4.0*A20*COSX*COSY*SIN3Z
 H(2,8)=-4.*A13* SINX*SINZ*COSY
 1+4.0*A19*COSY*(SIN3Z*SINX+SINZ*SIN3X)
 +4.0*A20*SINZ*SINX*COS3Y
```

```
G(2,8)=4.*A13*(COSX*COSZ*SINY)
1+4.0*A19*SINY*(COS3Z*COSX+COSZ*COS3X)
 +4.0*A20*COSZ*COSX*SIN3Y
H(3,3)=A7+4.0*A8* COS2Y*(COS2Z+COS2X)+4.0*A9*COS2Z*COS2X
G(3,3)=0.0
H(3,4)=-4.0*A10*SIN2Y*SIN2Z
G(3,4)=   -4.0*A11*SIN2X*(COS2Y-COS2Z)
H(3,5)=4.0*A6* COSY*SINZ*SINX
1-4.0*A15*COS3Y*SINZ*SINX-4.0*A16*COSY*(SINZ*SIN3X
 +SIN3Z*SINX)
G(3,5)=-4.0*A6*(SINY*COSZ*COSX)
1-4.0*A15*SIN3Y*COSZ*COSX-4.0*A16*SINY*(COSZ*COS3X
 +COS3Z*COSX)
H(3,6)=H(2,7)
G(3,6)=G(2,7)
H(3,7)=4.0*A12*(COSX*COSY*COSZ)
1+4.0*A17*COS3Y*COSZ*COSX+4.0*A18*COSY*(COSZ*COS3X
 +COS3Z*COSX)
G(3,7)=-4.0*A12* SINX*SINY*SINZ
1+4.0*A17*SIN3Y*SINZ*SINX+4.0*A18*SINY*(SINZ*SIN3X
 +SIN3Z*SINX)
H(3,8)=-4.0*A13* SINY*SINZ*COSX
1+4.0*A19*COSX*(SIN3Y*SINZ+SINY*SIN3Z)
 +4.0*A20*SINY*SINZ*COS3X
G(3,8)=4.0*A13*(COSY*COSZ*SINX)
1+4.0*A19*SINX*(COS3Y*COSZ+COSY*COS3Z)
 +4.0*A20*COSY*COSZ*SIN3X
H(4,4)=A7+4.0*A8*COS2Z*(COS2X+COS2Y)+4.0*A9*COS2X*COS2Y
G(4,4)=0.0
H(4,5)=4.0*A6* COSZ*SINX*SINY
1-4.0*A15*COS3Z*SINX*SINY-4.0*A16*COSZ*(SINX*SIN3Y
 +SIN3X*SINY)
G(4,5)=-4.0*A6*(SINZ*COSX*COSY)
1-4.0*A15*SIN3Z*COSX*COSY-4.0*A16*SINZ*(COSX*COS3Y
 +COS3X*COSY)
H(4,6)=-4.0*A13* SINX*SINZ*COSY
1+4.0*A19*COSY*(SIN3Z*SINX+SINZ*SIN3X)
 +4.0*A20*SINZ*SINX*COS3Y
G(4,6)=4.0*A13*(COSX*COSZ*SINY)
1+4.0*A19*SINY*(COS3Z*COSX+COSZ*COS3X)
 +4.0*A20*COSZ*COSX*SIN3Y
H(4,7)=-4.0*A13* SINY*SINZ*COSX
1+4.0*A19*COSX*(SIN3Y*SINZ+SINY*SIN3Z)
 +4.0*A20*SINY*SINZ*COS3X
G(4,7)=4.0*A13*(COSZ*COSY*SINX)
```

```
  1+4.0*A19*SINX*(COS3Y*COSZ+COSY*COS3Z)
  +4.0*A20*COSY*COSZ*SIN3X
  H(4,8)=4.0*A12*(COSX*COSY*COSZ)
  1+4.0*A17*COS3Z*COSX*COSY+4.0*A18*COSZ*(COSX*COS3Y
  +COS3X*COSY)
  G(4,8)=-4.0*A12* SINX*SINY*SINZ
  1+4.0*A17*SIN3Z*SINX*SINY+4.0*A18*SINZ*(SINX*SIN3Y
  +SIN3X*SINY)
  H(5,5)=H(1,1)
  G(5,5)=G(1,1)
  H(5,6)=4.0*A3*SIN2Z*SIN2Y
  G(5,6)=  4.0*A4*SIN2X*(COS2Y+COS2Z)
  H(5,7)=4.0*A3*SIN2X*SIN2Z
  G(5,7)= 4.0*A4*SIN2Y*(COS2Z+COS2X)
  H(5,8)=4.0*A3*SIN2Y*SIN2X
  G(5,8)= 4.0*A4*SIN2Z*(COS2X+COS2Y)
  H(6,6)=H(2,2)
  G(6,6)=G(2,2)
  H(6,7)=-4.0*A10*SIN2X*SIN2Y
  G(6,7)=  4.0*A11*SIN2Z*(COS2X-COS2Y)
  H(6,8)=-4.0*A10*SIN2X*SIN2Z
  G(6,8)=  -4.0*A11*SIN2Y*(COS2Z-COS2X)
  H(7,7)=H(3,3)
  G(7,7)=G(3,3)
  H(7,8)=-4.0*A10*SIN2Y*SIN2Z
  G(7,8)=  4.0*A11*SIN2X*(COS2Y-COS2Z)
  H(8,8)=H(4,4)
  G(8,8)=G(4,4)
  DO 310 L=1,N
  DO 310 K=L,N
  H(K,L)=H(L,K)
  G(K,L)=-G(L,K)
310 CONTINUE
  S(1,1)=1.0 +4.0*B2*(COS2X*COS2Y+COS2Y*COS2Z+COS2Z*COS2X)
  P(1,1)=0.0
  S(1,2)=-4.0*B3*SIN2Z*SIN2Y
  P(1,2)=      4.0*B4*SIN2X*(COS2Y+COS2Z)
  S(1,3)=-4.*B3*SIN2X*SIN2Z
  P(1,3)=     4.0*B4*SIN2Y*(COS2Z+COS2X)
  S(1,4)=-4.0*B3*SIN2Y*SIN2X
  P(1,4)=      4.0*B4*SIN2Z*(COS2X+COS2Y)
  S(1,5)=4.0*B5*(COSX*COSY*COSZ)
 1+4.*B14*(COS3X*COSY*COSZ+COSX*COS3Y*COSZ+COSX*COSY*COS3Z)
  P(1,5)=-4.*B5*(SINX*SINY*SINZ)
 1+4.*B14*(SIN3X*SINY*SINZ+SINX*SIN3Y*SINZ+SINX*SINY*SIN3Z)
```

```
S(1,6)=-4.*B6*COSX*SINY*SINZ
1+4.0*B15*COS3X*SINY*SINZ+4.0*B16*COSX*(SINY*SIN3Z
+SIN3Y*SINZ)
P(1,6)=4.*B6*(SINX*COSY*COSZ)
1+4.0*B15*SIN3X*COSY*COSZ+4.0*B16*SINX*(COSY*COS3Z
+COS3Y*COSZ)
S(1,7)=-4.0*B6*COSY*SINZ*SINX
1+4.0*B15*COS3Y*SINZ*SINX+4.0*B16*COSY*(SINZ*SIN3X
+SIN3Z*SINX)
P(1,7)=4.0*B6*(SINY*COSZ*COSX)
1+4.0*B15*SIN3Y*COSZ*COSX+4.0*B16*SINY*(COSZ*COS3X
+COS3Z*COSX)
S(1,8)=-4.0*B6* COSZ*SINX*SINY
1+4.0*B15*COS3Z*SINX*SINY+4.0*B16*COSZ*(SINX*SIN3Y
+SIN3X*SINY)
P(1,8)=4.0*B6*(SINZ*COSX*COSY)
1+4.0*B15*SIN3Z*COSX*COSY+4.0*B16*SINZ*(COSX*COS3Y
+COS3X*COSY)
S(2,2)=1.0  +4.0*B8*COS2X*(COS2Y+COS2Z)+4.0*B9*COS2Y*COS2Z
P(2,2)=0.0
S(2,3)=-4.0*B10*SIN2X*SIN2Y
P(2,3)=     -4.0*B11*SIN2Z*(COS2X-COS2Y)
S(2,4)=-4.0*B10*SIN2X*SIN2Z
P(2,4)=      4.0*B11*SIN2Y*(COS2Z-COS2X)
S(2,5)=-S(1,6)
P(2,5)=-P(1,6)
S(2,6)=4.0*B12*(COSX*COSY*COSZ)
1+4.0*B17*COS3X*COSY*COSZ+4.0*B18*COSX*(COSY*COS3Z
+COS3Y*COSZ)
P(2,6)=-4.0*B12*SINX*SINY*SINZ
1+4.0*B17*SIN3X*SINY*SINZ+4.0*B18*SINX*(SINY*SIN3Z
+SIN3Y*SINZ)
S(2,7)=-4.*B13*SINX*SINY*COSZ
1+4.0*B19*COSZ*(SIN3X*SINY+SINX*SIN3Y)
+4.0*B20*SINX*SINY*COS3Z
P(2,7)=4.*B13*(COSX*COSY*SINZ)
1+4.0*B19*SINZ*(COS3X*COSY+COSX*COS3Y)
+4.0*B20*COSX*COSY*SIN3Z
S(2,8)=-4.*B13*SINX*SINZ*COSY
1+4.0*B19*COSY*(SIN3Z*SINX+SINZ*SIN3X)
+4.0*B20*SINZ*SINX*COS3Y
P(2,8)=4.*B13*(COSX*COSZ*SINY)
1+4.0*B19*SINY*(COS3Z*COSX+COSZ*COS3X)
+4.0*B20*COSZ*COSX*SIN3Y
S(3,3)=1.0 +4.0*B8* COS2Y*(COS2Z+COS2X)+4.0*B9*COS2Z*COS2X
```

```
P(3,3)=0.0
S(3,4)=-4.0*B10*SIN2Y*SIN2Z
P(3,4)=    -4.0*B11*SIN2X*(COS2Y-COS2Z)
S(3,5)=4.0*B6*COSY*SINZ*SINX
1-4.0*B15*COS3Y*SINZ*SINX-4.0*B16*COSY*(SINZ*SIN3X
+SIN3Z*SINX)
P(3,5)=-4.0*B6*(SINY*COSZ*COSX)
1-4.0*B15*SIN3Y*COSZ*COSX-4.0*B16*SINY*(COSZ*COS3X
+COS3Z*COSX)
S(3,6)=S(2,7)
P(3,6)=P(2,7)
S(3,7)=4.0*B12*(COSX*COSY*COSZ)
1+4.0*B17*COS3Y*COSZ*COSX+4.0*B18*COSY*(COSZ*COS3X
+COS3Z*COSX)
P(3,7)=-4.0*B12*SINX*SINY*SINZ
1+4.0*B17*SIN3Y*SINZ*SINX+4.0*B18*SINY*(SINZ*SIN3X
+SIN3Z*SINX)
S(3,8)=-4.0*B13*SINY*SINZ*COSX
1+4.0*B19*COSX*(SIN3Y*SINZ+SINY*SIN3Z)
+4.0*B20*SINY*SINZ*COS3X
P(3,8)=4.0*B13*(COSY*COSZ*SINX)
1+4.0*B19*SINX*(COS3Y*COSZ+COSY*COS3Z)
+4.0*B20*COSY*COSZ*SIN3X
S(4,4)=1.0 +4.0*B8*COS2Z*(COS2X+COS2Y)+4.0*B9*COS2X*COS2Y
P(4,4)=0.0
S(4,5)=4.0*B6*COSZ*SINX*SINY
1-4.0*B15*COS3Z*SINX*SINY-4.0*B16*COSZ*(SINX*SIN3Y
+SIN3X*SINY)
P(4,5)=-4.0*B6*(SINZ*COSX*COSY)
1-4.0*B15*SIN3Z*COSX*COSY-4.0*B16*SINZ*(COSX*COS3Y
+COS3X*COSY)
S(4,6)=-4.0*B13*SINX*SINZ*COSY
1+4.0*B19*COSY*(SIN3Z*SINX+SINZ*SIN3X)
+4.0*B20*SINZ*SINX*COS3Y
P(4,6)=4.0*B13*(COSX*COSZ*SINY)
1+4.0*B19*SINY*(COS3Z*COSX+COSZ*COS3X)
+4.0*B20*COSZ*COSX*SIN3Y
S(4,7)=-4.0*B13*SINY*SINZ*COSX
1+4.0*B19*COSX*(SIN3Y*SINZ+SINY*SIN3Z)
+4.0*B20*SINY*SINZ*COS3X
P(4,7)=4.0*B13*(COSZ*COSY*SINX)
1+4.0*B19*SINX*(COS3Y*COSZ+COSY*COS3Z)
+4.0*B20*COSY*COSZ*SIN3X
S(4,8)=4.0*B12*(COSX*COSY*COSZ)
```

```
      1+4.0*B17*COS3Z*COSX*COSY+4.0*B18*COSZ*(COSX*COS3Y
      +COS3X*COSY)
      P(4,8)=-4.0*B12*SINX*SINY*SINZ
      1+4.0*B17*SIN3Z*SINX*SINY+4.0*B18*SINZ*(SINX*SIN3Y
      +SIN3X*SINY)
      S(5,5)=1.0 +4.0*B2*(COS2X*COS2Y+COS2Y*COS2Z+COS2Z*COS2X)
      P(5,5)=P(1,1)
      S(5,6)=4.0*B3*SIN2Z*SIN2Y
      P(5,6)=    4.0*B4*SIN2X*(COS2Y+COS2Z)
      S(5,7)=4.0*B3*SIN2X*SIN2Z
      P(5,7)=    4.0*B4*SIN2Y*(COS2Z+COS2X)
      S(5,8)=4.0*B3*SIN2Y*SIN2X
      P(5,8)=    4.0*B4*SIN2Z*(COS2X+COS2Y)
      S(6,6)=1.0  +4.0*B8*COS2X*(COS2Y+COS2Z)+4.0*B9*COS2Y*COS2Z
      P(6,6)=P(2,2)
      S(6,7)=-4.0*B10*SIN2X*SIN2Y
      P(6,7)=     4.0*B11*SIN2Z*(COS2X-COS2Y)
      S(6,8)=-4.0*B10*SIN2X*SIN2Z
      P(6,8)=    -4.0*B11*SIN2Y*(COS2Z-COS2X)
      S(7,7)=1.0 +4.0*B8* COS2Y*(COS2Z+COS2X)+4.0*B9*COS2Z*COS2X
      P(7,7)=P(3,3)
      S(7,8)=-4.0*B10*SIN2Y*SIN2Z
      P(7,8)=     4.0*B11*SIN2X*(COS2Y-COS2Z)
      S(8,8)=1.0  +4.0*B8*COS2Z*(COS2X+COS2Y)+4.0*B9*COS2X*COS2Y
      P(8,8)=P(4,4)
      DO 311 L=1,N
      DO 311 K=L,N
      S(K,L)=S(L,K)
      P(K,L)=-P(L,K)
  311 CONTINUE
c   WRITE(15,325) XX,YY,ZZ
  325 FORMAT(3(F8.5,2X))
      DO 101M=1,NS
      DO 101L=1,NS
  101 CMHG(M,L)=CMPLX(H(M,L),G(M,L))
C   DIAGONALIZES THE OVERLAP MATRIX S
      CALL CEIGEN(S,P,EVAL,EVR,EVI,NS)
       DO 8011 M=1,NS
       DO 8011 L=1,NS
        EVEC(M,L)=CMPLX(EVR(M,L),EVI(M,L))
 8011 CONTINUE
      DO 810 M=1,NS
      DO 810 J=1,M
      HNEW(M,J)=CMPLX(0.0,0.0)
      DO 820 K=1,NS
```

```
    DO 820 L=1,NS
    HNEW(M,J)=HNEW(M,J)+CONJG(EVEC(K,M))*CMHG(K,L)*EVEC(L,J)
 820 CONTINUE
    HNEW(M,J)=HNEW(M,J)/(SQRT(EVAL(M))*SQRT(EVAL(J)))
     HNEW(J,M)=CONJG(HNEW(M,J))
 810 CONTINUE
     DO 911 L=1,NS
     DO 911 M=1,NS
    HPR(L,M)=REAL(HNEW(M,L))
    HPI(L,M)=-AIMAG(HNEW(M,L))
 911  CONTINUE
C   DIAGONALIZES THE EFFECTIVE HAMILTONIAN HNEW
    CALL CEIGEN(HPR,HPI,EVAL1,EVR,EVI,NS)
     DO 9011 M=1,NS
     DO 9011 L=1,NS
      EVEC1(M,L)=CMPLX(EVR(M,L),EVI(M,L))
 9011 CONTINUE
    DO 850 I1=1,NS
    DO 850 I2=1,I1
    T(I1,I2)=CMPLX(0.0,0.0)
    DO 860 J1=1,NS
    T(I1,I2)=T(I1,I2)+EVEC(I1,J1)*EVEC1(J1,I2)/SQRT(EVAL(J1))
  860 CONTINUE
  850 CONTINUE
    DO 851 I1=1,NS
    DO 851 I2=I1,NS
    T(I1,I2)=CMPLX(0.0,0.0)
    DO 861 J1=1,NS
    T(I1,I2)=T(I1,I2)+EVEC(I1,J1)*EVEC1(J1,I2)/SQRT(EVAL(J1))
  861 CONTINUE
  851 CONTINUE
    DO 390 J=1,NS
    ANORM=0.0
    DO 865 LNORM=1,NS
    ANORM=ANORM+(CONJG(EVEC1(LNORM,J))*EVEC1(LNORM,J))/EVAL
     (LNORM)
  865 CONTINUE
    ANORM=SQRT(ANORM)
    DO 866 LNORM=1,NS
    T(LNORM,J)=T(LNORM,J)/ANORM
    U(LNORM,J)=CONJG(T(LNORM,J))
  866 CONTINUE
  390 CONTINUE
 7845 FORMAT (8F10.5)
  331 FORMAT(1X,8F10.5)
```

```
C    THE Q10,Q11,Q12,Q13 ARE THE SUMS OF THE SQUARES OF THE
C    EIGENVECTORS OVER 9 BANDS. THESE ARE USED TO FIND THE
C    ANGULAR MOMENTUM CHARACTER OF EACH STATE AND THUS
C    DECOMPOSE THE DOS
     DO 730 J=1,NS
     Q10(J,JK )=U(1,J)*T(1,J)
     Q11(J,JK )= U(2,J)*T(2,J) +U(3,J)*T(3,J)+U(4,J)*T(4,J)
     Q20(J,JK )= U(5,J)*T(5,J)
     Q21(J,JK )=U(6,J)*T(6,J)+U( 7,J)*T(7,J)+U( 8,J)*T(8,J)
 730 CONTINUE
c Note the conversion to Ry
     WRITE(15,332)
     1(EVAL1(J)/13.6d0,Q10(J,JK),Q11(J,JK),Q20(J,JK),Q21(J,
     JK),J=1,
    1N)
     WRITE(8,333)(EVAL1(J),
    1      Q10(J,JK),Q11(J,JK),Q20(J,JK),Q21(J,JK),XX,
    1YY,ZZ,J=1,N)
 333 FORMAT(F9.5,3X,4F8.5,16X,3F5.1)
 332 FORMAT(F9.5,3x,4F8.5)
 330 FORMAT(1X,8F10.5///)
 719 CONTINUE
 328 CONTINUE
 450 CONTINUE
 451 CONTINUE
 600 STOP
     END
```

```
 SILICON DIAMOND ORTHOGONAL THREE-CENTER
   2  3  20
-6.4704937935    1          SI THREE CENTER FIT 20 PARAMS
 2.1498334408    2          SI THREE CENTER FIT 20 PARAMS
-1.8440667391    3          SI THREE CENTER FIT 20 PARAMS
 1.0185183287    4          SI THREE CENTER FIT 20 PARAMS
 0.2994655669    5          SI THREE CENTER FIT 20 PARAMS
 1.4462300539    6          SI THREE CENTER FIT 20 PARAMS
 0.2114924043    7          SI THREE CENTER FIT 20 PARAMS
 0.0430580266    8          SI THREE CENTER FIT 20 PARAMS
-0.1434953809    9          SI THREE CENTER FIT 20 PARAMS
 0.0985795483   10          SI THREE CENTER FIT 20 PARAMS
-0.3289353251   11          SI THREE CENTER FIT 20 PARAMS
-0.1229215860   12          SI THREE CENTER FIT 20 PARAMS
-0.1824773401   13          SI THREE CENTER FIT 20 PARAMS
-0.1156223640   14          SI THREE CENTER FIT 20 PARAMS
-0.1248439848   15          SI THREE CENTER FIT 20 PARAMS
```

```
 0.0711577833   16          SI THREE CENTER  FIT 20 PARAMS
 0.0791977420   17          SI THREE CENTER  FIT 20 PARAMS
 0.0261967443   18          SI THREE CENTER  FIT 20 PARAMS
 0.0879656002   19          SI THREE CENTER  FIT 20 PARAMS
-0.0787469298   20          SI THREE CENTER  FIT 20 PARAMS
 0.0
 0.0
 0.0
 0.0
 0.0
 0.0
 0.0
 0.0
 0.0
 0.0
 0.0
 0.0
 0.0
 0.0
 0.0
 0.0
 0.0
```

```
0.0
```

D. **Hexagonal close-packed structure**

This program calculates the densities of states (IDOS = 1) or generates eigenvalues and eigenvectors for the 18 × 18 spd Hamiltonian (IDOS = 0); it diagonalizes a complex matrix. The user is expected to add a diagonalization subroutine for a complex matrix named CEIGEN which in turn calls the subroutines HTRIDI and HTRIBK not given here. The Slater–Koster parameters should be read in the order given, for example, in the titanium Tables.

```
      PROGRAM SLAKOS
C  HEXAGONAL STRUCTURE WITH OVERLAP 2-CENTER
C  THIS PROGRAM CALCULATES THE DENSITY OF STATES IF IDOS=1
C  IT CREATES AN INPUT TO ANOTHER DOS PROGRAM IF IDOS=0
C    CAUTION: THE DIMENSIONING OF Q10,Q11,Q12,Q13 MUST BE
C    CONSISTENT WITH THE NUMBER OF K-POINTS CREATED BY THE
C    ASSIGNED VALUE TO "MESH".
CC A diagonalization package needs to be added in order to
 diagonalze a
C complex matrix. CALL CEIGEN
      IMPLICIT REAL*8 (A-H,O-Z)
      DIMENSION WW(50000),DENSFL(8),DENS2(8)
      DIMENSION DENS(2000),DENT2G(2000),DENEG(2000),DENSS(2000),
      DENSP(
     12000),EL(2000),ENER(2000)
c     REAL*8 W,DENS,DENT2G,DENEG,DENSS,DENSP,SUM,DNORM,SWT
c    1,EL,ENER,DE,EFL,DNORFL,DENSFL,WW,SS
      DIMENSION ELS1(2000),ELP1(2000),ELT2G(2000),ELEG(2000)
      DIMENSION TITLE(20),EVR(18,18),EVI(18,18)
      DIMENSION A(3)
      DIMENSION  Z(18,18),ZI(18,18),U(18,18),UI(18,18)
     DIMENSION Q10(18,505),Q11(18,505 ),Q12(18,505 ),Q13(18,505 )
     DIMENSION Q20(18,505),Q21(18,505 ),Q22(18,505 ),Q23(18,505 )
      DIMENSION HPR(18,18),HPI(18,18)
      DIMENSION EVAL(18),EVAL1(18),EVEC(18,18),EVEC1(18,18)
      DIMENSION T(18,18),BAND(17),R(18,18)
      DIMENSION TT(18,18),RR(18,18)
      DIMENSION S(18,18),P(18,18),CMHG(18,18),HNEW(18,18)
      COMPLEX COM,EVEC,CMHG,HNEW,EVEC1,T,U
      COMMON/CD1/CX(320),CY(320),CZ(320),SX(320),SY(320),SZ(320),
     1CX2(320),CY2(320),CZ2(320),SX2(320),SY2(320),SZ2(320),
     2E(320),
     3H(18,18),G(18,18),PARAM(71),IFREEZE(71),N
      COMMON/CD2/COSX,COSY,COSZ,SINX,SINY,SINZ,COS2X,COS13Y,
      COS23Y,
     * SIN2X,SIN13Y,SIN23Y,COS2Z,COS43Y,SIN2Z,SIN43Y
```

```
   COMMON/CD3/CZ3(320),CY3(320),SZ3(320),SY3(320)
   COMMON /WINDO/ E1,E2,ELOW,EHIGH,IWON,IWIN
   COMMON/LIA/ IWRITE
   COMMON/PARAMS/     S0,P0,D0,D1,D2,PD0,SSS1,SPS1,SDS1,PPS1,
   PPP1,
  1PDS1,PDP1,DDS1,DDP1,DDD1,
  2 SSS2,SPS2,SDS2,PPS2,PPP2,
  2PDS2,PDP2,DDS2,DDP2,DDD2,
  3 SSS3,SPS3,SDS3,PPS3,PPP3,
  3PDS3,PDP3,DDS3,DDP3,DDD3
  4,OVPD0,OVSSS1,OVSPS1,OVSDS1,OVPPS1,OVPPP1,
  1OVPDS1,OVPDP1,OVDDS1,OVDDP1,OVDDD1
  2,OVSSS2,OVSPS2,OVSDS2,OVPPS2,OVPPP2,
  2OVPDS2,OVPDP2,OVDDS2,OVDDP2,OVDDD2
  3,OVSSS3,OVSPS3,OVSDS3,OVPPS3,OVPPP3,
  3OVPDS3,OVPDP3,OVDDS3,OVDDP3,OVDDD3
   REAL*8 NUELEC,INTERP
   OPEN(10,FILE='par.in',BLANK='ZERO')
   OPEN(15,FILE='hcp.out',BLANK='ZERO')
   READ(10,35) TITLE
 35 FORMAT(20A4)
   WRITE (15,335) TITLE
335 FORMAT (1H ,20A4)
C   MESH=K-POINT MESH GENERATOR; MUST BE AN ODD NUMBER
C   IPT=NUMBER OF DOS VALUES
C   IDOS=1 DOES A HISTOGRAM TYPE DOS
C  IDOS=0 GENERATES INPUT TO Other DOS Program such as TETRAHEDRON
   READ(10,9999)MESH,IPT,IDOS
   WRITE (15,9998)MESH,IPT,IDOS
9999 FORMAT (7I5)
9998 FORMAT (5X,'MESH=',I5,5X,'IPT=',I5,5X,'IDOS=',I5)
   DO 175II = 1,IPT
   DENSS(II)=0.0
   DENSP(II)=0.0
   DENT2G(II)=0.0
   DENEG(II)=0.0
 175 DENS(II)=0.0
C   NUELEC=NUMBER OF ELECTRONS AT EF
C   TOP=MAXIMUM ENERGY FOR WHICH DOS IS FOUND.CAUTION
C   REGARDING THE DIMENSIONS OF DOS FUNCTIONS
C   DEL=ENERGY STEP
C   START=LOWER ENERGY FOR WHICH DOS IS CALCULATED
   READ(10,1501) NUELEC,TOP,DEL,START
1501 FORMAT(4F10.5)
   WRITE(15,1503) NUELEC,TOP,DEL,START
```

```
    1503 FORMAT(1X,'NUELEC=',F10.5,5X,'TOP=',F10.5,5X,'DEL=',
    F10.5,5X,
   1'START=',F10.5)
C   THE ORDER OF THE SK-PARAMETERS HERE IS THE SAME AS IN THE TABLES
C   THE ORTHOGONAL FIT CONTAINS 36 PARAMETERS. SO IF WE WISH TO USE
C   ORTHOGONAL REPRESENTATION WE MUST SET THE PARAMETERS FROM
C   37-67 EQUAL TO ZERO
    NPAR=67
    READ(10,110) (PARAM(I),I=1,NPAR)
    SR3=SQRT(3.0)
    PI =3.14159265
  110 FORMAT(F13.10)
    WRITE(15,999) (PARAM(I),I=1,NPAR)
999 FORMAT(//10X,'VALUES OF THE SK-PARAMETERS',/67(25X,F15.8/))
    N=18
    NS=N
    DO 10000 I=1,NS
    DO 10000 J=1,NS
10000 HNEW(I,J)=CMPLX(0.0,0.0)
    S0 =PARAM(1)
    P0 =PARAM(2)
    D0=PARAM(3)
    D1=PARAM(4)
    D2 =PARAM(5)
    PD0=PARAM(6)
    SSS1=PARAM(7)
    SPS1=PARAM(8)
    SDS1=PARAM(9)
    PPS1=PARAM(10)
    PPP1=PARAM(11)
    PDS1=PARAM(12)
    PDP1=PARAM(13)
    DDS1=PARAM(14)
    DDP1=PARAM(15)
    DDD1=PARAM(16)
    SSS2=PARAM(17)
    SPS2=PARAM(18)
    SDS2=PARAM(19)
    PPS2=PARAM(20)
    PPP2=PARAM(21)
    PDS2=PARAM(22)
    PDP2=PARAM(23)
    DDS2=PARAM(24)
    DDP2=PARAM(25)
    DDD2=PARAM(26)
```

```
SSS3=PARAM(27)
SPS3=PARAM(28)
SDS3=PARAM(29)
PPS3=PARAM(30)
PPP3=PARAM(31)
PDS3=PARAM(32)
PDP3=PARAM(33)
DDS3=PARAM(34)
DDP3=PARAM(35)
DDD3=PARAM(36)
OVPD0=PARAM(37)
OVSSS1=PARAM(38)
OVSPS1=PARAM(39)
OVSDS1=PARAM(40)
OVPPS1=PARAM(41)
OVPPP1=PARAM(42)
OVPDS1=PARAM(43)
OVPDP1=PARAM(44)
OVDDS1=PARAM(45)
OVDDP1=PARAM(46)
OVDDD1=PARAM(47)
OVSSS2=PARAM(48)
OVSPS2=PARAM(49)
OVSDS2=PARAM(50)
OVPPS2=PARAM(51)
OVPPP2=PARAM(52)
OVPDS2=PARAM(53)
OVPDP2=PARAM(54)
OVDDS2=PARAM(55)
OVDDP2=PARAM(56)
OVDDD2=PARAM(57)
OVSSS3=PARAM(58)
OVSPS3=PARAM(59)
OVSDS3=PARAM(60)
OVPPS3=PARAM(61)
OVPPP3=PARAM(62)
OVPDS3=PARAM(63)
OVPDP3=PARAM(64)
OVDDS3=PARAM(65)
OVDDP3=PARAM(66)
OVDDD3=PARAM(67)
JK=0
NPY=MESH-1
IF (NPY.NE.2*(NPY/2)) GO TO 998
IMESH=MESH/2+1
```

```
      SCALE=1./FLOAT(2*MESH-2)
      DO 451 KK=1,MESH
      IZ=KK-1
      DO 451 JJ=1,IMESH
      IY=JJ-1
      IM=2*IY+1
      DO 451 II=IM,MESH
      IX=II-1
      JK=JK+1
      W=24.D0
      IF (IZ.EQ.0) W=W/2.D0
      IF (IY.EQ.0) W=W/2.D0
      IF (IX.EQ.0) W=W/3.D0
      IF (IX.EQ.2*IY) W=W/2.D0
      IF (IZ.EQ.NPY) W=W/2.D0
      IF (IX.EQ.NPY.AND.IY.EQ.0) W=W/3.D0
      IF (IX.EQ.NPY.AND.IY.NE.0) W=W/2.D0
      IF (IX+IY+IZ.EQ.0) W=1.D0
      IF (IX+IY.EQ.0.AND.IZ.EQ.NPY) W=1.D0
      WW(JK)=W
      SWT=0.0
      DO 25 I=1,JK
   25 SWT=SWT+WW(I)
      A(1)= FLOAT(IX)
      A(2)= FLOAT(IY)
      A(3)= FLOAT(IZ)
      XX =A(1)*SCALE
      YY =A(2)*SCALE
      ZZ =A(3)*SCALE
      ALFA=XX
      BETA=YY
      XX=ALFA+BETA
      YY=ALFA-BETA
C     CHECK UNITS HERE
      X2=PI*XX
      Y2=PI*YY
      Z2=PI*ZZ
      COSX=COS(X2)
      COSY=COS(Y2)
      COSZ=COS(Z2)
      SINX=SIN(X2)
      SINY=SIN(Y2)
      SINZ=SIN(Z2)
      COS2X=COS(2.*X2)
      COS2Z=COS(2.*Z2)
```

```
      COS13Y=COS(Y2/3.0)
      COS23Y=COS(2.*Y2/3.0)
      COS43Y=COS(4.*Y2/3.)
      SIN2X=SIN(2.*X2)
      SIN2Z=SIN(2.*Z2)
      SIN13Y=SIN(Y2/3.0)
      SIN23Y=SIN(2.*Y2/3.0)
      SIN43Y=SIN(4.*Y2/3.)
      CALL MATRI(S,P)
      IF (IDOS.NE.1)
     1WRITE(15,325) XX,YY,ZZ
  325 FORMAT(3(F8.5,2X))
      DO 101M=1,NS
      DO 101L=1,NS
 101  CMHG(M,L)=CMPLX(H(M,L),G(M,L))
       CALL CEIGEN(S,P,EVAL,EVR,EVI,NS)
       DO 8011 M=1,NS
       DO 8011 L=1,NS
        EVEC(M,L)=CMPLX(EVR(M,L),EVI(M,L))
 8011  CONTINUE
      DO 810 M=1,NS
      DO 810 J=1,M
      HNEW(M,J)=CMPLX(0.0,0.0)
      DO 820 K=1,NS
      DO 820 L=1,NS
      HNEW(M,J)=HNEW(M,J)+CONJG(EVEC(K,M))*CMHG(K,L)*EVEC(L,J)
 820  CONTINUE
      HNEW(M,J)=HNEW(M,J)/(SQRT((EVAL(M)))*SQRT((EVAL(J))))
       HNEW(J,M)=CONJG(HNEW(M,J))
 810  CONTINUE
       DO 911 L=1,NS
       DO 911 M=1,NS
      HPR(L,M)=REAL(HNEW(M,L))
      HPI(L,M)=-AIMAG(HNEW(M,L))
 911   CONTINUE
      CALL CEIGEN(HPR,HPI,EVAL1,EVR,EVI,NS)
       DO 9011 M=1,NS
       DO 9011 L=1,NS
        EVEC1(M,L)=CMPLX(EVR(M,L),EVI(M,L))
 9011  CONTINUE
      DO 850 I1=1,NS
      DO 850 I2=1,I1
      T(I1,I2)=CMPLX(0.0,0.0)
      DO 860 J1=1,NS
C     FIX FOR NEGATIVE EVAL
```

```
    T(I1,I2)=T(I1,I2)+EVEC(I1,J1)*EVEC1(J1,I2)/SQRT(ABS(EVAL
     (J1)))
 860 CONTINUE
 850 CONTINUE
     DO 851 I1=1,NS
     DO 851 I2=I1,NS
     T(I1,I2)=CMPLX(0.0,0.0)
     DO 861 J1=1,NS
C    FIX FOR NEGATIVE EVAL
     T(I1,I2)=T(I1,I2)+EVEC(I1,J1)*EVEC1(J1,I2)/SQRT(ABS(EVAL
      (J1)))
 861 CONTINUE
 851 CONTINUE
     DO 390 J=1,NS
     ANORM=0.0
     DO 865 LNORM=1,NS
     ANORM=ANORM+(CONJG(EVEC1(LNORM,J))*EVEC1(LNORM,J))/EVAL
      (LNORM)
 865 CONTINUE
C    CAUTION HERE WITH THE ABS
     ANORM=SQRT(ABS(ANORM))
     DO 866 LNORM=1,NS
     T(LNORM,J)=T(LNORM,J)/ANORM
     U(LNORM,J)=CONJG(T(LNORM,J))
 866 CONTINUE
 390 CONTINUE
 331 FORMAT(1X,8F10.5)
      IF (IDOS.EQ.1) GO TO 7730
     DO 730  J=1,NS
     Q10(J,JK )=U(1,J)*T(1,J)
     Q11(J,JK )= U(2,J)*T(2,J) +U(3,J)*T(3,J)+U(4,J)*T(4,J)
     Q12(J,JK)=U(5,J)*T(5,J)+U(6,J)*T(6,J)+U(7,J)*T(7,J)
     Q13(J,JK)=U(8,J)*T(8,J)+U(9,J)*T(9,J)
     Q20(J,JK )= U(10,J)*T(10,J)
    Q21(J,JK )=U(11,J)*T(11,J)+U(12,J)*T(12,J)+U(13,J)*T(13,J)
    Q22(J,JK )=U(14,J)*T(14,J)+U(15,J)*T(15,J)+U(16,J)*T(16,J)
     Q23(J,JK)=U(17,J)*T(17,J)+U(18,J)*T(18,J)
 730 CONTINUE
7730 CONTINUE
     IF (IDOS.NE.1) WRITE(15,333) xx,yy,zz,w, (EVAL1(J),j=1,16)
c 1       Q10(J,JK),Q11(J,JK),Q12(J,JK),Q13(J,JK),XX,YY,ZZ,
c 1        Q20(J,JK),Q21(J,JK),Q22(J,JK),Q23(J,JK),J=1,N)
c 333 FORMAT(F9.5,3X,4F10.5,16X,3F8.5,/4F10.5)
 333 FORMAT(4f9.3,/8F10.5/8F10.5)
 332 FORMAT(1X,9F10.5)
```

```
 330 FORMAT(1X,8F10.5///)
     IF (IDOS.NE.1) GO TO 451
     DO 391 J=1,NS
     IF(EVAL1(J).GE.TOP) GO TO 391
     NHJ=  (EVAL1(J)-START)/DEL
     IF (NHJ.GT.1999)
    1WRITE(6,666) NHJ,EVAL1(J),START,DEL,W
 666 FORMAT(1X,I5,4E15.8)
     DENS(NHJ+1) = DENS(NHJ+1) + W
     DENSS(NHJ+1)=DENSS(NHJ+1)+W*U( 1 ,J)*T(1,J)
     DENSP(NHJ+1)=DENSP(NHJ+1)+W*(U(2,J)*T(2,J)+U(3,J)*T(3,J)+
    1U(4,J)*T(4,J))
     DENT2G(NHJ+1)=DENT2G(NHJ+1)+W*(U(5,J)*T(5,J)+U(6,J)*T(6,J
    1)+U(7,J)*T(7,J))
     DENEG(NHJ+1)=DENEG(NHJ+1)+W*(U(8,J)*T(8,J)+U(9,J)*T(9,J))
 391 CONTINUE
 451 CONTINUE
     WRITE(15,4) SWT
 4   FORMAT('SUM OF WEIGHTS IS',E16.7)
     WRITE(15,452) JK
 452 FORMAT(1X,'NUMBER OF POINTS',2X,I5)
     DNORM=SWT*DEL
     SUM=0.0
     SUMS1=0.0
     SUMP1=0.0
     SUMT2G=0.0
     SUMEG =0.0
     DE=DEL
     DO 98 J=1,IPT
     ENER(J)=START+DE
     DE=DE+DEL
     SUM=SUM+DENS(J)
     SUMS1=SUMS1+DENSS(J)
     SUMP1=SUMP1+DENSP(J)
     SUMT2G=SUMT2G+DENT2G(J)
     SUMEG=SUMEG+DENEG(J)
     DENS(J)=DENS (J)/DNORM
     DENSS(J)=DENSS(J)/DNORM
     DENSP(J)=DENSP(J)/DNORM
     DENT2G(J)=DENT2G(J)/DNORM
     DENEG(J)=DENEG(J)/DNORM
     EL(J)=(SUM/DNORM)*DEL
     ELS1(J)=(SUMS1/DNORM)*DEL
     ELP1(J)=(SUMP1/DNORM)*DEL
     ELT2G(J)=(SUMT2G/DNORM)*DEL
```

```
      ELEG(J)=(SUMEG/DNORM)*DEL
      IF (IDOS.EQ.1)
      1WRITE(15,501) J,ENER(J), DENSS(J),DENSP(J),DENT2G(J),
      DENEG(J),
     1DENS(J),EL(J)
 501  FORMAT (1X,I5,F11.4,6E11.4)
  505 FORMAT(I5,4E15.4)
  506 FORMAT(I3,F5.2,4F9.4)
 504  FORMAT   (7F10.4)
 98   CONTINUE
      WRITE(15,601) NUELEC
      601 FORMAT(1X,46H THE NUMBER OF ELECTRONS AT THE FERMI LEVEL IS,
      F4.1/)
      MCOUNT=0
      IF (IDOS.NE.1) GO TO 500
      DO 177 J=1,IPT
      WRITE(15,501)J,ENER(J),ELS1(J),ELP1(J),ELT2G(J),ELEG(J),
      EL(J)
      IF(EL(J) .GT. NUELEC) GO TO 178
      GO TO 177
  178 IF(MCOUNT .NE. 0) GO TO 177
      MCOUNT=J
  177 CONTINUE
C     LAGRANGIAN INTERPOLATION OF THE FERMI LEVEL
      MC=MCOUNT
      MCM=MC-1
      SS=NUELEC
      EFL=INTERP(SS  ,EL,ENER(1),MCM,3)
      WRITE(15,844)EFL
      DNORFL=INTERP(SS  ,EL,DENS(1),MCM,3)
      WRITE(15,192)
      DENSFL(1)=INTERP(SS,EL,DENSS(1),MCM,3)
      DENSFL(2)=INTERP(SS,EL,DENSP(1),MCM,3)
      DENSFL(3)=INTERP(SS,EL,DENT2G(1),MCM,3)
      DENSFL(4)=INTERP(SS,EL,DENEG (1),MCM,3)
      WRITE(15,842)
  842 FORMAT(66X,10H---------- //)
      WRITE(15,839) EFL,NUELEC,DNORFL,(DENSFL(I),I=1,4)
  738 FORMAT(//10X,85H THE  F E R M I  L E V E L VIA LAGRANGIAN INTERPO
     1LATION                     )
  189 FORMAT(16X,20H-------------------- //)
  192 FORMAT(/72X,14HDECOMPOSED DOS )
      193 FORMAT(9X,6HENERGY,7X,9HELECTRONS,6X,9HTOTAL
      DOS,11X,1HS,14X,1HP,
     1 14X,4HDT2G,14X,3HDEG //)
```

```
194 FORMAT(72X,14H-------------- /)
839 FORMAT(2F10.5 ,3X,6F10.5//)
  843                                          FORMAT(10X,9HTOTAL
  DOS,11X,1HS,14X,1HP,14X,4HDT2G,14X,3HDEG,14X,
 11HS,14X,1HP//)
844 FORMAT(30X,7F15.5 //)
849 FORMAT(9F15.5)
446 FORMAT(//2X,'THE NUMBER OF RANDOM K-VECTORS IS',I10//)
500 CONTINUE
  STOP
998 WRITE(15,2222) MESH
2222 FORMAT(1H ,'*** WRONG MESH ,MESH=',I4,' ***')
  STOP
  END
  SUBROUTINE MATRI(S,P)
  IMPLICIT REAL*8 (A-H,O-Z)
  REAL*8 IWT
  COMMON/CD3/CZ3(320),CY3(320),SZ3(320),SY3(320)
  COMMON/CD1/     CX(320),CY(320),CZ(320),SX(320),SY(320),SZ
  (320),
 1CX2(320),CY2(320),CZ2(320),SX2(320),SY2(320),SZ2(320),
 2E(320),
 3H(18,18),G(18,18),PARAM(71),IFREEZE(71),NS
  COMMON/CD2/COSX,COSY,COSZ,SINX,SINY,SINZ,COS2X,COS13Y,
  COS23Y,
 * SIN2X,SIN13Y,SIN23Y,COS2Z,COS43Y,SIN2Z,SIN43Y
  COMMON/PARAMS/     S0,P0,D0,D1,D2,PD0,SSS1,SPS1,SDS1,PPS1,
  PPP1,
 1PDS1,PDP1,DDS1,DDP1,DDD1,
 2 SSS2,SPS2,SDS2,PPS2,PPP2,
 2PDS2,PDP2,DDS2,DDP2,DDD2,
 3 SSS3,SPS3,SDS3,PPS3,PPP3,
 3PDS3,PDP3,DDS3,DDP3,DDD3
 4,OVPD0,OVSSS1,OVSPS1,OVSDS1,OVPPS1,OVPPP1,
 1OVPDS1,OVPDP1,OVDDS1,OVDDP1,OVDDD1
 2,OVSSS2,OVSPS2,OVSDS2,OVPPS2,OVPPP2,
 2OVPDS2,OVPDP2,OVDDS2,OVDDP2,OVDDD2
 3,OVSSS3,OVSPS3,OVSDS3,OVPPS3,OVPPP3,
 3OVPDS3,OVPDP3,OVDDS3,OVDDP3,OVDDD3
  COMPLEX EVEC(18,18)
  DIMENSION EVAL(18),S(18,18),P(18,18)
  N=18
  SR2=SQRT(2.)
  SR3=SQRT(3.)
  H(1,1)=S0+2.*SSS1*(2.*COSX*COSY+COS2X)
```

```
1+2.*SSS3*COS2Z
 G(1,1)=0.0
 H(1,2)=0.0
 G(1,2)=2.*SPS1*(SINX*COSY+SIN2X)
 H(1,3)=0.0
 G(1,3)=2.*SR3*SPS1*COSX*SINY
 H(1,4)=0.0
 G(1,4)=2.0*SPS3*SIN2Z
 H(1,5)=-3.*SDS1*SINX*SINY
 G(1,5)=0.0
 H(1,6)=0.0
 G(1,6)=0.0
 H(1,7)=0.0
 G(1,7)=0.0
 H(1,8)=-SR3*SDS1*(COSX*COSY-COS2X)
 G(1,8)=0.0
 H(1,9)=-SDS1*(2.*COSX*COSY+COS2X)
1+2.0*SDS3*COS2Z
 G(1,9)=0.0
 H(2,2)=P0+(PPS1+3.*PPP1)*COSX*COSY+2.*PPS1*COS2X
1+2.0*PPP3*COS2Z
 G(2,2)=0.0
 H(2,3)=-SR3*(PPS1-PPP1)*SINX*SINY
 G(2,3)=0.0
 H(2,4)=0.0
 G(2,4)=0.0
 H(2,5)=PD0
 G(2,5)=(1.5*PDS1+SR3*PDP1)*COSX*SINY
 H(2,6)=0.0
 G(2,6)=0.0
 H(2,7)=0.0
 G(2,7)=2.0*PDP3*SIN2Z
 H(2,8)=0.0
 G(2,8)=-(0.5*SR3*PDS1-3.*PDP1)*SINX*COSY+SR3*PDS1*SIN2X
 H(2,9)=0.0
 G(2,9)=-PDS1*(SINX*COSY+SIN2X)
 H(3,3)=P0+(3.*PPS1+PPP1)*COSX*COSY+2.*PPP1*COS2X
1+2.0*PPP3*COS2Z
 G(3,3)=0.0
 H(3,4)=0.0
 G(3,4)=0.0
 H(3,5)=0.0
 G(3,5)=(1.5*SR3*PDS1-PDP1)*SINX*COSY+2.*PDP1*SIN2X
 H(3,6)=0.0
 G(3,6)=2.0*PDP3*SIN2Z
```

```
H(3,7)=0.0
G(3,7)=0.0
H(3,8)=PD0
G(3,8)=-(1.5*PDS1+SR3*PDP1)*COSX*SINY
H(3,9)=0.0
G(3,9)=-SR3*PDS1*COSX*SINY
H(4,4)=P0+2.*PPP1*(2.*COSX*COSY+COS2X)
1+2.0*PPS3*COS2Z
G(4,4)=0.0
H(4,5)=0.0
G(4,5)=0.0
H(4,6)=0.0
G(4,6)=2*SR3*PDP1*COSX*SINY
H(4,7)=0.0
G(4,7)=2.*PDP1*(SINX*COSY+SIN2X)
H(4,8)=0.0
G(4,8)=0.0
H(4,9)=0.0
G(4,9)=2.0*PDS3*SIN2Z
H(5,5)=D0+((9./4.)*DDS1+DDP1+0.75*DDD1)*COSX*COSY+2.
*DDP1*COS2X
1+2.0*DDD3*COS2Z
G(5,5)=0.0
H(5,6)=0.0
G(5,6)=0.0
H(5,7)=0.0
G(5,7)=0.0
H(5,8)=0.25*SR3*(3.*DDS1-4.*DDP1+DDD1)*SINX*SINY
G(5,8)=0.0
H(5,9)=1.5*(DDS1-DDD1)*SINX*SINY
G(5,9)=0.0
H(6,6)=D1+(3.*DDP1+DDD1)*COSX*COSY+2.*DDD1*COS2X
1+2.0*DDP3*COS2Z
G(6,6)=0.0
H(6,7)=-SR3*(DDP1-DDD1)*SINX*SINY
G(6,7)=0.0
H(6,8)=0.0
G(6,8)=0.0
H(6,9)=0.0
G(6,9)=0.0
H(7,7)=D1+(DDP1+3.*DDD1)*COSX*COSY+2.*DDP1*COS2X
1+2.0*DDP3*COS2Z
G(7,7)=0.0
H(7,8)=0.0
G(7,8)=0.0
```

```
H(7,9)=0.0
G(7,9)=0.0
H(8,8)=D0+(0.75*DDS1+3.*DDP1+0.25*DDD1)*COSX*COSY+0.5*(3.
*DDS1
1+DDD1)*COS2X
2+2.0*DDD3*COS2Z
G(8,8)=0.0
H(8,9)=0.5*SR3*(DDS1-DDD1)*(COSX*COSY-COS2X)
G(8,9)=0.0
H(9,9)=D2+0.5*(DDS1+3.*DDD1)*(2.*COSX*COSY+COS2X)
1+2.0*DDS3*COS2Z
G(9,9)=0.0
H(10,10)=H(1,1)
G(10,10)=0.0
H(10,11)=-H(1,2)
G(10,11)=G(1,2)
H(10,12)=-H(1,3)
G(10,12)=G(1,3)
H(10,13)=-H(1,4)
G(10,13)=G(1,4)
H(10,14)=H(1,5)
G(10,14)=-G(1,5)
H(10,15)=H(1,6)
G(10,15)=-G(1,6)
H(10,16)=H(1,7)
G(10,16)=-G(1,7)
H(10,17)=H(1,8)
G(10,17)=-G(1,8)
H(10,18)=H(1,9)
G(10,18)=-G(1,9)
H(11,11)=H(2,2)
G(11,11)=G(2,2)
H(11,12)=H(2,3)
G(11,12)=-G(2,3)
H(11,13)=H(2,4)
G(11,13)=-G(2,4)
H(11,14)=-H(2,5)
G(11,14)=G(2,5)
H(11,15)=-H(2,6)
G(11,15)=G(2,6)
H(11,16)=-H(2,7)
G(11,16)=G(2,7)
H(11,17)=-H(2,8)
G(11,17)=G(2,8)
H(11,18)=-H(2,9)
```

```
G(11,18)=G(2,9)
H(12,12)=H(3,3)
G(12,12)=G(3,3)
H(12,13)=H(3,4)
G(12,13)=-G(3,4)
H(12,14)=-H(3,5)
G(12,14)=G(3,5)
H(12,15)=-H(3,6)
G(12,15)=G(3,6)
H(12,16)=-H(3,7)
G(12,16)=G(3,7)
H(12,17)=-H(3,8)
G(12,17)=G(3,8)
H(12,18)=-H(3,9)
G(12,18)=G(3,9)
H(13,13)=H(4,4)
G(13,13)=G(4,4)
H(13,14)=-H(4,5)
G(13,14)=G(4,5)
H(13,15)=-H(4,6)
G(13,15)=G(4,6)
H(13,16)=-H(4,7)
G(13,16)=G(4,7)
H(13,17)=-H(4,8)
G(13,17)=G(4,8)
H(13,18)=-H(4,9)
G(13,18)=G(4,9)
H(14,14)=H(5,5)
G(14,14)=G(5,5)
H(14,15)=H(5,6)
G(14,15)=-G(5,6)
H(14,16)=H(5,7)
G(14,16)=-G(5,7)
H(14,17)=H(5,8)
G(14,17)=-G(5,8)
H(14,18)=H(5,9)
G(14,18)=-G(5,9)
H(15,15)=H(6,6)
G(15,15)=G(6,6)
H(15,16)=H(6,7)
G(15,16)=-G(6,7)
H(15,17)=H(6,8)
G(15,17)=-G(6,8)
H(15,18)=H(6,9)
G(15,18)=-G(6,9)
```

```
H(16,16)=H(7,7)
G(16,16)=G(7,7)
H(16,17)=H(7,8)
G(16,17)=-G(7,8)
H(16,18)=H(7,9)
G(16,18)=-G(7,9)
H(17,17)=H(8,8)
G(17,17)=G(8,8)
H(17,18)=H(8,9)
G(17,18)=-G(8,9)
H(18,18)=H(9,9)
G(18,18)=G(9,9)
H(1,10)=2.*SSS1*COSZ*(2.*COSX*COS13Y+COS23Y)
1    +2.*SSS2*COSZ*(2.*COS2X*COS23Y+COS43Y)
G(1,10)=2.*SSS1*COSZ*(2.*COSX*SIN13Y-SIN23Y)
1    +2.*SSS2*COSZ*(-2.*COS2X*SIN23Y+SIN43Y)
H(1,11)=-2.*SPS1*SINX*COSZ*SIN13Y
1    +2.*SR2*SPS2*SIN2X*COSZ*SIN23Y
G(1,11)=2.*SPS1*SINX*COSZ*COS13Y
1    +2.*SR2*SPS2*SIN2X*COSZ*COS23Y
H(1,12)=(2./3.)*SR3*SPS1*COSZ*(COSX*COS13Y-COS23Y)
1 +2.*SQRT(2./3.)*SPS2*COSZ*(-COS2X*COS23Y+COS43Y)
G(1,12)=(2./3.)*SR3*SPS1*COSZ*(COSX*SIN13Y+SIN23Y)
1 +2.*SQRT(2./3.)*SPS2*COSZ*(COS2X*SIN23Y+SIN43Y)
H(1,13)=-2.*SQRT(2./3.)*SPS1*SINZ*(2.*COSX*SIN13Y-SIN23Y)
1     +(2./3.)*SR3*SPS2*SINZ*(2.*COS2X*SIN23Y-SIN43Y)
G(1,13)=2.*SQRT(2./3.)*SPS1*SINZ*(2.*COSX*COS13Y+COS23Y)
1     +(2./3.)*SR3*SPS2*SINZ*(2.*COS2X*COS23Y+COS43Y)
H(1,14)=-SDS1*SINX*COSZ*SIN13Y
1-2.*SDS2*SIN2X*COSZ*SIN23Y
G(1,14)=SDS1*SINX*COSZ*COS13Y
1-2.*SDS2*SIN2X*COSZ*COS23Y
H(1,15)=-2.*SQRT(2./3.)*SDS1*SINZ*(COSX*SIN13Y+SIN23Y)
1-2.*SQRT(2./3.)*SDS2*SINZ*(COS2X*SIN23Y+SIN43Y)
G(1,15)=2.*SQRT(2./3.)*SDS1*SINZ*(COSX*COS13Y-COS23Y)
1-2.*SQRT(2./3.)*SDS2*SINZ*(COS2X*COS23Y-COS43Y)
H(1,16)=-2.*SR2*SDS1*SINX*SINZ*COS13Y
1-2.*SR2*SDS2*SIN2X*SINZ*COS23Y
G(1,16)=-2.*SR2*SDS1*SINX*SINZ*SIN13Y
1+2.*SR2*SDS2*SIN2X*SINZ*SIN23Y
H(1,17)=(1./3.)*SR3*SDS1*COSZ*(COSX*COS13Y-COS23Y)
1+(2./3.)*SR3*SDS2*COSZ*(COS2X*COS23Y-COS43Y)
G(1,17)=(1./3.)*SR3*SDS1*COSZ*(COSX*SIN13Y+SIN23Y)
1-(2./3.)*SR3*SDS2*COSZ*(COS2X*SIN23Y+SIN43Y)
H(1,18)=SDS1*COSZ*(2.*COSX*COS13Y+COS23Y)
```

```
G(1,18)=SDS1*COSZ*(2.*COSX*SIN13Y-SIN23Y)
H(2,11)=COSZ*((PPS1+3.*PPP1)*COSX*COS13Y+2.*PPP1*COS23Y)
1 +2.*COSZ*((PPS2+PPP2)*COS2X*COS23Y+PPP2*COS43Y)
G(2,11)=COSZ*((PPS1+3.*PPP1)*COSX*SIN13Y-2.*PPP1*SIN23Y)
1 -2.*COSZ*((PPS2+PPP2)*COS2X*SIN23Y-PPP2*SIN43Y)
H(2,12)=-(1./3.)*SR3*(PPS1-PPP1)*SINX*COSZ*SIN13Y
1 -(2./3.)*SR3*(PPS2-PPP2)*SIN2X*COSZ*SIN23Y
G(2,12)=(1./3.)*SR3*(PPS1-PPP1)*SINX*COSZ*COS13Y
1 -(2./3.)*SR3*(PPS2-PPP2)*SIN2X*COSZ*COS23Y
H(2,13)=-2.*SQRT(2./3.)*(PPS1-PPP1)*SINX*SINZ*COS13Y
1 -2.*SQRT(2./3.)*(PPS2-PPP2)*SIN2X*SINZ*COS23Y
G(2,13)=-2.*SQRT(2./3.)*(PPS1-PPP1)*SINX*SINZ*SIN13Y
1 +2.*SQRT(2./3.)*(PPS2-PPP2)*SIN2X*SINZ*SIN23Y
H(2,14)=COSZ*((0.5*PDS1+(1./3.)*SR3*PDP1)*COSX*COS13Y
1-(2./3.)*SR3*PDP1*COS23Y)
1 +SQRT(2./3.)*COSZ*(-SR3*PDS2*COS2X*COS23Y
1+2.*PDP2*COS43Y)
G(2,14)=COSZ*((0.5*PDS1+(1./3.)*SR3*PDP1)*COSX*SIN13Y
1+(2./3.)*SR3*PDP1*SIN23Y)
1 +SQRT(2./3.)*COSZ*(SR3*PDS2*COS2X*SIN23Y
1+2.*PDP2*SIN43Y)
H(2,15)=-(SQRT(2./3.)*PDS1-(2./3.)*SR2*PDP1)
*SINX*SINZ*COS13Y
1 +(2./9.)*SR3*(3.*PDS2-2.*SR3*PDP2)*SIN2X*SINZ*COS23Y
G(2,15)=-(SQRT(2./3.)*PDS1-(2./3.)*SR2*PDP1)
*SINX*SINZ*SIN13Y
1 -(2./9.)*SR3*(3.*PDS2-2.*SR3*PDP2)*SIN2X*SINZ*SIN23Y
H(2,16)=SINZ*(-(SR2*PDS1+2.*SQRT(2./3.)*PDP1)*COSX*SIN13Y
1+2.*SQRT(2./3.)*PDP1*SIN23Y)
1 +(2./3.)*SINZ*(3.*PDS2*COS2X*SIN23Y
1-SR3*PDP2*SIN43Y)
G(2,16)=SINZ*((SR2*PDS1+2.*SQRT(2./3.)*PDP1)*COSX*COS13Y
1+2.*SQRT(2./3.)*PDP1*COS23Y)
1 +(2./3.)*SINZ*(3.*PDS2*COS2X*COS23Y
1+SR3*PDP2*COS43Y)
H(2,17)=-(1./3.)*(0.5*SR3*PDS1+5.*PDP1)*SINX*COSZ*SIN13Y
1 +(1./3.)*SQRT(2./3.)*(4.*SR3*PDP2+3.*PDS2)
*SIN2X*COSZ*SIN23Y
G(2,17)=(1./3.)*(0.5*SR3*PDS1+5.*PDP1)*SINX*COSZ*COS13Y
1 +(1./3.)*SQRT(2./3.)*(4.*SR3*PDP2+3.*PDS2)
*SIN2X*COSZ*COS23Y
H(2,18)=-(PDS1-(4./3.)*SR3*PDP1)*SINX*COSZ*SIN13Y
1 -2.*SQRT(2./3.)*PDP2*SIN2X*COSZ*SIN23Y
G(2,18)=(PDS1-(4./3.)*SR3*PDP1)*SINX*COSZ*COS13Y
1 -2.*SQRT(2./3.)*PDP2*SIN2X*COSZ*COS23Y
```

```
H(3,12)=(1./3.)*COSZ*((PPS1+11.*PPP1)*COSX*COS13Y
1+2.*(PPS1+2.*PPP1)*COS23Y)
1 +(2./3.)*COSZ*((2.*PPS2+PPP2)*COS43Y
1+(PPS2+5.*PPP2)*COS2X*COS23Y)
G(3,12)=(1./3.)*COSZ*((PPS1+11.*PPP1)*COSX*SIN13Y
1-2.*(PPS1+2.*PPP1)*SIN23Y)
1 +(2./3.)*COSZ*((2.*PPS2+PPP2)*SIN43Y
1-(PPS2+5.*PPP2)*COS2X*SIN23Y)
H(3,13)=-(2./3.)*SR2*(PPS1-PPP1)*SINZ*(COSX*SIN13Y
+SIN23Y)
1 -(2./3.)*SR2*(PPS2-PPP2)*SINZ*(COS2X*SIN23Y+SIN43Y)
G(3,13)=(2./3.)*SR2*(PPS1-PPP1)*SINZ*(COSX*COS13Y-COS23Y)
1-(2./3.)*SR2*(PPS2-PPP2)*SINZ*(COS2X*COS23Y-COS43Y)
H(3,14)=H(2,17)
G(3,14)=G(2,17)
H(3,15)=(1./3.)*SINZ*(-(SR2*PDS1+10.*SQRT(2./3.)*PDP1)
*COSX*SIN13Y
1+2.*(SR2*PDS1+SQRT(2./3.)*PDP1)*SIN23Y)
1-(2./9.)*SINZ*((6.*PDS2-SR3*PDP2)*SIN43Y-
2(3.*PDS2+4.*SR3*PDP2)*COS2X*SIN23Y)
G(3,15)=(1./3.)*SINZ*((SR2*PDS1+10.*SQRT(2./3.)*PDP1)
*COSX*COS13Y
1+2.*(SR2*PDS1+SQRT(2./3.)*PDP1)*COS23Y)
1+(2./9.)*SINZ*((6.*PDS2-SR3*PDP2)*COS43Y+
2(3.*PDS2+4.*SR3*PDP2)*COS2X*COS23Y)
H(3,16)=H(2,15)
G(3,16)=G(2,15)
H(3,17)=(1./3.)*COSZ*((0.5*PDS1-(7./3.)*SR3*PDP1)
*COSX*COS13Y
1+2.*(0.5*PDS1+(2./3.)*SR3*PDP1)*COS23Y)
2-(1./3.)*SQRT(2./3.)*COSZ*(2.*(SR3*PDS2+PDP2)*COS43Y
3+(SR3*PDS2-8.*PDP2)*COS2X*COS23Y)
G(3,17)=(1./3.)*COSZ*((0.5*PDS1-(7./3.)*SR3*PDP1)
*COSX*SIN13Y
1-2.*(0.5*PDS1+(2./3.)*SR3*PDP1)*SIN23Y)
2-(1./3.)*SQRT(2./3.)*COSZ*(2.*(SR3*PDS2+PDP2)*SIN43Y
3-(SR3*PDS2-8.*PDP2)*COS2X*SIN23Y)
H(3,18)=(1./3.)*(SR3*PDS1-4.*PDP1)*COSZ*(COSX*COS13Y-
COS23Y)
1-(2./3.)*SR2*PDP2*COSZ*(COS43Y-COS2X*COS23Y)
G(3,18)=(1./3.)*(SR3*PDS1-4.*PDP1)*COSZ*(COSX*SIN13Y
+SIN23Y)
1-(2./3.)*SR2*PDP2*COSZ*(SIN43Y+COS2X*SIN23Y)
H(4,13)=(2./3.)*(2.*PPS1+PPP1)*COSZ*(2.*COSX*COS13Y
+COS23Y)
```

```
1+(2./3.)*(PPS2+2.*PPP2)*COSZ*(COS43Y+2.*COS2X*COS23Y)
G(4,13)=(2./3.)*(2.*PPS1+PPP1)*COSZ*(2.*COSX*SIN13Y-
SIN23Y)
1+(2./3.)*(PPS2+2.*PPP2)*COSZ*(SIN43Y-2.*COS2X*SIN23Y)
H(4,14)=H(2,15)
G(4,14)=G(2,15)
H(4,15)=(2./3.)*(2.*PDS1-(1./3.)*SR3*PDP1)*COSZ*
(COSX*COS13Y
1-COS23Y)
2+(2./3.)*SQRT(2./3.)*(SR3*PDS2+PDP2)*COSZ*(COS43Y-
COS2X*COS23Y)
G(4,15)=(2./3.)*(2.*PDS1-(1./3.)*SR3*PDP1)*COSZ*
(COSX*SIN13Y
1+SIN23Y)
2+(2./3.)*SQRT(2./3.)*(SR3*PDS2+PDP2)*COSZ*(SIN43Y
+COS2X*SIN23Y)
H(4,16)=-(2./3.)*(2.*SR3*PDS1-PDP1)*SINX*COSZ*SIN13Y
1+(2./3.)*SQRT(2./3.)*(3.*PDS2+SR3*PDP2)*SIN2X*COSZ*SIN23Y
G(4,16)=(2./3.)*(2.*SR3*PDS1-PDP1)*SINX*COSZ*COS13Y
1+(2./3.)*SQRT(2./3.)*(3.*PDS2+SR3*PDP2)*SIN2X*COSZ*COS23Y
H(4,17)=-(1./3.)*(SR2*PDS1-2.*SQRT(2./3.)*PDP1)*SINZ*
(COSX*SIN13Y
1+SIN23Y)
2+(2./9.)*(3.*PDS2-2.*SR3*PDP2)*SINZ*(SIN43Y+COS2X*SIN23Y)
G(4,17)=(1./3.)*(SR2*PDS1-2.*SQRT(2./3.)*PDP1)*SINZ*
(COSX*COS13Y
1-COS23Y)
2-(2./9.)*(3.*PDS2-2.*SR3*PDP2)*SINZ*(COS43Y-COS2X*COS23Y)
H(4,18)=-(SQRT(2./3.)*PDS1+(2./3.)*SR2*PDP1)*SINZ*(2.
*COSX*SIN13Y
1-SIN23Y)
2+(4./3.)*PDP2*SINZ*(2.*COS2X*SIN23Y-SIN43Y)
G(4,18)=(SQRT(2./3.)*PDS1+(2./3.)*SR2*PDP1)*SINZ*(2.
*COSX*COS13Y
1+COS23Y)
2+(4./3.)*PDP2*SINZ*(2.*COS2X*COS23Y+COS43Y)
H(5,14)=COSZ*(0.25*(DDS1+4.*DDP1+11.*DDD1)*COSX*COS13Y
1+(2./3.)*(DDP1+2.*DDD1)*COS23Y)
2+(1./3.)*COSZ*(2.*(2.*DDP2+DDD2)*COS43Y+(3.*DDS2+4.*DDP2
+5.*DDD2)
3*COS2X*COS23Y)
G(5,14)=COSZ*(0.25*(DDS1+4.*DDP1+11.*DDD1)*COSX*SIN13Y
1-(2./3.)*(DDP1+2.*DDD1)*SIN23Y)
2+(1./3.)*COSZ*(2.*(2.*DDP2+DDD2)*SIN43Y-(3.*DDS2+4.*DDP2
+5.*DDD2)
```

```
3*COS2X*SIN23Y)
H(5,15)=-SQRT(2./3.)*(0.5*DDS1+(4./3.)*DDP1-(11./6.)
*DDD1)*SINX*
1SINZ*COS13Y
2-(1./3.)*SQRT(2./3.)*(3.*DDS2+2.*DDP2-5.*DDD2)
*SIN2X*SINZ*COS23Y
G(5,15)=-SQRT(2./3.)*(0.5*DDS1+(4./3.)*DDP1-(11./6.)
*DDD1)*SINX*
1SINZ*SIN13Y
2+(1./3.)*SQRT(2./3.)*(3.*DDS2+2.*DDP2-5.*DDD2)
*SIN2X*SINZ*SIN23Y
H(5,16)=-SR2*SINZ*(0.5*(DDS1-DDD1)*COSX*SIN13Y+(2./3.)*
(DDP1-DDD1)
1*SIN23Y)
2-(1./3.)*SR2*SINZ*(2.*(DDP2-DDD2)*SIN43Y+(3.*DDS2-2.
*DDP2-DDD2)
3*COS2X*SIN23Y)
G(5,16)=SR2*SINZ*(0.5*(DDS1-DDD1)*COSX*COS13Y-(2./3.)*
(DDP1-DDD1)
1*COS23Y)
2+(1./3.)*SR2*SINZ*(2.*(DDP2-DDD2)*COS43Y-(3.*DDS2-2.
*DDP2-DDD2)
3*COS2X*COS23Y)
H(5,17)=-(1./3.)*SR3*(0.25*DDS1-(1./3.)*DDP1+(1./12.)
*DDD1)*SINX*
1COSZ*SIN13Y
2-(1./9.)*SR3*(3.*DDS2-4.*DDP2+DDD2)*SIN2X*COSZ*SIN23Y
G(5,17)=(1./3.)*SR3*(0.25*DDS1-(1./3.)*DDP1+(1./12.)
*DDD1)*SINX*
1COSZ*COS13Y
2-(1./9.)*SR3*(3.*DDS2-4.*DDP2+DDD2)*SIN2X*COSZ*COS23Y
H(5,18)=-(0.5*DDS1-(4./3.)*DDP1+(5./6.)*DDD1)
*SINX*COSZ*SIN13Y
1+(4./3.)*(DDP2-DDD2)*SIN2X*COSZ*SIN23Y
G(5,18)=(0.5*DDS1-(4./3.)*DDP1+(5./6.)*DDD1)
*SINX*COSZ*COS13Y
1+(4./3.)*(DDP2-DDD2)*SIN2X*COSZ*COS23Y
H(6,15)=(1./3.)*COSZ*((2.*DDS1+(19./3.)*DDP1+(11./3.)
*DDD1)*COSX
1*COS13Y+2.*(2.*DDS1+(1./3.)*DDP1+(2./3.)*DDD1)*COS23Y)
2+(2./9.)*COSZ*((6.*DDS2+DDP2+2.*DDD2)*COS43Y+(3.*DDS2+5.
*DDP2
3+10.*DDD2)*COS2X*COS23Y)
G(6,15)=(1./3.)*COSZ*((2.*DDS1+(19./3.)*DDP1+(11./3.)
*DDD1)*COSX
```

```
1*SIN13Y-2.*(2.*DDS1+(1./3.)*DDP1+(2./3.)*DDD1)*SIN23Y)
2+(2./9.)*COSZ*((6.*DDS2+DDP2+2.*DDD2)*SIN43Y-(3.*DDS2+5.
*DDP2
3+10.*DDD2)*COS2X*SIN23Y)
H(6,16)=-(1./3.)*SR3*(2.*DDS1-(5./3.)*DDP1-(1./3.)*DDD1)
*SINX*COSZ
1*SIN13Y
2-(2./9.)*SR3*(3.*DDS2-DDP2-2.*DDD2)*SIN2X*COSZ*SIN23Y
G(6,16)=(1./3.)*SR3*(2.*DDS1-(5./3.)*DDP1-(1./3.)*DDD1)
*SINX*COSZ
1*COS13Y
2-(2./9.)*SR3*(3.*DDS2-DDP2-2.*DDD2)*SIN2X*COSZ*COS23Y
H(6,17)=-(1./3.)*SR2*SINZ*((0.5*DDS1-(8./3.)*DDP1+(13./
6.)*DDD1)
1*COSX*SIN13Y-(DDS1+(2./3.)*DDP1-(5./3.)*DDD1)*SIN23Y)
2+(1./9.)*SR2*SINZ*(2.*(3.*DDS2-DDP2-2.*DDD2)*SIN43Y
3-(3.*DDS2-10.*DDP2+7.*DDD2)*COS2X*SIN23Y)
G(6,17)=(1./3.)*SR2*SINZ*((0.5*DDS1-(8./3.)*DDP1+(13./6.)
*DDD1)
1*COSX*COS13Y+(DDS1+(2./3.)*DDP1-(5./3.)*DDD1)*COS23Y)
2-(1./9.)*SR2*SINZ*(2.*(3.*DDS2-DDP2-2.*DDD2)*COS43Y
3+(3.*DDS2-10.*DDP2+7.*DDD2)*COS2X*COS23Y)
H(6,18)=-2.*SQRT(2./3.)*(0.5*DDS1-(1./3.)*DDP1-(1./6.)
*DDD1)*SINZ*
1(COSX*SIN13Y+SIN23Y)
2-(2./3.)*SQRT(2./3.)*(DDP2-DDD2)*SINZ*(SIN43Y
+COS2X*SIN23Y)
G(6,18)=2.*SQRT(2./3.)*(0.5*DDS1-(1./3.)*DDP1-(1./6.)
*DDD1)*SINZ*
1(COSX*COS13Y-COS23Y)
2+(2./3.)*SQRT(2./3.)*(DDP2-DDD2)*SINZ*(COS43Y-
COS2X*COS23Y)
H(7,16)=COSZ*((2.*DDS1+DDP1+DDD1)*COSX*COS13Y+(2./3.)*(2.
*DDP1
1+DDD1)*COS23Y)
2+(2./3.)*COSZ*((DDP2+2.*DDD2)*COS43Y+(3.*DDS2+DDP2+2.
*DDD2)
3*COS2X*COS23Y)
G(7,16)=COSZ*((2.*DDS1+DDP1+DDD1)*COSX*SIN13Y-(2./3.)*(2.
*DDP1
1+DDD1)*SIN23Y)
2+(2./3.)*COSZ*((DDP2+2.*DDD2)*SIN43Y-(3.*DDS2+DDP2+2.
*DDD2)
3*COS2X*SIN23Y)
H(7,17)=H(5,15)
```

```
 G(7,17)=G(5,15)
 H(7,18)=-2.*SR2*(0.5*DDS1-(1./3.)*DDP1-(1./6.)*DDD1)
 *SINX*SINZ*
1COS13Y
2-(2./3.)*SR2*(DDP2-DDD2)*SIN2X*SINZ*COS23Y
 G(7,18)=-2.*SR2*(0.5*DDS1-(1./3.)*DDP1-(1./6.)*DDD1)
 *SINX*SINZ*
1SIN13Y
2+(2./3.)*SR2*(DDP2-DDD2)*SIN2X*SINZ*SIN23Y
 H(8,17)=(1./3.)*COSZ*((0.25*DDS1+(11./3.)*DDP1+(97./12.)
 *DDD1)*
 1COSX*COS13Y+2.*(0.25*DDS1+(2./3.)*DDP1+(25./12.)*DDD1)
 *COS23Y)
2+(1./9.)*COSZ*(2.*(3.*DDS2+2.*DDP2+4.*DDD2)*COS43Y+
3(3.*DDS2+20.*DDP2+13.*DDD2)*COS2X*COS23Y)
 G(8,17)=(1./3.)*COSZ*((0.25*DDS1+(11./3.)*DDP1+(97./12.)
 *DDD1)*
 1COSX*SIN13Y-2.*(0.25*DDS1+(2./3.)*DDP1+(25./12.)*DDD1)
 *SIN23Y)
2+(1./9.)*COSZ*(2.*(3.*DDS2+2.*DDP2+4.*DDD2)*SIN43Y-
3(3.*DDS2+20.*DDP2+13.*DDD2)*COS2X*SIN23Y)
 H(8,18)=(2./3.)*SR3*(0.25*DDS1-(2./3.)*DDP1+(5./12.)
 *DDD1)*COSZ*
1(COSX*COS13Y-COS23Y)
2+(4./9.)*SR3*(DDP2-DDD2)*COSZ*(COS43Y-COS2X*COS23Y)
 G(8,18)=(2./3.)*SR3*(0.25*DDS1-(2./3.)*DDP1+(5./12.)
 *DDD1)*COSZ*
1(COSX*SIN13Y+SIN23Y)
2+(4./9.)*SR3*(DDP2-DDD2)*COSZ*(SIN43Y+COS2X*SIN23Y)
 H(9,18)=(0.5*DDS1+(4./3.)*DDP1+(1./6.)*DDD1)*COSZ*(2.
 *COSX*COS13Y
1+COS23Y)
2+(2./3.)*(2.*DDP2+DDD2)*COSZ*(COS43Y+2.*COS2X*COS23Y)
 G(9,18)=(0.5*DDS1+(4./3.)*DDP1+(1./6.)*DDD1)*COSZ*(2.
 *COSX*SIN13Y
1-SIN23Y)
2+(2./3.)*(2.*DDP2+DDD2)*COSZ*(SIN43Y-2.*COS2X*SIN23Y)
 H(2,10)=-H(1,11)
 G(2,10)=-G(1,11)
 H(3,10)=-H(1,12)
 G(3,10)=-G(1,12)
 H(4,10)=-H(1,13)
 G(4,10)=-G(1,13)
 H(5,10)=H(1,14)
 G(5,10)=G(1,14)
```

```
H(6,10)=H(1,15)
G(6,10)=G(1,15)
H(7,10)=H(1,16)
G(7,10)=G(1,16)
H(8,10)=H(1,17)
G(8,10)=G(1,17)
H(9,10)=H(1,18)
G(9,10)=G(1,18)
H(3,11)=H(2,12)
G(3,11)= G(2,12)
H(4,11)=H(2,13)
G(4,11)= G(2,13)
H(5,11)=-H(2,14)
G(5,11)=-G(2,14)
H(6,11)=-H(2,15)
G(6,11)=-G(2,15)
H(7,11)=-H(2,16)
G(7,11)=-G(2,16)
H(8,11)=-H(2,17)
G(8,11)=-G(2,17)
H(9,11)=-H(2,18)
G(9,11)=-G(2,18)
H(4,12)=H(3,13)
G(4,12)= G(3,13)
H(5,12)=-H(3,14)
G(5,12)=-G(3,14)
H(6,12)=-H(3,15)
G(6,12)=-G(3,15)
H(7,12)=-H(3,16)
G(7,12)=-G(3,16)
H(8,12)=-H(3,17)
G(8,12)=-G(3,17)
H(9,12)=-H(3,18)
G(9,12)=-G(3,18)
H(5,13)=-H(4,14)
G(5,13)=-G(4,14)
H(6,13)=-H(4,15)
G(6,13)=-G(4,15)
H(7,13)=-H(4,16)
G(7,13)=-G(4,16)
H(8,13)=-H(4,17)
G(8,13)=-G(4,17)
H(9,13)=-H(4,18)
G(9,13)=-G(4,18)
H(6,14)=H(5,15)
```

```
      G(6,14)=G(5,15)
      H(7,14)=H(5,16)
      G(7,14)=G(5,16)
      H(8,14)=H(5,17)
      G(8,14)=G(5,17)
      H(9,14)=H(5,18)
      G(9,14)=G(5,18)
      H(7,15)=H(6,16)
      G(7,15)=G(6,16)
      H(8,15)=H(6,17)
      G(8,15)=G(6,17)
      H(9,15)=H(6,18)
      G(9,15)=G(6,18)
      H(8,16)=H(7,17)
      G(8,16)=G(7,17)
      H(9,16)=H(7,18)
      G(9,16)=G(7,18)
      H(9,17)=H(8,18)
      G(9,17)=G(8,18)
      DO 310 L=1,N
      DO 310 K=L,N
      H(K,L)=H(L,K)
      G(K,L)=-G(L,K)
  310 CONTINUE
      S(1,1)=1.0+2.*OVSSS1*(2.*COSX*COSY+COS2X)
     1+2.*OVSSS3*COS2Z
      P(1,1)=0.0
      S(1,2)=0.0
      P(1,2)=2.*OVSPS1*(SINX*COSY+SIN2X)
      S(1,3)=0.0
      P(1,3)=2.*SR3*OVSPS1*COSX*SINY
      S(1,4)=0.0
      P(1,4)=2.0*OVSPS3*SIN2Z
      S(1,5)=-3.*OVSDS1*SINX*SINY
      P(1,5)=0.0
      S(1,6)=0.0
      P(1,6)=0.0
      S(1,7)=0.0
      P(1,7)=0.0
      S(1,8)=-SR3*OVSDS1*(COSX*COSY-COS2X)
      P(1,8)=0.0
      S(1,9)=-OVSDS1*(2.*COSX*COSY+COS2X)
     1+2.0*OVSDS3*COS2Z
      P(1,9)=0.0
      S(2,2)=1.0+(OVPPS1+3.*OVPPP1)*COSX*COSY+2.*OVPPS1*COS2X
```

```
1+2.0*OVPPP3*COS2Z
 P(2,2)=0.0
 S(2,3)=-SR3*(OVPPS1-OVPPP1)*SINX*SINY
 P(2,3)=0.0
 S(2,4)=0.0
 P(2,4)=0.0
 S(2,5)=OVPD0
 P(2,5)=(1.5*OVPDS1+SR3*OVPDP1)*COSX*SINY
 S(2,6)=0.0
 P(2,6)=0.0
 S(2,7)=0.0
 P(2,7)=2.0*OVPDP3*SIN2Z
 S(2,8)=0.0
 P(2,8)=-(0.5*SR3*OVPDS1-3.*OVPDP1)*SINX*COSY
 +SR3*OVPDS1*SIN2X
 S(2,9)=0.0
 P(2,9)=-OVPDS1*(SINX*COSY+SIN2X)
 S(3,3)=1.0+(3.*OVPPS1+OVPPP1)*COSX*COSY+2.*OVPPP1*COS2X
1+2.0*OVPPP3*COS2Z
 P(3,3)=0.0
 S(3,4)=0.0
 P(3,4)=0.0
 S(3,5)=0.0
 P(3,5)=(1.5*SR3*OVPDS1-OVPDP1)*SINX*COSY+2.*OVPDP1*SIN2X
 S(3,6)=0.0
 P(3,6)=2.0*OVPDP3*SIN2Z
 S(3,7)=0.0
 P(3,7)=0.0
 S(3,8)=OVPD0
 P(3,8)=-(1.5*OVPDS1+SR3*OVPDP1)*COSX*SINY
 S(3,9)=0.0
 P(3,9)=-SR3*OVPDS1*COSX*SINY
 S(4,4)=1.0+2.*OVPPP1*(2.*COSX*COSY+COS2X)
1+2.0*OVPPS3*COS2Z
 P(4,4)=0.0
 S(4,5)=0.0
 P(4,5)=0.0
 S(4,6)=0.0
 P(4,6)=2*SR3*OVPDP1*COSX*SINY
 S(4,7)=0.0
 P(4,7)=2.*OVPDP1*(SINX*COSY+SIN2X)
 S(4,8)=0.0
 P(4,8)=0.0
 S(4,9)=0.0
 P(4,9)=2.0*OVPDS3*SIN2Z
```

```
S(5,5)=1.0+((9./4.)*OVDDS1+OVDDP1+0.75*OVDDD1)*COSX*COSY
+2.*OVDDP1
1*COS2X+2.0*OVDDD3*COS2Z
P(5,5)=0.0
S(5,6)=0.0
P(5,6)=0.0
S(5,7)=0.0
P(5,7)=0.0
S(5,8)=0.25*SR3*(3.*OVDDS1-4.*OVDDP1+OVDDD1)*SINX*SINY
P(5,8)=0.0
S(5,9)=1.5*(OVDDS1-OVDDD1)*SINX*SINY
P(5,9)=0.0
S(6,6)=1.0+(3.*OVDDP1+OVDDD1)*COSX*COSY+2.*OVDDD1*COS2X
1+2.0*OVDDP3*COS2Z
P(6,6)=0.0
S(6,7)=-SR3*(OVDDP1-OVDDD1)*SINX*SINY
P(6,7)=0.0
S(6,8)=0.0
P(6,8)=0.0
S(6,9)=0.0
P(6,9)=0.0
S(7,7)=1.0+(OVDDP1+3.*OVDDD1)*COSX*COSY+2.*OVDDP1*COS2X
1+2.0*OVDDP3*COS2Z
P(7,7)=0.0
S(7,8)=0.0
P(7,8)=0.0
S(7,9)=0.0
P(7,9)=0.0
S(8,8)=1.0+(0.75*OVDDS1+3.*OVDDP1+0.25*OVDDD1)*COSX*COSY
+0.5*(3.* 1OVDDS1+OVDDD1)*COS2X
2+2.0*OVDDD3*COS2Z
P(8,8)=0.0
S(8,9)=0.5*SR3*(OVDDS1-OVDDD1)*(COSX*COSY-COS2X)
P(8,9)=0.0
S(9,9)=1.0+0.5*(OVDDS1+3.*OVDDD1)*(2.*COSX*COSY+COS2X)
1+2.0*OVDDS3*COS2Z
P(9,9)=0.0
S(10,10)=S(1,1)
P(10,10)=0.0
S(10,11)=-S(1,2)
P(10,11)=P(1,2)
S(10,12)=-S(1,3)
P(10,12)=P(1,3)
S(10,13)=-S(1,4)
P(10,13)=P(1,4)
```

```
S(10,14)=S(1,5)
P(10,14)=-P(1,5)
S(10,15)=S(1,6)
P(10,15)=-P(1,6)
S(10,16)=S(1,7)
P(10,16)=-P(1,7)
S(10,17)=S(1,8)
P(10,17)=-P(1,8)
S(10,18)=S(1,9)
P(10,18)=-P(1,9)
S(11,11)=S(2,2)
P(11,11)=P(2,2)
S(11,12)=S(2,3)
P(11,12)=-P(2,3)
S(11,13)=S(2,4)
P(11,13)=-P(2,4)
S(11,14)=-S(2,5)
P(11,14)=P(2,5)
S(11,15)=-S(2,6)
P(11,15)=P(2,6)
S(11,16)=-S(2,7)
P(11,16)=P(2,7)
S(11,17)=-S(2,8)
P(11,17)=P(2,8)
S(11,18)=-S(2,9)
P(11,18)=P(2,9)
S(12,12)=S(3,3)
P(12,12)=P(3,3)
S(12,13)=S(3,4)
P(12,13)=-P(3,4)
S(12,14)=-S(3,5)
P(12,14)=P(3,5)
S(12,15)=-S(3,6)
P(12,15)=P(3,6)
S(12,16)=-S(3,7)
P(12,16)=P(3,7)
S(12,17)=-S(3,8)
P(12,17)=P(3,8)
S(12,18)=-S(3,9)
P(12,18)=P(3,9)
S(13,13)=S(4,4)
P(13,13)=P(4,4)
S(13,14)=-S(4,5)
P(13,14)=P(4,5)
S(13,15)=-S(4,6)
```

```
      P(13,15)=P(4,6)
      S(13,16)=-S(4,7)
      P(13,16)=P(4,7)
      S(13,17)=-S(4,8)
      P(13,17)=P(4,8)
      S(13,18)=-S(4,9)
      P(13,18)=P(4,9)
      S(14,14)=S(5,5)
      P(14,14)=P(5,5)
      S(14,15)=S(5,6)
      P(14,15)=-P(5,6)
      S(14,16)=S(5,7)
      P(14,16)=-P(5,7)
      S(14,17)=S(5,8)
      P(14,17)=-P(5,8)
      S(14,18)=S(5,9)
      P(14,18)=-P(5,9)
      S(15,15)=S(6,6)
      P(15,15)=P(6,6)
      S(15,16)=S(6,7)
      P(15,16)=-P(6,7)
      S(15,17)=S(6,8)
      P(15,17)=-P(6,8)
      S(15,18)=S(6,9)
      P(15,18)=-P(6,9)
      S(16,16)=S(7,7)
      P(16,16)=P(7,7)
      S(16,17)=S(7,8)
      P(16,17)=-P(7,8)
      S(16,18)=S(7,9)
      P(16,18)=-P(7,9)
      S(17,17)=S(8,8)
      P(17,17)=P(8,8)
      S(17,18)=S(8,9)
      P(17,18)=-P(8,9)
      S(18,18)=S(9,9)
      P(18,18)=P(9,9)
      S(1,10)=2.*OVSSS1*COSZ*(2.*COSX*COS13Y+COS23Y)
     1   +2.*OVSSS2*COSZ*(2.*COS2X*COS23Y+COS43Y)
      P(1,10)=2.*OVSSS1*COSZ*(2.*COSX*SIN13Y-SIN23Y)
     1   +2.*OVSSS2*COSZ*(-2.*COS2X*SIN23Y+SIN43Y)
      S(1,11)=-2.*OVSPS1*SINX*COSZ*SIN13Y
     1   +2.*SR2*OVSPS2*SIN2X*COSZ*SIN23Y
      P(1,11)=2.*OVSPS1*SINX*COSZ*COS13Y
     1   +2.*SR2*OVSPS2*SIN2X*COSZ*COS23Y
```

```
S(1,12)=(2./3.)*SR3*OVSPS1*COSZ*(COSX*COS13Y-COS23Y)
1 +2.*SQRT(2./3.)*OVSPS2*COSZ*(-COS2X*COS23Y+COS43Y)
P(1,12)=(2./3.)*SR3*OVSPS1*COSZ*(COSX*SIN13Y+SIN23Y)
1 +2.*SQRT(2./3.)*OVSPS2*COSZ*(COS2X*SIN23Y+SIN43Y)
S(1,13)=-2.*SQRT(2./3.)*OVSPS1*SINZ*(2.*COSX*SIN13Y-
SIN23Y)
1    +(2./3.)*SR3*OVSPS2*SINZ*(2.*COS2X*SIN23Y-SIN43Y)
P(1,13)=2.*SQRT(2./3.)*OVSPS1*SINZ*(2.*COSX*COS13Y
+COS23Y)
1    +(2./3.)*SR3*OVSPS2*SINZ*(2.*COS2X*COS23Y+COS43Y)
S(1,14)=-OVSDS1*SINX*COSZ*SIN13Y
1-2.*OVSDS2*SIN2X*COSZ*SIN23Y
P(1,14)=OVSDS1*SINX*COSZ*COS13Y
1-2.*OVSDS2*SIN2X*COSZ*COS23Y
S(1,15)=-2.*SQRT(2./3.)*OVSDS1*SINZ*(COSX*SIN13Y+SIN23Y)
1-2.*SQRT(2./3.)*OVSDS2*SINZ*(COS2X*SIN23Y+SIN43Y)
P(1,15)=2.*SQRT(2./3.)*OVSDS1*SINZ*(COSX*COS13Y-COS23Y)
1-2.*SQRT(2./3.)*OVSDS2*SINZ*(COS2X*COS23Y-COS43Y)
S(1,16)=-2.*SR2*OVSDS1*SINX*SINZ*COS13Y
1-2.*SR2*OVSDS2*SIN2X*SINZ*COS23Y
P(1,16)=-2.*SR2*OVSDS1*SINX*SINZ*SIN13Y
1+2.*SR2*OVSDS2*SIN2X*SINZ*SIN23Y
S(1,17)=(1./3.)*SR3*OVSDS1*COSZ*(COSX*COS13Y-COS23Y)
1+(2./3.)*SR3*OVSDS2*COSZ*(COS2X*COS23Y-COS43Y)
P(1,17)=(1./3.)*SR3*OVSDS1*COSZ*(COSX*SIN13Y+SIN23Y)
1-(2./3.)*SR3*OVSDS2*COSZ*(COS2X*SIN23Y+SIN43Y)
S(1,18)=OVSDS1*COSZ*(2.*COSX*COS13Y+COS23Y)
P(1,18)=OVSDS1*COSZ*(2.*COSX*SIN13Y-SIN23Y)
S(2,11)=COSZ*((OVPPS1+3.*OVPPP1)*COSX*COS13Y+2.
*OVPPP1*COS23Y)
1 +2.*COSZ*((OVPPS2+OVPPP2)*COS2X*COS23Y+OVPPP2*COS43Y)
P(2,11)=COSZ*((OVPPS1+3.*OVPPP1)*COSX*SIN13Y-2.
*OVPPP1*SIN23Y)
1 -2.*COSZ*((OVPPS2+OVPPP2)*COS2X*SIN23Y-OVPPP2*SIN43Y)
S(2,12)=-(1./3.)*SR3*(OVPPS1-OVPPP1)*SINX*COSZ*SIN13Y
1 -(2./3.)*SR3*(OVPPS2-OVPPP2)*SIN2X*COSZ*SIN23Y
P(2,12)=(1./3.)*SR3*(OVPPS1-OVPPP1)*SINX*COSZ*COS13Y
1 -(2./3.)*SR3*(OVPPS2-OVPPP2)*SIN2X*COSZ*COS23Y
S(2,13)=-2.*SQRT(2./3.)*(OVPPS1-OVPPP1)*SINX*SINZ*COS13Y
1 -2.*SQRT(2./3.)*(OVPPS2-OVPPP2)*SIN2X*SINZ*COS23Y
P(2,13)=-2.*SQRT(2./3.)*(OVPPS1-OVPPP1)*SINX*SINZ*SIN13Y
1 +2.*SQRT(2./3.)*(OVPPS2-OVPPP2)*SIN2X*SINZ*SIN23Y
S(2,14)=COSZ*((0.5*OVPDS1+(1./3.)*SR3*OVPDP1)*COSX*COS13Y
1-(2./3.)*SR3*OVPDP1*COS23Y)
1 +SQRT(2./3.)*COSZ*(-SR3*OVPDS2*COS2X*COS23Y
```

```
1+2.*OVPDP2*COS43Y)
 P(2,14)=COSZ*((0.5*OVPDS1+(1./3.)*SR3*OVPDP1)*COSX*SIN13Y
1+(2./3.)*SR3*OVPDP1*SIN23Y)
1 +SQRT(2./3.)*COSZ*(SR3*OVPDS2*COS2X*SIN23Y
1+2.*OVPDP2*SIN43Y)
 S(2,15)=-(SQRT(2./3.)*OVPDS1-(2./3.)*SR2*OVPDP1)
 *SINX*SINZ*COS13Y
1 +(2./9.)*SR3*(3.*OVPDS2-2.*SR3*OVPDP2)*SIN2X*SINZ*COS23Y
 P(2,15)=-(SQRT(2./3.)*OVPDS1-(2./3.)*SR2*OVPDP1)
 *SINX*SINZ*SIN13Y
1 -(2./9.)*SR3*(3.*OVPDS2-2.*SR3*OVPDP2)*SIN2X*SINZ*SIN23Y
 S(2,16)=SINZ*(-(SR2*OVPDS1+2.*SQRT(2./3.)*OVPDP1)
 *COSX*SIN13Y
1+2.*SQRT(2./3.)*OVPDP1*SIN23Y)
1 +(2./3.)*SINZ*(3.*OVPDS2*COS2X*SIN23Y
1-SR3*OVPDP2*SIN43Y)
 P(2,16)=SINZ*((SR2*OVPDS1+2.*SQRT(2./3.)*OVPDP1)
 *COSX*COS13Y
1+2.*SQRT(2./3.)*OVPDP1*COS23Y)
1 +(2./3.)*SINZ*(3.*OVPDS2*COS2X*COS23Y
1+SR3*OVPDP2*COS43Y)
 S(2,17)=-(1./3.)*(0.5*SR3*OVPDS1+5.*OVPDP1)
 *SINX*COSZ*SIN13Y
 1 +(1./3.)*SQRT(2./3.)*(4.*SR3*OVPDP2+3.*OVPDS2)
 *SIN2X*COSZ*SIN23Y
 P(2,17)=(1./3.)*(0.5*SR3*OVPDS1+5.*OVPDP1)
 *SINX*COSZ*COS13Y
 1 +(1./3.)*SQRT(2./3.)*(4.*SR3*OVPDP2+3.*OVPDS2)
 *SIN2X*COSZ*COS23Y
 S(2,18)=-(OVPDS1-(4./3.)*SR3*OVPDP1)*SINX*COSZ*SIN13Y
1 -2.*SQRT(2./3.)*OVPDP2*SIN2X*COSZ*SIN23Y
 P(2,18)=(OVPDS1-(4./3.)*SR3*OVPDP1)*SINX*COSZ*COS13Y
1 -2.*SQRT(2./3.)*OVPDP2*SIN2X*COSZ*COS23Y
 S(3,12)=(1./3.)*COSZ*((OVPPS1+11.*OVPPP1)*COSX*COS13Y
1+2.*(OVPPS1+2.*OVPPP1)*COS23Y)
1 +(2./3.)*COSZ*((2.*OVPPS2+OVPPP2)*COS43Y
1+(OVPPS2+5.*OVPPP2)*COS2X*COS23Y)
 P(3,12)=(1./3.)*COSZ*((OVPPS1+11.*OVPPP1)*COSX*SIN13Y
1-2.*(OVPPS1+2.*OVPPP1)*SIN23Y)
1 +(2./3.)*COSZ*((2.*OVPPS2+OVPPP2)*SIN43Y
1-(OVPPS2+5.*OVPPP2)*COS2X*SIN23Y)
 S(3,13)=-(2./3.)*SR2*(OVPPS1-OVPPP1)*SINZ*(COSX*SIN13Y
 +SIN23Y)
1 -(2./3.)*SR2*(OVPPS2-OVPPP2)*SINZ*(COS2X*SIN23Y+SIN43Y)
```

```
P(3,13)=(2./3.)*SR2*(OVPPS1-OVPPP1)*SINZ*(COSX*COS13Y-
COS23Y)
1-(2./3.)*SR2*(OVPPS2-OVPPP2)*SINZ*(COS2X*COS23Y-COS43Y)
S(3,14)=S(2,17)
P(3,14)=P(2,17)
S(3,15)=(1./3.)*SINZ*(-(SR2*OVPDS1+10.*SQRT(2./3.)
*OVPDP1)*COSX*
1SIN13Y+2.*(SR2*OVPDS1+SQRT(2./3.)*OVPDP1)*SIN23Y)
1-(2./9.)*SINZ*((6.*OVPDS2-SR3*OVPDP2)*SIN43Y-
2(3.*OVPDS2+4.*SR3*OVPDP2)*COS2X*SIN23Y)
P(3,15)=(1./3.)*SINZ*((SR2*OVPDS1+10.*SQRT(2./3.)*OVPDP1)
*COSX*
1COS13Y+2.*(SR2*OVPDS1+SQRT(2./3.)*OVPDP1)*COS23Y)
1+(2./9.)*SINZ*((6.*OVPDS2-SR3*OVPDP2)*COS43Y+
2(3.*OVPDS2+4.*SR3*OVPDP2)*COS2X*COS23Y)
S(3,16)=S(2,15)
P(3,16)=P(2,15)
S(3,17)=(1./3.)*COSZ*((0.5*OVPDS1-(7./3.)*SR3*OVPDP1)
*COSX*COS13Y
1+2.*(0.5*OVPDS1+(2./3.)*SR3*OVPDP1)*COS23Y)
2-(1./3.)*SQRT(2./3.)*COSZ*(2.*(SR3*OVPDS2+OVPDP2)*COS43Y
3+(SR3*OVPDS2-8.*OVPDP2)*COS2X*COS23Y)
P(3,17)=(1./3.)*COSZ*((0.5*OVPDS1-(7./3.)*SR3*OVPDP1)
*COSX*SIN13Y
1-2.*(0.5*OVPDS1+(2./3.)*SR3*OVPDP1)*SIN23Y)
2-(1./3.)*SQRT(2./3.)*COSZ*(2.*(SR3*OVPDS2+OVPDP2)*SIN43Y
3-(SR3*OVPDS2-8.*OVPDP2)*COS2X*SIN23Y)
S(3,18)=(1./3.)*(SR3*OVPDS1-4.*OVPDP1)*COSZ*(COSX*COS13Y-
COS23Y)
1-(2./3.)*SR2*OVPDP2*COSZ*(COS43Y-COS2X*COS23Y)
P(3,18)=(1./3.)*(SR3*OVPDS1-4.*OVPDP1)*COSZ*(COSX*SIN13Y
+SIN23Y)
1-(2./3.)*SR2*OVPDP2*COSZ*(SIN43Y+COS2X*SIN23Y)
S(4,13)=(2./3.)*(2.*OVPPS1+OVPPP1)*COSZ*(2.*COSX*COS13Y
+COS23Y)
1+(2./3.)*(OVPPS2+2.*OVPPP2)*COSZ*(COS43Y+2.*COS2X*COS23Y)
P(4,13)=(2./3.)*(2.*OVPPS1+OVPPP1)*COSZ*(2.*COSX*SIN13Y-
SIN23Y)
1+(2./3.)*(OVPPS2+2.*OVPPP2)*COSZ*(SIN43Y-2.*COS2X*SIN23Y)
S(4,14)=S(2,15)
P(4,14)=P(2,15)
S(4,15)=(2./3.)*(2.*OVPDS1-(1./3.)*SR3*OVPDP1)*COSZ*
(COSX*COS13Y
1-COS23Y)+(2./3.)
```

```
2*SQRT(2./3.)*(SR3*OVPDS2+OVPDP2)*COSZ*(COS43Y-
COS2X*COS23Y)
P(4,15)=(2./3.)*(2.*OVPDS1-(1./3.)*SR3*OVPDP1)*COSZ*
(COSX*SIN13Y
1+SIN23Y)+(2.
2/3.)*SQRT(2./3.)*(SR3*OVPDS2+OVPDP2)*COSZ*(SIN43Y
+COS2X*SIN23Y)
S(4,16)=-(2./3.)*(2.*SR3*OVPDS1-OVPDP1)*SINX*COSZ*SIN13Y
1+(2./3.)*SQRT(2./3.)*(3.*OVPDS2+SR3*OVPDP2)
*SIN2X*COSZ*SIN23Y
P(4,16)=(2./3.)*(2.*SR3*OVPDS1-OVPDP1)*SINX*COSZ*COS13Y
1+(2./3.)*SQRT(2./3.)*(3.*OVPDS2+SR3*OVPDP2)
*SIN2X*COSZ*COS23Y
S(4,17)=-(1./3.)*(SR2*OVPDS1-2.*SQRT(2./3.)*OVPDP1)*SINZ*
(COSX*
1SIN13Y+SIN23Y)
2+(2./9.)*(3.*OVPDS2-2.*SR3*OVPDP2)*SINZ*(SIN43Y
+COS2X*SIN23Y)
P(4,17)=(1./3.)*(SR2*OVPDS1-2.*SQRT(2./3.)*OVPDP1)*SINZ*
(COSX
1*COS13Y-COS23Y)
2-(2./9.)*(3.*OVPDS2-2.*SR3*OVPDP2)*SINZ*(COS43Y-
COS2X*COS23Y)
S(4,18)=-(SQRT(2./3.)*OVPDS1+(2./3.)*SR2*OVPDP1)*SINZ*(2.
*COSX*
1SIN13Y-SIN23Y)
2+(4./3.)*OVPDP2*SINZ*(2.*COS2X*SIN23Y-SIN43Y)
P(4,18)=(SQRT(2./3.)*OVPDS1+(2./3.)*SR2*OVPDP1)*SINZ*(2.
*COSX
1*COS13Y+COS23Y)
2+(4./3.)*OVPDP2*SINZ*(2.*COS2X*COS23Y+COS43Y)
S(5,14)=COSZ*(0.25*(OVDDS1+4.*OVDDP1+11.*OVDDD1)
*COSX*COS13Y
1+(2./3.)*(OVDDP1+2.*OVDDD1)*COS23Y)
2+(1./3.)*COSZ*(2.*(2.*OVDDP2+OVDDD2)*COS43Y+(3.*OVDDS2
+4.*OVDDP2
3+5.*OVDDD2)*COS2X*COS23Y)
P(5,14)=COSZ*(0.25*(OVDDS1+4.*OVDDP1+11.*OVDDD1)
*COSX*SIN13Y
1-(2./3.)*(OVDDP1+2.*OVDDD1)*SIN23Y)
2+(1./3.)*COSZ*(2.*(2.*OVDDP2+OVDDD2)*SIN43Y-(3.*OVDDS2
+4.*OVDDP2
3+5.*OVDDD2)*COS2X*SIN23Y)
S(5,15)=-SQRT(2./3.)*(0.5*OVDDS1+(4./3.)*OVDDP1-(11./6.)
*OVDDD1)*
```

```
1SINX*SINZ*COS13Y
2-(1./3.)*SQRT(2./3.)*(3.*OVDDS2+2.*OVDDP2-5.*OVDDD2)
*SIN2X*SINZ*
1COS23Y
P(5,15)=-SQRT(2./3.)*(0.5*OVDDS1+(4./3.)*OVDDP1-(11./6.)
*OVDDD1)*
1SINX*SINZ*SIN13Y
2+(1./3.)*SQRT(2./3.)*(3.*OVDDS2+2.*OVDDP2-5.*OVDDD2)
*SIN2X*SINZ*
1SIN23Y
S(5,16)=-SR2*SINZ*(0.5*(OVDDS1-OVDDD1)*COSX*SIN13Y+(2./
3.)*(1OVDDP1-OVDDD1)*SIN23Y)
2-(1./3.)*SR2*SINZ*(2.*(OVDDP2-OVDDD2)*SIN43Y+(3.*OVDDS2-
2.*OVDDP2
3-OVDDD2)*COS2X*SIN23Y)
P(5,16)=SR2*SINZ*(0.5*(OVDDS1-OVDDD1)*COSX*COS13Y-(2./3.)
*(OVDDP1
1-OVDDD1)*COS23Y)
2+(1./3.)*SR2*SINZ*(2.*(OVDDP2-OVDDD2)*COS43Y-(3.*OVDDS2-
2.*OVDDP2
3-OVDDD2)*COS2X*COS23Y)
S(5,17)=-(1./3.)*SR3*(0.25*OVDDS1-(1./3.)*OVDDP1+(1./12.)
*OVDDD1)*
1SINX*COSZ*SIN13Y
2-(1./9.)*SR3*(3.*OVDDS2-4.*OVDDP2+OVDDD2)
*SIN2X*COSZ*SIN23Y
P(5,17)=(1./3.)*SR3*(0.25*OVDDS1-(1./3.)*OVDDP1+(1./12.)
*OVDDD1)*
1SINX*COSZ*COS13Y
2-(1./9.)*SR3*(3.*OVDDS2-4.*OVDDP2+OVDDD2)
*SIN2X*COSZ*COS23Y
S(5,18)=-(0.5*OVDDS1-(4./3.)*OVDDP1+(5./6.)*OVDDD1)
*SINX*COSZ
1*SIN13Y+(4./3.)*(OVDDP2-OVDDD2)*SIN2X*COSZ*SIN23Y
P(5,18)=(0.5*OVDDS1-(4./3.)*OVDDP1+(5./6.)*OVDDD1)
*SINX*COSZ*
1COS13Y+(4./3.)*(OVDDP2-OVDDD2)*SIN2X*COSZ*COS23Y
S(6,15)=(1./3.)*COSZ*((2.*OVDDS1+(19./3.)*OVDDP1+(11./3.)
*OVDDD1)
1*COSX*COS13Y+2.*(2.*OVDDS1+(1./3.)*OVDDP1+(2./3.)
*OVDDD1)*COS23Y)
2+(2./9.)*COSZ*((6.*OVDDS2+OVDDP2+2.*OVDDD2)*COS43Y+(3.
*OVDDS2
3+5.*OVDDP2+10.*OVDDD2)*COS2X*COS23Y)
```

```
P(6,15)=(1./3.)*COSZ*((2.*OVDDS1+(19./3.)*OVDDP1+(11./3.)
*OVDDD1)
1*COSX
1*SIN13Y-2.*(2.*OVDDS1+(1./3.)*OVDDP1+(2./3.)*OVDDD1)
*SIN23Y)
2+(2./9.)*COSZ*((6.*OVDDS2+OVDDP2+2.*OVDDD2)*SIN43Y-(3.
*OVDDS2
3+5.*OVDDP2+10.*OVDDD2)*COS2X*SIN23Y)
S(6,16)=-(1./3.)*SR3*(2.*OVDDS1-(5./3.)*OVDDP1-(1./3.)
*OVDDD1)*
1SINX*COSZ*SIN13Y
2-(2./9.)*SR3*(3.*OVDDS2-OVDDP2-2.*OVDDD2)
*SIN2X*COSZ*SIN23Y
P(6,16)=(1./3.)*SR3*(2.*OVDDS1-(5./3.)*OVDDP1-(1./3.)
*OVDDD1)
1*SINX*COSZ*COS13Y
2-(2./9.)*SR3*(3.*OVDDS2-OVDDP2-2.*OVDDD2)
*SIN2X*COSZ*COS23Y
S(6,17)=-(1./3.)*SR2*SINZ*((0.5*OVDDS1-(8./3.)*OVDDP1
+(13./6.)*
1OVDDD1)
1*COSX*SIN13Y-(OVDDS1+(2./3.)*OVDDP1-(5./3.)*OVDDD1)
*SIN23Y)
2+(1./9.)*SR2*SINZ*(2.*(3.*OVDDS2-OVDDP2-2.*OVDDD2)
*SIN43Y
3-(3.*OVDDS2-10.*OVDDP2+7.*OVDDD2)*COS2X*SIN23Y)
P(6,17)=(1./3.)*SR2*SINZ*((0.5*OVDDS1-(8./3.)*OVDDP1
+(13./6.)*
1OVDDD1)
1*COSX*COS13Y+(OVDDS1+(2./3.)*OVDDP1-(5./3.)*OVDDD1)
*COS23Y)
2-(1./9.)*SR2*SINZ*(2.*(3.*OVDDS2-OVDDP2-2.*OVDDD2)
*COS43Y
3+(3.*OVDDS2-10.*OVDDP2+7.*OVDDD2)*COS2X*COS23Y)
S(6,18)=-2.*SQRT(2./3.)*(0.5*OVDDS1-(1./3.)*OVDDP1-(1./
6.)*OVDDD1)
1*SINZ*(COSX*SIN13Y+SIN23Y)
2-(2./3.)*SQRT(2./3.)*(OVDDP2-OVDDD2)*SINZ*(SIN43Y
+COS2X*SIN23Y)
P(6,18)=2.*SQRT(2./3.)*(0.5*OVDDS1-(1./3.)*OVDDP1-(1./6.)
*OVDDD1)
1*SINZ*(COSX*COS13Y-COS23Y)
2+(2./3.)*SQRT(2./3.)*(OVDDP2-OVDDD2)*SINZ*(COS43Y-
COS2X*COS23Y)
```

```
S(7,16)=COSZ*((2.*OVDDS1+OVDDP1+OVDDD1)*COSX*COS13Y+(2./
3.)*(2.*
1OVDDP1+OVDDD1)*COS23Y)
2+(2./3.)*COSZ*((OVDDP2+2.*OVDDD2)*COS43Y+(3.*OVDDS2
+OVDDP2+2.*
3OVDDD2)*COS2X*COS23Y)
P(7,16)=COSZ*((2.*OVDDS1+OVDDP1+OVDDD1)*COSX*SIN13Y-(2./
3.)*(2.*
1OVDDP1+OVDDD1)*SIN23Y)
2+(2./3.)*COSZ*((OVDDP2+2.*OVDDD2)*SIN43Y-(3.*OVDDS2
+OVDDP2+2.*
3OVDDD2)*COS2X*SIN23Y)
S(7,17)=S(5,15)
P(7,17)=P(5,15)
S(7,18)=-2.*SR2*(0.5*OVDDS1-(1./3.)*OVDDP1-(1./6.)
*OVDDD1)*SINX*
1SINZ*COS13Y
2-(2./3.)*SR2*(OVDDP2-OVDDD2)*SIN2X*SINZ*COS23Y
P(7,18)=-2.*SR2*(0.5*OVDDS1-(1./3.)*OVDDP1-(1./6.)
*OVDDD1)*SINX*
1SINZ*SIN13Y
2+(2./3.)*SR2*(OVDDP2-OVDDD2)*SIN2X*SINZ*SIN23Y
S(8,17)=(1./3.)*COSZ*((0.25*OVDDS1+(11./3.)*OVDDP1+(97./
12.)*
1OVDDD1)*
1COSX*COS13Y+2.*(0.25*OVDDS1+(2./3.)*OVDDP1+(25./12.)
*OVDDD1)*
1COS23Y)
2+(1./9.)*COSZ*(2.*(3.*OVDDS2+2.*OVDDP2+4.*OVDDD2)*COS43Y+
3(3.*OVDDS2+20.*OVDDP2+13.*OVDDD2)*COS2X*COS23Y)
P(8,17)=(1./3.)*COSZ*((0.25*OVDDS1+(11./3.)*OVDDP1+(97./
12.)*
1OVDDD1)*
1COSX*SIN13Y-2.*(0.25*OVDDS1+(2./3.)*OVDDP1+(25./12.)
*OVDDD1)*
1SIN23Y)
2+(1./9.)*COSZ*(2.*(3.*OVDDS2+2.*OVDDP2+4.*OVDDD2)*SIN43Y-
3(3.*OVDDS2+20.*OVDDP2+13.*OVDDD2)*COS2X*SIN23Y)
S(8,18)=(2./3.)*SR3*(0.25*OVDDS1-(2./3.)*OVDDP1+(5./12.)
*OVDDD1)*
1COSZ*(COSX*COS13Y-COS23Y)
2+(4./9.)*SR3*(OVDDP2-OVDDD2)*COSZ*(COS43Y-COS2X*COS23Y)
P(8,18)=(2./3.)*SR3*(0.25*OVDDS1-(2./3.)*OVDDP1+(5./12.)
*OVDDD1)*
1COSZ*(COSX*SIN13Y+SIN23Y)
```

```
2+(4./9.)*SR3*(OVDDP2-OVDDD2)*COSZ*(SIN43Y+COS2X*SIN23Y)
 S(9,18)=(0.5*OVDDS1+(4./3.)*OVDDP1+(1./6.)*OVDDD1)*COSZ*
 (2.*COSX*
1COS13Y+COS23Y)
2+(2./3.)*(2.*OVDDP2+OVDDD2)*COSZ*(COS43Y+2.*COS2X*COS23Y)
 P(9,18)=(0.5*OVDDS1+(4./3.)*OVDDP1+(1./6.)*OVDDD1)*COSZ*
 (2.*COSX*
1SIN13Y-SIN23Y)
2+(2./3.)*(2.*OVDDP2+OVDDD2)*COSZ*(SIN43Y-2.*COS2X*SIN23Y)
 S(2,10)=-S(1,11)
 P(2,10)=-P(1,11)
 S(3,10)=-S(1,12)
 P(3,10)=-P(1,12)
 S(4,10)=-S(1,13)
 P(4,10)=-P(1,13)
 S(5,10)=S(1,14)
 P(5,10)=P(1,14)
 S(6,10)=S(1,15)
 P(6,10)=P(1,15)
 S(7,10)=S(1,16)
 P(7,10)=P(1,16)
 S(8,10)=S(1,17)
 P(8,10)=P(1,17)
 S(9,10)=S(1,18)
 P(9,10)=P(1,18)
 S(3,11)=S(2,12)
 P(3,11)= P(2,12)
 S(4,11)=S(2,13)
 P(4,11)= P(2,13)
 S(5,11)=-S(2,14)
 P(5,11)=-P(2,14)
 S(6,11)=-S(2,15)
 P(6,11)=-P(2,15)
 S(7,11)=-S(2,16)
 P(7,11)=-P(2,16)
 S(8,11)=-S(2,17)
 P(8,11)=-P(2,17)
 S(9,11)=-S(2,18)
 P(9,11)=-P(2,18)
 S(4,12)=S(3,13)
 P(4,12)= P(3,13)
 S(5,12)=-S(3,14)
 P(5,12)=-P(3,14)
 S(6,12)=-S(3,15)
 P(6,12)=-P(3,15)
```

```
      S(7,12)=-S(3,16)
      P(7,12)=-P(3,16)
      S(8,12)=-S(3,17)
      P(8,12)=-P(3,17)
      S(9,12)=-S(3,18)
      P(9,12)=-P(3,18)
      S(5,13)=-S(4,14)
      P(5,13)=-P(4,14)
      S(6,13)=-S(4,15)
      P(6,13)=-P(4,15)
      S(7,13)=-S(4,16)
      P(7,13)=-P(4,16)
      S(8,13)=-S(4,17)
      P(8,13)=-P(4,17)
      S(9,13)=-S(4,18)
      P(9,13)=-P(4,18)
      S(6,14)=S(5,15)
      P(6,14)= P(5,15)
      S(7,14)=S(5,16)
      P(7,14)= P(5,16)
      S(8,14)=S(5,17)
      P(8,14)= P(5,17)
      S(9,14)=S(5,18)
      P(9,14)= P(5,18)
      S(7,15)=S(6,16)
      P(7,15)= P(6,16)
      S(8,15)=S(6,17)
      P(8,15)= P(6,17)
      S(9,15)=S(6,18)
      P(9,15)= P(6,18)
      S(8,16)=S(7,17)
      P(8,16)= P(7,17)
      S(9,16)=S(7,18)
      P(9,16)= P(7,18)
      S(9,17)=S(8,18)
      P(9,17)= P(8,18)
      DO 311 L=1,N
      DO 311 K=L,N
      S(K,L)=S(L,K)
      P(K,L)=-P(L,K)
  311 CONTINUE
      RETURN
      END
      REAL FUNCTION INTERP*8 (XX,X,F,J1,N)
      IMPLICIT REAL*8 (A-H,O-Z)
```

```
      DIMENSION X(1),F(1)
      FX=0.
      ISTART=J1-N+N/2+1
      J2=ISTART+N-1
      DO 10 J=ISTART,J2
      P=F(J)
      DO 5 I=ISTART,J2
      IF (I.EQ.J) GO TO 5
      P=P*(XX-X(I))/(X(J)-X(I))
    5 CONTINUE
   10 FX=FX+P
      INTERP=FX
      RETURN
      END
 Titanium hcp Orthogonal
  5 220  0
 8.0   1.0   0.005   -0.1
 0.9040181041   1             TITANIUM HEXAGONAL
 1.1502143145   2             TITANIUM HEXAGONAL
 0.6853885055   3             TITANIUM HEXAGONAL
 0.6848548651   4             TITANIUM HEXAGONAL
 0.6946728230   5             TITANIUM HEXAGONAL
-0.0196536314   6             TITANIUM HEXAGONAL
-0.0610373951   7             TITANIUM HEXAGONAL
 0.0751746073   8             TITANIUM HEXAGONAL
 0.0473231450   9             TITANIUM HEXAGONAL
 0.0981194004  10             TITANIUM HEXAGONAL
-0.0223902948  11             TITANIUM HEXAGONAL
 0.0604609102  12             TITANIUM HEXAGONAL
-0.0242657419  13             TITANIUM HEXAGONAL
-0.0519485325  14             TITANIUM HEXAGONAL
 0.0308015794  15             TITANIUM HEXAGONAL
-0.0044235298  16             TITANIUM HEXAGONAL
-0.0055239233  17             TITANIUM HEXAGONAL
 0.0156004326  18             TITANIUM HEXAGONAL
 0.0062634605  19             TITANIUM HEXAGONAL
 0.0462687425  20             TITANIUM HEXAGONAL
-0.0076305936  21             TITANIUM HEXAGONAL
 0.0148749193  22             TITANIUM HEXAGONAL
-0.0053210435  23             TITANIUM HEXAGONAL
-0.0089383014  24             TITANIUM HEXAGONAL
 0.0018997233  25             TITANIUM HEXAGONAL
 0.0009837329  26             TITANIUM HEXAGONAL
-0.0043515223  27             TITANIUM HEXAGONAL
 0.0071796081  28             TITANIUM HEXAGONAL
```

```
-0.0017818911   29            TITANIUM HEXAGONAL
 0.0145518621   30            TITANIUM HEXAGONAL
 0.0034449229   31            TITANIUM HEXAGONAL
-0.0010188001   32            TITANIUM HEXAGONAL
-0.0019740488   33            TITANIUM HEXAGONAL
 0.0006534021   34            TITANIUM HEXAGONAL
 0.0002592382   35            TITANIUM HEXAGONAL
 0.0000696836   36            TITANIUM HEXAGONAL
 0.0000000000   37
 0.0000000000   38
 0.0000000000   39
 0.0000000000   40
 0.0000000000   41
 0.0000000000   42
 0.0000000000   43
 0.0000000000   44
 0.0000000000   45
 0.0000000000   46
 0.0000000000   47
 0.0000000000   48
 0.0000000000   49
 0.0000000000   50
 0.0000000000   51
 0.0000000000   52
 0.0000000000   53
 0.0000000000   54
 0.0000000000   55
 0.0000000000   56
 0.0000000000   57
 0.0000000000   58
 0.0000000000   59
 0.0000000000   60
 0.0000000000   61
 0.0000000000   62
 0.0000000000   63
 0.0000000000   64
 0.0000000000   65
 0.0000000000   66
 0.0000000000   67
```

Atomic Number Index

Z	Element	Page
1	Hydrogen	39–43
2	Helium	404–406
3	Lithium	45–50
4	Beryllium	82–87
5	Boron	306–308
6	Carbon	338–342
7	Nitrogen	360–362
8	Oxygen	376–378
9	Fluorine	392–393
10	Neon	407–411
11	Sodium	51–56
12	Magnesium	88–94
13	Aluminum	309–313
14	Silicon	343–347
15	Phosphorus	363–365
16	Sulfur	379–381
17	Chlorine	394–395
18	Argon	412–416
19	Potassium	57–62
20	Calcium	95–99
21	Scandium	116–122
22	Titanium	123–129
23	Vanadium	130–135
24	Chromium	136–141
25	Manganese	142–147
26	Iron	148–157
27	Cobalt	158–169
28	Nickel	170–178
29	Copper	179–183
30	Zinc	184–190
31	Gallium	314–318
32	Germanium	348–352
33	Arsenic	366–368
34	Selenium	382–384
35	Bromine	396–397
36	Krypton	417–421
37	Rubidium	63–68
38	Strontium	100–104
39	Yttrium	192–198
40	Zirconium	199–204

(continued)

Z	Element	Page
41	Niobium	205–210
42	Molybdenum	211–216
43	Technetium	217–222
44	Ruthenium	223–228
45	Rhodium	229–233
46	Palladium	234–238
47	Silver	239–243
48	Cadmium	244–250
49	Indium	319–324
50	Tin	353–358
51	Antimony	369–371
52	Tellurium	385–387
53	Iodine	398–399
54	Xenon	422–426
55	Cesium	69–74
56	Barium	105–110
57	Lanthanum	430–433
58	Cerium	434–435
59	Praseodymium	436–437
60	Neodymium	438–439
61	Promethium	440–441
62	Samarium	442–443
63	Europium	444–445
64	Gadolinium	446
65	Terbium	447
66	Dysprosium	448
67	Holmium	449
68	Erbium	450
69	Thulium	451
70	Ytterbium	452
71	Lutetium	453–455
72	Hafnium	252–258
73	Tantalum	259–264
74	Tungsten	265–270
75	Rhenium	271–277
76	Osmium	278–284
77	Iridium	285–289
78	Platinum	290–294
79	Gold	295–299
80	Mercury	300–303
81	Thallium	325–331
82	Lead	332–336
83	Bismuth	372–374

(continued)

Z	Element	Page
84	Polonium	388–390
85	Astatine	400–401
86	Radon	427–428
87	Francium	75–78
88	Radium	111–113
89	Actinium	458–462
90	Thorium	463–467
91	Protactinium	468
92	Uranium	469
93	Neptunium	470
94	Plutonium	471
95	Americium	472
96	Curium	473
97	Berkelium	474
98	Californium	475
99	Einsteinium	476
100	Fermium	477
101	Mendelevium	478
102	Nobelium	479–480
103	Lawrencium	481–482
104	Rutherfordium	484–486
105	Dubnium	487–489
106	Seaborgium	490–492
107	Bohrium	493–495
108	Hassium	496–498
109	Meitnerium	499–501
110	Darmstadtium	502–504
111	Roentgenium	505–507
112	Copernicium	508–510

Printed by Printforce, the Netherlands